H. Baumgärtel

Analytic Perturbation Theory for Matrices and Operators

Mathematische Lehrbücher und Monographien
Herausgegeben von der Akademie der Wissenschaften der DDR
Institut für Mathematik

II. Abteilung
Mathematische Monographien
Band 64

Analytic Perturbation Theory for Matrices and Operators

by H. Baumgärtel

Analytic Perturbation Theory for Matrices and Operators

by Hellmut Baumgärtel

with 8 figures

Akademie-Verlag · Berlin
1984

Verfasser:

Prof. Dr. habil. Hellmut Baumgärtel
Institut für Mathematik der AdW der DDR
1080 Berlin, Mohrenstr. 39

ISSN 0076-5430

Erschienen im Akademie-Verlag, DDR-1086 Berlin, Leipziger Str. 3—4

Lizenznummer: 202 · 100/417/84
Printed in the German Democratic Republic
Gesamtherstellung: VEB Druckhaus „Maxim Gorki", 7400 Altenburg
Lektor: Dipl.-Math. Gesine Reiher
LSV 1065
Bestellnummer: 763 352 6 (6823)
08000

Dedicated to
Izrail Cudikovič Gochberg

Preface

This book is an extended and partly modified English version of the author's book "Endlichdimensionale analytische Störungstheorie", which appeared in 1972. A systematic presentation of analytic perturbation theory for matrices (i.e., linear operators on finite-dimensional linear spaces) is given and the central problems are solved completely.

The implications of this theory for analytic perturbations of isolated eigenvalues of finite algebraic multiplicity of closed operators on Banach spaces are investigated.

Essential contributions to the perturbation theory for isolated eigenvalues of finite algebraic multiplicity are contained in the works of F. Rellich [1 to 5], B. v. Sz.-Nagy [1 and 2], T. Kato [1 to 5] and F. Wolf [1]. In 1967, the author realized that this theory could be extended and brought to completion. The results are presented in this book with detailed proofs. The book may be considered as a supplement to Kato's monograph on perturbation theory for linear operators [5], in particular Ch. II.

The main part of this book is devoted to the analytic perturbation theory for one complex variable. A supplement deals with the transfer of the essential results to the case of analytic perturbations depending on several complex variables. This supplement is based on a paper of the author in Mat. Issl. 1974 (see [7]).

Applications of perturbation theory receive minor emphasis in this book.

Of course, problems of perturbation theory occur in several parts of mathematics (for example, in the theory of matrices, ordinary and partial differential equations, and integral equations) and perturbation theory strongly influences areas beyond mathematics, for example, in theoretical physics.

For these reasons, the presentation is also intended for researchers and advanced students in the physical sciences. It is the author's hope that the book will be useful for those interested — for whatever reason — in a detailed and precise foundation of their ideas and knowledge of analytic perturbation theory.

The reader is assumed to have a basic knowledge of linear algebra and real and complex analysis. For the convenience of the reader, the basic concepts and propositions are presented, without proofs, in Chapter I and in the Appendix.

The structure of the book will be sketched in the introduction. There are ten Chapters, one Supplement and one Appendix. The Chapters are divided into sections and the sections into subsections. Furthermore, the text is organized by Definitions, Lemmas, Theorems, Corollaries and so on. These items are numbered consecutively within each section, as are the formulas. A theorem, lemma etc. from another section (or another chapter) is referred to by the section number (or the chapter number and the section number) followed by the number of the theorem, lemma etc. The symbol ■ will denote the end of a proof.

The essential results described in this book were presented at the Second Conference of Bulgarian Mathematicians at Družba near Varna in 1967. Earlier results were published by the author in 1964. At this conference, Professor I. C. Gochberg from Kišinev (USSR) suggested to present the theory in detailed form and complete this part of perturbation theory. The German original of the present text was the result of this suggestion.

At this occasion, it is a pleasure to say again that the author is much obliged to Professor Gochberg for his suggestions and constant encouragement during the late sixties and early seventies, when the German version was in preparation. He was kind enough to go through large parts of the manuscript, to discuss the content with the author, and to give many suggestions concerning mathematical as well as methodical problems. He also read the final draft of the manuscript and suggested several references.

The author also wishes to express once again his thanks, first, to the Institute of Mathematics at the Academy of Sciences of the MSSR (Kišinev) for the opportunity of lecturing at the Functional Analysis Seminar in June 1968 and January 1970, when parts of the material were presented, and, secondly, to all colleagues of the Department of "Functional Analysis and Approximation Methods" of this institute for discussions and suggestions.

Since that time, the theory of analytic matrix-valued functions (including their perturbation theory) was developed to a large extent by Gochberg, Lancaster, Langer, Krein, Rodman and others. The author did not contribute to this development and this topic is not presented in this book.

The author is grateful to the Akademie-Verlag Berlin, especially to Dipl.-Math. Renate Helle and to Dipl.-Math. Gesine Reiher for their careful work in preparing the manuscript.

Berlin, March 1984 Hellmut Baumgärtel

Contents

Introduction

Perturbation theory is the study of the behavior of mathematical objects under the influence of perturbations. It is not a well-defined mathematical topic with specific objects and methods, but rather a method of investigation.

In this book, perturbation theory will be developed for linear operators. First, interest focuses on the properties of spectral objects, such as eigenvalues, eigenprojections, eigenvectors and Jordan vectors, under perturbation of the underlying operator. This study encompasses some difficult problems. On the one hand, variations of the spectral objects need to be calculated quantitatively. The spectral objects are assumed known for the unperturbed operator, the determination (or at least the approximation) of the spectral objects for the perturbed operator is at issue. This is the starting point for the perturbation theory of L. Rayleigh [1] (see also R. Courant and D. Hilbert [1, p. 296sqq]). On the other hand, the spectral objects often undergo abrupt qualitative changes, even in the case of small perturbations. These changes cause significant complications.

Usually, the behavior of the spectral objects depends strongly on the assumptions about the nature of the perturbation. For example, one can assume continuity, differentiability, smoothness (i.e. arbitrary differentiability) or analyticity. In the following discussion, only the case of analytic (holomorphic) perturbations will be investigated, even if this strong restriction is applied, the problems remain difficult enough.

On the one hand, solving problems of perturbation theory is of conceptual interest. The study of intrinsic spectral properties of a linear operator undergoing perturbation leads to deeper insights and understanding of the structure of the operator. It also leads to the development of new tools for further investigations.

On the other hand, applications (inside and outside of mathematics) lead to new questions in perturbation theory.

One of the first calculations of perturbation theory was given by L. Rayleigh, who determined the eigenfrequencies and eigenmodes of an oscillating string, fixed at $x = 0$ and $x = \pi$, whose elasticity modulus is constant and whose mass density $\varrho(x)$ has only a small deviation from a constant value for all x, $0 \leqq x \leqq \pi$. (That is, the density $\varrho(x)$ is of the form $\varrho(x) = \varrho_0 + \varepsilon\sigma(x)$, where $\sigma(x)$ is a given function and where ε is a small perturbation parameter.)

Actually, as this example indicates, the starting point for the development of perturbation theory was the study of perturbations of spectral objects (eigenvalues and eigenvectors) for concrete classes of operators, for example, Fredholm integral operators or Sturm-Liouville differential operators (for example, see L. Lichtenstein [1]).

In each of these cases, the corresponding eigenvalues are always of finite multiplicity.

Of course, many other problems lead to the central problem of calculating eigenvalues and eigenvectors of selfadjoint, and also of nonselfadjoint, operators. Even in the case of selfadjoint operators, the exact calculation of eigenvalues and eigenvectors causes serious difficulties. A common method for the approximate calculation of these objects is the following. The given operator A is replaced by a "neighboring" operator A_0, whose eigenvalues and eigenvectors are known. Then the influence of the difference $B := A - A_0$ on the eigenvalues and eigenvectors is taken into account by "perturbation terms". For this purpose, one introduces a perturbation parameter ε and considers the operator $A_0 + \varepsilon B$. Its eigenvalues and eigenvectors are functions of ε and are assumed to converge to the corresponding eigenvalues and eigenvectors of A_0 as $\varepsilon \to 0$. The first problem is calculating the correction terms for the eigenvalues and eigenvectors if the transition from A_0 to $A_0 + \varepsilon B$ is performed. The second problem is justifying the considerations for $\varepsilon = 1$.

The method just sketched is well-known in quantum mechanics (for example, see G. Ludwig [1, Ch. VII], E. Fick [1, Teil 5, Kap. III]). In fact, strong impulses for the development of perturbation theory came from quantum mechanics. A quantum mechanical system is described by a Hilbert (state) space and a selfadjoint operator H (the Hamiltonian) in this space. An isolated eigenvalue λ_0 of H (e.g., of finite multiplicity), together with a corresponding normalized eigenvector x_0, is interpreted as a "bound state" x_0 with "energy level" λ_0. If an external field is switched on, or if one takes into account interactions neglected before, then H undergoes a perturbation and is replaced by $H + H_1$, where the perturbation H_1 describes the influences mentioned. The question arises whether there is at least one bound state in the neighborhood of the unperturbed bound state. The question is related to the stability of the eigenvalue λ_0. If λ_0 is stable, how do λ_0 and x_0 change under the perturbation, is there a splitting of eigenvalues, can one calculate the bound states and energy levels of $H + H_1$ in the neighborhood of λ_0, x_0? Sometimes, one introduces a perturbation parameter ε and studies the operator $H_0 + \varepsilon H_1$.

To investigate the quantitative dependence of the perturbed spectral objects, E. Schrödinger developed a formalism of perturbation theory, which strongly influenced the further purely mathematical development of the topic (see E. Schrödinger, 1926 [1, 2]). Schrödinger started with the Ansatz $H_0 + \varepsilon H_1$ for the perturbation, where ε is a real parameter, and *postulated* an analytic (holomorphic) dependence of the perturbed spectral objects on the perturbation parameter ε. Using this postulate, he obtained explicit formulas for the perturbed spectral objects. These are known as perturbation series in quantum mechanics. The splitting problem for the eigenvalues was also investigated by Schrödinger. Using a secular equation, he explicitly calculated the splitting which occurs in the first-order (linear) approximation.

If one generalizes the problem and turns from special classes of operators (as Schrödinger had in mind) to selfadjoint operators in a Hilbert space, then one has to verify Schrödinger's postulate; more precisely, one must investigate the conditions which imply a holomorphic dependence of the perturbed spectral objects on the perturbation parameter. More generally, the problem implied by Schrödinger's formalism can be formulated as follows:

Let an analytic perturbation of the unperturbed operator H_0 be given on a Hilbert space, i.e., a power series

$$H(\varepsilon) = H_0 + \varepsilon H_1 + \varepsilon^2 H_2 + \cdots, \tag{1}$$

where, for simplicity, the operators $H_0, H_1, H_2, \ldots$ are assumed bounded (and selfadjoint), and where the power series (1) is assumed convergent with respect to the operator norm. Let λ_0 be an isolated eigenvalue of H_0 of finite multiplicity. Then λ_0 is stable. How do the perturbed spectral objects depend on the perturbation parameter ε? Schrödinger's postulate can be justified in this case; i.e. the formal perturbation power series are convergent. This result is the main content of a famous series of articles by F. Rellich [1—5], who, beginning 1937, investigated the perturbations of spectral objects of selfadjoint operators in a Hilbert space. The following theorem is already contained in [1].

Theorem 1 (Rellich). *Let $A(\varepsilon)$ be an analytic and bounded selfadjoint perturbation of A_0,*

$$A(\varepsilon) = A_0 + \varepsilon A_1 + \varepsilon^2 A_2 + \cdots, \qquad A_i \in \mathcal{L}(H), \qquad \varepsilon \text{ real},$$

and let λ_0 be an isolated eigenvalue of A_0 of finite multiplicity n. Then, if $|\varepsilon|$ is sufficiently small, there are exactly n analytic eigenvalues (their multiplicities taken into account) in the neighborhood of λ_0,

$$\lambda_i(\varepsilon) = \lambda_0 + \varepsilon\lambda_{i1} + \varepsilon^2\lambda_{i2} + \cdots, \qquad i = 1, \ldots, n,$$

and n corresponding analytic eigenvectors,

$$x_i(\varepsilon) = x_i + \varepsilon x_{i1} + \varepsilon^2 x_{i2} + \cdots, \qquad i = 1, \ldots, n,$$

which can be chosen mutually orthogonal and normalized,

$$\big(x_i(\varepsilon), x_j(\varepsilon)\big) = \delta_{ij}.$$

It is a remarkable fact that this result is also nontrivial in the case that the underlying Hilbert space is finite-dimensional. This result was not known before Rellich. Rellich investigated the finite-dimensional problem as a first step for the general case.

In his articles [1—5], F. Rellich also studied unbounded operators $A(\varepsilon)$ and secondary problems, for example, the problem of estimating the radii of convergence of the perturbation series, error estimates and splitting problems.

In 1946/47, the method of Rellich, essentially going back to Poincaré and Lichtenstein, was replaced by the powerful method of the Cauchy-Riesz integral, where the Riesz projection

$$P(\varepsilon) = \frac{1}{2\pi i} \oint \big(\lambda - A(\varepsilon)\big)^{-1} \, d\lambda \tag{2}$$

plays a decisive role. This was done by B. v. Sz.-Nagy in [1]. (A first version of this paper had already appeared in 1943 in Hungarian.)

In the period following, the introduction of this new method brought some progress in perturbation theory, in particular in the generalization to nonselfadjoint operator-valued functions and complex perturbation parameters.

On the one hand, there are formal reasons for such a generalization: The assumption of analyticity (holomorphy) for $A(\varepsilon)$, ε real, leads naturally to the consideration of ε as a complex variable z. Obviously, in this case $A(z)$ cannot be selfadjoint in the whole domain of the complex z-plane (which is at least some disc centered at $z = 0$), even though $A(z)$ may be selfadjoint on the real axis. Thus, one is led to try to transfer the results to nonselfadjoint operator-valued functions $A(z)$ and to operator-valued functions acting on a Banach space.

On the other hand, applications lead to perturbation problems for nonselfadjoint operators. For example, in the theory of oscillations, perturbation problems occur naturally (see E. A. Coddington and N. Levinson [1, Ch. 14 and 15], R. Bellmann [1, Ch. 13]).

Let A be a complex $N \times N$-matrix and let $x'(t) = Ax(t)$ be a linear system (of ordinary differential equations) of order N with constant coefficients. A fundamental matrix is $\Phi(t) = \exp(tA)$, for example. Such a system has periodic solutions with period l if and only if 1 is an eigenvalue of $\exp(lA)$. Let the system be perturbed by the term εB, where B denotes an $N \times N$-matrix and ε a complex perturbation parameter. One obtains the system $x'(t) = (A + \varepsilon B)\, x(t)$, of which $\Phi(t, \varepsilon) = \exp\{t(A + \varepsilon B)\}$ is a fundamental matrix, i.e., a solution of the operator differential equation

$$X'(t) = AX(t) + \varepsilon BX(t), \qquad X(0) = 1. \tag{3}$$

From (3), it follows (cf. (1.16.1)) that $X(t)$ satisfies the following linear integral equation (see, e.g., R. Bellmann [1, p. 206 sqq.]):

$$X(t) = \exp(tA) + \varepsilon \int_0^t \exp\big((t - s)\, A\big)\, BX(s)\, ds. \tag{4}$$

Hence,

$$X(t) = \exp(tA) + \varepsilon \int_0^t \exp\big((t - s)\, A\big)\, B \exp(sA)\, ds + O(\varepsilon^2), \tag{5}$$

uniformly with respect to t on bounded intervals $[0, T]$; in particular

$$\Phi(l, \varepsilon) = \exp(lA) + \varepsilon \int_0^l \exp\big((l - s)\, A\big)\, B \exp(sA)\, ds + O(\varepsilon^2). \tag{6}$$

The investigation of the existence of periodic solutions of the perturbed system requires knowledge of the spectrum of $\Phi(l, \varepsilon)$ (see W. S. Loud [1], E. A. Coddington and N. Levinson [1]). Determining a first-order approximation of this spectrum is a typical problem of perturbation theory. Obviously, the operators appearing in this connection are generally nonselfadjoint.

The behavior of the eigenvalues and eigenvectors of a selfadjoint operator A on a finite-dimensional space under perturbation of A can be studied in terms of the behavior of the quadratic form (x, Ax). If the underlying space is real (a linear space is called real if the "scalars" are real, rather than complex numbers; cf. Subsect. 1.1.1), then the behavior of the lengths and directions of the main axes of this quadratic form is at issue.

Problems of perturbation theory are also of interest in numerical linear algebra. For example, one is interested in the variation of the eigenvalues of a matrix, if the coefficients of the matrix undergo a perturbation. Typically, the variation of an eigenvalue of an $N \times N$-matrix can not exceed the order of $\varepsilon^{1/N}$ if the perturbation of the coefficients is of the order of ε. Moreover, the order of $\varepsilon^{1/N}$ occurs, which is illustrated by the following example, which is due to Forsythe:

$$A(\varepsilon) = \begin{bmatrix} \lambda_0 & 1 & 0 & \dots & 0 & 0 \\ 0 & \lambda_0 & 1 & \dots & 0 & 0 \\ & & & \dots & & \\ 0 & 0 & 0 & \dots & \lambda_0 & 1 \\ \varepsilon & 0 & 0 & \dots & 0 & \lambda_0 \end{bmatrix};$$

see J. H. Wilkinson [1]. The characteristic polynomial of $A(\varepsilon)$ is $(\lambda - \lambda_0)^N - \varepsilon$; the perturbed eigenvalues are $\lambda(\varepsilon) = \lambda_0 + \varepsilon^{1/N}$.

Typical problems of oscillation theory are of the form

$$A(\varepsilon)\, x = \lambda B(\varepsilon)\, x, \tag{7}$$

where $A(\varepsilon)$ and $B(\varepsilon)$ are selfadjoint operators. Such problems appear if one has to determine the eigenfrequencies and eigenmodes of a beam whose physical constants are only known approximately (for example, see S. Birnbaum [1] and W. Greenlee [1, 2]). If $B(\varepsilon)$ is (boundedly) invertible, then (7) leads to a problem of the type $B(\varepsilon)^{-1} A(\varepsilon)\, x = \lambda x$, which is an eigenvalue problem for the operator-valued function $C(\varepsilon) = B(\varepsilon)^{-1} A(\varepsilon)$. This operator-valued function is generally not selfadjoint.

Results for the generalized perturbation problem are obtained with the resolvent integration method mentioned above. The theory can be presented more generally and in a very simple form. We cite a theorem, which was proved by B. v. Sz.-Nagy [2], T. Kato [4] and F. Wolf [1]. The same theorem can be formulated for unbounded operator-valued functions.

Theorem 2 (Sz.-Nagy, Kato, Wolf). *Let $A(z)$ be a function whose values are bounded operators acting on a Banach space, holomorphic at $z = 0$, and let λ_0 be an isolated eigenvalue of finite algebraic multiplicity n of $A_0 = A(0)$. Then the spectrum of $A(z)$ in the neighborhood of λ_0 consists, for sufficiently small $|z|$, of a finite number of eigenvalues $\lambda_{\alpha\sigma}(z)$, $\alpha = 1, \dots, r$, $\sigma = 1, \dots, g_\alpha$, where $\lambda_{\alpha 1}(z), \dots, \lambda_{\alpha g_\alpha}(z)$ are the g_α branches of a unique analytic function with an algebroid branching at $z = 0$ of order g_α, i.e., the eigenvalues can be expanded into power series of the form*

$$\lambda_{\alpha\sigma}(z) = \lambda_0 + \sum_{\mu=1}^{\infty} \lambda_{\alpha\mu} z^{\mu/g_\alpha}.$$

Similar function-theoretic properties are valid for the corresponding eigenprojections $P_{\alpha\sigma}(z)$ (where a pole at $z = 0$ is possible) and for the eigennilpotents $D_{\alpha\sigma}(z)$.

The result for selfadjoint operators mentioned above is implied by this theorem.

Theorem 2 has by no means the definitive character of Theorem 1 (for selfadjoint perturbations). Obviously, this is due to the lack of a complete proposition on the Jordan structure of that part of the operator $A(z)$ for $z \neq 0$ which belongs to the (isolated) part of the spectrum in the neighborhood of λ_0. Theorem 2 does not mention any properties of the perturbed Jordan vectors.

The unperturbed operator A_0, restricted to the part belonging to λ_0, has a well defined Jordan structure, as does each operator $A(z)$, $z \neq 0$, restricted to the part belonging to the spectrum in the neighborhood of λ_0. A generalization of Rellich's result should contain a complete proposition on the total set of perturbed spectral objects, including the perturbed Jordan vectors. In other words, one has to determine *the* (unique) perturbed Jordan structure, provided there is one. Furthermore, one has to clarify the function-theoretic dependence of all spectral objects on the perturbation parameter z.

This problem was never investigated, not even after 1952. Of course, some progress was made, for example, in obtaining estimates of radii of convergence and error estimates (see B. v. Sz.-Nagy [2], T. Kato [1, 3], J. Schröder [1, 2, 3], F. Schäfke [1, 2], [3], M. I. Višik and L. A. Lusternik [2], V. B. Lidskii [1], M. K. Gavurin [1, 2], B. V. Loginov [1], G. Porath [1]). Furthermore, some progress was made in the reduction theory (see F. Wolf [1], I. M. Višik and L. A. Ljusternik [2], V. B. Lidskii [1], G. Porath [1, 2]). These developments will be treated in Ch. 7. However, there have been no contributions to the perturbation theory of Jordan vectors.

As late as 1966, in Kato's monograph [5], the analytic perturbation theory of isolated eigenvalues of finite algebraic multiplicity was still presented in a form expressed by Theorem 2.

Recently, results became known which definitively solved the key problem for analytic perturbations of isolated eigenvalues of finite algebraic multiplicity, at least as far as the qualitative questions regarding the Jordan structure are concerned (see H. Baumgärtel [2 to 4]).

The problem left open by Theorem 2 is purely finite-dimensional. Namely, by means of the resolvent integration method, in particular the Riesz projection (2), the perturbation problem can be reduced to a purely finite-dimensional one (see Ch. 10).

The definitive development of the qualitative theory suggests a comprehensive and systematic presentation of the theory. Because the decisive step is purely finite-dimensional, it is preferable to give such a systematic presentation in a finite-dimensional framework. The corresponding results for analytic perturbations in Hilbert and Banach spaces are then obtained as corollaries.

The theory is based on considerations from the theory of holomorphic operator-valued functions on finite-dimensional (linear) spaces.

Let G be a region of the complex plane and let $A(z)$ be a holomorphic operator-valued function, acting on a finite-dimensional space (i.e., a holomorphic matrix-valued function). Let ζ be a point from G. We denote the eigenvalues of $A(\zeta)$ by $\lambda_i(\zeta)$ and the corresponding eigenprojections by $P_i(\zeta)$,

$$P_i(\zeta) = \frac{1}{2\pi i} \int_{\mathscr{C}_i} \left(\lambda - A(\zeta)\right)^{-1} \mathrm{d}\lambda, \qquad i = 1, 2, \ldots, r(\zeta);$$

each integration path $\mathscr{C}_i$ is a positively oriented closed contour (e.g., a circle with center $\lambda_i(\zeta)$) separating $\lambda_i(\zeta)$ from the rest of the spectrum of $A(\zeta)$. Then the operator

$$D(\zeta) = A(\zeta) - \sum_{i=1}^{r(\zeta)} \lambda_i(\zeta) P_i(\zeta)$$

is nilpotent. An operator-valued function $A(z)$ is said to be of *invariant Jordan structure* in G, if $r(\zeta) = r$ (constant) in G, if $\dim P_i(\zeta) = m_i$ (constant), where the ordering of the $P_i(\zeta)$ is suitably chosen, $\zeta \in G$ arbitrary, and if there is a constant nilpotent operator N such that $D(\zeta)$ and N are similar for each $\zeta \in G$. With this notion, the following theorem can be stated (see Sect. 5.7):

Theorem 3. *Let $A(z)$ be holomorphic in G. Then there is a set $A \subset G$, isolated and closed in G, such that $A(z)$ is of invariant Jordan structure in $G \setminus A$.*

A set $A \subset G$ is said to be *isolated and closed* in G if the accumulation points of A, if any, are found on ∂G (the boundary of G). Note that such a set $A \subset G$ is necessarily denumerable.

The importance of this theorem for perturbation theory is demonstrated by the following corollary:

Corollary 1. *Let $A(z)$ be a holomorphic perturbation of $A_0 = A(0)$, defined on $|z| < R$. Then there is a ϱ, $0 < \varrho \leqq R$, such that $A(z)$ is of invariant Jordan structure in $0 < |z| < \varrho$.*

A corresponding proposition is true for analytic perturbations in Banach spaces (see Subsect. 10.4.3). Corollary 1 shows that, immediately after a holomorphic perturbation is switched on, the Jordan structure is locally invariant. Therefore, it makes sense to speak of *the* perturbed Jordan structure, as opposed to the unperturbed Jordan structure.

Corollary 1 is closely connected with the following result: All perturbed spectral objects can be expanded into Puiseux series. The expansions of eigenvalues, eigenvectors and Jordan vectors are power series of the type

$$a_0 + \sum_{\varrho=1}^{\infty} a_\varrho z^{\varrho/p}, \qquad |z| < \varrho;$$

the expansions of eigenprojections and irreducible projections are Laurent series of the type

$$\sum_{\varrho=-N}^{\infty} P_\varrho z^{p/\varrho}, \qquad 0 < |z| < \varrho;$$

here, p denotes a natural number (from a finite set) and a_0 denotes an unperturbed spectral object (eigenvalue, eigenvector, Jordan vector).

These results supplement Theorem 2 definitively, because the propositions regarding the perturbed objects include propositions on the Jordan structure. Therefore, these results yield a true analog of the result of F. Rellich (Theorem 1) for arbitrary holomorphic (i.e., in general nonselfadjoint) perturbations of isolated eigenvalues of finite algebraic multiplicity.

As has been stated, the aim of the following presentation is a systematic development of the analytic perturbation theory of isolated eigenvalues of finite algebraic multiplicity. The theory will be presented in its finite-dimensional form, because the essential steps are purely finite-dimensional. The main results and their perturbation-theoretic implications are contained in Ch. 5 and 6.

The transfer of the results to perturbations on Banach spaces is described in Ch. 10.

A supplement to the English edition contains the transfer of the essential results to the case of analytic perturbations depending on several complex variables. This supplement is a translation of the author's article [7].

Ch. 7 contains the reduction theory, which is concerned with the relation between perturbed and unperturbed Jordan structures. Contributions of several authors have been included (F. Wolf [1], M. I. Višik and L. A. Ljusternik [2], M. G. Krein and I. C. Gochberg [1], V. B. Lidskii [1], G. Porath [1]).

The classical and recent results for estimates of the radii of convergence and error estimates are discussed in Ch. 8.

A special chapter (Ch. 9) is devoted to the implications of the theory for holomorphic perturbations of polynomial operator-valued functions. The results of V. Eni [2, 4] are improved by the results presented here.

Ch. 1 and 2 contain basic material from finite-dimensional linear analysis; in particular, the spectral theory and the Jordan structure of linear operators. There are no proofs in Ch. 1.

Ch. 3 contains the spectral theory of meromorphic operator-valued functions and pursues the development of the analytic perturbation theory up to Theorem 2.

Ch. 4 summarizes results for linear operators acting on finite-dimensional linear spaces over arbitrary commutative fields (see W. Graeub [1], in particular Ch. XI). These results are important for the main topic, which is presented later (in Ch. 5. and 6).

The appendix contains auxiliary material from complex function theory and algebra. There are no proofs in Sect. A 1.

1. Finite-dimensional linear analysis

The following investigations of analytic perturbations of an operator take place in the context of a complex linear space of finite dimension. Therefore, we collect here those parts of finite-dimensional linear analysis which are essential for perturbation theory.

Part A of this chapter deals with the basic algebraic concepts of the theory of finite-dimensional linear spaces. In **Part B** basic concepts of linear vector and operator analysis on these spaces have been summarized. (The term "vector analysis" denotes merely the set of concepts and propositions implied by the basic concept "convergence of a sequence of vectors" for vector-valued functions of a scalar argument, as e.g. continuity, derivation, integral etc.)

The concept of a holomorphic (meromorphic) operator-valued function of a scalar complex variable will play a central role.

In this chapter statements are given without proof. A basic reference for Part A is, for example, H. Boseck's "Introduction into the theory of linear vector spaces" [1]. Other references are: I. M. Gelfand [2], T. Kato [5] and, especially, I. M. Glazman and Ju. L. Ljubič [1].

A. Basic algebraic concepts

1.1. Finite-dimensional complex linear spaces

1.1.1. Complex linear spaces, linear combinations, linear dependence, dimension

A non-empty set X is called a *complex linear space* if there is an addition and a multiplication by complex numbers in X such that the following properties are true:

1.1. For any three elements $x, y, z \in X$ the relation $(x + y) + z = x + (y + z)$ is valid (*associativity*);

1.2. For any two elements $x, y \in X$, the relation $x + y = y + x$ is valid (*commutativity*);

1.3. There is an element in X, denoted by 0, such that $x + 0 = x$ for all $x \in X$. The element 0 is called the *zero element.*

1.4. To every vector $x \in X$ corresponds another vector, denoted by $-x$, such that $x + (-x) = 0$.

2.1. For any two complex numbers $\alpha, \beta \in \mathbb{C}$ and for all $x \in X$, the relation $\alpha(\beta x) = (\alpha\beta)\, x$ is valid;

2.2. $1x = x$ is valid for all $x \in X$;

3.1. For any two complex numbers $\alpha, \beta \in \mathbb{C}$ and for all $x \in X$, the relation $(\alpha + \beta)\, x = \alpha x + \beta x$ is valid;

3.2. For any two elements $x, y \in X$ and for all $\alpha \in \mathbb{C}$, the relation $\alpha(x + y) = \alpha x + \alpha y$ is valid.

The elements of X are called *vectors*.

The set $\mathbb{C}^n$ of all ordered n-tuples $\{z_1, z_2, \ldots, z_n\}$ of complex numbers is an important example of a complex linear space.

A vector of the form $\sum_{\varrho=1}^{k} \alpha_\varrho x_\varrho$ with complex numbers $\alpha_1, \alpha_2, \ldots, \alpha_k$ and vectors $x_1, x_2, \ldots, x_k$ is called a *linear combination* of $x_1, x_2, \ldots, x_k$.

Vectors $x_1, x_2, \ldots, x_k$ are said to be *linearly dependent* if there are complex numbers $\alpha_1, \alpha_2, \ldots, \alpha_k$, not all vanishing, such that $\sum_{\varrho=1}^{k} \alpha_\varrho x_\varrho = 0$.

Vectors $x_1, x_2, \ldots, x_k$, which are not linearly dependent, are called *linearly independent*.

The complex linear space X is said to be *n-dimensional* if there are n linearly independent vectors in X, and if any $n + 1$ vectors are linearly dependent. The dimension of X is denoted by $\dim X$, $n = \dim X$.

If there are arbitrarily many linearly independent vectors in X, then X is called *infinite-dimensional*.

In general, the spaces X considered in this book are finite-dimensional; the only exception occurs in Ch. 10.

The space $\mathbb{C}^n$ is n-dimensional.

1.1.2. Bases, coordinates, isomorphic spaces

A family of n linearly independent vectors $e_1, e_2, \ldots, e_n$ of an n-dimensional space X is called a *basis* in X.

Any family $x_1, \ldots, x_k$ of k linearly independent vectors in X, where $k < n$, can be enlarged to a basis in X by adding suitable vectors $x_{k+1}, \ldots, x_n$.

Each vector $x \in X$ can be represented uniquely as a linear combination of the vectors of a fixed basis, $x = \sum_{\varrho=1}^{n} \xi_\varrho e_\varrho$. The scalars $\xi_1, \xi_2, \ldots, \xi_n$ are called the *coordinates* of x with respect to the basis $e_1, e_2, \ldots, e_n$.

Any two bases $e_1, e_2, \ldots, e_n$ and $e_1', e_2', \ldots, e_n'$ are coupled by a system of linear relations,

$$e_i' = \sum_{j=1}^{n} \gamma_{ji} e_j, \qquad i = 1, 2, \ldots, n.$$

The $n \times n$-matrix $\Gamma = (\gamma_{ij})$ is called the ***transition matrix***, where $\det \Gamma \neq 0$. The coordinates $\xi_1, \xi_2, \ldots, \xi_n$ and $\xi_1', \xi_2', \ldots, \xi_n'$ of one and the same vector x with respect to $e_1, e_2, \ldots, e_n$ and $e_1', e_2', \ldots, e_n'$, respectively, are then related to each other by

$$\xi_i = \sum_{j=1}^{n} \gamma_{ij} \xi_j', \qquad i = 1, 2, \ldots, n.$$

In matrix notation, this is expressed by

$$\begin{pmatrix} \xi_1 \\ \xi_2 \\ \vdots \\ \xi_n \end{pmatrix} = \Gamma^{\mathsf{T}} \begin{pmatrix} \xi_1' \\ \xi_2' \\ \vdots \\ \xi_n' \end{pmatrix},$$

and one obtains the inverse formula

$$\begin{pmatrix} \xi_1' \\ \vdots \\ \xi_2' \end{pmatrix} = (\Gamma^{\mathsf{T}})^{-1} \begin{pmatrix} \xi_1 \\ \vdots \\ \xi_n \end{pmatrix}.$$

The spaces X and X' are said to be ***isomorphic*** if there is a one-to-one assignment of X onto X' which is compatible with the linear relations, i.e., if $x \leftrightarrow x'$ and $y \leftrightarrow y'$, then also $\alpha x + \beta y \leftrightarrow \alpha x' + \beta y'$ for all $\alpha, \beta \in \mathbb{C}$.

The assignment of the coordinates $(\xi_1, \xi_2, \ldots, \xi_n)$ to the corresponding vector $x \in X$ with respect to a fixed basis is an isomorphism between X and $\mathbb{C}^n$.

Spaces with the same dimension are always isomorphic; spaces with different dimensions are not isomorphic.

1.1.3. Subspaces

A subset M of X is called a ***subspace*** if M itself is a linear space with respect to the linear operations in X, i.e., $x, y \in M$ implies $x + y \in M$ and $\alpha x \in M$ for all $\alpha \in \mathbb{C}$.

The dimension of M is not greater than the dimension of X.

If S is a subset of X, then the family M of all linear combinations of vectors from S is a subspace; M is generated by S and is called the ***linear span*** of S. It is denoted by $M =: \operatorname{spa} S$.

The linear span of k vectors is at most k-dimensional; it is k-dimensional if these k vectors are linearly independent.

If $X = \operatorname{spa} \{y_1, y_2, \ldots, y_p\}$ and if $e_1, e_2, \ldots, e_k$ denote k linearly independent vectors from X, $k < \dim X$, then the family $e_1, e_2, \ldots, e_k$ can be enlarged to a basis of X by adding suitable vectors from the family $\{y_1, y_2, \ldots, y_p\}$.

If M_1 and M_2 are subspaces of X, then $M_1 + M_2$ denotes the subspace consisting of all vectors of the form $x_1 + x_2$, where $x_1 \in M_1$, $x_2 \in M_2$. It is called the (linear) ***sum***.

Similarly, the ***intersection*** $M_1 \cap M_2$ is a subspace. The relation $\dim M_1 + \dim M_2 = \dim (M_1 + M_2) + \dim (M_1 \cap M_2)$ is valid.

The operation $M_1 + M_2$ for subspaces is associative: $M_1 + (M_2 + M_3) = (M_1 + M_2) + M_3$. Hence, the notation $\sum_{\varrho=1}^{k} M_\varrho$ for the sum of k subspaces $M_1, M_2, \ldots, M_k$ contains no ambiguity.

X is said to be the *direct sum* of the subspaces $M_1, M_2, \ldots, M_k$ if $X = \sum_{\varrho=1}^{k} M_\varrho$, and if $\sum_{\varrho=1}^{k} u_\varrho = 0$, where $u_\varrho \in M_\varrho$, implies $u_1 = u_2 = \cdots = u_k = 0$. In this case, the notation $X = M_1 \oplus M_2 \oplus \cdots \oplus M_k = \sum_{\varrho=1}^{k} \oplus M_\varrho$ is used. Each vector can be uniquely represented in the form

$$x = \sum_{\varrho=1}^{k} u_\varrho, \qquad u_\varrho \in M_\varrho,$$

and

$$\dim X = \sum_{\varrho=1}^{k} \dim M_\varrho$$

is valid.

1.1.4. Factor spaces

Let M be a subspace of X. By $x + M$ one denotes the family of all vectors $x + u$, where $u \in M$ (inhomogeneous linear manifold). The family of all inhomogeneous linear manifolds is a linear space with respect to the linear operation

$$\alpha(x + M) + \beta(y + M) = (\alpha x + \beta y) + M.$$

This linear space is called the *factor space* of X with respect to M, it is denoted by $X \bmod M$. The zero vector of $X \bmod M$ is the subspace M. The equation $x + M = y + M$ is valid if and only if $x - y \in M$, this property is denoted by $x \equiv y (\bmod M)$.

The dimension of $X \bmod M$ is called the *codimension* or deficiency of M with respect to X, it is denoted by $\operatorname{codim} M$. The relation

$$\dim M + \operatorname{codim} M = \dim X$$

is valid.

Let M be a subspace of X. The family $x_1, x_2, \ldots, x_k$ of vectors from X is said to be *linearly dependent modulo* M (mod M) if there are scalars $\alpha_1, \alpha_2, \ldots, \alpha_k$, not all vanishing, such that $\sum_{\varrho=1}^{k} \alpha_\varrho x_\varrho \equiv 0 (\bmod M)$. Correspondingly, the family is called *linearly independent modulo* M if $\sum_{\varrho=1}^{k} \alpha_\varrho x_\varrho \in M$ implies $\alpha_1 = \alpha_2 = \cdots = \alpha_k = 0$.

The family $x_1, x_2, \ldots, x_k$ from X is said to form a *relative basis* of X with respect to M if $x_1 + M, x_2 + M, \ldots, x_k + M$ is a basis of $X \bmod M$, i.e., if the family $x_1, x_2, \ldots, x_k$ is linearly independent mod M and if this family, together with M, spans the whole space X. Then $k = \operatorname{codim} M$.

A relative basis of $X \bmod M$ and a basis of M jointly form a basis of X.

1.2. Antilinear forms and the adjoint space

1.2.1. Definition and basis representation, linear combinations

Let X be a complex linear space of finite dimension, $0 < \dim X = N < \infty$. A complex-valued function $f(x)$, defined for all $x \in X$, is called an *antilinear form* if, for any complex numbers $\alpha, \beta \in \mathbb{C}$ and for any vectors $x, y \in X$,

$$f(\alpha x + \beta y) = \bar{\alpha} f(x) + \bar{\beta} f(y). \tag{2.1}$$

Let $e_1, e_2, \ldots, e_N$ denote a basis of X. If $x = \sum \xi_j e_j$ is the basis representation of x, then because of (2.1), one obtains

$$f(x) = \sum \alpha_j \bar{\xi}_j, \qquad \alpha_j = f(e_j). \tag{2.2}$$

Thus, each antilinear form can be represented by an N-tuple $(\alpha_1, \alpha_2, \ldots, \alpha_N)$ with respect to a fixed basis and, conversely, each N-tuple $(\alpha_1, \alpha_2, \ldots, \alpha_N)$ determines, according to (2.2), an antilinear form.

Linear combinations are defined in the following way:

$$(\alpha f + \beta g)(x) = \alpha f(x) + \beta g(x). \tag{2.3}$$

1.2.2. Adjoint space, scalar product, adjoint basis

The family of all antilinear forms of X is a linear space with respect to (2.3), denoted by X^*. It is called the *adjoint* (or *dual*) *space*. The elements of X^* are sometimes called *covectors*. The complex number

$$f(x) = (f, x), \qquad f \in X^*, \quad x \in X,$$

is called the *scalar product* of f and x.

Let $e_1, e_2, \ldots, e_N$ be a basis of X. To every N-tuple $(\alpha_1, \alpha_2, \ldots, \alpha_N) \in \mathbb{C}^N$ corresponds an antilinear form $f \in X^*$ such that $(f, e_\varrho) = \alpha_\varrho$. In particular, to every subscript $i = 1, 2, \ldots, N$ corresponds an antilinear form $f_i \in X^*$ such that

$$(f_i, e_j) = \delta_{ij}, \qquad i, j = 1, 2, \ldots, N. \tag{2.4}$$

The covectors $f_1, f_2, \ldots, f_N$ are linearly independent and each $f \in X^*$ is a linear combination of $f_1, \ldots, f_N$. Therefore, the N covectors $f_1, \ldots, f_N$ form a basis of X^*, called the adjoint basis corresponding to the basis $e_1, \ldots, e_N$. Thus, $\dim X = \dim X^*$. The relations

$$x = \sum_{j=1}^{N} \overline{(f_j, x)}\, e_j, \qquad x \in X, \tag{2.5}$$

and

$$(f, x) = \sum_{j=1}^{N} (f, e_j)\,(f_j, x), \qquad f \in X^*, \quad x \in X, \tag{2.6}$$

are valid.

1.2.3. The space X^{**}, annihilator

$X^{**} = (X^*)^*$ is the linear space of all antilinear forms of X^*. For example

$$F_x(f) = \overline{(f, x)}$$

is such an antilinear form. Thus, to every $x \in X$ corresponds a unique antilinear form F_x of X^{**}. This correspondence is linear, i.e.,

$$F_{\alpha x+\beta y} = \alpha F_x + \beta F_y.$$

Because $\dim X^{**} = \dim X^* = \dim X$, the existence of other antilinear forms on X^* is impossible. The assignment $x \to F_x$ is an isomorphism. In the following we do not distinguish between x and F_x, that is, X and X^{**} are identified. Therefore, the relation

$$(x, f) = \overline{(f, x)}, \qquad f \in X^*, \quad x \in X,$$

is valid. If $f_1, \ldots, f_N$ is the adjoint basis corresponding to $e_1, \ldots, e_N$, then, conversely, $e_1, \ldots, e_N$ is the adjoint basis corresponding to $f_1, \ldots, f_N$.

If $(f, x) = 0$, then the notation $f \perp x$ is used and f is said to *annihilate* x (or conversely). If S is a subset of X and if $f \perp x$ for all $x \in S$, then f is said to annihilate S, $f \perp S$. The family of all f which annihilate S is denoted by $S^\perp$. $S^\perp$ is a subspace. It is called the *annihilator* of S. $(S^\perp)^\perp$ turns out to be the linear span spa S of S. If M is a subspace, then $(M^\perp)^\perp = M$.

1.3. Norms

In addition to antilinear forms (resp. linear forms), certain functions on X, called norms, play a special role. The introduction of norms is convenient, though not absolutely necessary, for the purposes of analysis on X. Usually, numerical considerations (estimates) in X are performed by means of a suitable norm.

If a fixed norm in X is introduced, then X is called a *normed* (linear) *space*. Every finite-dimensional space can be normed. Of course, many choices of norms are possible. The so-called unitary spaces are distinguished by a special choice of norms.

1.3.1. Definition and simple properties of norms

A real-valued function $\|x\|$ on X is called a *norm* if the following properties are satisfied:

1. $\|x\| \geqq 0$; $\|x\| = 0$ if and only if $x = 0$.
2. $\|\alpha x\| = |\alpha| \cdot \|x\|$ (*homogeneity*).
3. $\|x + y\| \leqq \|x\| + \|y\|$ (*triangle inequality*).

Therefore,

$$\big|\|x\| - \|y\|\big| \leqq \|x - y\|.$$

If $\|x\| = 1$, then x is called *normalized*. Normalizing a vector $x \neq 0$ means transition to $\|x\|^{-1} x$.

Let $e_1, e_2, \ldots, e_N$ be a basis in X and let $x = \sum_{\varrho=1}^{N} \xi_\varrho e_\varrho$. Then the functions

$$\|x\| = \max_{\varrho=1,\ldots,N} |\xi_\varrho|, \tag{3.1}$$

$$\|x\| = \left(\sum_{\varrho=1}^{N} |\xi_\varrho|^p \right)^{1/p}, \qquad 1 \leqq p < \infty, \tag{3.2}$$

are norms.

Any two norms $\|\cdot\|_1$ and $\|\cdot\|_2$ on X are equivalent, that is, there are constants $\alpha > 0$, $\beta > 0$ such that

$$\alpha \|x\|_1 \leqq \|x\|_2 \leqq \beta \|x\|_1, \qquad x \in X.$$

1.3.2. Norms on X^* and X^{**}

Let $\|\cdot\|$ be a fixed norm on X. Then X^* can be normed as follows:

$$\|f\| = \sup_{0 \neq x \in X} \frac{|(f, x)|}{\|x\|} = \sup_{\|f\|=1} |(f, x)|.$$

This norm $\|\cdot\|$ on X^* is uniquely assigned to the norm $\|\cdot\|$ on X. Obviously

$$|(f, x)| \leqq \|f\| \cdot \|x\| \quad (\textit{Schwarz's inequality})$$

and

$$\|x\| \geqq \frac{|(f, x)|}{\|f\|}, \qquad f \neq 0.$$

Moreover,

$$\|x\| = \sup_{0 \neq f \in X^*} \frac{|(f, x)|}{\|f\|} = \sup_{\|f\|=1} |(f, x)|.$$

This property is implied by the fact that to every $x_0 \in X$ corresponds an $f \in X^*$, where $\|f\| = 1$, such that

$$(f, x_0) = \|x_0\|.$$

Let F_x be the antilinear form on X^* assigned to $x \in X$. Then

$$\|x\| = \|F_x\|, \tag{3.3}$$

where the norm on X^{**} is defined in terms of the norm on X^*. Thus, the linear assignment $x \to F_x$ is isometric. Therefore, X and X^{**} may also be identified as normed spaces.

1.4. Unitary spaces

1.4.1. Inner product, norm and adjoint space

A complex linear space X of finite dimension is said to be ***unitary*** if there is a so-called inner product (x, y) on X satisfying the following properties:

1. $(x, y) = \overline{(y, x)}$.
2. $(\alpha x + \beta y, z) = \alpha(x, z) + \beta(y, z)$.
3. $(x, x) > 0$ if $x \neq 0$.

On a unitary space X one of the possible norms is distinguished, namely the so-called ***unitary norm***

$$\|x\| = (x, x)^{1/2}.$$

This norm satisfies Schwarz's inequality

$$|(x, y)| \leqq \|x\| \cdot \|y\|.$$

The space $\mathbb{C}^N$ is unitary with respect to

$$(x, y) = \sum_{i=1}^{N} x_i \bar{y}_i.$$

The unitary norm is

$$\|x\| = (\sum |x_i|^2)^{1/2}.$$

If X is unitary, then X and X^* may be identified. Let $x \in X$. Then $f_x(y) = (x, y)$ is an antilinear functional, uniquely assigned to the vector x. Because

$$\|f_x\| = \sup_{y \neq 0} \frac{|(x, y)|}{\|y\|} = \|x\|$$

and $\dim X = \dim X^*$, the assignment $x \leftrightarrow f_x$ from X onto X^* is one-to-one, linear and isometric. Thus, the identification of X and X^* is justified. The inner product (x, y) is then simultaneously the scalar product between vectors and covectors.

1.4.2. Orthogonality, orthonormal systems, Parseval's equation

If $(x, y) = 0$, one writes $x \perp y$ and x and y are called (mutually) ***orthogonal***. x is orthogonal to a subset S of X, in symbol $x \perp S$, if $x \perp y$ for all $y \in S$. The set of all $x \in X$ such that $x \perp S$ is called the ***annihilator*** of S and is denoted by $S^\perp$. Two subsets S and S' of X are called (mutually) ***orthogonal***, in symbol $S \perp S'$, if $x \perp y$ for all $x \in S$, $y \in S'$. $S^\perp$ is a subspace. If M is a subspace, then

$$X = M \oplus M^\perp. \tag{4.1}$$

$M^\perp$ is called the ***orthogonal complement*** of M in X.

A system of vectors $x_1, x_2, \ldots, x_n$ from X is called an *orthonormal system* if

$$(x_i, x_j) = \delta_{ij}.$$

The system is called an *orthonormal basis* if $k = N = \dim X$. An orthonormal basis coincides with its adjoint basis. Orthonormal systems satisfy Bessel's inequality

$$\sum_{i=1}^{k} |(u, x)|^2 \leqq \|u\|^2, \qquad u \in X.$$

For any linearly independent system $u_1, u_2, \ldots, u_n$, it is possible to construct an orthonormal system $x_1, x_2, \ldots, x_n$, such that the linear spans of $u_1, u_2, \ldots, u_k$ and $x_1, x_2, \ldots, x_k$ coincide for each $k = 1, 2, \ldots, n$.

If M_{k-1} denotes the linear spans of $u_1, \ldots, u_{k-1}$ and $x_1, \ldots, x_{k-1}$, then $X = M_{k-1} \oplus M_{k-1}^{\perp}$ and, correspondingly, $u_k = u_k' + u_k''$. Obviously, one has $u_k'' \neq 0$, therefore, x_k is defined by $x_k := \|u_k''\|^{-1} u_k''$. This construction is called the *Schmidt orthogonalization process*.

Thus, every subspace M, in particular X itself, has an orthonormal basis.

Let $e_1, e_2, \ldots, e_n$ be an orthonormal basis of X; then

$$x = \sum_{i=1}^{N} (x, e_i)\, e_i$$

and

$$(x, y) = \sum_i (x, e_i)\,(e_i, y) = \sum_i (x, e_i)\,\overline{(y, e_i)}.$$

In particular,

$$\|x\|^2 = \sum_i |(x, e_i)|^2 \quad (\textit{Parseval's identity}). \tag{4.2}$$

1.5. Linear operators

1.5.1. Definition and simple properties

Let X and Y be two linear spaces. A mapping A from X into Y is called *linear* or a *linear operator* of X into Y if A is compatible with the linear operations:

$$A(\alpha x + \beta y) = \alpha A x + \beta A y.$$

If $X = Y$, then A is called a linear operator or simply an *operator* on X.

In this book most operators are defined on a fixed linear space. The few exceptions are discussed separately.

The operator that sends every vector x into the zero vector is called the *zero operator*. One writes $0x = 0$.

The operator that sends every vector x into itself is called the *unit operator*. One writes $1x = x$.

If $y = Ax$, then y is called the *image* of x under A; x is called an *inverse image* of y under A.

The set of all images of A is a subspace ima A of X. It is called the *image* (or *range*) of A; dim ima A is called the *rank* of A and is denoted by ran A. The codimension codim ima A of ima A is called the *deficiency* of A and is denoted by def A. Thus

$$\operatorname{ran} A + \operatorname{def} A = \dim X. \tag{5.1}$$

The set of all inverse images of the zero vector is a subspace ker A, called the *kernel* or *null space* of A. One has

$$\dim \ker A = \operatorname{def} A.$$

1.5.2. Matrix elements

Let $e_1, e_2, \ldots, e_N$ be a basis in X, let $f_1, \ldots, f_N$ be the corresponding adjoint basis in X^*. Then

$$x = \sum_i (x, f_i)\, e_i$$

and

$$Ax = \sum_i (x, f_i)\, Ae_i.$$

Furthermore,

$$Ae_i = \sum_j (Ae_i, f_j)\, e_j.$$

The matrix

$$A = (\alpha_{ij}) = \big((Ae_j, f_i)\big)$$

is uniquely assigned to the operator A with respect to the basis $e_1, \ldots, e_N$. It is called the *matrix* of A with respect to the basis $e_1, \ldots, e_N$. The complex numbers (Ae_j, f_i) are called the *matrix elements* of A. One obtains

$$\begin{pmatrix} (Ax, f_1) \\ (Ax, f_2) \\ \vdots \\ (Ax, f_N) \end{pmatrix} = A \begin{pmatrix} (x, f_1) \\ (x, f_2) \\ \vdots \\ (x, f_N) \end{pmatrix}.$$

In general, each number (Ax, f), $x \in X$, $f \in X^*$ is called a *matrix element* of A.

If A denotes the matrix of A with respect to $e_1, \ldots, e_N$ and if there is a transition

$$e_i' = \sum_{i=1}^{N} \gamma_{ji} e_j, \qquad \Gamma = (\gamma_{ij}), \qquad \det \Gamma \neq 0,$$

to a new basis $e_1', \ldots, e_N'$, then the matrix A', assigned to A with respect to $e_1', \ldots, e_N'$, is given by the relation

$$A' = \Gamma^{-1} A \Gamma. \tag{5.2}$$

Thus, the family of all matrices B assigned to A is obtained from one of them by

$$B = \Gamma^{-1} A \Gamma, \tag{5.3}$$

where Γ runs through all regular matrices ($\det \Gamma \neq 0$).

Two matrices A and B which are related by (5.3) with some regular matrix Γ are called *similar*. One writes $A \sim B$. Obviously, this is an equivalence relation.

Thus, an operator $\mathbf{A}$ corresponds to an equivalence class of similar matrices. Sometimes these classes are called classes of similarity.

If $A \sim B$, then $\det A = \det B$ and $\operatorname{tr} A = \operatorname{tr} B$. Thus, these numbers are unique functions of the classes of similarity.

If A is a matrix of $\mathbf{A}$, then $\det A$ is called the *determinant* of $\mathbf{A}$; one writes $\det \mathbf{A} := \det A$. Similarly, $\operatorname{tr} A$ is called the *trace* of $\mathbf{A}$; one writes $\operatorname{tr} \mathbf{A} := \operatorname{tr} A$.

According to the foregoing definition, the similarity between two operators $\mathbf{A}$ and $\mathbf{B}$, $\mathbf{A} \sim \mathbf{B}$, is defined by the existence of an invertible operator $\mathbf{C}$, such that $\mathbf{A} = \mathbf{C}\mathbf{B}\mathbf{C}^{-1}$ (see Subsect. 1.6.2).

1.6. Linear operations for linear operators

1.6.1. Linear combinations, products, invertibility, determinant and trace

If $\mathbf{A}$ and $\mathbf{B}$ are two linear operators on X and if $\alpha, \beta \in \mathbb{C}$, then their linear combination $\alpha\mathbf{A} + \beta\mathbf{B}$ is defined by

$$(\alpha\mathbf{A} + \beta\mathbf{B})\, x = \alpha\mathbf{A}x + \beta\mathbf{B}x, \qquad x \in X.$$

With respect to these operations, the linear operators form a linear space of dimension N^2. It is denoted by $\mathscr{L}(X)$.

The product of two operators $\mathbf{A}, \mathbf{B}$ is defined by

$$(\mathbf{A}\mathbf{B})\, x = \mathbf{A}(\mathbf{B}x), \qquad x \in X.$$

The product is associative, $(\alpha\mathbf{A})\,\mathbf{B} = \mathbf{A}(\alpha\mathbf{B}) = \alpha(\mathbf{A}\mathbf{B})$, and both laws of distributivity are valid.

If $\mathbf{A}\mathbf{B} = \mathbf{B}\mathbf{A}$, then $\mathbf{A}$ and $\mathbf{B}$ are called *commuting*, sometimes one writes $\mathbf{A} \cup \mathbf{B}$. In general, $\mathbf{A}\mathbf{B} \neq \mathbf{B}\mathbf{A}$.

Summarizing, the set of all linear operators on X is said to be an *algebra*. The unit operator is the unit element of the algebra.

The operators of the form $\alpha\mathbf{1}$ are called *scalar operators*. They commute with all operators. Sometimes, instead of $\alpha\mathbf{1}$, one writes simply α.

If $\dim \ker \mathbf{A} = \operatorname{def} \mathbf{A} = 0$, then $\mathbf{A}$ is one-to-one and X is mapped onto itself by $\mathbf{A}$. In this case, the *inverse operator* $\mathbf{A}^{-1}$, which sends $\mathbf{A}x$ into x, is defined. One has

$$\mathbf{A}\mathbf{A}^{-1} = \mathbf{A}^{-1}\mathbf{A} = \mathbf{1}.$$

A^{-1} exists if and only if $\det A \neq 0$. Such operators are called *invertible* (or *regular*, or *nonsingular*). One has $(A^{-1})^{-1} = A$ and $(AB)^{-1} = B^{-1}A^{-1}$.

For the determinant and trace functions of an operator introduced above one obtains the relations

$$\det AB = \det A \cdot \det B,$$

$$\operatorname{tr}(\alpha A + \beta B) = \alpha \operatorname{tr} A + \beta \operatorname{tr} B,$$

$$\operatorname{tr}(AB) = \operatorname{tr}(BA).$$

1.6.2. Powers, operator polynomials

One writes

$$AA = A^2, \qquad AAA = A^3, \ldots, \text{etc},$$

and $A^0 = 1$. Thus

$$A^m A^n = A^{m+n}, \qquad (A^m)^n = A^{mn}, \qquad m, n = 0, 1, 2, \ldots \tag{6.1}$$

If $p(\lambda) = \alpha_0 + \alpha_1\lambda + \alpha_2\lambda^2 + \cdots + \alpha_n\lambda^n$ is a polynomial from $\mathbb{C}[\lambda]$, then the operator $p(A)$ is defined by

$$p(A) = \alpha_0 1 + \alpha_1 A + \alpha_2 A^2 + \cdots + \alpha_n A^n.$$

The mapping $p(\lambda) \mapsto p(A)$ is a homomorphism from $\mathbb{C}[\lambda]$ into $\mathscr{L}(X)$, because $p(\lambda) + q(\lambda) = r(\lambda)$ and $p(\lambda)\,q(\lambda) = r(\lambda)$ imply $p(A) + q(A) = r(A)$ and $p(A)\,q(A) = r(A)$, respectively.

Let $A \in \mathscr{L}(X)$. Then there is a number m, $0 \leqq m \leqq \dim X$, such that

$$\operatorname{ima}(1) \supset \operatorname{ima}(A) \supset \operatorname{ima}(A^2) \supset \cdots \supset \operatorname{ima}(A^m) = \operatorname{ima}(A^{m+1}) = \cdots \tag{6.2}$$

and

$$\ker(1) \subset \ker(A) \subset \ker(A^2) \subset \cdots \subset \ker(A^m) = \ker(A^{m+1}) = \cdots. \tag{6.3}$$

A is called *nilpotent* if there is a number m such that $A^m = 0$. Then $\ker A^m = X$ and (6.3) implies $A^{\dim X} = 0$. This relation is true for all nilpotent operators.

If A is invertible, then the negative powers of A are defined by $A^{-n} := (A^{-1})^n$. In this case, the relations (6.1) are valid for arbitrary integers.

1.7. Adjoint operator

1.7.1. Definition and simple properties

Let $A \in \mathscr{L}(X)$. Then (f, Ax) is an antilinear form on X. Thus, there is exactly one $g \in X^*$ such that

$$(f, Ax) = (g, x), \qquad x \in X. \tag{7.1}$$

The assignment $f \to g$ defines the adjoint operator A^*; setting $g = A^*f$ one has

$$(A^*f, x) = (f, Ax), \qquad f \in X^*, \quad x \in X. \tag{7.2}$$

A^* is a linear operator, $A^* \in \mathscr{L}(X^*)$. Moreover,

$$(\alpha A + \beta B)^* = \bar{\alpha} A^* + \bar{\beta} B^*,$$

$$(AB)^* = B^* A^*,$$

$$A^{**} = A.$$

If A is invertible, then A^* is invertible, and one has $(A^*)^{-1} = (A^{-1})^*$. Concerning kernel and image, one has the relations

$$\ker (A^*) = (\operatorname{ima} A)^{\perp}, \qquad \ker A = (\operatorname{ima} A^*)^{\perp}. \tag{7.3}$$

Because $\operatorname{def} A = \dim \ker A$, the relations

$$\dim \ker A^* = \operatorname{def} A, \qquad \dim \ker A = \operatorname{def} A^*,$$

$$\dim \ker A^* = \dim \ker A, \qquad \operatorname{def} A^* = \operatorname{def} A$$

are valid.

1.7.2. Matrix representations

Let $e_1, e_2, \ldots, e_N \in X$, $f_1, f_2, \ldots, f_N \in X^*$ be adjoint bases and let $A = (\alpha_{ij})$ be the assigned matrix of A with respect to the basis $e_1, \ldots, e_N$. Then, because

$$\alpha_{ij} = (Ae_j, f_i) = \overline{(f_i, Ae_j)} = \overline{(A^*f_i, e_j)},$$

the assigned matrix $A^* = (\alpha_{ij}^*)$ of A^* with respect to $f_1, \ldots, f_N$ satisfies the relation

$$\alpha_{ij}^* = (A^*f_j, e_i) = \bar{\alpha}_{ji}.$$

Hence

$$A^* = \bar{A}^{\mathsf{T}}.$$

Thus,

$$\det A^* = \overline{\det A}, \qquad \operatorname{tr} A^* = \overline{\operatorname{tr} A}.$$

1.8. Norms of operators

For the purpose of analysis in $\mathscr{L}(X)$, the introduction of operator norms is convenient. On the one hand, one can define arbitrary norms in $\mathscr{L}(X)$. On the other hand, as in the case of X^*, it is possible to define assigned operator norms, which have especially convenient properties.

1.8.1. Definition (assigned to a given vector norm)

Let $A \in \mathscr{L}(X)$. Then one defines

$$\|A\| = \sup_{0 \neq x \in X} \frac{\|Ax\|}{\|x\|} = \sup_{\|x\|=1} \|Ax\|.$$

$\|A\|$ is a real-valued function on $\mathscr{L}(X)$, which satisfies the characteristic norm properties:

$$\|A\| \geqq 0, \qquad \|A\| = 0 \quad \text{if and only if} \quad A = 0,$$

$$\|\alpha A\| = |\alpha| \cdot \|A\|,$$

$$\|A + B\| \leqq \|A\| + \|B\|.$$

Moreover,

$$\|Ax\| \leqq \|A\| \cdot \|x\|$$

and

$$\|AB\| \leqq \|A\| \cdot \|B\|.$$

Because of the second representation of the vector norm, the relation

$$\|A\| = \sup_{\substack{0 \neq x \in X \\ 0 \neq f \in X^*}} \frac{|(Ax, f)|}{\|f\| \cdot \|x\|} = \sup_{\substack{\|x\|=1 \\ \|f\|=1}} |(Ax, f)|$$

follows. One has

$$\|A\| = \|A^*\|.$$

1.8.2. Norms of operators as norms on the N^2-dimensional space

All the norms defined in Subsect. 1.8.1 are equivalent (when considered as norms in an N^2-dimensional linear space). A special norm in $\mathscr{L}(X)$ can be defined by a given basis $e_1, \ldots, e_N$ in X as follows:

$$\|A\|_1 = \max_{i,j=1,\ldots,N} |(Ae_i, f_j)|.$$

Because all norms are equivalent, this definition implies

$$|(Ae_i, f_j)| \leqq \gamma \|A\|, \tag{8.1}$$

$$\|A\| \leqq \delta \max_{i,j=1,\ldots,N} |(Ae_i, f_j)|, \tag{8.2}$$

where $\gamma > 0$, $\delta > 0$ are constants, which depend only on the norm $\|\cdot\|$ and the basis chosen, but not on A.

Concerning the norm of A^{-1}, the following estimate is true:

$$\|A^{-1}\| \leqq \gamma\, |\det A|^{-1}\, \|A\|^{N-1}, \qquad \gamma > 0, \tag{8.3}$$

where γ is independent of A, but depends on the norm used. (For example, $\gamma = 1$ for a unitary space X.)

1.8.3. Norm of A^n, spectral radius

The relations

$$\|A^n\| \leqq \|A\|^n, \qquad \|A^n\|^{1/n} \leqq \|A\|, \qquad n = 1, 2, \ldots,$$

are valid. The limit $\lim_{n\to\infty} \|A^n\|^{1/n}$ exists, and

$$\lim_{n\to\infty} \|A^n\|^{1/n} = \inf_n \|A^n\|^{1/n}.$$

This number is called the *spectral radius* of A. One writes spr A. The spectral radius of A is independent of the norm used.

1.9. Projections

1.9.1. Definition

Let $X = M \oplus N$. Then M and N are said to be *complementary*. In this case, every vector $x \in X$ can be uniquely represented in the form

$$x = m + n, \qquad m \in M, \qquad n \in N.$$

m is called the *projection* of x on M along N. The operator

$$Px = m$$

is linear, $P \in \mathscr{L}(X)$. P is called the projection on M along N. Then, because of the relation

$$(1 - P)\, x = n,$$

$1 - P$ is the projection on N along M.

$Px = x$ if and only if $x \in M$; $Px = 0$ if and only if $x \in N$. One has ima $P = M$ and ker $P = N$. The number dim ima P is called the *dimension* of the projection P; one writes dim $P :=$ dim ima P. One has $P(Px) = Px$ for all $x \in X$; hence,

$$P^2 = P. \tag{9.1}$$

Thus a projection P is idempotent. Conversely, if P is idempotent, then P is a projection, and

$$X = \text{ima}\, P \oplus \ker P.$$

Thus, P is the projection on ima P along ker P. One always has

$$\|P\| \geqq 1, \qquad P \neq 0. \tag{9.2}$$

If P and Q are projections and if $PQ = QP$, then PQ is also a projection, and ima PQ = ima $P \cap$ ima Q.

1.9.2. Disjoint systems of projections, completeness

Let $P_1, P_2, \ldots, P_r$ be a system of projections satisfying the property

$$P_i P_j = \delta_{ij} P_i; \qquad i, j = 1, 2, \ldots, r. \tag{9.3}$$

Then the projections $P_1, \ldots, P_r$ are called *pairwise* (or *mutually*) *disjoint* and the whole system is called *disjoint*. If $M_i = \text{ima } P_i$, then the sum of all $M_1, M_2, \ldots, M_r$ is a direct sum, $M_1 \oplus M_2 \oplus \cdots \oplus M_r$.

The disjoint system of projections $P_1, P_2, \ldots, P_r$ is called *complete* if $\sum_{i=1}^{r} P_i = 1$. Then,

$$X = M_1 \oplus M_2 \oplus \cdots \oplus M_r, \qquad M_i = \text{ima } (P_i). \tag{9.4}$$

A complete disjoint system of projections always corresponds to a decomposition of X into a direct sum of subspaces, namely the image spaces of the projections of the given system.

Conversely, if a decomposition (9.4) is given, then the projections assigned uniquely to the subspaces M_i form a complete disjoint system of projections. The projection P_i is defined by the projection on M_i along $M_1 \oplus M_2 \oplus \cdots \oplus M_{i-1} \oplus M_{i+1} \oplus \cdots \oplus M_r$.

1.9.3. Adapted bases

Let a decomposition (9.4) of X be given, where $m_i = \dim M_i$. A basis $x_\varrho^{(i)}$, $i = 1, 2, \ldots, r$; $\varrho = 1, 2, \ldots, m_i$, of X is called a basis *adapted to the decomposition* (9.4) if the vectors $x_\varrho^{(i)}$ form a basis of M_i for each index i.

If $r = 2$, i.e., $X = M \oplus N$, and if the basis $x_\varrho^{(i)}$, $i = 1, 2$, is adapted to this decomposition, then the assigned projections P and Q have a simple matrix with respect to this basis, namely

$$P \leftrightarrow \begin{pmatrix} E_{m_1} & 0 \\ 0 & 0 \end{pmatrix}, \qquad Q \leftrightarrow \begin{pmatrix} 0 & 0 \\ 0 & E_{m_2} \end{pmatrix},$$

where m_i denotes the $m_i \times m_i$-unit matrix. One always has $\dim P = \text{tr } P$.

1.9.4. Subprojections

Let P and Q be projections. P is called a *subprojection* of Q if $PQ = QP$ and if ima $P \subseteq$ ima Q. One writes $P \leqq Q$. If the stronger inclusion ima $P \subset$ ima Q holds, then one writes $P < Q$.

If $PQ = QP$, then $PQ \leqq P$ and $PQ \leqq Q$.

P is a subprojection of Q if and only if $PQ = QP = P$.

If $P \leqq Q$ and $Q \leqq P$, then $P = Q$.

The inequalities $P \leqq Q$ and $Q \leqq S$ (where S is a projection) imply $P \leqq S$.

A complete disjoint system of projections defines a hierarchy of subprojections in a natural way.

1.10. Linear operators on unitary spaces

1.10.1. Adjoint operators

Let H be a unitary space. Then, because $H = H^*$, the adjoint operator A^* of A is also an element of $\mathscr{L}(H)$. It is well-defined by the equation

$$(A^*x, y) = (x, Ay), \qquad x, y \in H. \tag{10.1}$$

If $e_1, e_2, \ldots, e_N$ is an orthonormal basis (i.e., a "selfadjoint" basis), then

$$(Ae_i, e_j) = (e_i, A^*e_j) = \overline{(A^*e_j, e_i)}.$$

Thus, the matrices A and A^* assigned to A and A^*, respectively, satisfy the relation $A^* = \overline{A}^\mathsf{T}$.

The operator A is called *selfadjoint* if $A = A^*$. For the assigned matrix A (with respect to an orthonormal basis) one has $A = \overline{A}^\mathsf{T}$ (Hermitean symmetric).

For a selfadjoint operator A, the assigned form (Ax, y) is Hermitean symmetric, i.e., one has

$$(Ax, y) = \overline{(Ay, x)}, \qquad x, y \in H.$$

In particular, (Ax, x) is real for all $x \in H$.

A is called *nonnegative*, $A \geqq 0$, if $(Ax, x) \geqq 0$ for all $x \in H$. A is called *positive*, $A > 0$, if $(Ax, x) > 0$ for all $x \neq 0$. This definition implies $(Ax, x) \geqq \gamma\|x\|^2$, where γ is a positive constant, $\gamma > 0$. If $A \geqq 0$, then

$$|(Ax, y)| \leqq (Ax, x)^{1/2} (Ay, y)^{1/2},$$

$$\big(A(x + y), x + y\big)^{1/2} \leqq (Ax, x)^{1/2} + (Ay, y)^{1/2}.$$

1.10.2. Normal and unitary operators, orthogonal projections

An operator $U \in \mathscr{L}(X)$ is called *unitary* if $\|Ux\| = \|x\|$ for all $x \in H$. Equivalently, $U^*U = UU^* = 1$. U^* is unitary whenever U is.

An operator A is called *normal* if $AA^* = A^*A$. Selfadjoint and unitary operators are special cases of normal operators. An equivalent formulation of $AA^* = A^*A$ is given by

$$\|A^*x\| = \|Ax\|, \qquad x \in H.$$

If A is normal, then

$$\|A^n\| = \|A\|^n, \qquad n = 1, 2, \ldots$$

In particular, one has

$$\operatorname{spr} A = \|A\|.$$

Let P be the projection on M along $M^\perp$. Then $P = P^*$ (and also $P \geqq 0$). Conversely, if P is a projection and $P = P^*$, then P is the projection on ima P along (ima $P)^\perp$.

Thus, selfadjoint projections, but no other projections, lead to orthogonal decompositions of H. Therefore, they are called *orthogonal projections*.

Disjoint systems of orthogonal projections project onto mutually orthogonal subspaces.

The sum of mutually disjoint orthogonal projections is again an orthogonal projection.

If the projection $P \neq 0$ is orthogonal, then $\|P\| = 1$.

If P is a normal projection, then P is orthogonal.

1.11. Operator algebras

The algebra $\mathcal{L}(X)$ of all linear operators on X will be introduced in Sect. 1.6. For some later applications (investigation of the splitting of eigenvalues by analytic perturbations) it is also useful to consider subalgebras of $\mathcal{L}(X)$, especially subalgebras connected with the concept of commutativity.

1.11.1. Definition, sum and intersection of algebras

A set of operators $\mathcal{A} \subset \mathcal{L}(X)$ is called a *subalgebra* (or simply an *algebra*) if it contains, together with A and B, $\alpha A + \beta B$ and AB.

$\mathcal{L}(X)$ is an algebra, as is the set of all scalar operators of $\mathcal{L}(X)$.

If an algebra $\mathcal{A}$ contains the unit operator, then it is called an *algebra with* 1.

The family of all polynomials $\alpha_0 + \alpha_1 A + \cdots + \alpha_n A^n$ of A is a commutative algebra with 1. It is denoted by $\mathcal{P}(A)$.

Let $\mathcal{A}$ and $\mathcal{B}$ be algebras. Then $\mathcal{A} \cap \mathcal{B}$ is again an algebra.

In general the union (in the sense of set theory) of two algebras is not an algebra. The algebra

$$\mathcal{C} = \bigcap_{\mathcal{R} \supset \mathcal{A} \cap \mathcal{B}} \mathcal{R}$$

is called the *sum* of $\mathcal{A}$ and $\mathcal{B}$ and is denoted by $\mathcal{C} = \mathcal{A} \cup \mathcal{B}$.

Let $A_1, A_2, \ldots, A_r$ be operators. Then the algebra $\mathcal{P}(A_1) \cup \mathcal{P}(A_2) \cup \cdots \cup \mathcal{P}(A_r)$ is called the *algebra generated by* $A_1, A_2, \ldots, A_r$ *and* 1. Every algebra $\mathcal{A}$ with 1 can be generated by finitely many operators.

1.11.2. Commutants

Let A be an operator. Then the set of all operators C which commute with A forms an algebra with 1, called the *commutant* of A. It contains $\mathcal{P}(A)$, but it is larger in general.

Let $\mathscr{A}$ be an algebra. Then the set of all operators C for which

$$CA = AC, \qquad A \in \mathscr{A},$$

forms an algebra with 1. It is denoted by com $\mathscr{A}$ and is called the *commutant* of $\mathscr{A}$. The relations

$$\text{com}\,(\mathscr{A} \cup \mathscr{B}) = \text{com}\,\mathscr{A} \cap \text{com}\,\mathscr{B}, \tag{11.1}$$

$$\text{com}\,(\mathscr{A} \cap \mathscr{B}) = \text{com}\,\mathscr{A} \cup \text{com}\,\mathscr{B} \tag{11.2}$$

are valid.

B. Analysis on finite-dimensional complex linear spaces

In this part the concepts of elementary real and complex analysis for vector- and operator-valued functions of a scalar (real or complex) argument are summarized.

1.12. Convergence

1.12.1. Definition

A sequence $x_1, x_2, x_3, \ldots$ from X is said to be *convergent to zero* (*zero sequence*), if

$$\lim_{\varrho\to\infty} (f, x_\varrho) = 0, \qquad f \in X^*, \tag{12.1}$$

is valid for all $f \in X^*$. The sequence $x_1, x_2, x_3, \ldots$ is said to be *convergent to the limit* x if the sequence $\{x_\varrho - x\}$ converges to zero. One writes $x = \lim\limits_{\varrho\to\infty} x_\varrho$ or $x_\varrho \to x$.

It is sufficient to require that the relation (12.1) is valid for a basis $f_1, \ldots, f_N$ of X^*. That is, $\{x_\varrho\}$ is a zero sequence if and only if each of the N sequences of coordinates of $\{x_\varrho\}$ converges to zero with respect to any basis in X.

Equivalently, the relation $x = \lim\limits_{\varrho\to\infty} x_\varrho$ can be expressed by means of an arbitrary norm on X,

$$\|x_\varrho - x\| \to 0.$$

Because of the equivalence of all norms, all norms lead to one and the same concept of convergence. Moreover, the norm $\|x\| = \max |\xi_\varrho|$ for $x = \sum\limits_\varrho \xi_\varrho e_\varrho$ shows that this concept of convergence coincides with the former.

1.12.2. Continuity, boundedness, theorem of Bolzano-Weierstrass

The linear operations are continuous with respect to convergence, i.e., $\alpha_n \to \alpha$, $\beta_n \to \beta$, $x_n \to x$, and $y_n \to y$ imply $\alpha_n x_n + \beta_n y_n \to \alpha x + \beta y$.

The scalar product is continuous with respect to both variables, i.e., $x_n \to x$ and $f_n \to f$ imply $(f_n, x_n) \to (f, x)$.

Any norm $\|x\|$ is continuous with respect to x because $\big|\|x\| - \|y\|\big| \leqq \|x - y\|$.

A sequence $x_1, x_2, x_3, \ldots$ is convergent if and only if it is a so-called *Cauchy-sequence*, i.e., if

$$\lim_{m,n\to\infty} \|x_n - x_m\| = 0.$$

A sequence $\{x_n\}$ from X is called *bounded* if the sequence $\{(f, x_n)\}$ is bounded for every $f \in X^*$. Equivalently, there exists an (open) ball $\|x\| < R$, such that each vector of the sequence lies inside the ball.

According to Bolzano-Weierstrass, every bounded sequence has a convergent subsequence (X is locally compact).

With respect to one (hence all) norms on X, the usual topological concepts (closed and open sets, etc.) are used.

1.12.3. Infinite series of vectors, absolute convergence

The infinite series $\sum_{n=1}^{\infty} x_n$ is said to be *convergent* if the sequence $\{s_n\}$ consisting of the partial sums $s_n = \sum_{\varrho=1}^{n} x_\varrho$ converges. If $s_n \to s$, then one writes $s = \sum_{\varrho=1}^{\infty} x_\varrho$. Thus, the convergence of the series $\sum_{n=1}^{\infty} x_n$ is equivalent with the convergence of the series $\sum_{n=1}^{\infty} (f, x_n)$ for all $f \in X^*$, or the convergence of each of the N coordinate series $\sum_{n=1}^{\infty} x_n^{(\varrho)}$, $\varrho = 1, 2, \ldots, N$, with respect to any basis $e_1, \ldots, e_N$, where $x_n = \sum x_n^{(\varrho)} e_\varrho$. Then,

$$\sum_{n=1}^{\infty} (f, x_n) = (f, s), \qquad f \in X^*.$$

The series $\sum_{n=1}^{\infty} x_n$ is said to be *absolutely convergent* if the series $\sum_{n=1}^{\infty} |(f, x_n)|$ is convergent for all $f \in X^*$, i.e., if the series $\sum_{n=1}^{\infty} (f, x_n)$ are absolutely convergent.

Again, it is sufficient to require this property for a certain basis in X^*, that is, for each of the N coordinate series with respect to some basis in X.

If $\sum_{n=1}^{\infty} \|x_n\| < \infty$ for some norm $\|\cdot\|$, then the series $\sum x_n$ is absolutely convergent.

Let $\sum_{n=1}^{\infty} |(f, x_n)| < \infty$ for all $f \in X^*$, in particular for a basis $f_1, f_2, \ldots, f_N$ of X^*. Then $\sum_{n=1}^{\infty} \sum_{\varrho=1}^{N} |(f_\varrho, x_n)| < \infty$; hence, $\sum_{n=1}^{\infty} \|x_n\| < \infty$, where $\|\cdot\|$ denotes the norm defined by (3.2) and $p = 1$ with respect to the basis which is adjoint to $f_1, \ldots, f_N$. Thus, $\sum \|x_n\| < \infty$ for all norms.

For absolutely convergent series one has

$$\|\sum x_n\| \leqq \sum \|x_n\|.$$

As in the scalar case, in an absolutely convergent series the order of the summands may be changed without changing the limit of the series.

A corresponding result is true for double series.

1.13. Vector-valued functions of a scalar argument

1.13.1. Vector-valued functions of a real argument, limit, continuity, differentiation

Let I denote an interval of the real line $(-\infty, \infty)$. Single-valued mappings $x(t)$ from I into X are called *vector-valued* functions on I.

$y \in X$ is called a *limit* of the vector-valued function $x(t)$ as $t \to a$, $y = \lim x(t)$, if $(f, y) = \lim\limits_{t \to a} \big(f, x(t)\big)$ for all $f \in X^*$. Equivalently, $\lim\limits_{t \to a} \|x(t) - y\| = 0$ for some norm.

The function $x(t)$ is called *continuous* at $t = a$ if $\lim\limits_{t \to a} x(t)$ exists and is equal to $x(a)$. $x(t)$ is said to be *continuous* on I if $x(t)$ is continuous at every point of I. The continuity of $x(t)$ implies the continuity of $\|x(t)\|$ on I.

$x(t)$ is said to be *differentiable* at $t = a$ with derivative $x'(a)$ at this point if

$$\lim_{h \to 0} \frac{x(a + h) - x(a)}{h} = x'(a)$$

exists. The usual rules for differentiation are satisfied. Moreover, if $\alpha(t)$ is a complex-valued scalar function and if $\alpha(t)$ is differentiable at $t = a$, then

$$\big(\alpha(t)\, x(t)\big)' = \alpha'(t)\, x(t) + \alpha(t)\, x'(t).$$

1.13.2. Riemann integrals

Let $x(t)$ be a continuous vector-valued function on the finite interval $[a, b]$. Then, the Riemann integral $\int\limits_a^b x(t)\, dt$ is defined as follows: For every $f \in X^*$, the scalar function $\big(f, x(t)\big)$ is continuous on $[a, b]$. Then the Riemann integral $\int\limits_a^b \big(x(t), f\big)\, dt$ exists. This expression is an antilinear form on X^*. Thus there exists exactly one vector $y \in X$, such that

$$\int\limits_a^b \big(x(t), f\big)\, dt = (y, f), \qquad f \in X^*,$$

One writes

$$y = \int\limits_a^b x(t)\, dt.$$

If $x(t) = \sum_{i=1}^{N} \alpha_i(t)\, e_i$, then $\int_a^b x(t)\, \mathrm{d}t = \sum_{i=1}^{N} \int_a^b \alpha_i(t)\, \mathrm{d}t e_i$. Thus, a continuous vector-valued function is integrated by means of its coordinates (with respect to any basis).

The integral is linear with respect to the integrand,

$$\int \big(\alpha x(t) + \beta y(t)\big)\, \mathrm{d}t = \alpha \int x(t)\, \mathrm{d}t + \beta \int y(t)\, \mathrm{d}t.$$

Furthermore,

$$\left\| \int_a^b x(t)\, \mathrm{d}t \right\| \leqq \int_a^b \|x(t)\|\, \mathrm{d}t$$

with respect to any norm.

Of course, $\int_a^b x(t)\, \mathrm{d}t$ may be defined as a limit of Riemann sums $\sum_\varrho x(\tau_\varrho)\, (t_{\varrho+1} - t_\varrho)$ as $\max_\varrho |t_{\varrho+1} - t_\varrho| \to 0$.

1.13.3. Vector-valued functions of a complex argument, Riemann integral, holomorphy, Cauchy's integral theorem

Let G be a region of the complex plane. Single-valued mappings $x(z)$ from G into X are called *vector-valued functions* on G. The concepts of limit, continuity and derivative for $x(z)$ are defined as in the case of a real argument.

Let $\mathscr{C}$ be a piecewise smooth bounded path in G and let $x(z)$ be a continuous vector-valued function in G. The Riemann integral $\int_{\mathscr{C}} x(z)\, \mathrm{d}z$ is again defined by

$$\left(f, \int_{\mathscr{C}} x(z)\, \mathrm{d}z\right) = \int_{\mathscr{C}} \big(f, x(z)\big)\, \mathrm{d}z, \qquad f \in X^*.$$

The integral is linear with respect to the integrand and

$$\left\| \int_{\mathscr{C}} x(z)\, \mathrm{d}z \right\| \leqq \int_{\mathscr{C}} \|x(z)\| \cdot |\mathrm{d}z|$$

for any norm.

$x(z)$ is said to be *holomorphic* in G if $x'(z)$ exists for all $z \in G$. If $x(z)$ is a holomorphic function in a simply connected region G, then Cauchy's integral theorem

$$\int_{\mathscr{C}} x(z)\, \mathrm{d}z = 0$$

is valid for arbitrary closed piecewise smooth paths lying in G.

If $\mathscr{C}$ denotes a simply closed positively oriented path, such that z lies inside $\mathscr{C}$, then Cauchy's integral formulas

$$x^{(n)}(z) = \frac{n!}{2\pi \mathrm{i}} \int_{\mathscr{C}} \frac{x(\zeta)}{(\zeta - z)^{n+1}}\, \mathrm{d}\zeta, \qquad n = 0, 1, 2, \ldots,$$

are valid.

1.13.4. Power series, uniform convergence

Let $\sum_{n=0}^{\infty} a_n(z - z_0)^n$ be a power series, $a_n \in X$. The convergence of this series at some point z is equivalent with the convergence of all series

$$\sum_{n=0}^{\infty} (a_n, f)\,(z - z_0)^n = \left(\sum_{n=0}^{\infty} a_n(z - z_0)^n, f\right), \qquad f \in X^*.$$

If f is fixed, then, according to Cauchy-Hadamard, this series has the *radius of convergence*

$$r(f) = \frac{1}{\limsup\limits_{n\to\infty} |(a_n, f)|^{1/n}}.$$

If $r = \inf\limits_{f \in X^*} r(f)$, then the radius of convergence of the original power series is r. In other words, the circle of convergence is the smallest circle centered at z_0, such that all power series of the coordinates converge with respect to some basis.

The sequence of functions $x_n(z)$, $z \in G$, $n = 1, 2, \ldots$, is said to *converge uniformly* in G to the limit $x(z)$ if the sequences $(f, x_n(z))$ converge uniformly to $(f, x(z))$, i.e., if

$$\sup_{z \in G} \left|\left(f, x_n(z) - x(z)\right)\right| \to 0$$

for all $f \in X^*$, or equivalently,

$$\sup_{z \in G} \|x_n(z) - x(z)\| \to 0$$

for any norm.

Power series converge absolutely uniformly in closed subregions of the circle of convergence. They represent holomorphic functions inside the circle of convergence. A vector-valued function $x(z)$ is holomorphic in G if and only if there is a power series expansion (Taylor's expansion) for $x(z)$ in the neighborhood of each point $\zeta \in G$.

1.13.5. Laurent series; polynomial, rational and meromorphic vector-valued functions

Let $x(z)$ be a holomorphic vector-valued function with an isolated singularity ζ, i.e., $x(z)$ is holomorphic in a dotted circle $0 < |z - \zeta| < r$. Then there is a Laurent series expansion

$$x(z) = \sum_{n=-\infty}^{\infty} a_n(z - \zeta)^n, \qquad a_n \in X, \qquad 0 < |z - \zeta| < r.$$

in the neighborhood of ζ. The classification of the singularities (removable singularities, poles and essential singularities) is done as usual. The expansion coefficients a_n are given by

$$a_n = \frac{1}{2\pi i} \int\limits_{\mathscr{C}} \frac{x(z)}{(z - \zeta)^{n+1}}\, dz,$$

where $\mathscr{C}$ is a suitable simply closed positively oriented piecewise smooth path, such that ζ lies inside $\mathscr{C}$.

The simplest holomorphic vector-valued functions are the polynomials

$$p(z) = \sum_{k=0}^{m} a_k z^k, \qquad a_k \in X,$$

and the rational functions, $r(z) = p(z)/\alpha(z)$, where $p(z)$ is a polynomial and $\alpha(z)$ denotes a nonvanishing scalar polynomial.

$p(z)$ is a polynomial if and only if $\big(p(z), f\big)$ is a scalar polynomial for all $f \in X^*$. $r(z)$ is a rational function if and only if $\big(r(z), f\big)$ is a scalar rational function for all $f \in X^*$. Thus, $p(z)$ and $r(z)$ are polynomial and rational if and only if all coordinates are polynomial and rational, respectively, with respect to some basis.

A vector-valued function $m(z)$ is called *meromorphic* if the only singularities of $m(z)$ in the complex plane are poles (which may accumulate at infinity). $m(z)$ is called *meromorphic in the region* G if the only singularities of $m(z)$ in G are poles (which may accumulate at the boundary ∂G of G). Thus, the set of all poles of a meromorphic function in G is isolated and closed in G (accumulation points of this set are to be found at most on the boundary ∂G).

1.14. Analysis of linear operators

1.14.1. Convergence; continuity of the linear operations, products, norms and Ax

The sequence $A_1, A_2, A_3, \ldots$ from $\mathscr{L}(x)$ is said to be *convergent to zero* (*zero sequence*) if $A_n x \to 0$ for all $x \in X$. It is said to be *convergent with limit* B if $A_n - B$ is a zero sequence. One writes $A_n \to B$.

$A_n x$ is a zero sequence for all x if and only if $A_n e_i$ is a zero sequence for all vectors $e_1, e_2, \ldots, e_N$ of a basis. But this is true if and only if $(A_n e_i, f_j)$ is a zero sequence for all $i, j = 1, 2, \ldots, N$, where $f_1, f_2, \ldots, f_N$ denotes the adjoint basis. Thus, A_n is a zero sequence if and only if all matrix elements of A_n are zero sequences with respect to some (hence all) bases. Furthermore, $A_n \to B$ is equivalent with $(A_n x, f) \to (Bx, f)$ for all $x \in X, f \in X^*$.

If some norm $\|\cdot\|$ in $\mathscr{L}(X)$ is fixed, then $A_n \to B$ is equivalent with $\|A_n - B\| \to 0$ (cf. Subsect. 1.8.2). The sequence A_n is convergent if and only if A_n is a Cauchy sequence, i.e., if $\lim\limits_{m,n\to\infty} \|A_n - A_m\| = 0$.

$\alpha A + \beta B$ is a continuous function of its variables, i.e. $\alpha_n \to \alpha$, $\beta_n \to \beta$, $A_n \to A$ and $B_n \to B$ imply $\alpha_n A_n + \beta_n B_n \to \alpha A + \beta B$.

The product AB is continuous in A and B jointly, i.e., $A_n \to A$ and $B_n \to B$ imply $A_n B_n \to AB$.

The norm $\|A\|$ is continuous because $\big|\|A\| - \|B\|\big| \leqq \|A - B\|$.

The vector-valued function Ax is continuous in both variables jointly, i.e., $x_n \to x$ and $A_n \to A$ imply $A_n x_n \to Ax$. In particular, $Ax_n \to Ax$ is true. Thus, the operator A

is continuous. Therefore,

$$A \int_a^b x(t)\, dt = \int_a^b Ax(t)\, dt$$

and

$$A \sum_{n=1}^{\infty} x_n = \sum_{n=1}^{\infty} Ax_n .$$

1.14.2. Infinite series of operators, absolute convergence

The expression $\sum_{n=1}^{\infty} A_n = B$ is equivalent with

$$\sum_{n=1}^{\infty} (A_n x, f) = (Bx, f)$$

for all $x \in X$, $f \in X^*$. This is implied by the usual definition of the convergence of the series.

$\sum_{n=1}^{\infty} A_n$ is said to be *absolutely convergent* if $\sum_{n=1}^{\infty} |(A_n x, f)| < \infty$ for all $x \in X$, $f \in X^*$. It is sufficient to require this condition for the matrix elements $(A_n e_i, f_j)$ of an arbitrarily chosen basis (together with its adjoint basis). Because $\sum_{n=1}^{\infty} \left(\sum_{i,j=1}^{N} |(A_n e_i, f_i)| \right) < \infty$, it is sufficient to require that $\sum_{n=1}^{\infty} \|A_n\| < \infty$ for some norm, hence for all norms (cf. Subsect. 1.8.2). Furthermore, $\left\| \sum_{n=1}^{\infty} A_n \right\| \leq \sum_{n=1}^{\infty} \|A_n\|$.

The continuity of the product implies

$$B \sum_{n=1}^{\infty} A_n = \sum_{n=1}^{\infty} BA_n , \qquad \left(\sum_{n=1}^{\infty} A_n \right) B = \sum_{n=1}^{\infty} A_n B .$$

If the series $\sum A_n$, $\sum B_n$ are absolutely convergent, then

$$\sum_{n=1}^{\infty} A_n \cdot \sum_{m=1}^{\infty} B_m = \sum_{n,m=1}^{\infty} A_n B_m$$

is valid, where the double series on the right hand side are absolutely convergent.

1.14.3. Exponential function, Neumann series, continuity of A^{-1}

Because $\sum_{n=0}^{\infty} (1/n!) \|A\|^n < \infty$, the series $\sum_{n=0}^{\infty} (1/n!) A^n$ is absolutely convergent; its sum is denoted by $\exp A$; one has $\|\exp A\| \leq e^{\|A\|}$. If A and B commute, then $\exp (A + B) = \exp A \exp B$.

The series $\sum_{n=0}^{\infty} A^n$ is absolutely convergent if $\|A\| < 1$, because $\sum_{n=0}^{\infty} \|A\|^n = 1/(1 - \|A\|) < \infty$. Then $1 - A$ is invertible, where $(1 - A)^{-1} = \sum_{n=0}^{\infty} A^n$. One has the estimate $\|(1 - A)^{-1}\| \leqq 1/(1 - \|A\|)$.

If the condition $\|A^m\| < 1$ is satisfied for some $m \geqq 1$ then

$$\sum_{n=0}^{\infty} \|A^n\| \leqq (1 + \|A\| + \cdots + \|A^{n-1}\|) \sum_{k=0}^{\infty} \|A^m\|^k \leqq \frac{1 + \|A\| + \cdots + \|A^{m-1}\|}{1 - \|A^m\|},$$

and the Neumann series is absolutely convergent. Note that $\|A^m\| < 1$ for some m is true if and only if $\operatorname{spr} A < 1$.

Let A be invertible, $C = B - A$ and let $\|C\| < 1/\|A^{-1}\|$. Then $1 + CA^{-1}$ is invertible; hence, B is also invertible. One has $B^{-1} = A^{-1}(1 + CA^{-1})^{-1}$ and the following estimates are valid

$$\|B^{-1}\| \leqq \frac{\|A^{-1}\|}{1 - \|C\| \cdot \|A^{-1}\|}$$

and

$$\|B^{-1} - A^{-1}\| \leqq \frac{\|C\| \cdot \|A^{-1}\|^2}{1 - \|C\| \cdot \|A^{-1}\|}.$$

This implies the continuity of A^{-1} as a function of A. Therefore, the set of all invertible operators is open.

1.15. Operator-valued functions of a scalar argument

1.15.1. Operator-valued functions of a real argument; limit, continuity, differentiation and Riemann integrals

Let I be an interval of the real line $(-\infty, \infty)$. Single-valued mappings $A(t)$ from I into $\mathscr{L}(X)$ are called *operator-valued functions.*

The operator B is called a *limit* of $A(t)$ for $t \to a$, $B = \lim_{t\to a} A(t)$, if $\lim_{t\to a} A(t)\, x = Bx$ for all $x \in X$. Equivalently, $\lim_{t\to a} \big(A(t)\, x, f\big) = (Bx, f)$ for all $x \in X$, $f \in X^*$; or $\lim_{t\to a} \|A(t) - B\| = 0$ for some norm.

The concepts of continuity and derivative are defined as in the case of vector-valued functions (cf. Subsect. 1.13.1). Moreover,

$$\big(A(t)\, x(t)\big)' = A'(t)\, x(t) + A(t)\, x'(t),$$

$$(AB)' = A'B + AB',$$

$$(A^{-1})' = -A^{-1}A'A^{-1}.$$

For continuous operator-valued functions $A(t)$ on $[a, b]$ the Riemann integral $\int_a^b A(t)\,dt$ is defined by

$$\left(\int_a^b A(t)\,dtx, f\right) = \int_a^b \left(A(t)\,x, f\right) dt;$$

that is, if a basis is chosen, the integral is calculated by integrating the matrix elements. An equivalent definition can be given by the limit of Riemann sums of the type $\sum_\varrho A(\tau_\varrho) \times (t_{\varrho+1} - t_\varrho)$, where $\max_\varrho |t_{\varrho+1} - t_\varrho| \to 0$. Hence,

$$\left\|\int_a^b A(t)\,dt\right\| \leq \int_a^b \|A(t)\|\,dt,$$

$$\int_a^b A(t)\,dtx = \int_a^b A(t)\,x\,dt,$$

$$B\int_a^b A(t)\,dt = \int_a^b BA(t)\,dt,$$

$$\left(\int_a^b A(t)\,dt\right) B = \int_a^b A(t)\,B\,dt.$$

1.15.2. Operator-valued functions of a complex argument; limit, continuity, differentiation and Riemann integrals

Let G be a region of the complex plane and let $A(z)$ be an operator-valued function on G (a single-valued mapping from G into $\mathscr{L}(X)$). Limit, continuity and derivative are defined as before (cf. Subsect. 1.15.1 and 1.13.1). As for vector-valued functions, the integral of continuous functions $A(z)$ along bounded piecewise smooth paths $\mathscr{C}$ in G is defined by

$$\left(\int_{\mathscr{C}} A(z)\,dz\,x, f\right) = \int_{\mathscr{C}} \left(A(z)\,x, f\right) dz, \qquad x \in X, \qquad f \in X^*;$$

that is, the integral is calculated by integrating the matrix elements. An equivalent definition can be given by the limit of Riemann sums $\sum_\varrho A(\zeta_\varrho)\,(z_{\varrho+1} - z_\varrho)$, where $\max_\varrho |z_{\varrho+1} - z_\varrho| \to 0$. One obtains the same properties as in Subsect. 1.15.1. In this case the estimate has the form

$$\left\|\int_{\mathscr{C}} A(z)\,dz\right\| \leq \int_{\mathscr{C}} \|A(z)\| \cdot |dz|.$$

1.15.3. Holomorphy and Cauchy's integral theorem

The operator-valued function $A(z)$ on G is said to be ***holomorphic*** on G if the derivative $A'(z)$ exists for all $z \in G$. Equivalently, all matrix elements $\big(A(z)\,x, f\big)$ are holomorphic.

If G is simply connected, then, for arbitrary holomorphic operator-valued functions $A(z)$, ***Cauchy's integral*** theorem

$$\int_{\mathscr{C}} A(z)\,\mathrm{d}z = 0$$

is valid for arbitrary closed piecewise smooth paths $\mathscr{C}$ lying in G.

As in the case of vector-valued functions (cf. Subsect. 1.13.1 to 1.13.5) one obtains for holomorphic operator-valued functions the well-known theorems of elementary complex analysis (complex function theory), such as Cauchy's integral formulas, the Taylor expansion of a holomorphic function in the neighborhood of holomorphic points, the theorem of identity for holomorphic functions, Liouville's theorem for entire functions, Laurent's expansion of holomorphic operator-valued functions in the neighborhood of isolated singular points, and classification of the singularities.

1.15.4. Power series; exponential functions, Neumann series; meromorphic operator-valued functions, projection-valued functions

The power series $A(z) = A_0 + zA_1 + z^2A_2 + \cdots$ converges at z if and only if

$$\big(A(z)\,x, f\big) = (A_0x, f) + z(A_1x, f) + z^2(A_2x, f) + \cdots \tag{15.1}$$

converges for all $x \in X$, $f \in X^*$. The radius of convergence of (15.1) is $r(x, f) = 1/\limsup |(A_nx, f)|^{1/n}$. The radius of convergence of $A(z)$ is $r = \inf\limits_{x\in X, f\in X^*} r(x, f)$, which is the radius of the smallest circle centered at 0, such that all matrix elements converge inside this circle. $A(z)$ is holomorphic in this circle.

The operator-valued function $\exp\{zA\} = \sum\limits_{n=0}^{\infty} (z^n/n!)\,A^n$ is an entire function, i.e., it is convergent in the whole complex plane.

The radius of convergence of the Neumann series $(1 - zA)^{-1} = \sum\limits_{n=0}^{\infty} z^nA^n$ is $r = 1/\mathrm{spr}\,A$.

An operator-valued function $A(z)$ is said to be *meromorphic* in G if its singularities in G are at most poles. The poles of $A(z)$ form an isolated and closed set in G. In the neighborhood of a pole ζ the function $A(z)$ has a Laurent expansion with a finite main part,

$$A(z) = \sum_{n=-M}^{\infty} A_n(z - \zeta)^n.$$

$A(z)$ is meromorphic in G if and only if all matrix elements $\big(A(z)\,x, f\big)$ are meromorphic in G.

A polynomial operator-valued function $A_0 + zA_1 + \cdots + z^mA_m$ is a special case of a meromorphic operator-valued function. $A(z)$ is polynomial if and only if all matrix elements are polynomial.

Let $P(z)$ be a holomorphic operator-valued function in G, the values of which are projections. Then, $\dim P(z)$ is constant in G because $\dim P(z) = \operatorname{tr} P(z)$. Together with $P(z)$, all matrix elements are holomorphic; hence, $\operatorname{tr} P(z)$ is holomorphic. In this case, the values of $\operatorname{tr} P(z)$ are natural numbers; hence, $\operatorname{tr} P(z)$ is constant.

1.16. Vector and operator differential equations

1.16.1. Real argument

Let the operator-valued function $A(t)$ be continuous for $t \geqq 0$. Then there exists a solution $x(t)$ of the vector differential equation

$$x'(t) = A(t)\, x(t), \qquad x(0) = c$$

for all $t \geqq 0$; it is uniquely determined and can be represented by

$$x(t) = X(t)\, c,$$

where $X(t)$ denotes an operator-valued function, uniquely determined by

$$X'(t) = A(t)\, X(t), \qquad X(0) = 1.$$

One has

$$\det X(t) = e^{\int_0^t \operatorname{tr} A(s)\, ds}.$$

If $A(t) \equiv A$ is constant, then $X(t) = \exp(tA)$.

The inhomogeneous equation

$$x'(t) = A(t)\, x(t) + f(t), \qquad x(0) = c,$$

has the solution

$$x(t) = X(t)\, c + \int_0^t X(t)\, X(s)^{-1} f(s)\, ds. \tag{16.1}$$

1.16.2. Complex argument

Let G be a simply connected region of the complex plane and let $A(z)$ be a holomorphic operator-valued function on G. Then the operator differential equation

$$X'(z) = A(z)\, X(z), \qquad X(\zeta) = 1, \qquad \zeta \in G,$$

has exactly one solution $U(z, \zeta)$, which is holomorphic with respect to z in G. It satisfies the relations

$$U(z_1, z_2)\, U(z_2, z_3) = U(z_1, z_3), \qquad z_1, z_2, z_3 \in G,$$
$$U(\zeta, \zeta) = 1.$$

One has

$$\det U(z, \zeta) = e^{\int_\zeta^z \operatorname{tr} A(u)\,du}.$$

$U(z, \zeta)$ can be represented by the series

$$U(z, \zeta) = 1 + \sum_{n=1}^{\infty} \int_\zeta^z d\zeta_1 A(\zeta_1) \int_\zeta^{\zeta_1} d\zeta_2 A(\zeta_2) \cdots \int_\zeta^{\zeta_{n-1}} A(\zeta_n)\, d\zeta_n, \qquad \zeta_0 = z, \tag{16.2}$$

which is absolutely uniformly convergent in compact subregions of G (successive approximation).

If $A(z) \equiv A$ is constant, then $U(z, \zeta) = \exp\big((z - \zeta) A\big)$.

The solution of the homogeneous vector differential equation

$$x'(z) = A(z)\, x(z), \qquad x(\zeta) = c, \tag{16.3}$$

is

$$x(z) = U(z, \zeta)\, c,$$

the solution of the corresponding inhomogeneous equation

$$x'(z) = A(z)\, x(z) + f(z), \qquad x(\zeta) = c,$$

can be expressed in the form

$$x(z) = U(z, \zeta)\, c + \int_\zeta^z U(z, s)\, f(s)\, ds.$$

The operator-valued function $U(z, \zeta)$ is also holomorphic with respect to ζ. With respect to ζ it satisfies the operator differential equation

$$\frac{\partial}{\partial \zeta} U(z, \zeta) = -U(z, \zeta)\, A(\zeta).$$

This function $U(z, \zeta)$, which is holomorphic in both variables $(z, \zeta) \in G \times G$ jointly, is sometimes called the *evolution operator* of the equation (16.3).

2. Spectral theory and Jordan structure of linear operators

The determination of the Jordan structure (also called Jordan normal form) for holomorphic (or meromorphic) operator-valued functions and its function-theoretic behavior is one of the main problems of analytic perturbation theory.

In this chapter, the Jordan structure is determined for a single operator. This will be done in two steps: First the spectral representation of an operator A is derived. It is developed by a thorough discussion of the eigenvalue problem. The function-theoretic approach in this discussion is essentially based on the resolvent and the Cauchy-Riesz integral, which are basic concepts for the functional calculus of an operator A (cf. for example T. Kato [5, Ch. I]). The spectral representation is uniquely determined and yields a decomposition of the basic space X into a direct sum of reducing subspaces.

The second step deals with a decomposition of these reducing subspaces into a direct sum of irreducible subspaces. Furthermore, the structure of A is determined with respect to these irreducible subspaces.

There are alternative approaches to the Jordan structure, for example: Weierstrass's theory of elementary divisors or the theory of annihilating polynomials. (See, for example, F. Gantmacher [1]; W. Graeub [1, in particular Ch. XI].)

Finally, the commutant of A is determined and some properties of algebras and their commutants are proved for later use (in Ch. 7 for bifurcation problems).

Ch. 2 contains complete proofs.

2.1. Eigenvalue problem

2.1.1. Basic concepts

Let A be an operator from $\mathscr{L}(X)$. The complex number λ is called an *eigenvalue* of A if there is a vector $x \neq 0$ from X such that

$$Ax = \lambda x.$$

x is called an *eigenvector* corresponding to the eigenvalue λ; one has $x \in \ker(A - \lambda)$. (x is sometimes called a *geometric eigenvector*.)

λ is an eigenvalue of A if and only if $\ker(A - \lambda) \supset \{0\}$. In this case $\ker(A - \lambda)$ consists of all eigenvectors belonging to λ (except the zero vector).

$\ker(A - \lambda)$ is called the *geometric eigenspace* of λ and $\dim \ker(A - \lambda)$ the *geometric multiplicity* of λ.

For an eigenvalue λ the operator $\lambda - A$ is not invertible, i.e., singular. If λ is not an eigenvalue, then $\ker(A - \lambda) = \{0\}$; that is, $\lambda - A$ is invertible.

$\lambda - A$ is singular if and only if $\det(\lambda - A) = 0$. This equation is called the *characteristic equation* and the polynomial $\chi(\lambda, A) := \det(\lambda - A)$ is called the *characteristic polynomial* of A. This polynomial is of degree $N = \dim X$ and

$$\chi(\lambda, A) = \lambda^N - \operatorname{tr} A\lambda^{N-1} \pm \cdots + (-1)^N \det A.$$

Thus, there are at most N different eigenvalues of A and there is at least one.

2.1.2. Spectrum, resolvent

The *spectrum* $\operatorname{spec} A$ of A is defined as the set of all eigenvalues. The spectrum is nonempty and consists of at most N different points.

If $\lambda \notin \operatorname{spec} A$, then $\lambda - A$ is invertible, i.e., $(\lambda - A)^{-1}$ exists. In this case, λ is called a *regular point*. The set of all regular points is called the *resolvent set* and is denoted by $\operatorname{res} A$. In other words, $\operatorname{res} A$ consists of all points λ such that the equation $\lambda x - Ax = y$ is uniquely solvable for arbitrary vectors $y \in X$. One writes

$$R(\lambda) = R(\lambda, A) = (\lambda - A)^{-1}, \qquad \lambda \in \operatorname{res} A.$$

This operator-valued function, defined on $\operatorname{res} A$, is called the *resolvent* of A. The relation

$$R(\lambda_1) - R(\lambda_2) = -(\lambda_1 - \lambda_2) R(\lambda_1) R(\lambda_2) \tag{1.1}$$

is valid. It is called *Hilbert's identity* (or first resolvent equation). For its proof, note that

$$R(\lambda_1) = R(\lambda_1)(\lambda_2 - A) R(\lambda_2)$$

and

$$R(\lambda_2) = R(\lambda_1)(\lambda_1 - A) R(\lambda_2).$$

Hence, (1.1) follows by subtraction. From (1.1)

$$\frac{R(\lambda_1) - R(\lambda_2)}{\lambda_1 - \lambda_2} = -R(\lambda_1) R(\lambda_2) \tag{1.2}$$

is obtained. Therefore,

$$R(\lambda_1) R(\lambda_2) = R(\lambda_2) R(\lambda_1), \qquad \lambda_1, \lambda_2 \in \operatorname{res} A.$$

An immediate consequence of (1.2) is:

The resolvent $R(\lambda)$ is differentiable at each point $\lambda \in \operatorname{res} A$, that is, $R(\lambda)$ is holomorphic in $\operatorname{res} A$ and the equations

$$R'(\lambda) = -R(\lambda)^2$$

and

$$R^{(n)}(\lambda) = n!(-1)^n R(\lambda)^{n+1}, \qquad n = 1, 2, 3, \ldots$$

are valid. Thus the Taylor expansion at $\lambda_0 \in \operatorname{res} A$ *has the form*

$$R(\lambda) = \sum_{n=0}^{\infty} (\lambda - \lambda_0)^n (-1)^n R(\lambda_0)^{n+1} = \big(1 + (\lambda - \lambda_0) R(\lambda_0)\big)^{-1} R(\lambda_0) \tag{1.3}$$

and is convergent (cf. Subsect. 1.14.3) *at least for*

$$|\lambda - \lambda_0| < \|R(\lambda_0)\|^{-1}.$$

(1.3) is called the *first Neumann series* of the resolvent. The exact radius of convergence of (1.3) is $1/\operatorname{spr} R(\lambda_0) = \big(\lim\limits_{n\to\infty} \|R(\lambda_0)^n\|^{1/n}\big)^{-1}$ (cf. Subsect. 1.15.4).

For large $|\lambda|$, one has

$$R(\lambda) = (\lambda - A)^{-1} = \lambda^{-1}(1 - \lambda^{-1}A)^{-1} = \lambda^{-1} \sum_{n=0}^{\infty} \lambda^{-n}A^n = \sum_{n=0}^{\infty} \frac{A^n}{\lambda^{n+1}}. \tag{1.4}$$

This expansion converges if and only if $|\lambda| > \operatorname{spr} A$ (again, cf. Subsect. 1.15.4). Therefore, *the resolvent is holomorphic at* $\lambda = \infty$ *and has the value* 0 *there.*

The resolvent is a rational function of λ, because (according to the inversion of $\lambda - A$)

$$R(\lambda) = \frac{1}{\chi(\lambda, A)} B_{N-1}(\lambda),$$

where $B_{N-1}(\lambda)$ is a polynomial operator-valued function of degree at most $N - 1$. Hence, the eigenvalues are at most poles of the resolvent.

If we assume that the eigenvalue λ_0 is a removable singularity of $R(\lambda)$, the limit $\lim\limits_{\lambda\to\lambda_0} R(\lambda) = R$ exists. The relations $(\lambda - A) R(\lambda) = R(\lambda) (\lambda - A) = 1$ imply $(\lambda_0 - A) R = R(\lambda_0 - A) = 1$, which contradicts $\lambda_0 \in \operatorname{spec} A$. Therefore:

Eigenvalues are poles of $R(\lambda)$.

At the boundary of the region of convergence of (1.4) there is at least one singularity of $R(\lambda)$. Therefore,

$$\operatorname{spr} A = \max_{\lambda \in \operatorname{spec} A} |\lambda|. \tag{1.5}$$

2.2. Invariant, reducing and irreducible subspaces (the generalized eigenvalue problem)

The eigenvalue problem involves the calculation of all eigenvalues and eigenvectors of A. An eigenvector x spans a one-dimensional subspace M such that $y \in M$ implies $Ay \in M$. The operator acts on M as multiplication with the corresponding eigenvalue. Obviously, the property $y \in M$ implies $Ay \in M$ does not depend on the dimension of M. Taking into account subspaces M of higher dimension with this property, one obtains a generalized eigenvalue problem, namely, the determination of all such subspaces M and the action of A in M. The original (one-dimensional) eigenvalue problem also yields a suitable approach to solve the generalized problem.

2.2.1. Invariant subspaces

Let $A \in \mathscr{L}(X)$. Then one defines:

Definition 1. The subspace $M \subset X$ is said to be *invariant with respect to* A if $x \in M$ implies $Ax \in M$.

In this case, A can be considered as an operator on M. This operator is called the *restriction* of A to M. The *part of A in M* is denoted by $A_M = A \upharpoonright M$.

The simplest examples of invariant subspaces are given by the one-dimensional subspaces spanned by an eigenvector, or by $\ker (A - \lambda)$ for an eigenvalue λ.

Lemma 1. *Let M be a subspace and P a projection, where* ima $P = M$. *Then M is an invariant subspace of A if and only if*

$$PAP = AP. \tag{2.1}$$

Proof. Let M be invariant. Then $PAPx = APx$ for all $x \in X$; (2.1) follows. Conversely, from (2.1) it follows that ima $P = M$ is invariant with respect to A. ■

Note that P is by no means uniquely determined by M. In the following we also use the notation PX for ima P.

2.2.2. Reducing subspaces

Definition 2. Let M be invariant with respect to A. M is said to be *reducing* if there is a second invariant subspace N of A such that $M \oplus N = X$.

In this case A splits into two completely separated parts A_M and A_N. M and N are reducing subspaces.

The *complementary* invariant subspaces M and N decompose the operator A. This is called a *direct decomposition* of A into the direct sum

$$A = A_M \oplus A_N.$$

Therefore such a decomposition exists if and only if there are two complementary invariant (hence, reducing) subspaces.

The simplest example of a direct decomposition of A is given by an operator acting as a "generalized block-diagonal matrix"

$$A \leftrightarrow \begin{pmatrix} A_1 & 0 \\ 0 & A_2 \end{pmatrix}$$

with respect to some basis, where A_1 and A_2 are the matrix representations on the complementary invariant subspaces.

Lemma 2. *Let M be a reducing subspace and N a corresponding complementary invariant subspace. If P is the projection on M along N, then*

$$AP = PA. \tag{2.2}$$

Proof. $PAP = AP$, because M is invariant. Further, $(1 - P) A(1 - P) = A(1 - P)$, because N is invariant. The second equation yields $PAP = PA$. (2.2) follows. ∎

Lemma 3. *Let P be a projection. If $PA = AP$, then* ima P *is a reducing subspace.*

Proof. ima $P = PX$ is invariant because $PAP = AP$, $(1 - P) M$ is invariant because $(1 - P) A(1 - P) = A(1 - P)$. PX and $(1 - P) X$ are complementary. ∎

A projection P with the property (2.2) is called a *reducing projection* of A.

2.2.3. Complete disjoint systems of reducing projections

Let

$$X = M_1 \oplus M_2 \oplus \cdots \oplus M_r \tag{2.3}$$

be a direct decomposition of X and let every subspace M_i be invariant with respect to A. Then every M_i is also reducing (the complementary invariant subspace is $M_1 \oplus M_2 \oplus \cdots \oplus M_{i-1} \oplus M_{i+1} \oplus \cdots \oplus M_r$). Therefore, the corresponding projection P_i, where $P_iX = M_i$, is reducing, that is,

$$AP_i = P_iA, \qquad i = 1, 2, \ldots, r. \tag{2.4}$$

The system $P_1, P_2, \ldots, P_r$ is disjoint and complete.

Conversely, if every projection P_i of a complete disjoint system $P_1, \ldots, P_r$ has the property (2.4), then the corresponding decomposition (2.3) of X consists of invariant, hence reducing subspaces. If $A_\varrho := A \restriction M_\varrho$, then one writes

$$A = A_1 \oplus A_2 \oplus \cdots \oplus A_r$$

and this equation is called a *decomposition* of A into a *direct sum of operators which are parts of A.*

2.2.4. Irreducible subspaces

Definition 3. A reducing subspace M of A is said to be *irreducible* with respect to A if there is no nontrivial reducing subspace of $A \restriction M$.

If N is complementary with respect to M and if P denotes the projection on M along N, then P is called an *irreducible projection.*

Lemma 4. *A projection P is irreducible with respect to A if an only if $AP = PA$ and there is no proper subprojection Q, $0 < Q < P$, which commutes with A, $QA = AQ$.*

Proof. Let P be irreducible. Then $X = PX \oplus (1 - P) X$ and $PX = M$ is irreducible with respect to A. Furthermore, $(1 - P) X = N$ is invariant with respect to A. Therefore, $AP = PA$. If Q is a proper subprojection of P with the property $AQ = QA$, then, according to Subsect. 1.9.4, Q satisfies $\{0\} \subset QX \subset M$, where QX is a nontrivial reducing subspace.

Conversely, let $AP = PA$ and assume that there is no proper subprojection Q of P, $0 < Q < P$, with the property $AQ = QA$. Then we have to show that $PX = M$ is irreducible. Let $(1 - P) X = N$.

If M_1 is a nontrivial reducing subspace of M with respect to A, $\{0\} \subset M_1 \subset M$, that is, if

$$M = M_1 \oplus M_2,$$

where M_2 is also invariant with respect to $A \restriction M$, then

$$X = M_1 \oplus M_2 \oplus N.$$

The corresponding equation for the assigned projections is

$$1 = Q_1 + Q_2 + (1 - P), \qquad Q_1 \neq 0, \qquad Q_1 \neq P.$$

From

$$(Q_1 \restriction M)(A \restriction M) = (A \restriction M)(Q_1 \restriction M)$$

one obtains

$$Q_1AP = APQ_1P$$

and, because $Q_1P = PQ_1 = Q_1$, also

$$Q_1A = AQ_1$$

which contradicts the assumption. ■

Definition 4. An operator A from $\mathscr{L}(X)$ is called *irreducible* if X is irreducible with respect to A, that is, if the identity 1 is an irreducible projection with respect to A.

For example, the matrix

$$\begin{pmatrix} 0 & 1 & 0 & \dots & 0 & 0 \\ 0 & 0 & 1 & \dots & 0 & 0 \\ & & & \dots & & 0 \\ 0 & 0 & 0 & \dots & 0 & 1 \\ 0 & 0 & 0 & \dots & 0 & 0 \end{pmatrix}$$

is irreducible in $\mathscr{L}(\mathbb{C}^N)$.

Obviously the central problem in the investigation of the structure of A is finding a complete disjoint system of irreducible projections of A and studying the structure of A restricted to the irreducible subspaces, that is, determining the invariant subspaces, etc. Clearly such a system exists. This problem will be solved in the following subsection by the spectral representation and the Jordan representation of A.

2.2.5. Spectral radius and nilpotent operators

This subsection is based on Subsect. 2.1.2. According to (1.5), if $\operatorname{spr} A = 0$ then $\lambda = 0$ is the only point of $\operatorname{spec} A$, hence, A is noninvertible.

If $\operatorname{spr} A = 0$ and M is an invariant subspace of A, then $\lambda = 0$ is also the only eigenvalue of A in M; hence, $\operatorname{spr}(A \restriction M) = 0$. Thus:

Lemma 5. *The equation* spr $A = 0$ *is valid if and only if* A *is nilpotent.*

Proof. Let spr $A = 0$. According to Subsect. 1.6.2. one considers the chain of inclusions

$$X \supset \text{ima } A \supset \text{ima } A^2 \supset \cdots \supset \text{ima } A^m = \text{ima } A^{m+1} = \cdots.$$

Let k be a natural number. The subspace ima A^k is invariant with respect to A. The relation ima $A^k \supset \{0\}$ implies ima $A^k \supset$ ima A^{k+1}, because, if one assumes that ima A^k = ima $A^{k+1} \supset \{0\}$, then A is invertible on ima A^k, which is impossible according to the remarks preceding this lemma. Hence, the dimension of ima A^k decreases by at least one as k increases by one. Necessarily therefore, ima $A^N = \{0\}$, that is, $A^N = 0$, so A is nilpotent. The converse is obvious. ■

2.3. Cauchy-Riesz functional calculus

2.3.1. Motivation

Let $\mathscr{C}$ be a positively oriented simply closed and piecewise smooth contour lying in the region $|z| >$ spr A.

Lemma 1. *The formulas*

$$\frac{1}{2\pi i} \int_{\mathscr{C}} \lambda^m R(\lambda, A)\, d\lambda = A^m, \quad m = 0, 1, 2, \ldots, \tag{3.1}$$

are valid.

Proof. According to (1.4),

$$\lambda^m R(\lambda, A) = \sum_{n=0}^{\infty} \lambda^{m-n-1} A^n;$$

hence,

$$\int_{\mathscr{C}} \lambda^m R(\lambda, A)\, d\lambda = \sum_{n=0}^{\infty} A^n \int_{\mathscr{C}} \lambda^{m-n-1}\, d\lambda = 2\pi i A^m.$$

Therefore, if $p(\lambda) = \alpha_0 + \alpha_1\lambda + \alpha_2\lambda^2 + \cdots + \alpha_m\lambda^m$ is an arbitrary polynomial, then

$$p(A) = \frac{1}{2\pi i} \int_{\mathscr{C}} p(\lambda) R(\lambda, A)\, d\lambda. \tag{3.2}$$

This formula is an operator analog of Cauchy's integral formula for holomorphic functions.

The contour $\mathscr{C}$ may be replaced by other contours, for example by sufficiently small positively oriented spheres which are centered at the eigenvalues and do not contain other eigenvalues in their interior.

Formula (3.2) will be used to define $f(A)$ for an enlarged class of functions, but it will turn out that all such functions $f(A)$ can be expressed by operator polynomials in A.

According to R. Bellman [1, p. 132] H. Poincaré was the first to use the path integral $(2\pi i)^{-1} \int f(\lambda)\,(\lambda - A)^{-1}\,d\lambda$ for the representation of functions $f(A)$ of a matrix A; cf. H. Poincaré [1, p. 193]. F. Riesz [1] proposed the calculus for the infinite-dimensional case and for bounded operators (see F. Riesz and B. v. Sz.-Nagy [1, p. 414]). The calculus was worked out in full generality by N. Dunford [1, 2]. For the following, cf. N. Dunford and J. T. Schwartz [1, Ch. VII].

2.3.2. Cauchy-Riesz integral

Let $f(\lambda)$ be a complex-valued function, which is holomorphic on spec A, that is, $f(\lambda)$ is holomorphic in small discs at centered the eigenvalues. Thus, in general, $f(\lambda)$ is piecewise holomorphic. Let $\mathscr{C}$ be a suitable system of positively oriented simply closed contours of integration which consist of regular points and lie inside the holomorphic region of $f(\lambda)$, where the eigenvalues are inside these contours. Then one defines

$$f(A) = \frac{1}{2\pi i} \int\limits_{\mathscr{C}} f(\lambda)\,R(\lambda, A)\,d\lambda. \tag{3.3}$$

$f(A)$ is independent of the special choice of $\mathscr{C}$. Some properties of the assignment $f(\lambda) \mapsto f(A)$ are collected in the following theorem:

Theorem 1. *Let f_1, f_2 be defined as described above and let $f_1(A), f_2(A)$ etc. be defined according to* (3.3). *Then*

I. $g(\lambda) = \alpha f_1(\lambda) + \beta f_2(\lambda)$ *implies* $g(A) = \alpha f_1(A) + \beta f_2(A)$.

II. $g(\lambda) = f_1(\lambda)\,f_2(\lambda)$ *implies* $g(A) = f_1(A)\,f_2(A)$.

III. *If* $f(\lambda) = \sum_{\varrho=0}^{\infty} \alpha_\varrho \lambda^\varrho$ *converges for* $|\lambda| < \operatorname{spr} A + \varepsilon$, $\varepsilon > 0$, *then* $f(A) = \sum_{\varrho=0}^{\infty} \alpha_\varrho A^\varrho$.

Proof. I. Obvious.

II. Let U be a joint neighborhood of holomorphy for spec A with respect to f_1 and f_2. Then the contours of integration in the neighborhood of each eigenvalue λ_0 can be chosen such that the $\mathscr{C}_1$-part lies inside the $\mathscr{C}_2$-part:

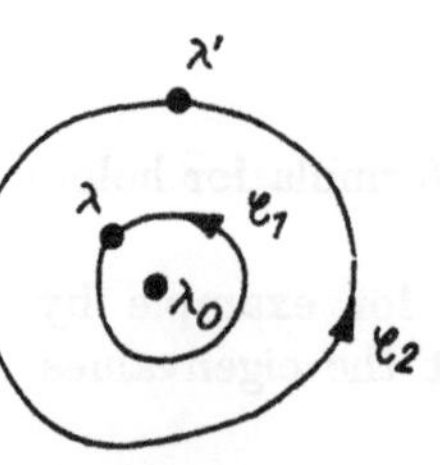

Fig. 1

Then,

$$f_1(A)\, f_2(A) = \frac{1}{(2\pi i)^2} \int\limits_{\mathscr{C}_1} f_1(\lambda)\, R(\lambda, A)\, d\lambda \int\limits_{\mathscr{C}_2} f_2(\lambda')\, R(\lambda', A)\, d\lambda'$$

$$= \frac{1}{(2\pi i)^2} \int\limits_{\mathscr{C}_1}\int\limits_{\mathscr{C}_2} f_1(\lambda)\, f_2(\lambda')\, R(\lambda, A)\, R(\lambda', A)\, d\lambda\, d\lambda'$$

$$= \frac{1}{(2\pi i)^2} \int\limits_{\mathscr{C}_1}\int\limits_{\mathscr{C}_2} f_1(\lambda)\, f_2(\lambda')\, \frac{R(\lambda, A) - R(\lambda', A)}{\lambda' - \lambda}\, d\lambda\, d\lambda'$$

$$= \frac{1}{(2\pi i)^2} \int\limits_{\mathscr{C}_1} f_1(\lambda)\, R(\lambda, A) \left[\int\limits_{\mathscr{C}_2} \frac{f_2(\lambda')}{\lambda' - \lambda}\, d\lambda'\right] d\lambda + \frac{1}{(2\pi i)^2} \int\limits_{\mathscr{C}_2} f_2(\lambda')\, R(\lambda', A) \left[\int\limits_{\mathscr{C}_1} \frac{f_1(\lambda)}{\lambda' - \lambda}\, d\lambda\right] d\lambda'$$

$$= \frac{1}{2\pi i} \int\limits_{\mathscr{C}_1} f_1(\lambda)\, f_2(\lambda)\, R(\lambda, A)\, d\lambda.$$

III. Let $\mathscr{C}$ be a positively oriented sphere with radius r, let $\operatorname{spr} A < r < \operatorname{spr} A + \varepsilon$, $\varepsilon > 0$. Then

$$f(A) = \frac{1}{2\pi i} \int\limits_{\mathscr{C}} \left[\sum_{\varrho=0}^{\infty} \alpha_\varrho \lambda^\varrho\right] R(\lambda, A)\, d\lambda = \frac{1}{2\pi i} \sum_{\varrho=0}^{\infty} \alpha_\varrho \int\limits_{\mathscr{C}} \lambda^\varrho\, R(\lambda, A)\, d\lambda = \sum_{\varrho=0}^{\infty} \alpha_\varrho A^\varrho.$$

This concludes the proof. ■

The assignment $f(\lambda) \mapsto f(A)$ is called a *homomorphism* of the algebra of all piecewise holomorphic functions on spec A into the algebra $\mathscr{L}(X)$.

2.3.3. Spectral mapping theorem

The spectrum of $f(A)$ can be expressed by spec A by a simple formula.

Theorem 2. *For piecewise holomorphic functions $f(\lambda)$ on* spec A,

$$\operatorname{spec}\big(f(A)\big) = f(\operatorname{spec} A) \tag{3.4}$$

where $f(\operatorname{spec} A)$ denotes the image (range) of $f(\cdot)$ on spec A.

Proof. First the inclusion $f(\operatorname{spec} A) \subseteq \operatorname{spec}\big(f(A)\big)$ will be proven. Let $\lambda_0 \in \operatorname{spec} A$. Consider

$$g(\lambda) = \frac{f(\lambda) - f(\lambda_0)}{\lambda - \lambda_0},$$

$g(\lambda)$ is piecewise holomorphic on spec A. One has

$$f(\lambda_0) - f(\lambda) = (\lambda_0 - \lambda)\, g(\lambda).$$

Hence,

$$f(\lambda_0)\, 1 - f(A) = (\lambda_0 - A)\, g(A).$$

Now, if $f(\lambda_0) \in \operatorname{res}\big(f(A)\big)$, i.e., if $\big(f(\lambda_0) - f(A)\big)^{-1}$ exists, then

$$(\lambda_0 - A)\, g(A) \big(f(\lambda_0) - f(A)\big)^{-1} = 1.$$

Hence, $(\lambda_0 - A)^{-1}$ exists, which contradicts $\lambda_0 \in \operatorname{spec} A$. Therefore, $f(\lambda_0) \in \operatorname{spec}\big(f(A)\big)$. Conversely, if $\lambda_0 \in \operatorname{spec}\big(f(A)\big)$ and $\lambda_0 \notin f(\operatorname{spec} A)$, then

$$g(\lambda) = \frac{1}{f(\lambda) - \lambda_0}$$

is piecewise holomorphic on spec A and

$$g(A)\big(f(A) - \lambda_0\big) = \big(f(A) - \lambda_0\big)\, g(A) = 1$$

which contradicts $\lambda_0 \in \operatorname{spec}\big(f(A)\big)$. ∎

2.4. Riesz eigenprojections

2.4.1. Definition and simple properties

Let $\lambda_1, \lambda_2, \ldots, \lambda_r$ denote the different eigenvalues of A, where $1 \leqq r \leqq N$ and where

$$\chi(\lambda_\varrho, A) = 0, \qquad \varrho = 1, 2, \ldots, r.$$

Let $\mathscr{C}_\varrho$ be a positively oriented sphere with center λ_ϱ, which consists of regular points and is such that no other eigenvalue lies inside $\mathscr{C}_\varrho$. Then one defines the operators $P_1, \ldots, P_r$ by

$$P_\varrho = \frac{1}{2\pi \mathrm{i}} \int\limits_{\mathscr{C}_\varrho} R(\lambda, A)\, \mathrm{d}\lambda. \tag{4.1}$$

P_ϱ is a function of A; namely $P_\varrho = k_\varrho(A)$, where

$$k_\varrho(\lambda) = \begin{cases} 1, & |\lambda - \lambda_\varrho| < \delta,\ \delta > 0, \\ 0, & \text{elsewhere.} \end{cases}$$

Basic properties of $P_1, \ldots, P_r$ are collected in

Lemma 1. *The operators $P_1, \ldots, P_r$ form a complete disjoint system of nonzero projections reducing A; that is, the relations*

$$P_\varrho P_\sigma = \delta_{\varrho\sigma} P_\varrho, \qquad \varrho, \sigma = 1, 2, \ldots, r, \tag{4.2}$$

$$\sum_{\varrho=1}^{r} P_\varrho = 1, \tag{4.3}$$

$$A P_\varrho = P_\varrho A, \qquad \varrho = 1, 2, \ldots, r, \tag{4.4}$$

$$P_\varrho \neq 0, \qquad \varrho = 1, 2, \ldots, r, \tag{4.5}$$

hold.

Proof. The properties (4.2), (4.3) and (4.4) follow immediately from (4.1) and Theorem 3.1. According to Theorem 3.2., one has spec $P_\varrho = k_\varrho(\operatorname{spec} A) \supseteqq \{1\}$. This implies $P_\varrho \neq 0$, because the zero operator does not have a nonzero eigenvalue. ■

The projection P_ϱ is called the *(Riesz) eigenprojection* of A belonging to the eigenvalue λ_ϱ. If $M_\varrho = P_\varrho X$ and $A_\varrho = A \upharpoonright M_\varrho$, then

$$X = M_1 \oplus M_2 \oplus \cdots \oplus M_r \tag{4.6}$$

and

$$A = A_1 \oplus A_2 \oplus \cdots \oplus A_r. \tag{4.7}$$

Thus, for the investigation of the structure of A it is sufficient to determine the structure of the parts A_ϱ.

2.4.2. Eigennilpotent (operator) and index

Let the operators D_ϱ be defined by

$$D_\varrho = (A - \lambda_\varrho) P_\varrho, \qquad \varrho = 1, 2, \ldots, r. \tag{4.8}$$

D_ϱ is a function of A, $D_\varrho = f_\varrho(A)$, where

$$f_\varrho(\lambda) = (\lambda - \lambda_\varrho) k_\varrho(\lambda). \tag{4.9}$$

From Theorem 3.2., the relation

$$\operatorname{spec} D_\varrho = \{0\} \tag{4.10}$$

follows, because $f_\varrho(\lambda)$ maps spec A onto zero. (4.10) is equivalent with $\operatorname{spr} D_\varrho = 0$ (see (1.5)).

Lemma 2. *The operators $D_1, D_2, \ldots, D_r$ are nilpotent.*

Proof. Obvious, because of Lemma 2.5 and $\operatorname{spr} D_\varrho = 0$. ■

The operator D_ϱ is called the *eigennilpotent (operator)*, belonging to the eigenvalue λ_ϱ. P_ϱ reduces D_ϱ. Hence, $D_\varrho \upharpoonright M_\varrho$ is also nilpotent, i.e., $(A - \lambda_\varrho) \upharpoonright M_\varrho$ is nilpotent. Therefore, spec $\{(A - \lambda_\varrho) \upharpoonright M_\varrho\} = \{0\}$; that is, *the operator A has exactly one eigenvalue in M_ϱ, namely λ_ϱ.*

Because D_ϱ is nilpotent, to every D_ϱ corresponds a natural number $m_\varrho \geqq 1$ such that

$$D_\varrho^{m_\varrho} = 0, \qquad D_\varrho^{m_\varrho - 1} \neq 0, \qquad \varrho = 1, 2, \ldots, r. \tag{4.11}$$

The number m_ϱ is called the *index* of the eigenvalue λ_ϱ; one writes $m_\varrho = \nu(\lambda_\varrho)$.

2.4.3. Jordan vectors

Given the decomposition (4.6), the following questions arise: How can one give an inner characterization of the subspace M_ϱ; what properties have the vectors of M_ϱ? To answer these questions, the concept of Jordan vectors will be introduced.

Definition 1. A vector $x \in X$ is called a *Jordan vector* with respect to A belonging to the eigenvalue λ_0 if there exists a natural number $m \geqq 0$ such that

$$(A - \lambda_0)^m x = 0. \tag{4.12}$$

The name for this concept varies in the literature. In German articles, "Hauptvektor" is sometimes used (for example: G. Porath [1]), in Russian articles always "root vector", in English articles also "principal vector". The name "algebraic eigenvector" seems also possible, as distinguished from "geometric eigenvector".

For example, all eigenvectors of A belonging to λ_0 are also Jordan vectors belonging to λ_0. For nonzero Jordan vectors, one distinguishes Jordan vectors of degree 1, 2, 3, ...
x is called a *Jordan vector of degree g* if

$$(A - \lambda_0)^g x = 0,$$

but

$$(A - \lambda_0)^{g-1} x \neq 0.$$

Thus eigenvectors are Jordan vectors of degree 1.

Lemma 3. *The subspace M_ϱ consists exactly of the Jordan vectors belonging to the eigenvalue λ_ϱ.*

Proof. Let $x \in M_\varrho$, i.e., $P_\varrho x = x$. Then, because of (4.11),

$$D_\varrho^{m_\varrho} x = (A - \lambda_\varrho)^{m_\varrho} x = 0;$$

that is, x is a Jordan vector belonging to λ_ϱ of degree at most m_ϱ. Conversely, let x be a Jordan vector belonging to λ_ϱ, i.e., let $(A - \lambda_\varrho)^n x = 0$ for some n. According to the decomposition (4.6),

$$x = x_1 + x_2 + \cdots + x_r;$$

hence,

$$(A - \lambda_\varrho)^n x = 0 = (A - \lambda_\varrho)^n x_1 + (A - \lambda_\varrho)^n x_2 + \cdots + (A - \lambda_\varrho)^n x_r.$$

Because of $x_\sigma \in M_\sigma$ and because of the invariance of M_ϱ with respect to A, $(A - \lambda_\varrho)^n x_\sigma \in M_\sigma$, therefore,

$$(A - \lambda_\varrho)^n x_\sigma = 0, \qquad \sigma = 1, 2, \ldots, r.$$

Now, $(A - \lambda_\varrho)^n \upharpoonright M_\sigma$ is invertible for $\varrho \neq \sigma$, because spec $\{A \upharpoonright M_\sigma\} = \{\lambda_\sigma\}$; hence, spec $\{(A - \lambda_\varrho)^n \upharpoonright M_\sigma\} = \{(\lambda_\sigma - \lambda_\varrho)^n\}$, and $(\lambda_\sigma - \lambda_\varrho)^n$ is different from zero for $\sigma \neq \varrho$. Therefore, $x_\sigma = 0$ follows for $\sigma \neq \varrho$ and, finally, $x = x_\varrho \in M_\varrho$. ■

The proof shows at the same time that *the maximal degree of Jordan vectors belonging to λ_ϱ is equal to $\nu(\lambda_\varrho)$.*

$M_\varrho = P_\varrho X$ is called the *algebraic eigenspace* of λ_ϱ and $\dim M_\varrho = \dim P_\varrho$ is called the *algebraic multiplicity* of λ_ϱ.

There are some inequalities between the algebraic multiplicity n_ϱ, the geometric multiplicity m_ϱ and the index ν_ϱ of λ_ϱ:

$$1 \leqq m_\varrho \leqq n_\varrho, \qquad \varrho = 1, 2, \ldots, r, \tag{4.13}$$

$$1 \leqq \nu_\varrho \leqq n_\varrho, \qquad \varrho = 1, 2, \ldots, r. \tag{4.14}$$

The last inequality (4.14) is valid, because the eigennilpotent D_ϱ of λ_ϱ satisfies the equation $D_\varrho^{n_\varrho} = 0$, since dim $P_\varrho = n_\varrho$ (cf. the proof of Lemma 2.5).

The eigenvalue λ_ϱ is said to be *simple* if $n_\varrho = 1$, *semisimple* if $\nu_\varrho = 1$ and *prime* if $m_\varrho = 1$.

2.4.4. Algebraic multiplicity and the characteristic polynomial

The algebraic multiplicity n_ϱ of λ_ϱ is nothing else than its multiplicity as a root of the characteristic polynomial. One has

Lemma 4. *Let $\lambda_1, \lambda_2, \ldots, \lambda_r$ be the different eigenvalues of A with the algebraic multiplicities $n_1, n_2, \ldots, n_r$, respectively. Then*

$$\chi(\lambda, A) = \prod_{\varrho=1}^{r} (\lambda - \lambda_\varrho)^{n_\varrho}. \tag{4.15}$$

Proof. Let $P_1, \ldots, P_r$ be the corresponding eigenprojections and let $P_\varrho X = M_\varrho$. Then $n_\varrho = \dim M_\varrho$. Let $A_\varrho = A \restriction M_\varrho$. Recall that A_ϱ has only one eigenvalue, namely λ_ϱ. Hence, $\chi(\lambda, A_\varrho) = (\lambda - \lambda_\varrho)^{n_\varrho}$, $\varrho = 1, 2, \ldots, r$. Furthermore,

$$\lambda - A = (\lambda - A_1) \oplus (\lambda - A_2) \oplus \cdots \oplus (\lambda - A_r).$$

Thus,

$$\det(\lambda - A) = \det(\lambda - A_1) \det(\lambda - A_2) \cdots \det(\lambda - A_r);$$

that is, $\chi(\lambda, A) = \prod_{\varrho=1}^{r} \chi(\lambda, A_\varrho)$, which implies the assertion. ∎

2.5. Spectral representation

2.5.1. Existence

Let $\lambda_1, \ldots, \lambda_r$ be the (different) eigenvalues of A, $P_1, \ldots, P_r$ the eigenprojections and $D_1, \ldots, D_r$ the eigennilpotents. Then,

$$AP_\varrho = P_\varrho A = \lambda_\varrho P_\varrho + D_\varrho, \qquad \varrho = 1, 2, \ldots, r,$$

and

$$A = \sum_{\varrho=1}^{r} \lambda_\varrho P_\varrho + \sum_{\varrho=1}^{r} D_\varrho.$$

Setting

$$S = \sum_{\varrho=1}^{r} \lambda_\varrho P_\varrho$$

and

$$D = \sum_{\varrho=1}^{r} D_\varrho.$$

one obtains

$$A = S + D. \tag{5.1}$$

According to Theorem 3.2, S is an operator with the eigenvalues $\lambda_1, \lambda_2, \ldots, \lambda_r$. The corresponding eigenprojections are $P_1, \ldots, P_r$, because

$$(\lambda - S)^{-1} = \sum_{\varrho=1}^{r} \frac{1}{\lambda - \lambda_\varrho} P_\varrho, \qquad \lambda \in \operatorname{res} S,$$

which implies

$$\frac{1}{2\pi i} \int_{\mathscr{C}_\varrho} (\lambda - S)^{-1} \, d\lambda = P_\varrho.$$

Moreover, because

$$SP_\varrho - \lambda_\varrho P_\varrho = 0,$$

the corresponding eigennilpotents are zero, i.e.,

$$\nu(\lambda_\varrho, S) = 1, \qquad \varrho = 1, 2, \ldots, r,$$

where $\nu(\lambda_\varrho, S)$ denotes the index of λ_ϱ with respect to S; hence, all eigenvalues of S are semisimple.

An operator S is called a *diagonal operator* if its eigenvalues are semisimple. S is given by a diagonal matrix with respect to a basis adapted to the system $P_1, \ldots, P_r$ of eigenprojections. The elements on the main diagonal are the eigenvalues, where the algebraic multiplicity has to be taken into account. (Here the algebraic and geometric multiplicities coincide.)

The operator D is nilpotent. One has

$$DS = SD. \tag{5.2}$$

A representation (5.1) of A is called a *spectral representation* of A if S is diagonal and D is nilpotent and if (5.2) is satisfied.

2.5.2. Uniqueness

First we note: *Let $f(\lambda)$ be piecewise holomorphic on* spec A *and $f(A)$ be given by* (3.3). *Furthermore, let $AB = BA$. Then $f(A)\,B = Bf(A)$, because $AB = BA$ implies $R(\lambda, A)\,B = BR(\lambda, A)$, hence also*

$$\int_{\mathscr{C}} f(\lambda)\, R(\lambda, A) \, d\lambda \, B = B \int_{\mathscr{C}} f(\lambda)\, R(\lambda, A)\, d\lambda.$$

Next we prove the uniqueness of the spectral representation (5.1).

Theorem 1 (Spectral representation). *An arbitrary operator A can be decomposed into a sum of the form $A = S + D$, where $S = \sum_{\varrho=1}^{r} \lambda_\varrho P_\varrho$ is a diagonal operator, D is a nilpotent* *operator and* $SD = DS$. *This representation is uniquely determined.* $\lambda_1, \ldots, \lambda_r$ *are the eigenvalues of* A *and* $P_1, \ldots, P_r$ *the corresponding eigenprojections. If* $D_\varrho := DP_\varrho = P_\varrho D$, *then*

$$A = \sum_{\varrho=1}^{r} (\lambda_\varrho P_\varrho + D_\varrho).$$

Proof. The existence has been proved in Subsect. 2.5.1. Now the uniqueness will be proved. Let

$$A = S' + D'$$

be a second representation, where S' is diagonal, D' nilpotent and $D'S' = S'D'$. First one has

$$S' = \sum_{\varrho=1}^{p} \lambda'_\varrho P'_\varrho,$$

where $\lambda'_1, \ldots, \lambda'_p$ denote the eigenvalues and $P'_1, \ldots, P'_p$ the corresponding eigenprojections. Then

$$P'_\varrho = \frac{1}{2\pi i} \int_{\mathscr{C}_\varrho} R(\lambda, S')\, d\lambda = k_\varrho(S'),$$

where $\mathscr{C}_\varrho$ and k_ϱ are defined as usual (cf. Subsect. 4.1). Hence,

$$P'_\varrho D' = D'P'_\varrho, \qquad \varrho = 1, 2, \ldots, p.$$

Let

$$P'_\varrho D' = D'P'_\varrho = D'_\varrho.$$

Then

$$A = \sum_{\varrho=1}^{p} \lambda'_\varrho P'_\varrho + \sum_{\varrho=1}^{p} D'_\varrho = \sum_{\varrho=1}^{p} (\lambda'_\varrho P'_\varrho + D'_\varrho).$$

Therefore

$$AP'_\varrho = P'_\varrho A = \lambda'_\varrho P'_\varrho + D'_\varrho,$$

i.e., the projections P'_ϱ reduce A and $(A - \lambda'_\varrho) \upharpoonright P'_\varrho X$ is nilpotent for all $\varrho = 1, 2, \ldots, p$.

This result implies that $\lambda'_1, \lambda'_2, \ldots, \lambda'_p$ are the eigenvalues of A; that is, one has $p = r$ and $\lambda_\varrho = \lambda'_\varrho$, $\varrho = 1, 2, \ldots, r$, if the subscripts are suitably chosen.

Furthermore, because $P_\varrho = k_\varrho(A)$,

$$P_\varrho P'_\sigma = P'_\sigma P_\varrho, \qquad \varrho, \sigma = 1, 2, \ldots, r. \tag{5.3}$$

Let $Q_{\varrho\sigma} = P_\varrho P'_\sigma$. The projections $Q_{\varrho\sigma}$ reduce A. Moreover, $Q_{\varrho\sigma} \leqq P_\varrho$ and $Q_{\varrho\sigma} \leqq P'_\varrho$. If $\dim Q_{\varrho\sigma} = m_{\varrho\sigma}$, then

$$\chi(\lambda, A/Q_{\varrho\sigma}X) = (\lambda - \lambda_\varrho)^{m_{\varrho\sigma}} = (\lambda - \lambda_\sigma)^{m_{\varrho\sigma}};$$

that is, $m_{\varrho\sigma} = 0$ if $\varrho \neq \sigma$. Thus

$$P_\varrho P'_\sigma = P'_\sigma P_\varrho = 0, \qquad \varrho \neq \sigma.$$

From $\sum_{\varrho=1}^{r} P_\varrho = \sum_{\varrho=1}^{r} P'_\varrho = 1$ it follows that $P'_\varrho P_\varrho = P'_\varrho$ and $P_\varrho P'_\varrho = P_\varrho$, i.e., $P_\varrho = P'_\varrho$. Therefore, $D'_\varrho = (A - \lambda_\varrho) P_\varrho = D_\varrho$, $S = S'$ and $D = D'$. ■

The uniqueness of the spectral representation can be proved also by other approaches (for example by the uniqueness of the partial fraction decomposition of the resolvent, see T. KATO [5, p. 42]). This decomposition will be investigated in the following section. In this chapter (which is devoted to the investigation of the Jordan structure of A), this decomposition is not used, but later it will play an essential role in establishing the so-called perturbation series (see Sect. 3.8 and 3.9).

2.6. Partial fraction decomposition of the resolvent

2.6.1. Principal parts

Let λ_0 be one of the eigenvalues of A. Then λ_0 is a pole of $R(\lambda)$. Hence, there is a Laurent expansion

$$R(\lambda) = \sum_{n=-M}^{\infty} A_n (\lambda - \lambda_0)^n \tag{6.1}$$

in the neighborhood of $\lambda = \lambda_0$, where

$$A_n = \frac{1}{2\pi i} \int_{\mathscr{C}} \frac{1}{(\lambda - \lambda_0)^{n+1}} R(\lambda)\, d\lambda \tag{6.2}$$

and where $\mathscr{C}$ denotes a suitable sufficiently small positively oriented sphere centered at λ_0. From (6.2) it follows that

$$A_{-1} = \frac{1}{2\pi i} \int_{\mathscr{C}} R(\lambda)\, d\lambda = P_0,$$

$$A_{-2} = \frac{1}{2\pi i} \int_{\mathscr{C}} (\lambda - \lambda_0) R(\lambda)\, d\lambda = (A - \lambda_0) P_0 = D_0,$$

and, finally,

$$A_{-k-1} = \frac{1}{2\pi i} \int_{\mathscr{C}} (\lambda - \lambda_0)^k R(\lambda)\, d\lambda = D_0^k, \qquad k = 1, 2, 3, \ldots,$$

where P_0 denotes the eigenprojection and D_0 the eigennilpotent belonging to λ_0. Hence, (note Subsect. 2.4.2) *λ_0 is a pole of $R(\lambda)$ of order $\nu(\lambda_0)$ with the principal part*

$$\frac{P_0}{\lambda - \lambda_0} + \frac{D_0}{(\lambda - \lambda_0)^2} + \frac{D_0^2}{(\lambda - \lambda_0)^3} + \cdots + \frac{D_0^{\nu(\lambda_0)-1}}{(\lambda - \lambda_0)^{\nu(\lambda_0)}}. \tag{6.3}$$

2.6.2. Reduced resolvents

Consider (6.2) for $n = 0, 1, 2, \ldots$ The integrals cannot be calculated as functions of A by Theorem 3.1, because $1/(\lambda - \lambda_0)^{n+1}$ is not piecewise holomorphic on spec A in this case. But one can derive a product formula for the integrals (6.2) which is the analog of the product formula for the Cauchy-Riesz integral (see Theorems 3.1 and 3.2).

Again, we choose two different simply closed positively oriented piecewise smooth paths $\mathscr{C}$ and $\mathscr{C}'$ consisting of regular points, where the only eigenvalue inside $\mathscr{C}$ is λ_0 and where $\mathscr{C}$ lies inside $\mathscr{C}'$:

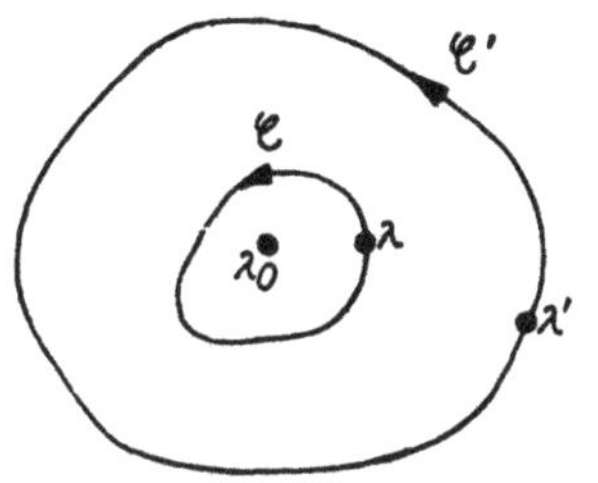

Fig. 2

Then,

$$A_n A_m = \frac{1}{(2\pi i)^2} \int_{\mathscr{C}} (\lambda - \lambda_0)^{-n-1} R(\lambda)\, d\lambda \int_{\mathscr{C}'} (\lambda' - \lambda_0)^{-m-1} R(\lambda')\, d\lambda'$$

$$= \frac{1}{(2\pi i)^2} \int_{\mathscr{C}} \int_{\mathscr{C}'} (\lambda - \lambda_0)^{-n-1} (\lambda' - \lambda_0)^{-m-1} R(\lambda) R(\lambda')\, d\lambda\, d\lambda'$$

$$= \frac{1}{(2\pi i)^2} \int_{\mathscr{C}} \int_{\mathscr{C}'} (\lambda - \lambda_0)^{-n-1} (\lambda' - \lambda_0)^{-m-1} \frac{R(\lambda) - R(\lambda')}{\lambda' - \lambda}\, d\lambda\, d\lambda'$$

$$= \frac{1}{(2\pi i)^2} \int_{\mathscr{C}} (\lambda - \lambda_0)^{-n-1} R(\lambda)\, d\lambda \int_{\mathscr{C}'} \frac{d\lambda'}{(\lambda' - \lambda_0)^{m+1} (\lambda' - \lambda)}$$

$$- \frac{1}{(2\pi i)^2} \int_{\mathscr{C}'} (\lambda' - \lambda_0)^{-m-1} R(\lambda')\, d\lambda' \int_{\mathscr{C}} \frac{d\lambda}{(\lambda - \lambda_0)^{n+1} (\lambda' - \lambda)}.$$

Now one has the identities

$$\int_{\mathscr{C}'} \frac{d\lambda'}{(\lambda' - \lambda_0)^{m+1} (\lambda' - \lambda)} = 0, \qquad m = 0, 1, 2, \ldots,$$

and

$$\frac{1}{2\pi i} \int_{\mathscr{C}} \frac{d\lambda}{(\lambda - \lambda_0)^{n+1} (\lambda' - \lambda)} = (\lambda' - \lambda_0)^{-n-1}, \qquad n = 0, 1, 2, \ldots$$

Thus,

$$A_n A_m = -\frac{1}{2\pi i} \int_{\mathscr{C}'} (\lambda' - \lambda_0)^{-m-n-2} R(\lambda')\, d\lambda' = -A_{n+m+1}, \tag{6.4}$$

for $n, m \geqq 0$. For $n = -1$,

$$\int_{\mathscr{C}} \frac{d\lambda}{\lambda' - \lambda} = 0;$$

hence,

$$A_{-1}A_m = A_m A_{-1} = 0, \qquad m = 0, 1, 2, \ldots \tag{6.5}$$

If

$$A_0 = \frac{1}{2\pi i} \int_{\mathscr{C}} \frac{1}{\lambda - \lambda_0} R(\lambda)\, d\lambda = S_0 \tag{6.6}$$

then it follows from (6.4) that

$$A_n = (-1)^n S_0^{n+1}, \qquad n = 0, 1, 2, \ldots,$$

and from (6.5) that

$$P_0 S_0 = S_0 P_0 = 0, \tag{6.7}$$

where P_0 is the eigenprojection of λ_0. Therefore, *the Laurent expansion of the resolvent in the neighborhood of $\lambda = \lambda_0$ takes the form*

$$R(\lambda) = \frac{P_0}{\lambda - \lambda_0} + \sum_{\varrho=2}^{\nu(\lambda_0)} \frac{D_0^{\varrho-1}}{(\lambda - \lambda_0)^\varrho} + \sum_{n=0}^{\infty} (-1)^n S_0^{n+1} (\lambda - \lambda_0)^n. \tag{6.8}$$

The holomorphic term

$$S(\lambda) = \sum_{n=0}^{\infty} (-1)^n S_0^{n+1} (\lambda - \lambda_0)^n \tag{6.9}$$

of the resolvent is called the *reduced resolvent* of A with respect to λ_0.

Sometimes, the operator $S_0 = S(\lambda_0)$ is also called the *reduced resolvent* of A at λ_0; (note that the resolvent $R(\lambda)$ has a pole at λ_0).

This ambiguous notation arises also in the case of the resolvent itself: In this book resolvent means the resolvent function $R(\lambda)$. Sometimes the value of the resolvent function at a point λ_0 is called the resolvent (at λ_0).

For S_0 one obtains

$$A S_0 = \frac{1}{2\pi i} \int_{\mathscr{C}} \frac{A R(\lambda)}{\lambda - \lambda_0}\, d\lambda = S_0 A, \tag{6.10}$$

i.e., S_0 and A commute. Because

$$A R(\lambda) = R(\lambda)\, A = -1 + \lambda R(\lambda),$$

one obtains

$$AS_0 = S_0A = \frac{1}{2\pi i}\int_{\mathscr{C}} \frac{\big(-1 + \lambda R(\lambda)\big)}{\lambda - \lambda_0}\, d\lambda$$

$$= -1 + \frac{1}{2\pi i}\int_{\mathscr{C}} R(\lambda)\, d\lambda + \frac{\lambda_0}{2\pi i}\int_{\mathscr{C}} \frac{1}{\lambda - \lambda_0} R(\lambda)\, d\lambda = -1 + P_0 + \lambda_0 S_0;$$

that is,

$$(\lambda_0 - A)\, S_0 = S_0(\lambda_0 - A) = 1 - P_0. \tag{6.11}$$

Moreover, according to (6.7), P_0 reduces S_0, and $S_0 \restriction P_0X = 0$.

From (6.11) it can be derived that $(\lambda_0 - A)$ is invertible in $(1 - P_0)\,X$ and

$$[(\lambda_0 - A) \restriction (1 - P_0)\,X]^{-1} = S_0 \restriction (1 - P_0)\,X. \tag{6.12}$$

Therefore, *the reduced resolvent S_0 at λ_0 is the "true" resolvent of $A \restriction (1 - P_0)\,X$ at $\lambda = \lambda_0$.*

Furthermore, from (6.7) and (6.9),

$$S(\lambda)\, P_0 = P_0S(\lambda) = 0, \tag{6.13}$$

i.e., $S(\lambda)$ is reduced by P_0 and is equal to zero in P_0X. Again,

$$\begin{aligned}(\lambda - A)\, S(\lambda) &= (\lambda - \lambda_0)\, S(\lambda) + (\lambda_0 - A)\, S(\lambda)\\ &= (\lambda - \lambda_0)\, S(\lambda) + (\lambda_0 - A)\, S_0 + \sum_{n=1}^{\infty} (\lambda_0 - A)^n\, (-1)^n\, S_0^{n+1}(\lambda - \lambda_0)^n\\ &= (\lambda - \lambda_0)\, S(\lambda) + 1 - P_0 + \sum_{n=1}^{\infty} (1 - P_0)\, (-1)^n\, S_0^n(\lambda - \lambda_0)^n\\ &= (\lambda - \lambda_0)\, S(\lambda) + 1 - P_0 - (\lambda - \lambda_0) \sum_{n=0}^{\infty} (-1)^{n-1}\, S_0^n(\lambda - \lambda_0)^{n-1}\\ &= 1 - P_0;\end{aligned}$$

i.e., *the operator $\lambda - A$ is invertible in $(1 - P_0)\,X$ and*

$$[(\lambda - A) \restriction (1 - P_0)\,X]^{-1} = S(\lambda) \restriction (1 - P_0)\,X. \tag{6.14}$$

The relation (6.14) justifies the name "reduced resolvent" for $S(\lambda)$. One has

$$S(\lambda) \restriction (1 - P_0)\,X = R\big(\lambda; A \restriction (1 - P_0)\,X\big).$$

2.6.3. Decomposition into partial fractions. Corollaries

The decomposition of $R(\lambda)$ into a sum of partial fractions follows from (6.3) and Liouville's theorem (note Subsect. 2.1.2, in particular $R(\infty) = 0$).

Lemma 1. *The resolvent $R(\lambda)$ can be decomposed into a sum of partial fractions thus:*

$$R(\lambda) = \sum_{\varrho=1}^{r} \left(\frac{P_\varrho}{\lambda - \lambda_\varrho} + \sum_{n=2}^{\nu(\lambda_\varrho)} \frac{D_\varrho^{n-1}}{(\lambda - \lambda_\varrho)^n}\right),$$

where $\lambda_1, \dots, \lambda_r$ denote the eigenvalues, $P_1, \dots, P_r$ the eigenprojections, $D_1, \dots, D_r$ the eigennilpotents and $\nu(\lambda_1), \dots, \nu(\lambda_r)$ the indices of the eigenvalues.

Proceeding from Lemma 1 one can obtain some deeper insight into the Cauchy-Riesz integral (cf. Subsect. 2.1.1 and 2.1.2). In particular, all functions of A are necessarily operator polynomials of A.

Theorem 1 (Formula of Lagrange-Sylvester). *Let $f(\lambda)$ be piecewise holomorphic on* spec $A = \{\lambda_1, \lambda_2, \dots, \lambda_r\}$. *Then*

$$f(A) = \sum_{\varrho=1}^{r} \left(f(\lambda_\varrho) P_\varrho + \frac{f'(\lambda_\varrho)}{1!} D_\varrho + \cdots + \frac{f^{(\nu(\lambda_\varrho)-1)}(\lambda_\varrho)}{(\nu(\lambda_\varrho)-1)!} D_\varrho^{\nu(\lambda_\varrho)-1} \right). \tag{6.15}$$

Proof. From Lemma 1 it follows that

$$\frac{1}{2\pi i} \operatorname*{Res}_{\lambda_\varrho} f(\lambda) R(\lambda) = f(\lambda_\varrho) P_\varrho + \frac{f'(\lambda_\varrho)}{1!} D_\varrho + \cdots + \frac{f^{(\nu(\lambda_\varrho)-1)}(\lambda_\varrho)}{(\nu(\lambda_\varrho)-1)!} D_\varrho^{\nu(\lambda_\varrho)-1}.$$

Thus, because of (3.3) one obtains (6.15). ∎

(6.15) is called the **Lagrange-Sylvester formula.** It shows that $f(A)$ is an operator polynomial of A in all cases. Each polynomial $p_f(\lambda)$ with the properties

$$p_f^{(j)}(\lambda_\varrho) = f^{(j)}(\lambda_\varrho), \quad j = 0, 1, 2, \dots, \nu(\lambda_\varrho) - 1; \quad \varrho = 1, \dots, r, \tag{6.16}$$

yields the same operator polynomial as f does. The relations (6.16) define the so-called Hermitean interpolation polynomials.

Corollary 1 (Cayley-Hamilton). *If $p(\lambda) = \chi(\lambda, A)$ then $p(A) = 0$.*

Proof. One has $\chi(\lambda, A) = \prod_{\varrho=1}^{r} (\lambda - \lambda_\varrho)^{n_\varrho}$. Therefore, $\chi(\lambda, A) R(\lambda)$ has no poles in the complex plane. This implies the assertion. ∎

2.7. Spectral representation of normal and other special operators

2.7.1. Adjoint operators

Let the spectral representation of A be given by $A = \sum_{\varrho=1}^{r} (\lambda_\varrho P_\varrho + D_\varrho)$. Then,

$$A^* = \sum_{\varrho=1}^{r} (\bar{\lambda}_\varrho P_\varrho^* + D_\varrho^*); \tag{7.1}$$

hence,

$$A^* = S^* + D^*, \tag{7.2}$$

where

$$S^* = \sum_{\varrho=1}^{r} \bar{\lambda}_\varrho P_\varrho^*, \qquad D^* = \sum_{\varrho=1}^{r} D_\varrho^*.$$

Because of Theorem 5.1, (7.2.) is the spectral representation of A^*; $\bar{\lambda}_1, \ldots, \bar{\lambda}_r$ are the eigenvalues, $P_1^*, \ldots, P_r^*$ the corresponding eigenprojections and $D_1^*, \ldots, D_r^*$ the eigennilpotents of A^*.

2.7.2. Normal operators

Let $X = H$ be unitary and A normal. Let M_0 be an invariant subspace of A, i.e., let $P_0AP_0 = AP_0$ be valid for any projection P_0 on M_0, in particular for the orthogonal projection $P_0 = P_0^*$ on M_0. Then $M_0^\perp$ is an invariant subspace of A^*, i.e., $1 - P_0$ is an invariant (and orthogonal) projection of A^*, because $(1 - P_0) A^*(1 - P_0) - A^*(1 - P_0) = P_0A^*P_0 - P_0A^* = 0$.

Furthermore, for any complex λ_0,

$$\|(A - \lambda_0)^n x\| = \|(A^* - \bar{\lambda}_0)^n x\|, \qquad x \in H, \qquad n = 1, 2, 3, \ldots,$$

because

$$\begin{aligned} \|(A - \lambda_0)^n x\|^2 &= \big((A - \lambda_0)^n x, (A - \lambda_0)^n x\big) = \big((A^* - \bar{\lambda}_0)^n (A - \lambda_0)^n x, x\big) \\ &= \big((A - \lambda_0)^n (A^* - \bar{\lambda}_0)^n x, x\big) = \big((A^* - \bar{\lambda}_0)^n x, (A^* - \bar{\lambda}_0) x\big) \\ &= \|(A^* - \bar{\lambda}_0)^n x\|^2. \end{aligned}$$

Consequently, the algebraic and also the geometric eigenspaces of A and A^* belonging to the eigenvalues λ_0 and $\bar{\lambda}_0$, respectively, coincide.

From the spectral representations $A = \sum_{\varrho=1}^{r} (\lambda_\varrho P_\varrho + D_\varrho)$ and $A^* = \sum_{\varrho=1}^{r} (\bar{\lambda}_\varrho P_\varrho^* + D_\varrho^*)$, it follows that

$$P_\varrho = P_\varrho^*, \qquad \varrho = 1, 2, \ldots, r,$$

i.e., the eigenprojections of a normal operator A are orthogonal. Moreover, the eigennilpotents $D_1, \ldots, D_r$ vanish. First note that the geometric eigenspaces of A and A^* in $P_\varrho H$ coincide, i.e.,

$$\ker\big((A - \lambda_\varrho) \restriction P_\varrho H\big) = \ker\big((A^* - \bar{\lambda}_\varrho) \restriction P_\varrho H\big) = K_\varrho, \qquad \varrho = 1, 2, \ldots, r.$$

Next assume $D_\varrho \neq 0$, i.e., $\nu(\lambda_\varrho) > 1$. Then $\{0\} \subset K_\varrho \subset P_\varrho H$, so there is a nontrivial orthogonal complement L_ϱ of K_ϱ in $P_\varrho H$. Because L_ϱ is invariant with respect to $A \restriction P_\varrho H$ and $A^* \restriction P_\varrho H$, there is a (common) eigenvector of A and A^* in L_ϱ. But this is impossible, so $D_\varrho = 0$.

Thus we have shown

Lemma 1. *The eigenvalues $\lambda_1, \lambda_2, \ldots, \lambda_r$ of a normal operator A are semisimple, i.e., $\nu(\lambda_\varrho) = 1$, the eigenprojections are orthogonal, $P_\varrho = P_\varrho^*$, and the spectral representations of A and A^* take the form*

$$A = \sum_{\varrho-1}^{r} \lambda_\varrho P_\varrho, \qquad A^* = \sum_{\varrho-1}^{r} \bar{\lambda}_\varrho P_\varrho, \qquad P_\varrho = P_\varrho^*, \qquad \varrho = 1, \ldots, r. \tag{7.3}$$

Choosing orthonormal bases in $P_\varrho H$, one obtains an orthonormal basis $e_1, e_2, \ldots, e_N$ of H, such that

$$Ae_\varrho = \mu_\varrho e_\varrho, \qquad \varrho = 1, 2, \ldots, N,$$

where $\mu_1, \mu_2, \ldots, \mu_N$ denote the eigenvalues of A, where their multiplicities are taken into account.

Furthermore, (7.3) implies

$$A^*A = AA^* = \sum_{\varrho=1}^{r} |\lambda_\varrho|^2 P_\varrho. \tag{7.4}$$

Hence,

$$\|Au\|^2 = (A^*Au, u) = \sum_\varrho |\lambda_\varrho|^2 (P_\varrho u, u) = \sum_\varrho |\lambda_\varrho|^2 \|P_\varrho u\|^2 \leqq \max_\varrho |\lambda_\varrho|^2 \cdot \|u\|^2,$$

which implies the inequality

$$\|A\| \leqq \max_\varrho |\lambda_\varrho|.$$

On the other hand, one has $\|Au\| = |\lambda_\varrho| \cdot \|u\|$ if $u \in P_\varrho H$. Thus,

$$\|A\| = \max_\varrho |\lambda_\varrho|. \tag{7.5}$$

Together with A, $R(\lambda, A)$ is also normal and so is the reduced resolvent S_ϱ at the point λ_ϱ (see (6.6)). From (7.3) it follows that

$$R(\lambda, A) = \sum_{\sigma=1}^{r} \frac{1}{\lambda - \lambda_\sigma} P_\sigma$$

and, according to (6.6),

$$S_\varrho = \sum_{\substack{\sigma=1 \\ \sigma \neq \varrho}}^{r} \frac{1}{\lambda_\varrho - \lambda_\sigma} P_\sigma.$$

Using (7.5), one obtains

$$\|R(\lambda, A)\| = 1/\min_\sigma |\lambda - \lambda_\sigma| \tag{7.6}$$

and

$$\|S_\varrho\| = 1/\min_{\sigma \neq \varrho} |\lambda_\varrho - \lambda_\sigma|. \tag{7.7}$$

The right hand side of (7.6) is the inverse of the distance of λ from spec A; the right hand side of (7.7) is the inverse of the distance of λ_ϱ from spec $A \setminus \{\lambda_\varrho\}$, i.e., from the "rest spectrum". If $\operatorname{dis}(\lambda, M) = \inf_{z \in M} |\lambda - z|$, one has

Lemma 2. *If A is normal, then*

$$\|R(\lambda, A)\| = 1/\operatorname{dis}(\lambda, \operatorname{spec} A), \tag{7.8}$$

$$\|S_\varrho\| = 1/\operatorname{dis}(\lambda_\varrho, \operatorname{spec} A \setminus \{\lambda_\varrho\}), \qquad \varrho = 1, \ldots, r, \tag{7.9}$$

where $R(\lambda, A)$ denotes the resolvent and $S_1, S_2, \ldots, S_r$ are the reduced resolvents at the points $\lambda_1, \ldots, \lambda_r$.

2.7.3. Unitary and selfadjoint operators

If U is unitary and $U = \sum_{\varrho=1}^{r} \lambda_\varrho P_\varrho$, then $U^*U = \sum_\varrho |\lambda_\varrho|^2 P_\varrho = 1$, that is, $|\lambda_1| = |\lambda_2| = \cdots = |\lambda_r| = 1$. Conversely, if this condition is satisfied, then $U = \sum \lambda_\varrho P_\varrho$ is unitary:

A normal operator U is unitary if and only if its spectrum lies on the unit sphere $|\lambda| = 1$.

If A is selfadjoint and $A = \sum \lambda_\varrho P_\varrho$, then $\lambda_\varrho = \bar{\lambda}_\varrho$, $\varrho = 1, \ldots, r$, that is, the eigenvalues are real. Conversely, if this condition is satisfied, then $A = \sum \lambda_\varrho P_\varrho$ is selfadjoint:

A normal operator A is selfadjoint if and only if its spectrum lies on the real axis.

For a selfadjoint operator $A = \sum \lambda_\varrho P_\varrho$, one has $(Ax, x) = \sum \lambda_\varrho (P_\varrho x, x) = \sum \lambda_\varrho \|P_\varrho x\|^2$; therefore,

$$\min_\varrho \lambda_\varrho \cdot \|x\|^2 \leqq (Ax, x) \leqq \max_\varrho \lambda_\varrho \cdot \|x\|^2, \qquad x \in H. \tag{7.10}$$

2.8. Continuous perturbations of an operator

In this section, certain basic qualitative properties of spectrum, resolvent and eigenprojections are proved, which hold if an operator A is subjected to small continuous perturbations. Later, if a perturbation parameter is introduced, (see Ch. 3), we shall also obtain quantitative results.

2.8.1. Spectrum and resolvent

Let $\|\cdot\|$ be an arbitrary but fixed norm in $\mathscr{L}(X)$. Then the spectrum spec A and the resolvent $R(\lambda, A)$ depend continuously on the operator A and on the pair (λ, A) with $\lambda \in \operatorname{res} A$, respectively. More precisely, one has

Lemma 1. *Let $\lambda_0 \in \operatorname{res} A_0$. Then there are positive numbers $\alpha > 0$, $\beta > 0$ such that $\lambda \in \operatorname{res} A$ for all λ and all A, where $|\lambda - \lambda_0| < \alpha$, $\|A - A_0\| < \beta$.*

Proof. A formal calculation yields

$$\begin{aligned}(\lambda - A)^{-1} &= [\lambda_0 - A_0 + (\lambda - \lambda_0) - (A - A_0)]^{-1} \\ &= \left[\left\{1 - \left((A - A_0) - (\lambda - \lambda_0)\right)(\lambda_0 - A_0)^{-1}\right\}(\lambda_0 - A_0)\right]^{-1} \\ &= (\lambda_0 - A_0)^{-1}\,[1 - \{(A - A_0) - (\lambda - \lambda_0)\}\,(\lambda_0 - A_0)^{-1}]^{-1}.\end{aligned}$$

Now, if

$$|\lambda - \lambda_0| + \|A - A_0\| < \|R(\lambda_0, A_0)\|^{-1}, \tag{8.1}$$

then the operator

$$[1 - \{(A - A_0) - (\lambda - \lambda_0)\}\, R(\lambda_0, A_0)]^{-1}$$

exists, because the corresponding Neumann series is absolutely convergent (see Subsect. 1.14.3). Hence $(\lambda - A)^{-1}$ exists. Therefore, one has to choose $\alpha > 0$ and $\beta > 0$ such that $\alpha + \beta < \|R(\lambda_0, A_0)\|^{-1}$. ■

From the proof of Lemma 1, in particular from the equation

$$(\lambda - A)^{-1} - (\lambda_0 - A_0)^{-1} = (\lambda_0 - A_0)^{-1} \sum_{\varrho=1}^{\infty} [\{(A - A_0) - (\lambda - \lambda_0)\} (\lambda_0 - A_0)^{-1}]^\varrho \tag{8.2}$$

(see also Subsect. 1.14.3), one obtains

Lemma 2. *$R(\lambda, A)$ is a continuous function in the variables λ and A jointly, where $\lambda \in \operatorname{res} A$.*

Lemma 3. $\operatorname{spec} A$ *depends continuously on A, i.e., if U is an open set of complex numbers with $U \supset \operatorname{spec} A_0 = \{\lambda_1, \lambda_2, \ldots, \lambda_r\}$, for example U is a union of small open discs $|\lambda - \lambda_\varrho| < \varepsilon$, $\varrho = 1, 2, \ldots, r$, then there exists a number $\alpha > 0$ such that $U \supset \operatorname{spec} A$ if $\|A - A_0\| < \alpha$.*

Proof. If $\mathbb{C} \setminus U$ denotes the complement of U, then $\mathbb{C} \setminus U \subset \operatorname{res} A_0$. $\mathbb{C} \setminus U$ is closed, so (cf. Subsect. 2.1.2),

$$\inf_{\lambda \in \mathbb{C} \setminus U} \|R(\lambda, A_0)\|^{-1} = \alpha > 0.$$

Because

$$(\lambda - A)^{-1} = (\lambda - A_0)^{-1} [1 - \{A - A_0\} (\lambda - A_0)^{-1}]^{-1}$$

for $\|A - A_0\| < \alpha$ and $\lambda \in \mathbb{C} \setminus U$, $\lambda \in \operatorname{res} A$ always follows. Therefore, $\mathbb{C} \setminus U \subset \operatorname{res} A$, or $U \supset \operatorname{spec} A$. ■

2.8.2. Eigenprojections

The eigenprojections also depend continuously on A. More precisely, one has

Lemma 4. *Let $\lambda_0 \in \operatorname{spec} A_0$ and let $\varepsilon > 0$ be chosen such that the dotted disc $0 < |\lambda - \lambda_0| \leqq \varepsilon$ belongs to $\operatorname{res} A_0$. Then the Riesz projection*

$$P(A) = \frac{1}{2\pi i} \int_{|\lambda - \lambda_0| = \varepsilon} (\lambda - A)^{-1} \, d\lambda \tag{8.3}$$

(where the sphere $|\lambda - \lambda_0| = \varepsilon$ is positively oriented) depends continuously on A in a certain neighborhood $\|A - A_0\| < \delta$.

Proof. According to the proof of Lemma 3, $|\lambda - \lambda_0| = \varepsilon$ belongs to $\operatorname{res} A$ for a certain neighborhood $\|A - A_0\| < \delta$. Furthermore, according to Lemma 2, $R(\lambda, A)$ depends continuously on (λ, A) jointly, where $(\lambda, A) \in \{\lambda : |\lambda - \lambda_0| = \varepsilon\} \times \{A : \|A - A_0\| < \delta\}$. Note that $M \times N$ denotes the cartesian product of the sets M and N. Then, because of (8.3), $P(A)$ is continuous in the neighborhood $\|A - A_0\| < \delta$. ■

Corollary 1. *If $\|A - A_0\| < \delta$, then*

$$\dim P(A) = \dim P(A_0). \tag{8.4}$$

Proof. One has $\dim P(A) = \operatorname{tr} P(A)$. But $\operatorname{tr} P(A)$ is a continuous scalar function of A on $\|A - A_0\| < \delta$, because together with $P(A)$ all matrix elements are continuous functions of A. The special scalar function $\operatorname{tr} P(A)$ necessarily has integer values, hence it is constant. ■

Corollary 2. *The relation*

$$n(\lambda_0) = \sum_{\lambda \in \operatorname{spec} A \cap \{\lambda : |\lambda - \lambda_0| < \varepsilon\}} n(\lambda),$$

is valid, where $n(\lambda)$, $n(\lambda_0)$ denote the algebraic multiplicities of the corresponding eigenvalues.

Proof. According to the definition of the eigenprojections (cf. Subsect. 2.4.1), one has

$$P(A) = \sum_{\lambda \in \operatorname{spec} A \cap \{\lambda : |\lambda - \lambda_0| < \varepsilon\}} P(\lambda),$$

where $P(\lambda)$ denotes the eigenprojection belonging to the eigenvalue λ. This implies the assertion. ■

The operator A is said to have a *simple spectrum* if all eigenvalues of A are simple.

Corollary 3. *The set of all operators with a simple spectrum is an open set in $\mathcal{L}(X)$.*

Proof. Let A_0 be an operator which has a simple spectrum. Thus, A has $N = \dim X$ different eigenvalues $\lambda_1, \ldots, \lambda_N$. Then, according to Lemma 3, the spectrum $\operatorname{spec} A$ of any A from a certain neighborhood $\|A - A_0\| < \delta$ is contained in small discs $|\lambda - \lambda_\varrho| < \varepsilon$ and, according to Corollary 2, in every disc $|\lambda - \lambda_\varrho| < \varepsilon$ there is exactly one eigenvalue. Therefore, A has a simple spectrum. ■

2.9. Jordan structure of a linear operator

In Sect. 2.5 the spectral representation of A was derived. That is, A can be decomposed by $A = A_1 \oplus A_2 \oplus \cdots \oplus A_r$ with respect to $X = M_1 \oplus M_2 \oplus \cdots \oplus M_r$, where M_ϱ denotes the algebraic eigenspace belonging to the eigenvalue λ_ϱ. The next problem is the investigation of the structure of A on M_ϱ, or, in other words, of the structure of the eigennilpotents D_ϱ on $P_\varrho X = M_\varrho$. Obviously, it is sufficient to consider one of these subspaces. Therefore, without loss of generality, one can assume $A - \lambda_0$ to be nilpotent on X, i.e., A has only one eigenvalue λ_0 on X. Let $\nu(\lambda_0) = m$, then $1 \leqq m \leqq N$.

2.9.1. Eigennilpotent kernels

Because $(A - \lambda_0)^m = 0$, $(A - \lambda_0)^{m-1} \neq 0$, one obtains the following chain of inclusions for the kernels of the powers of the eigennilpotents:

$$\{0\} \subset \ker (A - \lambda_0) \subset \ker (A - \lambda_0)^2 \subset \cdots \subset \ker (A - \lambda_0)^{m-1} \subset \ker (A - \lambda_0)^m = X. \quad (9.1)$$

We abbreviate:

$$K_\varrho = K_\varrho(\lambda_0) = \ker (A - \lambda_0)^\varrho, \qquad \varrho = 1, 2, \ldots, m.$$

K_ϱ is called the *eigennilpotent kernel of degree* ϱ with respect to A and λ_0. K_ϱ is an *invariant subspace* of A, which consists of those Jordan vectors of A with respect to λ_0, where the degree is at most ϱ. K_1 is the geometric eigenspace; K_m is the algebraic eigenspace with respect to λ_0. One has

Lemma 1. *Let the vectors* $x_1, x_2, \ldots, x_l \in K_{\varrho+1}$ *be linearly independent* $\operatorname{mod} K_\varrho$ (i.e., linearly independent with respect to the factor space $K_{\varrho+1} \operatorname{mod} K_\varrho$, cf. Subsect. 1.1.10 and 1.1.11). *Then the vectors* $(A - \lambda_0)\, x_1, (A - \lambda_0)\, x_2, \ldots, (A - \lambda_0)\, x_l \in K_\varrho$ *are linearly independent* $\operatorname{mod} K_{\varrho-1}$.

Proof. Let $\sum \alpha_\sigma (A - \lambda_0)\, x_\sigma \in K_{\varrho-1}$, that is, let

$$\sum \alpha_\sigma (A - \lambda_0)\, x_\sigma = y, \qquad (A - \lambda_0)^{\varrho-1}\, y = 0.$$

Then

$$(A - \lambda_0)^\varrho \sum \alpha_\sigma x_\sigma = 0,$$

i.e., $\sum \alpha_\sigma x_\sigma \in K_\varrho$. Hence, $\alpha_1 = \alpha_2 = \cdots = \alpha_l = 0$. ■

2.9.2. Construction of a Jordan basis in X

Now a special basis in X will be constructed with essential properties with respect to A. This basis appears in the form of a certain table, which will be derived in this section.

First one considers a basis of $X \operatorname{mod} K_{m-1}$. The vectors of this basis are denoted by $y_m^{(1)}, y_m^{(2)}, \ldots, y_m^{(p_m)}$, where $p_m = \dim X \operatorname{mod} K_{m-1} = \dim K_m \operatorname{mod} K_{m-1} \geqq 1$. Then, according to Lemma 1, the vectors

$$(A - \lambda_0)\, y_m^{(\sigma_m)}, \qquad \sigma_m = 1, 2, \ldots, p_m, \qquad p_m \geqq 1, \tag{9.2}$$

are linearly independent in $K_{m-1} \operatorname{mod} K_{m-2}$. Next the vectors (9.2) can be enlarged to a basis of $K_{m-1} \operatorname{mod} K_{m-2}$ through the addition of suitable vectors $y_{m-1}^{(\sigma_{m-1})}$, $\sigma_{m-1} = 1, 2, \ldots, p_{m-1}$. Of course, the case $p_{m-1} = 0$ is not excluded. In this case, the vectors (9.2) already form a basis of $K_{m-1} \operatorname{mod} K_{m-2}$. Furthermore, the vectors

$$(A - \lambda_0)^2\, y_m^{(\sigma_m)}, \quad (A - \lambda_0)\, y_{m-1}^{(\sigma_{m-1})}, \quad \sigma_m = 1, 2, \ldots, p_m; \quad \sigma_{m-1} = 1, 2, \ldots, p_{m-1}, \tag{9.3}$$

are linearly independent in $K_{m-2} \operatorname{mod} K_{m-3}$. Again, they are enlarged to a basis of $K_{m-2} \operatorname{mod} K_{m-3}$ through the addition of suitable vectors $y_{m-2}^{(\sigma_{m-2})}$. After finitely many steps, one obtains the following table (each row contains the basis of the space given in the last column

$$\left.\begin{array}{lllll}
y_m^{(\sigma_m)} & \ldots\ldots\ldots & & & X/K_{m-1} \\
(A - \lambda_0)\, y_m^{(\sigma_m)} & y_{m-1}^{(\sigma_{m-1})} & \ldots\ldots & & K_{m-1}/K_{m-2} \\
(A - \lambda_0)^2\, y_m & (A - \lambda_0)\, y_{m-1}^{(\sigma_{m-1})} & y_{m-2}^{(\sigma_{m-1})} & \ldots\ldots & K_{m-2}/K_{m-3} \\
\ldots\ldots & \ldots\ldots & \ldots\ldots & \ldots\ldots & \\
(A - \lambda_0)^{m-1}\, y_m^{(\sigma_m)} & (A - \lambda_0)^{m-2}\, y_{m-1}^{(\sigma_{m-1})} & (A - \lambda_0)^{m-3}\, y_{m-2}^{(\sigma_{m-2})} & \cdots\ (A - \lambda_0)\, y_2^{(\sigma_2)},\ y_1^{(\sigma_1)} \cdots & K_1.
\end{array}\right\} \tag{9.4}$$

The upper indices $\sigma_1, \sigma_2, \ldots, \sigma_m$ run through the following values:

$$\sigma_m = 1, 2, \ldots, p_m, \qquad p_m \geqq 1,$$

$$\sigma_{m-1} = 1, 2, \ldots, p_{m-1}, \qquad p_{m-1} \geqq 0,$$

$$\vdots$$

$$\sigma_1 = 1, 2, \ldots, p_1, \qquad p_1 \geqq 0.$$

Therefore: *The vectors in Table* (9.4) *form a basis in X* (see Subsect. 1.1.11). Consider the subspaces K_ϱ^σ of X spanned by the vectors of the columns of Table (9.4). They are defined by

$$K_\varrho^\sigma = \operatorname{spa}\left(y_\varrho^{(\sigma)}, (A - \lambda_0)\, y_\varrho^{(\sigma)}, \ldots, (A - \lambda_0)^{\varrho-1}\, y_\sigma^{(\sigma)}\right), \tag{9.5}$$

where $\varrho = 1, 2, \ldots, m$ and $\sigma = 1, 2, \ldots, p_\varrho$. The properties of these subspaces are collected in

Lemma 2. *The subspaces K_ϱ^σ of X defined by* (9.5) *have the following properties:*

I. $\dim K_\varrho^\sigma = \varrho$, $\varrho = 1, 2, \ldots, m$; $\sigma = 1, 2, \ldots, p_\varrho$.

II. *If $x \in K_\varrho^\sigma$, then x is a Jordan vector with respect to A and λ_0 of degree at most ϱ.*

III. *The identity $X = \sum_{\varrho=1}^{m} \sum_{\sigma=1}^{p_\varrho} \oplus K_\varrho^\sigma$ holds.*

IV. *K_ϱ^σ reduces A.*

V. *K_ϱ^σ is irreducible with respect to A.*

Proof. I follows from the linear independence of the vectors $y_\varrho^{(\sigma)}, (A - \lambda_0)\, y_\varrho^{(\sigma)}, \ldots, (A - \lambda_0)^{\varrho-1}\, y_\varrho^{(\sigma)}$ in X.

II follows from $y_\varrho^{(\sigma)} \in K_\varrho$.

III is obtained from the fact that Table (9.4) forms a basis in X.

IV. An immediate consequence of the structure of the basis $y_\varrho^{(\sigma)}, (A - \lambda_0)\, y_\varrho^{(\sigma)}, \ldots, (A - \lambda_0)^{\varrho-1}\, y_\varrho^{(\sigma)}$ is that K_ϱ^σ is invariant with respect to $A - \lambda_0$, hence, also with respect to A. Thus, because of III, IV follows.

V. K_ϱ^σ is irreducible because $y_\varrho^{(\sigma)}$ is a Jordan vector of degree ϱ, i.e., $(A - \lambda_0)^\varrho\, y_\varrho^{(\sigma)} = 0$, but $(A - \lambda_0)^{\varrho-1}\, y_\varrho^{(\sigma)} \neq 0$. Let $K_\varrho^\sigma = H_1 \oplus H_2$ be a proper decomposition of K_ϱ^σ into the sum of two reducing subspaces with respect to A, i.e., $\max\{\dim H_1, \dim H_2\} = m < \varrho$. Then, $(A - \lambda_0)^m\, x_1 = 0$ for all $x_1 \in H_1$, and $(A - \lambda_0)^m\, x_2 = 0$ for all $x_2 \in H_2$; hence, $(A - \lambda_0)^m\, x = 0$ for all $x \in K_\varrho^\sigma$, in contradiction to the existence of $y_\varrho^{(\sigma)} \in K_\varrho^\sigma$. ■

A basis in X formed according to table (9.4) is called a *Jordan basis* in X with respect to A.

2.9.3. Jordan operators, Jordan projections

The considerations of Subsect. 2.9.2 suggest the following definitions:

Definition 1. A is said to be a *Jordan operator* in X if there exists a basis $e_1, e_2, \ldots, e_N$

of X such that

$$Ae_1 = \lambda_0 e_1,$$

$$Ae_2 = \lambda_0 e_2 + e_1,$$

$$\dots$$

$$Ae_{N-1} = \lambda_0 e_{N-1} + e_{N-2},$$

$$Ae_N = \lambda_0 e_N + e_{N-1},$$

that is, if A has the assigned matrix

$$\Gamma_N = \begin{pmatrix} \lambda_0 & 1 & 0 & \dots & 0 & 0 \\ 0 & \lambda_0 & 1 & \dots & 0 & 0 \\ & & & \dots & & \\ 0 & 0 & 0 & \dots & \lambda_0 & 1 \\ 0 & 0 & 0 & \dots & 0 & \lambda_0 \end{pmatrix} \quad (9.6)$$

with respect to the basis $e_N, e_{N-1} = (A - \lambda_0) e_N, e_{N-2} = (A - \lambda_0)^2 e_N, \dots, e_2 = (A - \lambda_0)^{N-2} e_N, e_1 = (A - \lambda_0)^{N-1} e_N$. Here λ_0 denotes a complex number.

Γ_N is an $N \times N$-matrix and is called a *Jordan block of degree N* with respect to the eigenvalue λ_0.

If A is a Jordan operator in X, then λ_0 is the only eigenvalue of A, $A - \lambda_0$ is nilpotent and $\nu(\lambda_0) = N$, because $(A - \lambda_0)^{N-1} e_N \neq 0$.

Definition 2. Let A be any operator in X. A reducing projection P_0 of A is called a *Jordan projection* if $A \restriction P_0X$ is a Jordan operator.

Note that one has to be careful to distinguish between the concepts "Jordan operator" and "Jordan projection". Only the one-dimensional Jordan projections are Jordan operators.

Lemma 3. *A is a Jordan operator in X if and only if A is irreducible in X.*

Proof. Let A be a Jordan operator. Then from Subsect. 2.9.2, in particular from Lemma 2.V the irreducibility of A follows. Conversely, let A be irreducible in X. Obviously, there exists only one eigenvalue for A, because otherwise the spectral representation yields a nontrivial decomposition of A. Thus one can construct a Jordan basis of A according to Table (9.4). In this case the decomposition III of Lemma 2 contains only one subspace, because A is irreducible. In this subspace, i.e., in X, A acts as a Jordan operator. ■

In other words: *A reducing projection P_0 of A is irreducible if and only if P_0 is a Jordan projection.*

In this case therefore, the concepts "irreducible operator" and "Jordan operator" and also the concepts "irreducible projection" and "Jordan projection" are synonymous, because the linear spaces considered are linear spaces over $\mathbb{C}$, the complex numbers. In general (see Ch. 4), the situation is much more complicated.

The result of Subsect. 2.9.2 can now be expressed as follows:

Lemma 4. *Let A be an operator in X with only one eigenvalue λ_0. Then to A there corresponds a complete disjoint system of reducing Jordan projections P_ϱ^σ, and $A \upharpoonright P_\varrho^\sigma X$ is a Jordan operator with respect to λ_0.*

The matrix assigned to A with respect to the Jordan basis of Table (9.4) is a matrix formed by diagonal blocks, namely, by p_1 Jordan blocks of degree 1, p_2 Jordan blocks of degree 2, ..., p_m Jordan blocks of degree m:

$$A \leftrightarrow \{\underbrace{\Gamma_1, \Gamma_1, \ldots, \Gamma_1}_{p_1 \text{times}}; \underbrace{\Gamma_2, \ldots, \Gamma_2}_{p_2 \text{times}}; \ldots ; \underbrace{\Gamma_m, \ldots, \Gamma_m}_{p_m \text{times}}\}. \tag{9.7}$$

A matrix formed by diagonal blocks according to (9.7), i.e., where the blocks are Jordan blocks, is called a *Jordan normal form*. The structure of A is best expressed by the Jordan normal form. One refers to it as the *Jordan structure of A*.

2.9.4. Uniqueness of the Jordan structure, Frobenius formulas

A Jordan basis of A (with only one eigenvalue λ_0) in X is not uniquely determined. This is readily seen in the case of the unit operator 1, where every basis in X is a Jordan basis. But the Jordan normal form, hence the Jordan structure of A in X, is uniquely determined up to the order of the Jordan blocks. In other words:

Lemma 5. *The number and the dimensions of the Jordan projections of A are uniquely determined.*

Proof. Again, let $\nu(\lambda_0) = m$. Then the assertion is that the numbers $p_1, p_2, \ldots, p_m$ are uniquely determined. The proof of this assertion will be based on a calculation of the p_i. Because m is not known a priori either, the numbers $p_1, p_2, \ldots, p_N$ are calculated. Then m is the maximal index with $p_m \neq 0$. Again one denotes the eigennilpotent kernels by $K_1, K_2, \ldots, K_N$. Then K_ϱ is the subspace of all Jordan vectors of degree at most ϱ. Note that these eigennilpotent kernels are defined invariantly, i.e., without reference to a basis; thus the numbers $\dim K_\varrho$, $\varrho = 1, 2, \ldots, N$, are invariants of A, which are uniquely defined by A. The following equations hold:

$$\dim K_\varrho = p_1 + 2p_2 + \cdots + \varrho p_\varrho + \varrho p_{\varrho+1} + \cdots + \varrho p_N, \qquad \varrho = 1, \ldots, N. \tag{9.8}$$

From (9.8)

$$p_\varrho = 2 \dim K_\varrho - \dim K_{\varrho-1} - \dim K_{\varrho+1}, \qquad \varrho = 1, \ldots, N, \tag{9.9}$$

follows, where $K_0 = \{0\}$ and $K_{N+1} = X$. Thus the numbers p_ϱ have been calculated. In particular, according to (9.9), these numbers are uniquely determined by A. The inequalities

$$2 \dim K_\varrho - \dim K_{\varrho-1} - \dim K_{\varrho+1} \geqq 0, \qquad \varrho = 1, \ldots, N, \tag{9.10}$$

which follow from (9.9), are often called the *Frobenius inequalities* (see, for example, I. M. Glazman and Ju. L. Ljubič [1, p. 110]). In addition, $\dim K_\varrho = \dim \ker (A - \lambda_0)^\varrho = \operatorname{def} (A - \lambda_0)^\varrho$. Hence, it is also possible to write the equations (9.8) in the form

$$\operatorname{def} (A - \lambda_0)^\varrho = p_1 + 2p_2 + \cdots + \varrho p_\varrho + \varrho p_{\varrho+1} + \cdots + \varrho p_N. \tag{9.11}$$

2.9.5. Jordan structure of an arbitrary operator

Let A be an arbitrary operator with the spectral representation $A = \sum_{\varrho=1}^{r} (\lambda_\varrho P_\varrho + D_\varrho)$. Then the foregoing considerations are true for every algebraic eigenspace $P_\varrho X$ of A. In every algebraic eigenspace $P_\varrho X$ of A one can choose a Jordan basis according to Table (9.4) with all the implications following from this fact. Thus one obtains a basis in the whole space X which is called a *Jordan basis of A in X*. The assigned matrix of A with respect to this basis is called a *Jordan normal form of A*. Because of the uniqueness of the spectral representation and Lemma 5, the Jordan normal form is uniquely determined up to the order of the Jordan blocks. It takes the form

$$\{\underbrace{\Gamma_1(\lambda_1), \dots, \Gamma_1(\lambda_1)}_{\varkappa(1,1)\text{times}}, \dots, \underbrace{\Gamma_{\nu_1}(\lambda_1), \dots, \Gamma_{\nu_1}(\lambda_1)}_{\varkappa(1,\nu_1)\text{times}}; \underbrace{\Gamma_1(\lambda_2), \dots, \Gamma_1(\lambda_2)}_{\varkappa(2,1)\text{times}}, \dots, \underbrace{\Gamma_{\nu_r}(\lambda_r), \dots, \Gamma_{\nu_r}(\lambda_r)}_{\varkappa(r,\nu_r)\text{times}}\},$$

where $\Gamma_m(\lambda_0) = \Gamma_m$ denotes a Jordan block of degree m belonging to the eigenvalue λ_0. $\nu_1, \dots, \nu_r$ are the indices of $\lambda_1, \dots, \lambda_r$, respectively. The numbers $\varkappa(\varrho, \sigma)$, $\sigma = 1, 2, \dots, \nu_\varrho$; $\varrho = 1, 2, \dots, r$, are uniquely determined.

The results obtained can be collected in the following theorem:

Theorem 1 (Jordan structure). *Let A be an arbitrary operator on X with the spectral representation $A = \sum_{\varrho=1}^{r} (\lambda_\varrho P_\varrho + D_\varrho)$. Then P_ϱ can be decomposed by*

$$P_\varrho = \sum_{\sigma=1}^{\nu_\varrho} \sum_{\varkappa=1}^{\varkappa(\varrho,\sigma)} P_\varrho^{\sigma\varkappa}, \tag{9.12}$$

where the projections $P_\varrho^{\sigma\varkappa}$ denote a disjoint system of reducing Jordan projections of A, i.e., $A \restriction P_\varrho^{\sigma\varkappa} X$ is a Jordan operator with the eigenvalue λ_ϱ. One has $\dim P_\varrho^{\sigma\varkappa} = \sigma$ *for all $\varkappa = 1, 2, \dots, \varkappa(\varrho, \sigma)$ and the numbers $\varkappa(\varrho, \sigma)$ of Jordan projections with fixed dimension σ belonging to the eigenvalue λ_ϱ are uniquely determined.*

The numbers $\varkappa(\varrho, \sigma)$ can be determined according to Subsect. 2.9.4. Because

$$\operatorname{def} (A - \lambda_\varrho)^\sigma \restriction P_\varrho X = \operatorname{def} (A - \lambda_\varrho)^\sigma, \qquad \sigma = 1, 2, 3, \dots,$$

the equation (9.11) implies

$$\operatorname{def} (A - \lambda_\varrho)^\sigma = \varkappa(\varrho, 1) + 2\varkappa(\varrho, 2) + \dots + \sigma\varkappa(\varrho, \sigma)$$
$$+ \sigma\varkappa(\varrho, \sigma + 1) + \dots + \sigma\varkappa(\varrho, n_\varrho), \tag{9.13}$$

where σ runs from 1 to n_ϱ, the algebraic multiplicity of λ_ϱ. Thus one obtains

$$\varkappa(\varrho, \sigma) = 2 \operatorname{def} (A - \lambda_\varrho)^\sigma - \operatorname{def} (A - \lambda_\varrho)^{\sigma-1} - \operatorname{def} (A - \lambda_\varrho)^{\sigma+1}. \tag{9.14}$$

2.9.6. Complete system of invariants for a class of similar matrices, elementary divisors

The result obtained in Subsect. 2.9.5 can also be interpreted as follows: There is a one-to-one correspondence between operators A and classes of similar matrices (see Subsect. 1.5.2). As we have seen, in such a class $\{A\}$ of similar matrices there is exactly one Jordan normal form (apart from the order of the Jordan blocks). The Jordan normal form is uniquely fixed by the numbers $\varkappa(\lambda_0, \sigma)$, $\lambda_0 \in \operatorname{spec} A$, $\sigma = 1, 2, 3, \ldots$ Therefore, one can formulate the following statement: *The system of numbers* $\varkappa(\lambda_0, \sigma)$, $\lambda_0 \in \operatorname{spec} A$, $\sigma = 1, 2, \ldots$, *forms a complete system of invariants for the classes of similar matrices*, that is, such a class can be uniquely represented by this system.

If λ_0 is an eigenvalue of A and P_0 a Jordan projection of A belonging to λ_0, then $\chi(\lambda; A \upharpoonright P_0X) = (\lambda - \lambda_0)^{\dim P_0}$ is called the *corresponding elementary divisor*. Of course, the system of all elementary divisors is also a complete system of invariants for A.

2.9.7. Alternate notation for the system of Jordan projections

Sometimes it is convenient to use another indexing notation for the system of Jordan projections than the one given in Theorem 1. The new notation can be described as follows:

Let $\lambda_1, \lambda_2, \ldots, \lambda_r$ be the system of all eigenvalues. Let $\nu(\lambda_\varrho) = m_\varrho^{(1)}$. Let the dimensions of the relevant Jordan projections belonging to λ_ϱ be ordered in decreasing order

$$m_\varrho^{(1)} > m_\varrho^{(2)} > \cdots > m_\varrho^{(k_\varrho)} > 0. \tag{9.15}$$

Furthermore, let the number of Jordan projections $P_\varrho^{\sigma\varkappa}$ with $\dim P_\varrho^{\sigma\varkappa} = m_\varrho^{(\sigma)}$ be denoted by $l_\varrho^{(\sigma)}$. Then the decomposition (9.12) can be written in the form

$$P_\varrho = \sum_{\sigma=1}^{k_\varrho} \sum_{\varkappa=1}^{l_\varrho^{(\sigma)}} P_\varrho^{\sigma\varkappa}, \tag{9.16}$$

where

$$n_\varrho = \sum_{\sigma=1}^{k_\varrho} m_\varrho^{(\sigma)} l_\varrho^{(\sigma)}, \tag{9.17}$$

where n_ϱ is the algebraic multiplicity of λ_ϱ. The geometric multiplicity can be expressed by $\sum_{\sigma=1}^{k_\varrho} l_\varrho^{(\sigma)}$.

2.9.8. Adjoint operator

Together with the decomposition of A into a direct sum of Jordan operators according to the decomposition of the projections (9.12), the corresponding problem for A^* is also solved. If $A = \sum \lambda_\varrho P_\varrho + \sum D_\varrho$, then $A^* = \sum \bar{\lambda}_\varrho P_\varrho^* + \sum D_\varrho^*$ is the spectral representation of A^* (see Subsect. 2.7.1), where $\bar{\lambda}_\varrho$ are the eigenvalues, P_ϱ^* the eigenprojections and D_ϱ^* the eigennilpotents of A^*. Furthermore, from (9.12) it follows that

$$P_\varrho^* = \sum_{\sigma,\varkappa} (P_\varrho^{\sigma\varkappa})^*.$$

The system of all $(P_\varrho^{\sigma\varkappa})^*$ is a system of disjoint Jordan projections of A^* belonging to $\bar{\lambda}_\varrho$. (Note that from Subsect. 2.9.3 it follows that A^* is a Jordan operator in X^*

if A is a Jordan operator in X.) Because $\dim P_\varrho^{\sigma *} = \dim (P_\varrho^{\sigma *})^*$, the numbers and dimensions of the Jordan blocks remain invariant.

If $e_1, \ldots, e_N$ is a Jordan basis of A, then the corresponding adjoint basis $f_1, \ldots, f_N$ is a Jordan basis of A^*.

One can assume without loss of generality that A is a Jordan operator in X (with a single eigenvalue λ_0). Let $A - \lambda_0 = D_0$, then $D_0 e_1 = 0$, $D_0 e_i = e_{i-1}$, $i = 2, 3, \ldots, N$. Therefore,

$$D_0^* f_N = 0$$

and

$$D_0^* f_j = f_{j+1}, \qquad j = 1, 2, \ldots, N-1,$$

so $f_1, \ldots, f_N$ is a Jordan basis with the eigenvector f_N.

Sometimes, the *generating* vector e_N of a Jordan basis is called a *cyclic* vector. Then f_1 is a cyclic vector with respect to the adjoint basis.

2.9.9. Lagrange-Sylvester formula

By means of the Jordan normal form of A, the Lagrange-Sylvester formula (see Subsect. 2.6.3) can be strengthened. First one has: *If $A = \sum \oplus A_\varrho$, then $f(A) = \sum \oplus f(A_\varrho)$.* If P_ϱ denote the projections belonging to the decomposition of A, then $R(\lambda, A) P_\varrho = P_\varrho R(\lambda, A)$ and, moreover, $R(\lambda, A_\varrho) = R(\lambda, A) \upharpoonright P_\varrho X$. Then $R(\lambda, A) = \sum \oplus R(\lambda, A_\varrho)$ and therefore,

$$\int f(\lambda)\, R(\lambda, A)\, d\lambda = \sum \oplus \int f(\lambda)\, R(\lambda, A_\varrho)\, d\lambda$$

hence, $f(A) = \sum \oplus f(A_\varrho)$.

For this reason it is sufficient to take the function $f(A)$ for all Jordan operators belonging to A.

Let A be a Jordan operator in X with the eigenvalue λ_0. Then, if $m = \nu(\lambda_0)$ and $D_0 = A - \lambda_0$, then

$$f(A) = f(\lambda_0)\, 1 + \frac{f'(\lambda_0)}{1!} D_0 + \frac{f''(\lambda_0)}{2!} D_0^2 + \cdots + \frac{f^{(m-1)}(\lambda_0)}{(m-1)!} D_0^{m-1}. \tag{9.18}$$

Hence, the matrix of $f(A)$ assigned to a Jordan basis of A has the form

$$f(A) \leftrightarrow \begin{pmatrix} f(\lambda_0) & f'(\lambda_0)/1! & f''(\lambda_0)/2! & \ldots & f^{(m-1)}(\lambda_0)/(m-1)! \\ 0 & f(\lambda_0) & f'(\lambda_0)/1! & \ldots & f^{(m-2)}(\lambda_0)/(m-2)! \\ & & & \ldots & \\ 0 & 0 & 0 & \ldots & f'(\lambda_0)/1! \\ 0 & 0 & 0 & \ldots & f(\lambda_0) \end{pmatrix}.$$

2.10. Commutant of A

In the following discussion A is a fixed operator. The problem is determining the commutant of A (see Subsect. 1.11.2). This problem is called the *problem of Frobenius* (see, for example, F. Gantmacher [1, p. 204]). For example, all functions $f(A)$, i.e., all operator polynomials in A commute with A. But, as is shown, in general there are many more operators with the intertwining property.

2.10.1. Reduction of the problem

If $AB = BA$, then also $p(A)\,B = Bp(A)$ for all polynomials $p(\lambda)$. This is obvious, because $A^nB = BA^n$ for all $n = 1, 2, 3, \ldots$ Let $P_1, P_2, \ldots, P_r$ denote the eigenprojections of A. These eigenprojections are polynomials of A. Hence,

$$P_\varrho B = BP_\varrho, \qquad \varrho = 1, \ldots, r, \tag{10.1}$$

i.e., the projections P_ϱ reduce B. If $B_\varrho = B \restriction P_\varrho X$, then,

$$B = B_1 \oplus B_2 \oplus \cdots \oplus B_r. \tag{10.2}$$

If $A_\varrho = A \restriction P_\varrho X$, then, because $AB = BA$,

$$A_\varrho B_\varrho = B_\varrho A_\varrho, \qquad \varrho = 1, \ldots, r. \tag{10.3}$$

Conversely, if the operators B_ϱ are defined on $P_\varrho X$ with the property (10.3), then the operator defined by (10.2) commutes with A. Thus one obtains

Lemma 1. *Let $P_1, \ldots, P_r$ be the eigenprojections of A. Then $AB = BA$ if and only if $B = B_1 \oplus B_2 \oplus \cdots \oplus B_r$ with respect to the system $P_1, P_2, \ldots, P_r$, and if $A_\varrho B_\varrho = B_\varrho A_\varrho$, $\varrho = 1, 2, \ldots, r$, where $A_\varrho = A \restriction P_\varrho X$.*

Therefore, for the investigation of the structure of all operators B commuting with A, one can assume, without loss of generality, that A has only one eigenvalue. In other words: The investigation may be done for every algebraic eigenspace separately.

For a diagonal operator S (see Subsect. 2.5.1) the commutant is already determined by Lemma 1.

Corollary 1. *Let $S = \sum_{\varrho=1}^{r} \lambda_\varrho P_\varrho$ be a diagonal operator. Then $BS = SB$ if and only if $B = B_1 \oplus B_2 \oplus \cdots \oplus B_r$ with respect to the system $P_1, \ldots, P_r$, where B_ϱ can be chosen arbitrarily from $\mathscr{L}(P_\varrho X)$.*

Proof. In this case, $A_\varrho = \lambda_\varrho$ is a scalar operator and all operators from $\mathscr{L}(P_\varrho X)$ commute with λ_ϱ. ■

2.10.2. Commutant for operators with only one eigenvalue

Let A be an operator with only one eigenvalue λ_0. Assume that the projections P_1, $P_2, \ldots, P_r$ form a complete disjoint system of Jordan projections with respect to A, i.e., the operators $A \restriction P_\varrho X$ are Jordan operators. Let $\dim P_\varrho = m_\varrho$ and assume that these dimensions are ordered in a non-increasing sequence:

$$m_1 \geqq m_2 \geqq \cdots \geqq m_r \geqq 1.$$

Without loss of generality one may take $\lambda_0 = 0$, i.e., one may assume A to be nilpotent.

Furthermore, let B be an arbitrary operator. Then,

$$B = \sum_{\varrho,\sigma=1}^{r} P_\varrho B P_\sigma. \tag{10.4}$$

Let

$$B_{\varrho\sigma} = P_\varrho B P_\sigma \tag{10.5}$$

then, obviously,

Lemma 2. *The equation $AB = BA$ holds if and only if $AB_{\varrho\sigma} = B_{\varrho\sigma}A$ for all $\varrho, \sigma = 1, 2, \ldots, r$.*

To determine the commutant of A it is sufficient to investigate all equations

$$AB_{\varrho\sigma} = B_{\varrho\sigma}A, \qquad \varrho, \sigma = 1, \ldots, r. \tag{10.6}$$

According to (10.5), the operator $B_{\varrho\sigma}$ vanishes on the subspace $(1 - P_\sigma) X$ and maps $P_\sigma X$ into $P_\varrho X$, i.e., ima $B_{\varrho\sigma} \subseteqq P_\varrho X$. Therefore, it is sufficient to consider the operator $B_{\varrho\sigma}$ on $P_\sigma X$.

The subscripts ϱ and σ are arbitrary but fixed. Let dim $P_\sigma = m_\sigma$, dim $P_\varrho = m_\varrho$ and choose a Jordan basis in every space $P_\sigma X$ and $P_\varrho X$:

$$P_\varrho X = \text{spa}\,(f_1, f_2, \ldots, f_{m_\varrho}),$$

$$P_\sigma X = \text{spa}\,(e_1, e_2, \ldots, e_{m_\sigma}).$$

If $f := f_{m_\varrho}$ and $e := e_{m_\sigma}$, then,

$$A^{m_\varrho-1}f = f_1, \qquad A^{m_\varrho-2}f = f_2, \ldots, Af = f_{m_\varrho-1}, \tag{10.7}$$

$$A^{m_\sigma-1}e = e_1, \qquad A^{m_\sigma-2}e = e_2, \ldots, Ae = e_{m_\sigma-1}. \tag{10.8}$$

Furthermore, let $k = \min\,(m_\varrho, m_\sigma)$. One defines k operators $C_1, C_2, \ldots, C_k$ as follows: On $(1 - P_\sigma) X$ they are defined by zero. On $P_\sigma X$ they are defined by the formulas

$$C_\mu e_{m_\sigma} = f_\mu, \qquad C_\mu e_{m_\sigma-1} = f_{\mu-1}, \ldots, C_\mu e_{m_\sigma-\mu+1} = f_1, \tag{10.9}$$

$$C_\mu e_{m_\sigma-\mu} = C_\mu e_{m_\sigma-\mu-1} = \cdots = C_\mu e_1 = 0, \tag{10.10}$$

where $\mu = 1, 2, \ldots, k$. Then the following lemma can be proved.

Lemma 3. *The equation $AB_{\varrho\sigma} = B_{\varrho\sigma}A$ holds if and only if $B_{\varrho\sigma} = \sum_{\mu=1}^{k} \alpha_\mu C_\mu$ where $\alpha_1, \alpha_2, \ldots, \alpha_k$ are arbitrary complex numbers.*

Proof. Let

$$B_{\varrho\sigma} e_\varkappa = \sum_{\mu=1}^{m_\varrho} b_{\mu\varkappa} f_\mu, \qquad \varkappa = 1, 2, \ldots, m_\sigma.$$

Let $AB_{\varrho\sigma} = B_{\varrho\sigma}A$. Then $AB_{\varrho\sigma}e_\varkappa = B_{\varrho\sigma}Ae_\varkappa$ and, because of (10.8) and $Ae_1 = 0$,

$$\sum_{\mu=1}^{m_\varrho} b_{\mu\varkappa}Af_\mu = \sum_{\mu=2}^{m_\varrho} b_{\mu,\varkappa-1}f_\mu, \qquad \varkappa = 2, \ldots, m_\sigma. \tag{10.11}$$

Moreover, for $\varkappa = 1$,

$$\sum_{\mu=1}^{m_\varrho} b_{\mu 1}Af_\mu = 0 = \sum_{\mu=2}^{m_\varrho} b_{\mu 1}Af_\mu.$$

Hence,

$$b_{21} = b_{31} = \cdots = b_{m_\varrho 1} = 0. \tag{10.12}$$

Furthermore, from (10.11) and because

$$\sum_{\mu=1}^{m_\varrho} b_{\mu\varkappa}Af_\mu = \sum_{\mu=2}^{m_\varrho} b_{\mu\varkappa}f_{\mu-1}, \qquad \varkappa = 2, \ldots, m_\sigma,$$

immediately one obtains

$$\sum_{\mu=2}^{m_\varrho} b_{\mu\varkappa}f_{\mu-1} = \sum_{\mu=1}^{m_\varrho} b_{\mu,\varkappa-1}f_\mu$$

or

$$\sum_{\mu=1}^{m_\varrho-1} b_{\mu+1,\varkappa}f_\mu = \sum_{\mu=1}^{m_\varrho} b_{\mu,\varkappa-1}f_\mu, \qquad \varkappa = 2, \ldots, m_\sigma. \tag{10.13}$$

From (10.13) it follows that

$$b_{m_\varrho 1} = b_{m_\varrho 2} = \cdots = b_{m_\varrho, m_\sigma-1} = 0. \tag{10.14}$$

A further implication of (10.13) is the relation

$$b_{\mu+1,\varkappa+1} = b_{\mu\varkappa}, \qquad \mu = 1, 2, \ldots, m_\varrho - 1; \qquad \varkappa = 1, 2, \ldots, m_\sigma - 1. \tag{10.15}$$

In particular,

$$b_{1,m_\sigma-1} = b_{2,m_\sigma},$$

$$b_{1,m_\sigma-2} = b_{2,m_\sigma-1} = b_{3,m_\sigma} \cdots.$$

If $m_\sigma \geqq m_\varrho$, (i.e., if the number m_σ of columns of the matrix $B = (b_{\mu\varkappa})$ is greater than or equal to the number m_ϱ of rows), then finally,

$$b_{1,m_\sigma-m_\varrho+1} = b_{2,m_\sigma-m_\varrho+2} = \cdots = b_{m_\varrho,m_\sigma} \tag{10.16}$$

and

$$b_{11} = b_{12} = \cdots = b_{1,m_\sigma-m_\varrho} = 0. \tag{10.17}$$

On the other hand, if $m_\sigma < m_\varrho$, then

$$b_{11} = b_{22} = \cdots = b_{m_\sigma,m_\sigma} \tag{10.18}$$

and

$$b_{m_\sigma+1,m_\sigma} = b_{m_\sigma+2,m_\sigma} = \cdots = b_{m_\varrho,m_\sigma} = 0. \tag{10.19}$$

The situation can be described by the following diagrams:

$$m_\sigma \geqq m_\varrho: \quad \begin{array}{cccccccc} 0 & 0 & \cdots\cdots & 0 & * & * & \cdots & * \\ \vdots & \vdots & & \vdots & 0 & \ddots & \ddots & \vdots \\ \vdots & \vdots & & \vdots & \vdots & \ddots & \ddots & * \\ 0 & 0 & \cdots\cdots & 0 & 0 & \cdots & 0 & * \end{array} \tag{10.20}$$

$\underbrace{\quad}_{1^{\text{st}}\text{column}}$ $\qquad$ $\underbrace{\quad}_{(m_\sigma - m_\varrho)^{\text{st}}\text{column}}$

and

$$m_\sigma < m_\varrho: \quad \begin{array}{cccc} * & * & \cdots & * \\ 0 & \bullet & \ddots & \vdots \\ \vdots & & \ddots & * \\ 0 & \cdots & 0 & * \\ 0 & \cdots & 0 & 0 \quad (m_\sigma+1)^{\text{st}}\text{row} \\ \vdots & & \vdots & \vdots \\ 0 & \cdots & 0 & 0 \quad m_\varrho^{\text{st}}\text{row} \end{array} \tag{10.21}$$

Note that, according to (10.15), one has to put the same number on each of the upper diagonals of the matrices.

The conditions (10.20) and (10.21) are also sufficient, that is, from these conditions one obtains again the equation $AB_{\varrho\sigma} = B_{\varrho\sigma}A$. If $k = \min\{m_\varrho, m_\sigma\}$ and

$$\alpha_1 = b_{1,m_\sigma}, \qquad \alpha_2 = b_{2,m_\sigma}, \ldots, \alpha_k = b_{k,m_\sigma}, \tag{10.22}$$

Lemma 3 has been proved. ■

Matrices of the type (10.20) and (10.21) are called *upper triangular matrices.* Thus *the matrix of an operator B commuting with A takes the form of a diagonal block matrix (according to the spectral representation), where every diagonal block again forms a diagonal block matrix (according to the Jōrdan decomposition) and where every diagonal block is an upper triangular matrix. The converse is also true.*

The three Lemmata 1, 2 and 3 solve the Frobenius problem. In Ch. 7 these results will be applied to perturbation theory.

2.10.3. Special case

A characteristic commutant of A is obtained if all eigenvalues $\lambda_1, \lambda_2, \ldots, \lambda_r$ are prime, i.e., if all geometric multiplicities are 1, i.e., if the eigenprojections $P_1, \ldots, P_r$ are already Jordan projections. One has (note Subsect. 1.11.1).

Lemma 4. *The commutant of A coincides with $\mathscr{P}(A)$ if and only if all eigenvalues of A are prime.*

Proof. 1. Let all eigenvalues of A be prime. According to Lemma 1, without loss of generality, A can be assumed to be a Jordan operator. Let $AB = BA$. In this case there is only one operator $B_{\varrho\sigma}$ (cf. the decomposition (10.4)), namely B itself. One has

$m_\varrho = m_\sigma = k = N = \dim X$. From the equations (10.9) and (10.10), which define the operators $C_1, C_2, \ldots, C_N$, one obtains

$$C_N = 1, \quad C_{N-1} = A - \lambda_0, \quad C_{N-2} = (A - \lambda_0)^2, \ldots, C_1 = (A - \lambda_0)^{N-1}, \tag{10.23}$$

that is, the commutant of A consists of the polynomials $\sum_{\varrho=1}^{N-1} \alpha_\varrho (A - \lambda_0)^\varrho$, i.e., of the family of all polynomials $\sum_{\varrho=\sigma}^{N-1} \beta_\varrho A^\varrho$, i.e., of $\mathscr{P}(A)$.

2. Conversely, let λ_0 be an eigenvalue that is not prime. Then the corresponding eigenprojection P_0 has a nontrivial decomposition $P_0 = Q_1 + Q_2$, where Q_1, Q_2 are reducing projections with respect to A. Then, according to Subsect. 2.10.2, there is an operator commuting with A which is not an operator polynomial in A. ∎

2.11. Algebras and their commutants

Later, in Sect. 7.2, properties of linear operators, of algebras of linear operators and their commutants are used which are connected with *Schur's lemma*. In this section the corresponding results are formulated and proved (see, for example, I. M. Glazman and Ju. L. Ljubič [1, p. 123 sqq.]).

2.11.1. Schur's lemma

First we formulate the original form of Schur's lemma as a special case of more general results to be discussed later.

Lemma 1 (Schur). *An operator commuting with all operators of $\mathscr{L}(X)$ is necessarily a scalar operator.*

Proof. Let A be any operator with $AC = CA$ for all $C \in \mathscr{L}(X)$, in particular for all one-dimensional projections $P: AP = PA$. Hence, $Ax = \lambda(x)\, x$ for all $x \in X$, where $\lambda(x)$ denotes a complex number with the property $\lambda(\mu x) = \lambda(x)$ for all complex numbers μ. From the linearity of A one obtains $\lambda(x + y)\, x + \lambda(x + y)\, y = \lambda(x)\, x + \lambda(y)\, y$ for all $x, y \in X$. This implies $\lambda(x) = \lambda(y) = \lambda(x + y)$ if x and y are linearly independent. Therefore, $\lambda(x)$ is a constant. ∎

Schur's lemma may be written in the form

$$\operatorname{com} \mathscr{L}(X) \subseteqq \{\alpha 1\}$$

Together with the trivial relations $\{\alpha 1\} \subseteqq \operatorname{com} \mathscr{L}(X)$ and $\operatorname{com} \{\alpha 1\} \subseteqq \mathscr{L}(X)$, one obtains

$$\operatorname{com}\operatorname{com} \mathscr{L}(X) = \mathscr{L}(X); \qquad \operatorname{com}\operatorname{com} \{\alpha 1\} = \{\alpha 1\}. \tag{11.1}$$

In this Schur's lemma is a special case of more general propositions.

Lemma 2. *Let A be a Jordan operator. Furthermore, let B commute with all operators from the commutant of A. Then B is an operator polynomial in A.*

Proof. Because A is a Jordan operator, there is only one eigenvalue λ_0 of A, which in addition is prime. Then, according to Lemma 10.4, the commutant of A coincides with $\mathscr{P}(A)$. Hence, for B one obtains $BC = CB$ for all $C \in \mathscr{P}(A)$; in particular one has $BA = AB$, that is, B belongs to the commutant of A, which implies that B is an operator polynomial in A. ■

Sometimes one writes $B \cup\!\cup A$ if B commutes with all operators of the commutant of A. If the commutant of A is denoted by $\mathscr{C}$, then the family of all B with $B \cup\!\cup A$ coincides with the commutant com $\mathscr{C}$ of $\mathscr{C}$ (see Subsect. 1.11.2).

Thus, the essential content of Lemma 2 is the statement that for Jordan operators the commutant com $\mathscr{C}$ of the commutant $\mathscr{C}$ of A coincides with $\mathscr{P}(A)$, com $\mathscr{C} = \mathscr{P}(A)$, because polynomials $p(A)$ obviously commute with all operators C which commute with A.

The following Lemma extends this proposition to operators with only one eigenvalue.

Lemma 3. *Let A be an operator with only one eigenvalue λ_0. Furthermore, let $\mathscr{C}$ be the commutant of A and let $B \in$ com $\mathscr{C}$. Then B is an operator polynomial in A; that is,* com $\mathscr{C} = \mathscr{P}(A)$.

Proof. Let $P_1, P_2, \ldots, P_r$ be a complete disjoint system of Jordan projections with respect to A, i.e., $AP_\varrho = P_\varrho A$, $\varrho = 1, 2, \ldots, r$, and $A_\varrho = A \upharpoonright P_\varrho X$ is a Jordan projection. Then $P_\varrho \in \mathscr{C}$. B commutes with all operators of $\mathscr{C}$, in particular with P_ϱ, i.e.,

$$P_\varrho B = BP_\varrho, \qquad \varrho = 1, \ldots, r, \tag{11.2}$$

that is, the projections P_ϱ reduce B. Let $B_\varrho = B \upharpoonright P_\varrho X$. The relation $B \cup\!\cup A$ implies $B_\varrho \cup\!\cup A_\varrho$, i.e., B_ϱ commutes with all operators from $\mathscr{L}(P_\varrho X)$ which commute with A_ϱ. But A_ϱ is a Jordan operator. Hence, Lemma 2 is applicable and B_ϱ is an operator polynomial in A_ϱ,

$$B_\varrho = p_\varrho(A_\varrho), \qquad \varrho = 1, 2, \ldots, r. \tag{11.3}$$

Now $BC = CB$ for all $C \in \mathscr{C}$. If $C_{\varrho\sigma} = P_\varrho C P_\sigma$, then,

$$B_\varrho C_{\varrho\sigma} = C_{\varrho\sigma} B_\sigma. \tag{11.4}$$

$C_{\varrho\sigma}$ maps $P_\sigma X$ into $P_\varrho X$. Because of (11.3),

$$p_\varrho(A_\varrho)\, C_{\varrho\sigma} = C_{\varrho\sigma} p_\sigma(A_\sigma). \tag{11.5}$$

Now one has $CA = AC$ for all $C \in \mathscr{C}$, hence, $C_{\varrho\sigma} A = AC_{\varrho\sigma}$ and

$$C_{\varrho\sigma} A_\sigma = A_\varrho C_{\varrho\sigma}. \tag{11.6}$$

From this equation it follows that

$$p_\varrho(A_\varrho)\, C_{\varrho\sigma} = C_{\varrho\sigma} p_\varrho(A_\sigma) = C_{\varrho\sigma} p_\sigma(A_\sigma). \tag{11.7}$$

(11.7) holds for all $\varrho, \sigma = 1, 2, \ldots, r$. Setting $\dim P_\varrho = m_\varrho$ and choosing $m_1 \geqq m_2 \geqq \cdots \geqq m_r \geqq 1$, one obtains

$$C_{1\sigma} p_1(A_\sigma) = C_{1\sigma} p_\sigma(A_\sigma), \qquad \sigma = 1, 2, \ldots, r. \tag{11.8}$$

Because $\dim P_\sigma \leqq \dim P_1$ for all $\sigma = 1, 2, \ldots, r$, the mapping $C_{1\sigma}$ can be chosen one-to-one (see (10.9) and (10.10)). Then,

$$p_1(A_\sigma) = p_\sigma(A_\sigma), \qquad \sigma = 1, 2, \ldots, r, \tag{11.9}$$

and

$$B = \sum_{\varrho=1}^{r} \oplus B_\varrho = \sum_{\varrho=1}^{r} \oplus p_\varrho(A_\varrho) = \sum_{\varrho=1}^{r} \oplus p_1(A_\varrho) = p_1 \left(\sum_{\varrho=1}^{r} \oplus A_\varrho \right) = p_1(A).$$

Obviously, $\mathscr{P}(A) \subseteqq \operatorname{com} \mathscr{C}$. This concludes the proof. ■

One can extend the proposition of Lemma 3 to arbitrary operators.

Theorem 1. *If A is an arbitrary operator and $\mathscr{C}$ the commutant of A, then* $\operatorname{com} \mathscr{C} = \mathscr{P}(A)$.

Proof. The eigenvalues and eigenprojections of A are denoted by $\lambda_1, \ldots, \lambda_r$ and $P_1, \ldots, P_r$, respectively. Let $B \cup\!\cup A$. Then $BP_\varrho = P_\varrho B$, $\varrho = 1, \ldots, r$. If $B_\varrho = B \upharpoonright P_\varrho X$ and $A_\varrho = A \upharpoonright P_\varrho X$, one obtains: A_ϱ is an operator with exactly one eigenvalue and the relation $B_\varrho \cup\!\cup A_\varrho$ holds. Then, according to Lemma 3, the existence of polynomials $p_\varrho(\lambda)$ with $B_\varrho = p_\varrho(A_\varrho)$ follows. Let

$$p_\varrho(\lambda) = \alpha_\varrho^{(0)} + \alpha_\varrho^{(1)}(\lambda - \lambda_\varrho) + \cdots + \alpha_\varrho^{(\nu(\lambda_\varrho)-1)}(\lambda - \lambda_\varrho)^{(\nu(\lambda_\varrho)-1)}.$$

Now, if $p(\lambda)$ is a polynomial with

$$p(\lambda_\varrho) = \alpha_\varrho^{(0)}, \quad \frac{p'(\lambda_\varrho)}{1!} = \alpha_\varrho^{(1)}, \ldots, \frac{p^{(\nu(\lambda_\varrho)-1)}(\lambda_\varrho)}{(\nu(\lambda_\varrho) - 1)!} = \alpha_\varrho^{(\nu(\lambda_\varrho)-1)}, \qquad \varrho = 1, 2, \ldots, r,$$

then $p(A_\varrho) = p_\varrho(A_\varrho)$, hence, $B = p(A)$. Again, the converse $\mathscr{P}(A) \subseteqq \operatorname{com} \mathscr{C}$ is obvious. ■

From Theorem 1 it follows in particular that the equation $\operatorname{com} \operatorname{com} \mathscr{P}(A) = \mathscr{P}(A)$ holds for all operators A, because $\operatorname{com} \mathscr{P}(A)$ is the commutant of A.

Let $\mathscr{A}$ be any subalgebra with 1 of $\mathscr{L}(X)$. Then (see Subsect. 1.11.1)

$$\mathscr{A} = \mathscr{P}(A_1) \cup \mathscr{P}(A_2) \cup \cdots \cup \mathscr{P}(A_r),$$

where $A_1, A_2, \ldots, A_r$ denote certain finitely many operators. An immediate consequence of the commutation relations (Subsect. 1.11.2) is

$$\operatorname{com} \operatorname{com} \mathscr{A} = \operatorname{com} \operatorname{com} \mathscr{P}(A_1) \cup \operatorname{com} \operatorname{com} \mathscr{P}(A_2) \cup \cdots \cup \operatorname{com} \operatorname{com} \mathscr{P}(A_r).$$

Thus one can formulate

Theorem 2. *If $\mathscr{A}$ is a subalgebra with 1 of $\mathscr{L}(X)$, then* $\operatorname{com} \operatorname{com} \mathscr{A} = \mathscr{A}$.

This theorem contains Schur's lemma (see relation (11.1)).

2.11.2. Reducing and irreducible projections of algebras

These concepts are defined analogously as in the case of a single operator (see Sect. 2.2). Let $\mathscr{A}$ be a subalgebra with 1 of $\mathscr{L}(X)$.

Definition 1. A projection P that reduces each operator A from $\mathscr{A}$ is called a *reducing projection* of the algebra.

Definition 2. A projection P is called *irreducible* with respect to the algebra $\mathscr{A}$ if P is a reducing projection of $\mathscr{A}$ and if no proper subprojection of P exists which is reducing with respect to $\mathscr{A}$.

The question of the decomposition of an algebra $\mathscr{A}$ by a complete disjoint system of irreducible projections arises analogously as in the case of a single operator. Obviously, such decompositions exist. Concerning uniqueness the following lemma is sufficient for our purposes.

Lemma 4. *Let $\mathscr{A}$ be a subalgebra with* 1 *of $\mathscr{L}(X)$ and let* com $\mathscr{A} \subseteq \mathscr{A}$. *Then there is exactly one complete disjoint system $P_1, \ldots, P_r$ of irreducible projections with respect to $\mathscr{A}$.*

Proof. Assume $P_1, P_2, \ldots, P_r$ and $Q_1, Q_2, \ldots, Q_s$ to be complete disjoint systems of projections which are irreducible with respect to $\mathscr{A}$. Thus one has the relations

$$1 = P_1 + P_2 + \cdots + P_r, \qquad P_iP_j = \delta_{ij}P_i, \qquad i, j = 1, \ldots, r, \tag{11.10}$$

$$P_iA = AP_i, \qquad A \in \mathscr{A}, \qquad i = 1, \ldots, r, \tag{11.11}$$

and, moreover, the P_i are irreducible with respect to $\mathscr{A}$. The corresponding relations hold for the projections $Q_1, \ldots, Q_s$. From (11.11) it follows that

$$P_i \in \operatorname{com} \mathscr{A}, \qquad i = 1, \ldots, r.$$

Correspondingly,

$$Q_j \in \operatorname{com} \mathscr{A}, \qquad j = 1, \ldots, s.$$

Now, com $\mathscr{A} \subseteq \mathscr{A}$ which implies the commutativity of the algebra com $\mathscr{A}$. Hence,

$$P_iQ_j = Q_jP_i, \qquad i = 1, \ldots, r; \qquad j = 1, \ldots, s.$$

Therefore, the operators $P_iQ_j =: R_{ij}$ are also projections and because $R_{ij} \in \operatorname{com} \mathscr{A}$, one obtains

$$R_{ij}A = AR_{ij}, \qquad i = 1, \ldots, r; \qquad j = 1, \ldots, s.$$

R_{ij} is a subprojection of P_i. If the subscript i is fixed, then,

$$R_{ij} = \begin{cases} P_i, \\ 0, \end{cases} \qquad j = 1, 2, \ldots, s.$$

(This means that only P_i or 0 are possible values for R_{ij}.) Because $\sum_{j=1}^{s} R_{ij} = P_i$, there is exactly one subscript $j = j(i)$ such that $R_{ij} = P_i$. For all other subscripts j one has $R_{ij} = 0$.

On the other hand, R_{ij} is a subprojection of Q_j. If j is fixed, then,

$$R_{ij} = \begin{cases} Q_j, \\ 0, \end{cases} \qquad i = 1, \ldots, r.$$

(Again this means that only Q_j or 0 are possible values for R_{ij}.) Again, the existence of exactly one subscript $i = i(j)$ follows, such that $R_{ij} = Q_j$. For all other subscripts i one obtains $R_{ij} = 0$. Thus there is a one-to-one assignment of the subscripts $i \leftrightarrow j$, where $i = i(j)$ and $j = j(i)$, such that

$$P_i = R_{ij} = Q_j.$$

This implies the relation $r = s$ and the coincidence of both systems of projections. ■

The following insight results:

Corollary 1. *Let A be any operator and let $\mathscr{C}$ be the commutant of A, i.e., $\mathscr{C} = \mathrm{com}\,\mathscr{P}(A)$. Then to $\mathscr{C}$ there corresponds a uniquely determined complete disjoint system of irreducible projections.*

Proof. If $\mathscr{C} = \mathrm{com}\,\mathscr{P}(A)$, then $\mathrm{com}\,\mathscr{C} = \mathscr{P}(A)$. Because $\mathscr{P}(A) \subseteqq \mathscr{C}$, one has $\mathrm{com}\,\mathscr{C} \subseteqq \mathscr{C}$. Application of Lemma 4 yields the assertion. ■

Of course, the question arises how the uniquely determined complete disjoint system of irreducible projections of $\mathrm{com}\,\mathscr{P}(A)$, which exists according to Corollary 1, can be characterized by A. The answer is given by

Lemma 5. *Let A be any operator. The projection P is irreducible with respect to* $\mathrm{com}\,\mathscr{P}(A)$ *if and only if P is an eigenprojection of A, that is, the system of irreducible projections of* $\mathrm{com}\,\mathscr{P}(A)$ *and the system of eigenprojections of A coincide.*

Proof. Let $P_1, P_2, \ldots, P_r$ denote the system of all eigenprojections of A. Since P_ϱ is a polynomial in A, P_ϱ belongs to $\mathscr{P}(A)$. It follows from $\mathscr{P}(A) = \mathrm{com}\,\mathrm{com}\,\mathscr{P}(A)$ that

$$P_\varrho C = CP_\varrho, \qquad C \in \mathrm{com}\,\mathscr{P}(A), \qquad \varrho = 1, \ldots, r,$$

that is, the eigenprojection P_ϱ reduces the algebra $\mathrm{com}\,\mathscr{P}(A)$. The investigation of Subsect. 2.9.2 shows that no proper subprojection of P_ϱ exists which reduces all operators C of $\mathrm{com}\,\mathscr{P}(A)$. ■

3. Spectral theory of meromorphic operator-valued functions

As mentioned in the introduction to Ch. 2, the determination of the Jordan structure of meromorphic operator-valued functions and their function-theoretic behavior is one of the central problems of analytic perturbation theory. In this chapter a first essential step will be taken toward the solution of this problem.

In the analysis of the Jordan structure of a single operator in Ch. 2, the spectral representation of the operator appeared, as it were, as an intermediate step. Here one also first investigates the analytic properties of the spectral representation of meromorphic operator-valued functions. This amounts to an investigation of the analytic properties of eigenvalues and eigenprojections.

Let $A(z)$ be a meromorphic operator-valued function defined on a region G of the complex plane (see Subsect. 1.15.4). First, one investigates the global properties of eigenvalues and eigenprojections in G. Then, the question arises how eigenvalues and eigenprojections of $A(z)$ vary as a function of z in the neighborhood of a holomorphic point or a pole z_0 of $A(z)$. The second, local, question is the actual topic of (analytic) perturbation theory.

A priori, the local question is not obviously connected with a perturbation parameter, not even with a holomorphic one. (The reason for almost exclusively investigating analytic perturbations is the following: First, particularly, deep and strong results are obtained; second, in the applications of the finite-dimensional perturbation theory mostly analytic perturbations occur; see the Introduction.) The general question, of the variation of eigenvalues and eigenprojections of an operator under small perturbations (small in norm), was treated in Sect. 2.8; that section is the basis for the following more special investigations.

The central topic of this chapter is the investigation of the spectral representation of meromorphic operator-valued functions with special consideration of the local implications. Further topics are: the transformation of holomorphic disjoint systems of projections (Sect. 3.4), the investigation of normal and selfadjoint holomorphic perturbations (Sect. 3.5 and 3.6) in particular the results of Rellich, and the determination of the perturbation series for the perturbed objects (Sect. 3.7 and 3.8). This rough summary already shows that the normal and selfadjoint holomorphic perturbations are treated as special cases after the discussion of the general case. Anticipating the treatment of these special perturbations hardly yields advantages. On the one hand, the applications require the general case (see the Introduction). On the other hand, the study of the Riesz eigenprojections, for example, (see Sect. 3.3) can hardly be simplified for normal and selfadjoint perturbations; that is, the discussions would have to be repeated once more later.

The essential results of this chapter are due to F. Rellich [1 to 5], B. v. Sz.-Nagy [2], T. Kato [4], F. Wolf [1] (in particular, T. Kato [5, Ch. II]).

3.1. Formal perturbation theory

3.1.1. Selfadjoint perturbations in the case $N = 2$

Let $X = \mathbb{C}^2$ and let A and B be selfadjoint with respect to $(x, y) = x_1\bar{y}_1 + x_2\bar{y}_2$. For example, let

$$A = \begin{pmatrix} \lambda_1 & 0 \\ 0 & \lambda_2 \end{pmatrix}, \qquad B = \begin{pmatrix} \alpha & b \\ \bar{b} & \gamma \end{pmatrix}, \qquad \lambda_1, \lambda_2, \alpha, \gamma \text{ real}.$$

Consider a perturbation $A + \varepsilon B$ of A, where ε denotes a complex perturbation parameter. One easily calculates that

$$\chi(\lambda; A + \varepsilon B) = \left(\lambda - \frac{\lambda_1 + \lambda_2 + \varepsilon(\alpha + \gamma)}{2}\right)^2 - \frac{(\lambda_1 - \lambda_2)^2}{4}$$

$$- \frac{\varepsilon}{2}(\lambda_1 - \lambda_2)(\alpha - \gamma) - \varepsilon^2\left(|b|^2 + \frac{1}{4}(\alpha - \gamma)^2\right).$$

Hence,

$$\chi(\lambda; A + \varepsilon B) = \big(\lambda - \lambda_1(\varepsilon)\big)\big(\lambda - \lambda_2(\varepsilon)\big),$$

where

$$\lambda_{1,2}(\varepsilon) = \frac{\lambda_1 + \lambda_2 + \varepsilon(\alpha + \gamma)}{2}$$

$$\pm \left[\frac{(\lambda_1 - \lambda_2)^2}{4} + \frac{\varepsilon}{2}(\lambda_1 - \lambda_2)(\alpha - \gamma) + \varepsilon^2\left(|b|^2 + \frac{1}{4}(\alpha - \gamma)^2\right)\right]^{1/2}.$$

Thus, $\lambda_{1,2}(\varepsilon)$ can be expanded in a convergent power series with respect to ε in a neighborhood of $\varepsilon = 0$. This is obvious for $\lambda_1 = \lambda_2$; for $\lambda_1 \neq \lambda_2$ this is implied by the holomorphy of both branches of $\zeta^{1/2}$ for $\zeta \neq 0$. Similarly, one obtains convergent power series in ε for the eigenvectors.

As will be seen (in Sect. 3.5), the existence of convergent power series with respect to the perturbation parameter for eigenvalues and eigenvectors is already implied by the selfadjointness of A and B (also for arbitrary N).

3.1.2. Perturbation theory (according to E. Schrödinger)

For the following calculations see E. Schrödinger [1, 2]; also I. M. Gelfand [1]. It is assumed that the spectrum of A is simple.

Let $A + \varepsilon B$ be given, where ε denotes the perturbation parameter. The eigenvalues and (orthonormal) eigenvectors of A are assumed to be known and are denoted by $\lambda_1, \lambda_2, \ldots, \lambda_N$, resp. $e_1, e_2, \ldots, e_N$.

According to Corollary 2.8.3, for sufficiently small $|\varepsilon|$, the operator $A + \varepsilon B$ also has a simple spectrum. Therefore, its eigenvalues and orthonormal eigenvectors may be denoted without ambiguity by $\lambda_1(\varepsilon), \lambda_2(\varepsilon), \ldots, \lambda_N(\varepsilon)$, resp. $e_1(\varepsilon), e_2(\varepsilon), \ldots, e_N(\varepsilon)$. According to Lemma 2.8.3 and Lemma 2.8.4 these functions are continuous with respect to ε and

$$\lambda_\varrho(0) = \lambda_\varrho, \qquad e_\varrho(0) = e_\varrho, \qquad \varrho = 1, \ldots, N. \tag{1.1}$$

(Note that $e_\varrho(\varepsilon) = P_\varrho(\varepsilon)\, e_\varrho / \|P_\varrho(\varepsilon)\, e_\varrho\|$, where $P_\varrho(\varepsilon)$ denotes the eigenprojection belonging to $\lambda_\varrho(\varepsilon)$.) The question arises, whether these functions $\lambda_\varrho(\varepsilon)$, $e_\varrho(\varepsilon)$ have stronger properties, as continuity by itself is not useful for calculations.

E. Schrödinger has postulated and F. Rellich has proved that the eigenvalues and eigenvectors depend holomorphically on ε (see Sect. 3.5). If this is assumed, then one can start a priori with convergent power series

$$\lambda_\varrho(\varepsilon) = \lambda_\varrho^{(0)} + \varepsilon \lambda_\varrho^{(1)} + \varepsilon^2 \lambda_\varrho^{(2)} + \cdots, \qquad \lambda_\varrho^{(0)} = \lambda_\varrho, \tag{1.2}$$

$$e_\varrho(\varepsilon) = e_\varrho^{(0)} + \varepsilon e_\varrho^{(1)} + \varepsilon^2 e_\varrho^{(2)} + \cdots, \qquad e_\varrho^{(0)} = e_\varrho, \tag{1.3}$$

and one knows a priori that the corrections for λ_ϱ and e_ϱ arise as a sequence of corrections to the first, second and higher orders.

The determination of these corrections is a purely calculational process. First one has

$$(A + \varepsilon B)\, e_\varrho(\varepsilon) = \lambda_\varrho(\varepsilon)\, e_\varrho(\varepsilon), \qquad \varrho = 1, \ldots, N, \tag{1.4}$$

or

$$(A + \varepsilon B) \sum_{i=0}^{\infty} e_\varrho^{(i)} \varepsilon^i = \sum_{i=0}^{\infty} \lambda_\varrho^{(i)} \varepsilon^i \sum_{j=0}^{\infty} e_\varrho^{(j)} \varepsilon^j. \tag{1.5}$$

Comparing the coefficients of the terms of first order ε^1 with respect to ε, one obtains the equation

$$A e_\varrho^{(1)} + B e_\varrho = \lambda_\varrho e_\varrho^{(1)} + \lambda_\varrho^{(1)} e_\varrho. \tag{1.6}$$

After multiplication of (1.6) by e_ϱ (inner product), the relation

$$(A e_\varrho^{(1)}, e_\varrho) + (B e_\varrho, e_\varrho) = \lambda_\varrho (e_\varrho^{(1)}, e_\varrho) + \lambda_\varrho^{(1)}$$

follows. Because

$$(A e_\varrho^{(1)}, e_\varrho) = (e_\varrho^{(1)}, A e_\varrho) = \lambda_\varrho (e_\varrho^{(1)}, e_\varrho),$$

one obtains:

$$\lambda_{(1)}^\varrho = (B e_\varrho, e_\varrho). \tag{1.7}$$

For the calculation of the $e_\varrho^{(1)}$, (1.6) is multiplied by e_σ, where $\sigma \neq \varrho$. Then

$$(Ae_\varrho^{(1)}, e_\sigma) + (Be_\varrho, e_\sigma) = \lambda_\varrho(e_\varrho^{(1)}, e_\sigma).$$

Again

$$(Ae_\varrho^{(1)}, e_\sigma) = (e_\varrho^{(1)}, Ae_\sigma) = \lambda_\sigma(e_\varrho^{(1)}, e_\sigma);$$

thus one obtains

$$(e_\varrho^{(1)}, e_\sigma) = \frac{(Be_\varrho, e_\sigma)}{\lambda_\varrho - \lambda_\sigma}, \qquad \varrho \neq \sigma.$$

For the determination of $(e_\varrho^{(1)}, e_\varrho)$, one uses the normalization condition

$$\left(\sum_{i=0}^{\infty} e_\varrho^{(i)}\varepsilon^i, \sum_{j=0}^{\infty} e_\varrho^{(j)}\varepsilon^j\right) = 1. \tag{1.8}$$

By comparison of the coefficients, this condition yields, for real ε, the relation

$$(e_\varrho^{(1)}, e_\varrho) + (e_\varrho, e_\varrho^{(1)}) = 2\,\mathrm{Re}\,(e_\varrho^{(1)}, e_\varrho) = 0.$$

The vector $e_\varrho(\varepsilon)$ is not yet uniquely determined by the normalization condition (1.8) (it is possible to multiply with a holomorphic scalar function $\alpha(\varepsilon)$, where $|\alpha(\varepsilon)| = 1$). If one requires additionally

$$(e_\varrho, e_\varrho^{(i)}) \text{ is real}, \qquad i = 1, 2, 3, \ldots,$$

then $e_\varrho(\varepsilon)$ is unique up to multiplication with a complex constant α with $|\alpha| = 1$ (cf., for example, F. Rellich [4, p. 358]). This new "normalization condition" implies

$$(e_\varrho^{(1)}, e_\varrho) = 0. \tag{1.9}$$

Collecting the results, one obtains

$$e_\varrho^{(1)} = \sum_{\substack{i=1\\ i\neq\varrho}}^{N} \frac{(Be_\varrho, e_i)}{\lambda_\varrho - \lambda_i}\, e_i. \tag{1.10}$$

Thus the corrections of the first order are determined; that means, they are expressed by the given data.

Now the expressions $\lambda_\varrho^{(2)}$, i.e., the second-order corrections of the eigenvalues will be determined. Again, the starting points are the equations (1.4) and (1.5). Comparison of the coefficients of the terms of second order ε^2 yields the equation

$$Be_\varrho^{(1)} + Ae_\varrho^{(2)} = \lambda_\varrho^{(2)}e_\varrho + \lambda_\varrho^{(1)}e_\varrho^{(1)} + \lambda_\varrho e_\varrho^{(2)}. \tag{1.11}$$

This equation is multiplied by e_ϱ (inner product). Furthermore, one uses $(Ae_\varrho^{(2)}, e_\varrho) = (e_\varrho^{(2)}, Ae_\varrho) = \lambda_\varrho(e_\varrho^{(2)}, e_\varrho)$. Then one obtains

$$(Be_\varrho^{(1)}, e_\varrho) = \lambda_\varrho^{(2)} + \lambda_\varrho^{(1)}(e_\varrho^{(1)}, e_\varrho).$$

Because of (1.9)

$$\lambda_\varrho^{(2)} = (Be_\varrho^{(1)}, e_\varrho)$$

or, if one replaces $e_\varrho^{(1)}$ by the expression (1.10), the relation

$$\lambda_\varrho^{(2)} = \sum_{\substack{i=1\\ i\neq\varrho}}^{N} \frac{(Be_\varrho, e_i)\,(Be_i, e_\varrho)}{\lambda_\varrho - \lambda_i}$$

follows; that is,

$$\lambda_\varrho^{(2)} = \sum_{\substack{i=1\\ i\neq\varrho}}^{N} \frac{|(Be_\varrho, e_i)|^2}{\lambda_\varrho - \lambda_i}. \tag{1.12}$$

Similarly, one can determine $e_\varrho^{(2)}$. This calculation is omitted because the perturbation series will be calculated by another method later (see Sect. 3.7 and 3.8).

3.1.3. General perturbation problem

If X is not necessarily unitary, and if A and B are arbitrary operators from $\mathscr{L}(X)$, then the problem corresponding to the problem in Subsect. 3.1.2 takes the following form: Determination of eigenvalues and eigenprojections of the operator $A + \varepsilon B$ as functions of ε. Here eigenvectors are replaced by eigenprojections, because, for non-selfadjoint operators $A + \varepsilon B$, the situation within the algebraic eigenspace may be complicated, due to the existence of Jordan vectors of higher degrees.

Simple examples already show that in this case Schrödinger's postulate is not satisfied in general, but a more complicated dependence on ε may arise. For example, let $X = \mathbb{C}^2$ and let

$$A + \varepsilon B = \begin{pmatrix} 0 & 1\\ 0 & 0\end{pmatrix} + \varepsilon \begin{pmatrix} 0 & 0\\ 1 & 0\end{pmatrix} = \begin{pmatrix} 0 & 1\\ \varepsilon & 0\end{pmatrix}.$$

Then $\chi(\lambda; A + \varepsilon B) = \lambda^2 - \varepsilon$, i.e., one has $\lambda_{1/2}(\varepsilon) = \varepsilon^{1/2}$. Therefore, in this case the eigenvalues change much more rapidly than the matrix itself if ε deviates from $\varepsilon = 0$.

In most applications it is sufficient to assume that the perturbation εB of A is of the first order with respect to ε. It is also possible to treat perturbation terms of second and higher order with respect to ε. Thus one is led to consider a *perturbation* of an operator A_0 given by a convergent operator-valued power series

$$A(z) = A_0 + zA_1 + z^2A_2 + \cdots, \qquad |z| < r, \quad r > 0,$$

with respect to the *perturbation parameter* z, where $A(0) = A_0$ and where there is a positive radius of convergence $r > 0$ (see Subsect. 1.14.3 and 1.15.4). $A(z)$ represents a holomorphic operator-valued function, defined on the disc $|z| < r$. Note that $z = 0$ is a distinguished point; to this point the *unperturbed operator* A_0 is assigned. One says: *the unperturbed operator* A_0 *is perturbed holomorphically* (or *analytically*) *by* $A(z)$. The holomorphic function $A(z)$ on $|z| < r$ (see Subsect. 1.15.3) determines a global analytic operator-valued function. That is, for global considerations it is not necessary (and makes no sense) to distinguish the point $z = 0$ and to restrict oneself to a small disc $|z| < r$. For global purposes one uses a certain region G of meromorphy for the operator-valued function $A(z)$ (see Subsect. 1.15.4). If $z_0 \in G$ is a holomorphic point for $A(z)$, then $A(z_0)$ has a spectral representation $A(z_0) = \sum_\varrho \big(\lambda_\varrho(z_0) P_\varrho(z_0) + D_\varrho(z_0)\big)$. As has been pointed out in the introduction of this chapter, one first studies the global and local analytic properties of the spectral representation.

3.2. Spectrum

In the following, $A(z)$ denotes a meromorphic operator-valued function defined on a region G of complex numbers. The global behavior of the spectrum of $A(z)$ on G, and the local behavior of the spectrum in the neighborhood of holomorphic points z_0 of $A(z)$ will be investigated. If z_0 is a pole of $A(z)$ of order k, then $A(z)$ is replaced by $(z - z_0)^k A(z) =: B(z)$, such that $B(z_0) \neq 0$, and z_0 is a holomorphic point with respect to $B(z)$. In the following, theorems which have been proved in the Appendix, are often cited and applied. The corresponding concepts and notations are also explained and discussed in the Appendix. Let $\dim X = N > 0$.

3.2.1. Characteristic polynomial

Let $A(z)$ be meromorphic in G. Then

$$\chi\big(\lambda, A(z)\big) = \lambda^N + a_1(z)\, \lambda^{N-1} + \cdots + a_N(z),$$

where the coefficients $a_1(z), \ldots, a_N(z)$ are meromorphic scalar functions in G. One writes $\chi(\lambda, A(z)) \in \mathscr{M}_G[\lambda]$; χ is a normalized polynomial (see Subsect. A 2.1 and A 2.2).

If $A(z)$ is holomorphic in G, then $\chi \in \mathscr{H}_G[\lambda]$ (see Subsect. A 2.1). If, additionally, the equation $a_1(z_0) = a_2(z_0) = \cdots = a_N(z_0) = 0$ holds at a fixed point z_0, then one writes $\chi \in \mathscr{H}_G^{z_0}[\lambda]$. The following theorem holds (see Subsect. A 2.6).

Theorem 1. *Let q be a normalized polynomial from $\mathscr{M}_G[\lambda]$. Then there exists a unique decomposition into irreducible (or prime) factors*

$$q = \prod_{\varrho=1}^{r} p_\varrho^{m_\varrho}, \qquad p_\varrho \in \mathscr{M}_G[\lambda], \tag{2.1}$$

where the polynomials $p_1, p_2, \ldots, p_r$ are normalized and irreducible. Poles of coefficients of $p_1, \ldots, p_r$ are necessarily also poles of the coefficients of q.

If $q \in \mathscr{H}_G[\lambda]$, then all prime factors p_ϱ also belong to $\mathscr{H}_G[\lambda]$; if $q \in \mathscr{H}_G^{z_0}[\lambda]$, then all prime factors p_ϱ also belong to $\mathscr{H}_G^{z_0}[\lambda]$.

Hence, for the characteristic polynomial χ, one obtains the following proposition.

Corollary 1. *The equation $\chi\big(\lambda; A(z)\big) = \prod_{\varrho=1}^{r} p_\varrho(\lambda, z)^{m_\varrho}$ holds, where $p_1, p_2, \ldots, p_r \in \mathscr{M}_G[\lambda]$ and where the polynomials p_ϱ are normalized and irreducible. The poles of their coefficients* *necessarily belong to S_A, where S_A denotes the set of all poles of $A(z)$. Thus, if $\chi \in \mathscr{H}_G[\lambda]$, then $p_\varrho \in \mathscr{H}_G[\lambda]$, and if $\chi \in \mathscr{H}_G^{z_0}[\lambda]$, then $p_\varrho \in \mathscr{H}_G^{z_0}[\lambda]$, $\varrho = 1, 2, \ldots, r$.*

If $\deg p_\varrho = g_\varrho$, $\varrho = 1, \ldots, r$ (deg means the degree of the corresponding polynomial), one obtains the equation

$$N = \sum_{\varrho=1}^{r} m_\varrho g_\varrho . \tag{2.2}$$

3.2.2. Simple and multiple points

The set $G \setminus S_A$ of all holomorphic points with respect to $A(z)$ is divided into the set of all simple points and the set of all multiple points. For the preparation of the corresponding definitions a theorem from the Appendix (Subsect. A 3.2) is cited.

Theorem 2. *Let p be a normalized and irreducible polynomial from $\mathscr{M}_G[\lambda]$. The set of all poles of the coefficients of p is denoted by S_p. For all $z_0 \in G \setminus S_p$, the polynomial $p = p(\lambda, z_0)$ can be considered as a polynomial from $\mathbb{C}[\lambda]$. Then: with the exception of an isolated and in G closed set, $p(\lambda, z_0)$ has only simple roots for all $z_0 \in G \setminus S_p$.*

Hence one obtains

Corollary 2. *The prime factors $p_1(\lambda, z), \ldots, p_r(\lambda, z)$ of $\chi(\lambda, A(z))$ can be considered as polynomials from $\mathbb{C}[\lambda]$ for all $z \in G \setminus S_A$. With the exception of an isolated and in G closed set, all polynomials $p_1(\lambda, z), \ldots, p_r(\lambda, z)$ have only simple roots for all $z \in G \setminus S_A$.*

In other words, the polynomials $p_1(\lambda, z_0), \ldots, p_r(\lambda, z_0)$ have multiple roots with respect to λ for at most a denumerable set of points z_0 from G, whose accumulation points, if any, are to be found on ∂G. One defines the polynomial q by

$$q(\lambda, z) = \prod_{\varrho=1}^{r} p_\varrho(\lambda, z) .$$

Definition 1. The point $z_0 \in G \setminus S_A$ is called a *multiple point* if $q(\lambda, z_0)$ has at least one multiple root with respect to λ.

From Subsect. A 3.1 and from Corollary 2, one obtains

Lemma 1. *The set M of all multiple points is isolated and in G closed, i.e., M is denumerable and any accumulation point of M is to be found on ∂G.*

Definition 2. The point $z_0 \in G \setminus S_A$ is called a *simple point* if $q(\lambda, z_0)$ has only simple roots with respect to λ.

Let $s(z)$ denote the number of different roots of $q(\lambda, z)$, that is, the number of different eigenvalues of $A(z)$. Then z is simple if and only if $s(z) = s = \sum_{\varrho=1}^{r} g_\varrho = \deg q$. Correspondingly, z is multiple if and only if $s(z) < s$. Therefore, the number $s(z)$ is constant and equal to s on $G \setminus (S_A \cup M)$, but it decreases at the points of M.

Because of (2.2), the relation $s = N$ holds if and only if all $m_\varrho = 1$, that is, if the decomposition (2.1) does not contain a multiple factor. In this case for all $z \in G \setminus (S_A \cup M)$, the operator $A(z)$ has exactly N different eigenvalues, i.e., $A(z)$ is an operator with simple spectrum and, in particular, a diagonal operator. The relation $s < N$ is valid if and only if there is an m_ϱ with $m_\varrho > 1$, i.e., if there is at least one multiple prime factor in (2.1). In this case, the operator $A(z)$ is called *permanently degenerated.*

3.2.3. Function theory of the eigenvalues

Let $z \in G \setminus S_A$. Then, the relation $\lambda \in \operatorname{spec} A(z)$ holds if and only if one of the equations $p_\varrho(\lambda, z) = 0$, $\varrho = 1, \ldots, r$, is satisfied. Preparing the characterization of the spectrum according to this remark, two theorems from the Appendix (Subsect. A 5.2 and A 5.3) are cited.

Theorem 3. *Let $\mathfrak{P}$ denote an n-valued algebroid function over G. Then there is exactly one normalized polynomial p of degree n from $\mathcal{M}_G[\lambda]$ such that $\mathfrak{P}$ satisfies the equation $p(\lambda, z) = 0$. This polynomial is irreducible and its coefficients have poles at most at the poles and singularities of $\mathfrak{P}$.*

Theorem 4. *Let p be a normalized and irreducible polynomial of degree n from $\mathcal{M}_G[\lambda]$. Then there is exactly one n-valued algebroid function $\mathfrak{P}$ over G such that $\mathfrak{P}$ satisfies the equation $p(\lambda, z) = 0$. Poles and singularities of $\mathfrak{P}$ are found at most at the poles of the coefficients of p and at points z_0 such that $p(\lambda, z_0)$ has at least one multiple root with respect to λ.*

Applying this theorem to the present situation, one obtains

Corollary 3. *The equation $p_\varrho(\lambda, z) = 0$, $\varrho = 1, \ldots, r$, defines (in the sense of Theorem 4) an analytic function $\mathfrak{P}_\varrho$ over G. $\mathfrak{P}_\varrho$ is algebroid and g_ϱ-valued. If $z_0 \in G \setminus S_A$, then only holomorphic points (i.e., no poles) or branching points with finite limits (i.e., no branching poles) can lie over z_0. The family $\{\mathfrak{P}_1(z_0), \mathfrak{P}_2(z_0), \ldots, \mathfrak{P}_r(z_0)\}$ of all sets of values $\mathfrak{P}_\varrho(z_0)$ of $\mathfrak{P}_1, \ldots, \mathfrak{P}_r$ at z_0 coincides with the spectrum* $\operatorname{spec} A(z_0)$.

For the concept of the set of values of an analytic function at a point, see Subsect. A 1.3.

Note that the last proposition in Corollary 3 is true whether z_0 is regular or singular with respect to the functions $\mathfrak{P}_1, \mathfrak{P}_2, \ldots, \mathfrak{P}_r$. *If $z_0 \in G \setminus S_A$ is a simple point, then z_0 is regular for every function $\mathfrak{P}_1, \ldots, \mathfrak{P}_r$,* for the simplicity of z_0 implies the simplicity of all roots of $p_\varrho(\lambda, z_0)$. Therefore, according to Theorem 4, z_0 cannot be singular for $\mathfrak{P}_\varrho$ in this case. Conversely, it is possible that z_0 is regular for all functions $\mathfrak{P}_1, \ldots, \mathfrak{P}_r$ but nevertheless multiple; this can be shown by the following example:

Let $A(z) = \begin{pmatrix} z & 0 \\ 0 & z^2 \end{pmatrix}$. Then, $\chi(\lambda; A(z)) = (\lambda - z)(\lambda - z^2)$. $z = 0$ is regular for both functions z and z^2; nevertheless, $z = 0$ is a multiple point because $q(\lambda, 0) = \lambda^2$.

According to Corollary 3, one can speak of those eigenvalues of $A(z)$ that belong to the function $\mathfrak{P}_\varrho$; this set of eigenvalues is defined by the set of values $\mathfrak{P}_\varrho(z)$ of the function $\mathfrak{P}_\varrho$ at z (where $z \in G \setminus S_A$).

The eigenvalues of $A(z)$ belonging to $\mathfrak{P}_\varrho$ (for brevity: the eigenvalues of $\mathfrak{P}_\varrho$) form a *G-cycle* of eigenvalues. Every two eigenvalues $\lambda(z)$ and $\lambda'(z')$ from a fixed G-cycle $\mathfrak{P}_\varrho$ (where z and z' are from $G \setminus S_A$) are connected by a suitable mutual analytic continuation in G. Namely, one has to choose function elements lying over z and z' with the values $\lambda(z)$ and $\lambda'(z')$ at z and z', respectively. Furthermore, these function elements can be analytically continued mutually into each other, because they are function elements of a unique analytic function $\mathfrak{P}_\varrho$.

The points (function elements) of $\mathfrak{P}_\varrho$ are denoted by ζ_ϱ (for the purposes pursued here, one can identify $\mathfrak{P}_\varrho$ with the Riemann surface of $\mathfrak{P}_\varrho$, as is often done; then it is possible to speak of the points of the Riemann surface). Thus, on $\mathfrak{P}_\varrho$ there is defined a single-valued "eigenvalue function" $\lambda(\zeta_\varrho)$ that assignes a specific value to the point ζ_ϱ with trace point z_0. Note that z_0 is called *trace point* of ζ_ϱ if ζ_ϱ lies over z_0. Note further that for $z_0 \in S_A$ the value $\lambda(\zeta_\varrho) = \infty$ may appear.

With the use of the functions $\lambda(\zeta_\varrho)$, the analytic functions $\mathfrak{P}_1, \mathfrak{P}_2, \ldots, \mathfrak{P}_r$ can be mutually "stitched together" even further. If $\zeta_\varrho \in \mathfrak{P}_\varrho$ and $\zeta_\sigma \in \mathfrak{P}_\sigma$ lie over one and the same trace point z_0 and if $\lambda(\zeta_\varrho) = \lambda(\zeta_\sigma)$, then the points ζ_ϱ and ζ_σ are identified (with respect to their values over z_0; of course, this does not mean an identification of the function elements ζ_ϱ and ζ_σ, which is impossible).

Let $z \in G \setminus S_A$ and let z_0 be the trace point of $\zeta_\varrho \in \mathfrak{P}_\varrho$. Then $\lambda(\zeta_\varrho)$ is an eigenvalue of $A(z_0)$. The family of all points $\zeta_1, \zeta_2, \ldots, \zeta_r$ of the functions $\mathfrak{P}_1, \mathfrak{P}_2, \ldots, \mathfrak{P}_r$, respectively, lying over z_0, then yields the spectrum spec $A(z_0)$ via $\lambda(\zeta_1), \lambda(\zeta_2), \ldots, \lambda(\zeta_r)$ (according to Corollary 3).

3.2.4. Holomorphic and polynomial operator-valued functions

If $A(z)$ is holomorphic in G, then, according to Theorem 4, the functions $\mathfrak{P}_1, \mathfrak{P}_2, \ldots, \mathfrak{P}_r$ do not have any poles or branching poles. The exceptional set S_A is empty in this case and the sets of values $\mathfrak{P}_1(z_0) \cup \mathfrak{P}_2(z_0) \cup \cdots \cup \mathfrak{P}_r(z_0)$ represent spec $A(z_0)$ for any $z_0 \in G$.

If $A(z)$ is polynomial, then one can choose G as the closed z-plane. In this case, $\mathfrak{P}_1, \ldots, \mathfrak{P}_r$ are algebraic functions which do not have any pole in a finite region. Only $z = \infty$ may be a pole in this case.

3.2.5. Local behavior of the eigenvalues in the neighborhood of simple points

Let z_0 be a holomorphic point of $A(z)$, i.e., $z_0 \in G \setminus S_A$. Furthermore, let z_0 be simple. Then, according to Lemma 1, there exists a neighborhood of z_0, for example a disc $|z - z_0| < \varepsilon$, $\varepsilon > 0$, such that all points of this disc are simple holomorphic points of $A(z)$. Thus all points z of this disc are also regular with respect to all functions $\mathfrak{P}_1, \ldots, \mathfrak{P}_r$ (see Subsect. 3.2.3). Hence, for every $\mathfrak{P}_\alpha$, $\alpha = 1, 2, \ldots, r$, there are exactly g_α different holomorphic function elements $\lambda_{\alpha\sigma}(z)$, $\sigma = 1, 2, \ldots, g_\alpha$, lying over $|z - z_0| < \varepsilon$ and

$$p_\alpha(\lambda, z) = \prod_{\sigma=1}^{g_\alpha} \left(\lambda - \lambda_{\alpha\sigma}(z)\right), \qquad |z - z_0| < \varepsilon; \qquad \alpha = 1, 2, \ldots, r. \tag{2.3}$$

Altogether, one obtains different holomorphic function elements over $|z - z_0| < \varepsilon$, and the numbers $\lambda_{\alpha\sigma}(z)$ are mutually different for arbitrary but fixed z with $|z - z_0| < \varepsilon$, in particular for $z = z_0$. One has

$$\chi(\lambda, A(z)) = \prod_{\alpha=1}^{r} \prod_{\sigma=1}^{g_\alpha} (\lambda - \lambda_{\alpha\sigma}(z)). \tag{2.4}$$

Obviously, $\lambda_{\alpha\sigma}(z) \to \lambda_{\alpha\sigma}(z_0)$ as $z \to z_0$. That is, considering $A(z)$ on $|z - z_0| < \varepsilon$ as a holomorphic perturbation of $A(z_0)$, if $A(z_0)$ changes to $A(z)$, then the eigenvalue $\lambda_{\alpha\sigma}(z_0)$ changes to the only eigenvalue $\lambda_{\alpha\sigma}(z)$ and one obtains a holomorphic dependence on the perturbation parameter z.

3.2.6. Local (splitting) behavior of the eigenvalues in the neighborhood of multiple points

Let z_0 be a holomorphic point of $A(z)$, i.e., let $z_0 \in G \setminus S_A$ and let z_0 be multiple; that is, $q(\lambda, z_0)$ has at least one multiple root. In this case, the behavior of the eigenvalues of $A(z)$ in the neighborhood of z_0 can be very complicated. Splitting of eigenvalues may occur.

Let λ_0 be a fixed root of $q(\lambda, z_0)$ with multiplicity $m > 1$. The multiplicity of λ_0 with respect to $p_\varrho(\lambda, z_0)$ is denoted by n_ϱ. One has $\sum_{\varrho=1}^{r} n_\varrho = m$. There is at least one polynomial $p_\varrho(\lambda, z_0)$ with $n_\varrho > 0$. Now every polynomial $p_\varrho(\lambda, z_0)$ with $n_\varrho > 0$ will be "*related to the point* z_0". What is meant by this notation results from the following theorem (Weierstrass preparation theorem for polynomials), which has been cited from Subsect. A 4.2.

Theorem 5. *Let* $p \in \mathcal{M}_G[\lambda]$, *where* p *is normalized, let* $z_0 \in G \setminus S_P$ *and* $p(\lambda, z_0) = \prod_{\varrho=1}^{r} (\lambda - \lambda_\varrho)^{k_\varrho}$. *Then there is a neighborhood of* z_0, *for example a disc* $K = \{z : |z - z_0| < \varepsilon, \varepsilon > 0\}$, *such that the following properties hold: There exist* r *polynomials* $q_1(\mu, z)$, $q_2(\mu, z), \ldots, q_r(\mu, z)$ *belonging to* $\mathcal{H}_K^{z_0}[\mu]$ *and with* $\deg q_\varrho = k_\varrho$, $\varrho = 1, \ldots, r$, *such that*

$$p(\lambda, z) = \prod_{\varrho=1}^{r} q_\varrho(\lambda - \lambda_\varrho, z), \qquad z \in K. \tag{2.5}$$

(2.5) *is a decomposition of* p *within* $\mathcal{H}_K[\lambda]$.

Relating the polynomial p to the point z_0 means constructing the decomposition (2.5).

Furthermore, Theorem 5 will be applied to the polynomial p_ϱ with $n_\varrho > 0$. The disc K occuring in this theorem is chosen as a common disc for all p_ϱ and, moreover, such that the dotted disc $0 < |z - z_0| < \varepsilon$ consists only of simple points (this is possible because of Lemma 1). The decomposition (2.5) for the polynomial $p_\varrho(\lambda, z)$ contains a factor corresponding to the root λ_0. This factor is denoted by $q_\varrho(\lambda - \lambda_0, z)$, where $\deg q_\varrho = n_\varrho$ and $q_\varrho(\mu, z) \in \mathcal{H}_K^{z_0}[\mu]$. This polynomial has a decomposition into prime factors with respect to $\mathscr{H}_K[\mu]$,

$$q_\varrho(\lambda - \lambda_0, z) = \prod_{\alpha=1}^{k_\varrho} q_{\varrho\alpha}(\lambda - \lambda_0, z), \qquad q_{\varrho\alpha}(\mu, z) \in \mathscr{H}_K^{z_0}[\mu],$$

where the polynomials $q_{\varrho\alpha}$ are irreducible with respect to $\mathscr{H}_K[\mu]$. Note that no multiple factors can occur, because the polynomials q_ϱ are factors of p_ϱ and this polynomial is irreducible with respect to $\mathscr{M}_G[\lambda]$; that is, the polynomials $q_{\varrho 1}, \ldots, q_{\varrho k_\varrho}$ are mutually different. One puts $\deg q_{\varrho\alpha} = g_{\varrho\alpha}$. Then $\sum_{\alpha=1}^{k_\varrho} g_{\varrho\alpha} = n_\varrho$.

Further properties of the polynomials $q_{\varrho\alpha}$ are obtained from the following theorem (cited from Subsect. A 5.4).

Theorem 6. *Let $p(\mu, z) \in \mathscr{H}_K^{z_0}[\mu]$ where K is a disc $|z - z_0| < \varepsilon$, $\deg p = g$ and where p is irreducible in $\mathscr{H}_K^{z_0}[\mu]$ (hence, also irreducible with respect to $\mathscr{H}_K[\mu]$). Furthermore, for all $z \in K$ with $z \neq z_0$, $p(\mu, z)$ is assumed to have only simple roots with respect to μ.*

Then p defines a unique g-valued algebroid function $\mathfrak{P}$ over K, where z_0 is the only singular point, namely a branching point of order g. As $z \to z_0$, the function $\mathfrak{P}$ assumes the unique finite limit 0.

$\mathfrak{P}$ is represented over K by a convergent Puiseux series of the form

$$\sum_{\varrho=0}^{\infty} a_\varrho[(z - z_0)^{1/g}]^\varrho, \qquad a_0 = 0, \qquad |z - z_0| < \varepsilon; \tag{2.6}$$

that is, if one denotes the values of $\mathfrak{P}$ over z for $z \neq z_0$ by $\lambda_1(z), \lambda_2(z), \ldots, \lambda_g(z)$, then (for suitable ordering),

$$\lambda_\sigma(z) = \sum_{\varrho=0}^{\infty} a_\varrho[\zeta^\sigma(z - z_0)^{1/g}]^\varrho, \qquad \sigma = 1, 2, \ldots, g,$$

where $(z - z_0)^{1/g}$ denotes a fixed root and ζ a primitive g-th root of unity.

Therefore, the situation can be described as follows: Every polynomial $q_{\varrho\alpha}(\lambda - \lambda_0, z)$ represents a branch $\mathfrak{H}_{\varrho\alpha}$ of $\mathfrak{P}_\varrho$ of valuedness $g_{\varrho\alpha}$ over K with the only singularity z_0, which is a branching point of $\mathfrak{H}_{\varrho\alpha}$ of order $g_{\varrho\alpha}$. $\mathfrak{H}_{\varrho\alpha}$ assumes the unique finite limit λ_0 as $z \to z_0$. For $z \neq z_0$, $z \in K$, the values of the branch $\mathfrak{H}_{\varrho\alpha}$ over z are denoted by $\lambda_{\varrho\alpha 1}(z)$, $\lambda_{\varrho\alpha 2}(z), \ldots, \lambda_{\varrho\alpha g_{\alpha\varrho}}(z)$. These values simultaneously represent holomorphic function elements of $\mathfrak{H}_{\varrho\alpha}$ and $\mathfrak{P}_\varrho$ over z, respectively.

The numbers $\lambda_{\varrho\alpha\sigma}(z)$ are mutually distinct if the subscripts vary arbitrarily,

$$(\varrho = 1, \ldots, r;\ \alpha = 1, \ldots, k_\varrho;\ \sigma = 1, \ldots, g_{\varrho\alpha}),$$

because z is simple for $0 < |z - z_0| < \varepsilon$. Because $m = \sum_{\varrho=1}^{r} \sum_{\alpha=1}^{k_\varrho} g_{\varrho\alpha}$, one obtains exactly m different holomorphic function elements over z with the property $\lambda_{\varrho\alpha\sigma}(z) \to \lambda_0$ as $z \to z_0$.

There is no other function element of $\mathfrak{P}_1, \mathfrak{P}_2, \ldots, \mathfrak{P}_r$ in the neighborhood of z_0 with this property, because otherwise a higher multiplicity of λ_0 as a root of $q(\lambda, z_0)$ would follow.

The family $\{\lambda_{\varrho\alpha\sigma}(z)\}$ of all eigenvalues $\lambda(z)$ with $\lambda(z) \to \lambda_0$ as $z \to z_0$, i.e., the roots of $\prod_{\varrho=1}^{r} q_\varrho(\lambda - \lambda_0, z)$, is called *the* (z_0, λ_0)*-group* of perturbed eigenvalues.

In the case discussed before, the (z_0, λ_0)-group consists of at least two different eigenvalues. Because $m > 1$, either there is a $g_{\varrho\alpha} \geqq 2$ or there are two different $g_{\varrho\alpha}$ with $g_{\varrho\alpha} \geqq 1$.

The (z_0, λ_0)-group represents exactly those eigenvalues of $A(z)$ for $z \in K$, which are the *perturbed* and/or *splitted* eigenvalues of λ_0 if $A(z_0)$ undergoes the perturbation $A(z_0) \to A(z)$, i.e., if z_0 is changed to z.

One can formulate the following results:

An eigenvalue λ_0 *of* $A(z_0)$ *always splits under the perturbation* $A(z)$ *if it is a multiple root of* $q(\lambda, z_0)$.

If z_0 *is multiple, then there is at least one eigenvalue* λ_0 *of* $A(z_0)$ *which splits under the perturbation* $A(z)$.

If z_0 *is simple, then there is no splitting for the eigenvalues of* $A(z_0)$ *under the perturbation* $A(z)$.

If the operator-valued function $A(z)$ is considered only over the restricted region K, the eigenvalues of each branch $\mathfrak{H}_{\varrho\alpha}$ form a K-cycle of eigenvalues (see Subsect. 3.2.3). The number $g_{\varrho\alpha}$ of different eigenvalues belonging to the K-cycle given by $\mathfrak{H}_{\varrho\alpha}$ is called its *period*. In other words, the period of the K-cycle $\mathfrak{H}_{\varrho\alpha}$ is equal to the valuedness of $\mathfrak{H}_{\varrho\alpha}$ and to the order of the branching point z_0 with respect to $\mathfrak{H}_{\varrho\alpha}$, respectively.

In this way one obtains a *decomposition of the* (z_0, λ_0)*-group into several* K*-cycles. The eigenvalues* $\lambda_{\varrho\alpha\sigma}(z)$, $\sigma = 1, \ldots, g_{\varrho\alpha}$, *of a fixed* K*-cycle* $\mathfrak{H}_{\varrho\alpha}$ *are given over* K, *according to Theorem* 6, *by a Puiseux series of the form*

$$\lambda_{\varrho\alpha\sigma}(z) = \lambda_0 + \sum_{\varkappa=1}^{\infty} a_{\varrho\alpha\varkappa}[(z - z_0)^{1/g_{\varrho\alpha}}]^\varkappa, \qquad |z - z_0| < \varepsilon, \tag{2.7}$$

where the $g_{\varrho\alpha}$ *different eigenvalues of the cycle are obtained if* $(z - z_0)^{1/g_{\varrho\alpha}}$ *runs through the different values of the root.*

Thus the knowledge of the periods of the K-cycles is essential for the determination of the Puiseux series for the eigenvalues. Within a (z_0, λ_0)-group these periods are extremely important characteristics of the perturbation $A(z)$ of $A(z_0)$ with respect to the unperturbed eigenvalue λ_0.

The different (z_0, λ_0)-groups and, within a (z_0, λ_0)-group, the different K-cycles of $A(z)$ over K can be determined if one calculates the corresponding Puiseux series from the characteristic equation $\chi(\lambda, A(z)) = 0$ with the Puiseux-Newton diagram (this method will be presented in Sect. A 7).

By applying this method, one obtains in each (z_0, λ_i)-group cycles $\mathfrak{Z}_{i1}, \mathfrak{Z}_{i2}, \ldots, \mathfrak{Z}_{ir_i}$ which occur with certain multiplicities $m_{i1}, m_{i2}, \ldots, m_{ir_i}$, where the different eigenvalues of $A(z_0)$ are denoted by $\lambda_1, \lambda_2, \ldots, \lambda_p$. The different global multiplicities $m_1, \ldots, m_r$ are then obtained by the different values of the numbers $m_{i1}, m_{i2}, \ldots, m_{ir_i}$. The subscripts ϱ and α appearing above (for example in (2.7)) cannot be distinguished by this method, because they indicate global properties (the cycles where ϱ is fixed belong to the global analytic function $\mathfrak{P}_\varrho$). Therefore, the local behavior of the eigenvalues in the neighborhood of multiple points z_0 can be characterized, on the one hand, by the "splitting effect" and, on the other hand, by the fact that the eigenvalues can, in general, no longer be represented by power series with respect to $(z - z_0)$, but merely by power series with respect to fractional powers $(z - z_0)^{1/g}$. These two effects are connected in the following way: the existence of a cycle with a period > 1 always implies a splitting for λ_0, whereas the converse need not be true.

3.2.7. Splitting points and normal points

The results obtained in Subsect. 3.2.5 and 3.2.6 suggest the following definitions. Let $\zeta_\varrho \in \mathfrak{P}_\varrho$ and let the trace point z_0 of ζ_ϱ be from $G \setminus S_A$, i.e., $A(z_0)$ is well-defined. Then ζ_ϱ can be uniquely described by z_0 and some eigenvalue λ_0 of $A(z_0)$, namely by $\lambda_0 = \lambda(\zeta_\varrho)$. One writes $\zeta_\varrho = (z_0, \lambda_0)$, for short.

Definition 3. The point $\zeta_\varrho = (z_0, \lambda_0)$ of $\mathfrak{P}_\varrho$ is called a *splitting point* if λ_0 is a multiple root of $q(\lambda, z_0)$.

In other words: $\zeta_\varrho = (z_0, \lambda_0)$ is called a splitting point, if λ_0 splits under the perturbation $z_0 \to z$, i.e., if there is a "stitching", either within $\mathfrak{P}_\varrho$ or with respect to other functions $\mathfrak{P}_\sigma$, at ζ_ϱ (see Subsect. 3.2.3).

All other points $\zeta_\varrho \in \mathfrak{P}_\varrho$ with trace points from $G \setminus S_A$ are called *normal*. That is, $\zeta_\varrho = (z_0, \lambda_0)$ is called normal, if λ_0 is a simple root of $q(\lambda, z_0)$, i.e., if λ_0 does not split under the perturbation $z_0 \to z$.

If (z_0, λ_0) is normal, then the (z_0, λ_0)-group consists of exactly one perturbed eigenvalue $\lambda(z)$ given by

$$\lambda(z) = \lambda_0 + \lambda_1(z - z_0) + \lambda_2(z - z_0)^2 + \cdots,$$

that is, $\lambda(z)$ is holomorphic in the neighborhood of z_0.

This statement is valid because each normal point (z_0, λ_0) of $\mathfrak{P}_\varrho$ is necessarily a holomorphic point of $\mathfrak{P}_\varrho$; in the other case λ_0 would necessarily be a multiple root of the irreducible polynomial belonging to $\mathfrak{P}_\varrho$ at z_0.

If $A(z)$ is holomorphic in G, then, by the foregoing distinctions, all points of the functions $\mathfrak{P}_1, \mathfrak{P}_2, \ldots, \mathfrak{P}_r$ are divided into splitting points and normal points. The branching points are necessarily splitting points but the converse need not be true.

3.2.8. The case of a simple eigenvalue λ_0 of $A(z_0)$

As mentioned above, the periods of the eigenvalue cycles of a (z_0, λ_0)-group are important characteristics of the perturbation $A(z)$ of $A(z_0)$ if $z_0 \in G \setminus S_A$. Determination of these periods is of great importance. In this connection, the point is to get propositions on the periods without using the complicated Puiseux series expansions of the roots of the characteristic equation (see Subsect. 3.2.6). Later this problem plays an essential role in the so-called reduction theory (see Ch. 7).

Obviously, for simple eigenvalues λ_0 of $A(z_0)$ one can state the following proposition:

Lemma 2. *If λ_0 is a simple eigenvalue of $A(z_0)$, where $z_0 \in G \setminus S_A$, then (z_0, λ_0) is a normal point, i.e., the (z_0, λ_0)-group consists of exactly one eigenvalue $\lambda(z)$ which is holomorphic in the neighborhood of $z = z_0$.*

Proof. Let λ_0 be simple. Then λ_0 is a simple root of $\chi(\lambda, A(z_0)) = \prod_{\varrho=1}^{r} p_\varrho(\lambda, z_0)^{m_\varrho}$. Hence, one of the multiplicities m_ϱ, say m_1, must be equal to 1 and λ_0 is a simple root of $p_1(\lambda, z_0)$. Relating $p_1(\lambda, z)$ to the point $z = z_0$, one finds that the factor $q(\lambda - \lambda_0, z)$ belonging to λ_0 is of degree 1, that is, the (z_0, λ_0)-group consists of only one cycle of period 1; thus λ_0 is also a simple root of $q(\lambda, z_0) = \prod_{\varrho=1}^{r} p_\varrho(\lambda, z_0)$. ■

For semisimple eigenvalues λ_0 there is in general no holomorphic dependence of the perturbed eigenvalues. For example, let $X = \mathbb{C}^2$ and $A(z) = \begin{pmatrix} 0 & z \\ z^2 & 0 \end{pmatrix}$. $\lambda_0 = 0$ is a semi-simple eigenvalue of $A(0)$. However, $\chi(\lambda, A(z)) = \lambda^2 - z^3$, i.e., the perturbed eigenvalues are $\lambda_{1,2}(z) = z^{3/2}$, that is, the eigenvalue $\lambda_0 = 0$ splits into only one cycle of period 2.

Also for prime eigenvalues λ_0, one obtains, in general, not a holomorphic dependence for the perturbed eigenvalues, as is shown by the example $A(z) = \begin{pmatrix} 0 & 1 \\ z & 0 \end{pmatrix}$. $\lambda_0 = 0$ is a prime eigenvalue of $A(0)$, but $\lambda_0 = 0$ splits into only one cycle of period 2.

3.2.9. Local behavior of the eigenvalues in the neighborhood of poles of $A(z)$

If $z_0 \in S_A$, i.e., if z_0 is a pole of $A(z)$, then $A(z) = (z - z_0)^{-m} B(z)$, where m denotes the order of the pole z_0 and where $B(z)$ is meromorphic in G, holomorphic at $z = z_0$ and $B(z_0) \neq 0$. If μ_0 is an eigenvalue of $B(z_0)$, then the local considerations of Subsect. 3.2.5 and 3.2.6 and also the classifications of (z_0, μ_0), given in Subsect. 3.2.7, are applicable to (z_0, μ_0). Therefore, (z_0, μ_0) is either normal or a splitting point for the perturbation $B(z)$ of $B(z_0)$, and the perturbed eigenvalues have local expansions of the form

$$\mu(z) = \mu_0 + \sum_{\varrho=1}^{\infty} b_\varrho (z - z_0)^{\varrho/k}$$

in the neighborhood of $z = z_0$. Therefore, in the neighborhood of $z = z_0$, one obtains expansions for the eigenvalues $\lambda(z)$ of $A(z)$ of the form

$$\lambda(z) = \mu_0 (z - z_0)^{-m} + \sum_{\varrho=1}^{\infty} b_\varrho (z - z_0)^{\frac{\varrho}{k} - m},$$

such that the "first term" of the Puiseux series expansion of the eigenvalues $\lambda(z)$ is at most $\mu_0(z - z_0)^{-m}$, where m is the order of the pole z_0 of $A(z)$ and where μ_0 is an eigenvalue of $B(z_0)$.

A similar result is obtained, when z_0 is equal to infinity, $z_0 = \infty$. In this case one has to replace $1/(z - z_0)$ by z.

3.2.10. Examples

The following examples are illustrations of the concepts introduced above.

(a) $$A(z) = \begin{pmatrix} -z & 0 & z \\ 3z^2+1 & 0 & -z^2 \\ 1 & -1 & 0 \end{pmatrix}.$$

$A(z)$ is considered as a perturbation of $A(0)$ in the neighborhood of $z_0 = 0$. The only eigenvalue of $A(0)$ is $\lambda_0 = 0$. On has

$$\chi(\lambda, A(z)) = \lambda^3 + z\lambda^2 - (z + z^2)\,\lambda + z^2 + 2z^3.$$

One applies the Puiseux diagram (see Sect. A 7) to the equation $\chi = 0$. In this way, one obtains the local cycles:

$$\lambda_1(z) = z + \cdots; \qquad \lambda_2(z) = z^{1/2} + \cdots, \qquad \lambda_3(z) = -z^{1/2} + \cdots, \tag{2.8}$$

where $\cdots$ indicates the existence of terms of higher order. (2.8) implies that there is one cycle of period 1: $\lambda_1(z)$, and one cycle of period 2: $\lambda_2(z)$, $\lambda_3(z)$. The Puiseux series of the eigenvalues of this cycle are formed by powers of $z^{1/2}$. The "global multiplicities" are 1; there is no permanent degeneration (see Subsect. 3.2.2).

$z_0 = 0$ is multiple. The points $z_0 \neq 0$ are simple. $(z_0, \lambda_0) = (0, 0)$ is a splitting point (the only splitting point over $z_0 = 0$). The points $(z, \lambda_i(z))$, $z \neq 0$, $i = 1, 2, 3$, are normal.

The Jordan normal form of $A(0)$ is

$$\begin{pmatrix} 0 & 1 & 0 \\ 0 & 0 & 1 \\ 0 & 0 & 0 \end{pmatrix},$$

i.e., $A(0)$ is a Jordan operator and there exists only one linearly independent eigenvector corresponding to $\lambda_0 = 0$. For $z \neq 0$ the operator $A(z)$ is diagonal.

(b) $$A(z) = \begin{pmatrix} 1+z & 0 & 0 \\ 0 & 0 & 1 \\ 0 & z & 0 \end{pmatrix}.$$

$A(z)$ is considered as a perturbation of $A(0)$ in the neighborhood of $z_0 = 0$. There are two different eigenvalues $\lambda_1 = 1$, $\lambda_2 = 0$ for $A(0)$. One has

$$\chi(\lambda, A(z)) = (\lambda - 1 - z)\,(\lambda^2 - z).$$

Hence there is a (0, 1)-group and a (0, 0)-group of perturbed eigenvalues. Each group consists of only one cycle, the periods are 1 and 2. $z_0 = 0$ is multiple. The points $z \neq 0$ are simple. $(z_0, \lambda_2) = (0, 0)$ is a splitting point, $(z_0, \lambda_1) = (0, 1)$ is normal.

Example (b) shows that over a multiple trace point normal points can also occur.

(c) $$A(z) = \begin{pmatrix} 0 & z \\ -z & 2z \end{pmatrix}.$$

$A(z)$ is considered in the complex plane. One has

$$\chi(\lambda, A(z)) = (\lambda - z)^2.$$

In this case $s = 2$, i.e., there is a permanent degeneration. All points are simple, so is $z = 0$. There are no multiple points; hence, no splitting points either.

(d) $$A(z) = \begin{pmatrix} 0 & z & 0 \\ -z & 2z & 0 \\ 0 & 0 & -z \end{pmatrix}.$$

$A(z)$ is considered in the complex plane. One has

$$\chi(\lambda, A(z)) = (\lambda - z)^2 (\lambda + z).$$

In this case $s = 2$, i.e., there is a permanent degeneration. The relation $q(\lambda, z) = (\lambda - z)(\lambda + z)$ holds. The point $z = 0$ is multiple; the points $z \neq 0$ are simple. The only splitting point over $z = 0$ is $(0, 0)$. There are two cycles of period 1 within the $(0, 0)$-group.

(e) $$A(z) = \begin{pmatrix} 1 & z^3 - z^2 \\ -1 & 1 \end{pmatrix}.$$

$A(z)$ is considered in the complex plane. One has

$$\chi(\lambda, A(z)) = \lambda^2 - 2\lambda + z^3 - z^2 + 1.$$

χ is irreducible with respect to $\mathbb{C}$. Thus there is no permanent degeneration. The eigenvalues are

$$\lambda_{1,2}(z) = 1 \pm z(z - 1)^{1/2}.$$

The points $z_1 = 0$ and $z_2 = 1$ are multiple. All other points are simple. Let $\mathfrak{P}$ denote the analytic function given by χ. The multiple point $z_1 = 0$ is regular for $\mathfrak{P}$, the multiple point $z_2 = 1$ is singular for $\mathfrak{P}$. $(0, 1)$ is the only splitting point over $z_1 = 0$; $(1, 1)$ is the only splitting point over $z_2 = 1$. The $(0, 1)$-group consists of two cycles of period 1, the $(1, 1)$-group consists of one cycle of period 2.

3.3. Eigenprojections of meromorphic operator-valued functions

In this section, the global properties of the eigenprojections of an operator-valued function $A(z)$, meromorphic in a region G, are investigated. The local behavior of the eigenprojections in the neighborhood of holomorphic points and poles of $A(z)$ is also discussed. As in the foregoing section, theorems are cited which have been proved in the Appendix.

3.3.1. Resolvent

First we refer to Subsect. 2.8.1. There the following proposition has been proved: *Let $\lambda \in \operatorname{res} A_0$ and $\|A - A_0\| < \|R(\lambda, A_0)\|^{-1}$. Then the series $\sum_{k=0}^{\infty} \{(A - A_0) R(\lambda, A_0)\}^k$ is absolutely convergent, the relation $\lambda \in \operatorname{res} A$ holds and one has*

$$R(\lambda, A) = R(\lambda, A_0) \sum_{k=0}^{\infty} [\{A - A_0\} R(\lambda, A_0)]^k. \tag{3.1}$$

The series (3.1) is called the *second Neumann series* for the resolvent. (The first Neumann series for the resolvent coincides with its Taylor expansion with respect to finite points λ or $\lambda = \infty$; see Subsect. 2.1.2).

Now A_0 is replaced by $A(z_0)$ and A by $A(z)$. Then, as a condition of convergence for the series (3.1), one obtains the estimate $\|A(z) - A(z_0)\| < \|R(\lambda, A(z_0))\|^{-1}$. A sufficient condition that this estimate is satisfied is given by

$$\sum_{m=1}^{\infty} |z - z_0|^m \|A_m\| < \|R(\lambda, A(z_0))\|^{-1}, \tag{3.2}$$

if $A(z) = \sum_{m=0}^{\infty} A_m(z - z_0)^m$ in a neighborhood of z_0 (i.e., let $z_0 \in G \setminus S_A$, z_0 a holomorphic point for $A(z)$).

Now let K be a compact subregion of $\operatorname{res} A(z_0)$ and let L be a sufficiently small closed disc $|z - z_0| \leqq r$ such that r satisfies the condition

$$\sum_{m=1}^{\infty} r^m \|A_m\| < \inf_{\lambda \in K} \|R(\lambda, A(z_0))\|^{-1} \tag{3.3}$$

(cf. Subsect. 2.8.1). Then one obtains:

The series (3.1) *converges uniformly absolutely for* $(z, \lambda) \in L \times K$*; in particular, the inclusion* $K \subset \operatorname{res} A(z)$ *holds for all* $z \in L$.

The terms in the series (3.1) are holomorphic jointly in both variables $(z, \lambda) \in L \times K$. Hence one obtains

Lemma 1. *The resolvent $R(\lambda, A(z))$ is holomorphic in both variables $(z, \lambda) \in L \times K$ jointly.*

In particular, the series (3.1) can be arranged as a power series with respect to $z - z_0$ and it can be integrated termwise with respect to the variable $\lambda \in K$. One obtains

$$R(\lambda, A(z)) = \sum_{m=0}^{\infty} R_m(\lambda) (z - z_0)^m, \qquad |z - z_0| \leqq r, \tag{3.4}$$

where

$$R_0(\lambda) = R(\lambda, A(z_0)),$$

$$R_1(\lambda) = R_0(\lambda) A_1 R_0(\lambda),$$

$$R_2(\lambda) = R_0(\lambda) A_2 R_0(\lambda) + R_0(\lambda) A_1 R_0(\lambda) A_1 R_0(\lambda),$$

etc. The general expression is given by the formula

$$R_p(\lambda) = \sum_{\substack{\mu_1+\mu_2+\cdots+\mu_k=p \\ \mu_1,\mu_2,\ldots,\mu_k\geqq 1 \\ k\geqq 1}} R_0 A_{\mu_1} R_0 A_{\mu_2} \cdots R_0 A_{\mu_k} R_0, \quad R_0 = R_0(\lambda), \quad p = 1, 2, 3, \ldots \tag{3.5}$$

If, in particular, $\mathscr{C}$ denotes a piecewise smooth bounded path in res $A(z_0)$, then

$$\int_{\mathscr{C}} R(\lambda, A(z))\, d\lambda = \sum_{m=0}^{\infty} (z - z_0)^m \int_{\mathscr{C}} R_m(\lambda)\, d\lambda, \tag{3.6}$$

where again (3.6) converges uniformly absolutely with respect to $|z - z_0| \leqq r$. That is, (3.6) represents an operator-valued function, holomorphic in $|z - z_0| \leqq r$.

The relation (3.3) will be applied later (see Ch. 8) in connection with quantitative estimates of the radii of convergence for the perturbation series. The formulas (3.5) play a role in the calculation of the coefficients of the perturbation series (see Sect. 3.8 and 3.9).

3.3.2. The (z_0, λ_0)-groups

Let $A(z)$ be meromorphic in G and $z_0 \in G \setminus S_A$, i.e., $A(z_0)$ is well-defined. Let λ_0 be an eigenvalue of $A(z_0)$. Furthermore, let $\varepsilon > 0$ be chosen such that *the sphere* $|\lambda - \lambda_0| = \varepsilon$ *separates the eigenvalue* λ_0 *from the rest spectrum*, that is, the sphere consists only of regular points with respect to $A(z_0)$ and

$$\{\lambda: |\lambda - \lambda_0| < \varepsilon\} \cap \operatorname{spec}(A(z_0)) = \{\lambda_0\}.$$

The open disc $|\lambda - \lambda_0| < \varepsilon$ is denoted by K. Then one has

Lemma 2. *There exists a* $\delta > 0$ *such that* spec $A(z) \cap K$ *is exactly the* (z_0, λ_0)*-group of perturbed eigenvalues of* $A(z)$ *if* $|z - z_0| < \delta$.

Proof. The (z_0, λ_0)-group consists of a certain number of branches, say $\mathfrak{P}_1, \mathfrak{P}_2, \ldots, \mathfrak{P}_s$ over a neighborhood of z_0, i.e., $\mathfrak{P}_1(z_0) = \mathfrak{P}_2(z_0) = \cdots = \mathfrak{P}_s(z_0) = \lambda_0$ and for all other eigenvalue branches $\mathfrak{Q}$ of $A(z)$ over a neighborhood of z_0 one has $\mathfrak{Q}(z_0) = \{\mu(\mathfrak{Q})\}$, where $\mu(\mathfrak{Q}) \neq \lambda_0$. The continuity of the eigenvalue functions $\lambda(\zeta)$, defined on the several branches, implies the existence of a $\delta > 0$ with $\mathfrak{P}_\varrho(z) \subset K$, $\varrho = 1, 2, \ldots, s$, and $\mathfrak{Q}(z) \cap \operatorname{clo} K = \emptyset$ if $|z - z_0| < \delta$, where $\mathfrak{Q}$ runs through all other eigenvalue branches of $A(z)$. Hence, for $|z - z_0| < \delta$ one obtains $\mathfrak{P}_1(z) \cup \mathfrak{P}_2(z) \cup \cdots \cup \mathfrak{P}_s(z) = \operatorname{spec} A(z) \cap K$. ■

If one takes into account the results of Subsect. 3.3.1, then Lemma 2 can be strengthened as follows:

Lemma 3. *If* $r > 0$ *satisfied the inequality*

$$\sum_{m=1}^{\infty} r^m \|A_m\| < \inf_{|\lambda-\lambda_0|=\varepsilon} \|R(\lambda, A(z_0))\|^{-1}, \tag{3.7}$$

then spec $A(z) \cap K$ *is exactly the* (z_0, λ_0)*-group of perturbed eigenvalues of* $A(z)$ *for all* z *with* $|z - z_0| \leqq r$.

Proof. From (3.7), according to Subsect. 3.3.1, follows that the relation $\{\lambda: |\lambda - \lambda_0| = \varepsilon\} \subset \operatorname{res} A(z)$ holds for all z with $|z - z_0| \leqq r$. Therefore, the sphere $|\lambda - \lambda_0| = \varepsilon$ separates the part of the spectrum $\operatorname{spec} A(z)$ inside the sphere from the part of the spectrum $\operatorname{spec} A(z)$ outside the sphere, for all z with $|z - z_0| \leqq r$. According to Lemma 2, there is a $\delta > 0$ with $\operatorname{spec} A(z) \cap K = \mathfrak{P}_1(z) \cup \mathfrak{P}_2(z) \cup \cdots \cup \mathfrak{P}_s(z)$ for $|z - z_0| < \delta$, where the branches $\mathfrak{P}_1, \mathfrak{P}_2, \ldots, \mathfrak{P}_s$ represent the (z_0, λ_0)-group. For the other eigenvalue branches $\mathfrak{Q}$ of $A(z)$ one has $\operatorname{spec} A(z) \cap \operatorname{clo} K = \emptyset$ if $|z - z_0| < \delta$. We consider the branches $\mathfrak{P}_\varrho, \mathfrak{Q}$ over $|z - z_0| \leqq r$. Assume that $|\lambda(\zeta_\varrho) - \lambda_0| > \varepsilon$ for some point $\zeta_\varrho \in \mathfrak{P}_\varrho$, where the trace point z of ζ_ϱ satisfies $|z - z_0| \leqq r$. Furthermore, $\lambda(\zeta_\varrho^0) = \lambda_0$, where $\zeta_\varrho^0 \in \mathfrak{P}_\varrho$ and where the trace point of ζ_ϱ^0 is just z_0, i.e., $|\lambda(\zeta_\varrho^0) - \lambda_0| = 0$. Now we consider an analytic continuation from ζ_ϱ^0 into ζ_ϱ over $|z - z_0| \leqq r$; the corresponding continuation path is denoted by $\zeta_\varrho(t)$, $0 \leqq t \leqq 1$, where $\zeta_\varrho(0) = \zeta_\varrho^0$ and $\zeta_\varrho(1) = \zeta_\varrho$. The trace point $z(\zeta_\varrho(t))$ of $\zeta_\varrho(t)$ satisfies $|z(\zeta_\varrho(t)) - z_0| \leqq r$. Then the function $f(t) = |\lambda(\zeta_\varrho(t)) - \lambda_0|$ is continuous on $0 \leqq t \leqq 1$ with $f(0) = 0$ and $f(1) > \varepsilon$. Therefore, there is a t_0 with $0 < t_0 < 1$ such that $f(t_0) = \varepsilon$, i.e., there is a point $\xi_\varrho = \zeta_\varrho(t_0) \in \mathfrak{P}_\varrho$ over $|z - z_0| \leqq r$ such that $|\lambda(\xi_\varrho) - \lambda_0| = \varepsilon$. Because $\lambda(\xi_\varrho) \in \operatorname{spec} A(\tilde{z})$, where $\tilde{z}$ denotes the trace point of ξ_ϱ and where $|\tilde{z} - z_0| \leqq r$, this contradicts $\lambda(\xi_\varrho) \in \operatorname{res} A(\tilde{z})$. Therefore, over $|z - z_0| \leqq r$ the inequality $|\lambda(\zeta_\varrho) - \lambda_0| < \varepsilon$ always holds for all $\zeta_\varrho \in \mathfrak{P}_\varrho$, $\varrho = 1, 2, \ldots, s$. Similar conclusions are used for the treatment of the other eigenvalue branches. ■

Lemma 3 is also true if $|\lambda - \lambda_0| = \varepsilon$ is replaced by another simply closed piecewise smooth path in the eigenvalue plane, separating λ_0 from the rest spectrum.

3.3.3. Group projections and eigenprojections

Let z_0 be a holomorphic point of $A(z)$ and let $\lambda_0 \in \operatorname{spec} A(z_0)$. Let L denote a positively oriented simply closed piecewise smooth path, lying in $\operatorname{res} A(z_0)$ and separating λ_0 from the rest spectrum of $A(z_0)$. Let $r > 0$ be chosen according to

$$\sum_{m=1}^{\infty} r^m \|A_m\| < \inf_{\lambda \in L} \|R(\lambda, A(z_0))\|^{-1}. \tag{3.8}$$

Then an operator-valued function $P(z; z_0, \lambda_0)$ can be defined on $|z - z_0| \leqq r$ by

$$P(z; z_0, \lambda_0) = \frac{1}{2\pi i} \int_L R(\lambda, A(z))\, d\lambda, \qquad |z - z_0| \leqq r. \tag{3.9}$$

The inner region of L is denoted by K. Then, according to Lemma 3, for $|z - z_0| \leqq r$ the (z_0, λ_0)-group of $A(z)$ is given by $\operatorname{spec} A(z) \cap K$. This group consists of some branches $\mathfrak{P}_1, \ldots, \mathfrak{P}_s$ with the valuednesses $g_1, \ldots, g_s$, respectively. The values of $\mathfrak{P}_i$ over z, where $0 < |z - z_0| \leqq r$, are denoted by $\lambda_{i1}(z), \ldots, \lambda_{ig_i}(z)$. The properties of $P(z; z_0, \lambda_0)$ are collected in the following lemma.

Lemma 4. *The operator-valued function $P(z; z_0, \lambda_0)$ satisfies the following properties:*

I. *$P(z; z_0, \lambda_0)$ is holomorphic on $|z - z_0| \leqq r$.*

II. *$P(z; z_0, \lambda_0)$ is a projection for every z with $|z - z_0| \leqq r$; i.e., the function $P(z; z_0, \lambda_0)$ is projection-valued.*

III. *The relation $A(z)\, P(z; z_0, \lambda_0) = P(z, z_0, \lambda_0)\, A(z)$ holds for all z with $|z - z_0| \leqq r$; i.e., $P(z; z_0, \lambda_0)$ reduces $A(z)$.*

IV. *$P(z_0; z_0, \lambda_0)$ is the eigenprojection of λ_0 with respect to $A(z_0)$.*

V. *If the algebraic multiplicity of λ_0 with respect to $A(z_0)$ is denoted by $n(\lambda_0)$, then*

$$\dim P(z; z_0, \lambda_0) = n(\lambda_0).$$

VI. $\operatorname{spec}\big(A(z) \restriction P(z; z_0, \lambda_0)\, X\big) = \operatorname{spec} A(z) \cap K$.

VII. *If the eigenprojection of $\lambda_{i\alpha}(z)$ with respect to $A(z)$ is denoted by $Q\big(\lambda_{i\alpha}(z)\big)$, then*

$$P(z; z_0, \lambda_0) = \sum_{i=1}^{s} \sum_{\alpha=1}^{g_i} Q\big(\lambda_{i\alpha}(z)\big).$$

VIII. *The power series expansion of $P(z; z_0, \lambda_0)$ takes the form*

$$P(z; z_0, \lambda_0) = P(z_0; z_0, \lambda_0) + \sum_{m=1}^{\infty} (z - z_0)^m\, P_m(z_0, \lambda_0),$$

where

$$P_m(z_0, \lambda_0) = \frac{1}{2\pi \mathrm{i}} \int_L R_m(\lambda)\, \mathrm{d}\lambda.$$

Proof. I follows from Lemma 1. II follows from Theorem 2.3.1 and Theorem 2.3.2; the same is true for III. IV immediately follows from the definition (3.9) if one takes $z = z_0$. Note that L separates the point λ_0 from the rest spectrum. V follows from $\dim P(z_0; z_0, \lambda_0) = n(\lambda_0)$ and Subsect. 1.15.4. VII follows, for example, from Theorem 2.6.1 (or directly from Theorem 2.3.1). From VII one obtains VI; note Lemma 1. Finally, VIII follows from Subsect. 3.3.1. ■

Property VII of this Lemma expresses the fact that $P(z; z_0, \lambda_0)$ is the sum of all eigenprojections belonging to eigenvalues of the (z_0, λ_0)-group. Therefore, $P(z; z_0, \lambda_0)$ is called the *group projection* of the (z_0, λ_0)-group. $P(z; z_0, \lambda_0)\, X$ is called the *algebraic group eigenspace*; it is the direct sum of all algebraic eigenspaces belonging to eigenvalues of the (z_0, λ_0)-group.

Consider the special case where (z_0, λ_0) is *normal*. In this case there is exactly one eigenvalue $\lambda(z)$ in the (z_0, λ_0)-group, and it is holomorphic in $|z - z_0| \leqq r$. Because of Lemma 3.7, $P(z; z_0, \lambda_0)$ is the eigenprojection of $\lambda(z)$ with respect to $A(z)$. Hence one obtains

Corollary 1. *Let (z_0, λ_0) be normal, where the only eigenvalue of the (z_0, λ_0)-group is denoted by $\lambda(z)$. $\lambda(z)$ is holomorphic in $|z - z_0| \leqq r$. Then $P(z; z_0, \lambda_0)$ is the eigenprojection of $\lambda(z)$ with respect to $A(z)$ for all z with $|z - z_0| \leqq r$.*

If (z_0, λ_0) is normal, then there is exactly one irreducible factor $p(\lambda, z)$ belonging to the decomposition of $\chi\big(\lambda, A(z)\big)$ into prime factors, such that the relation $p(\lambda_0, z_0) = 0$

holds. Moreover, λ_0 is a simple root of $p(\lambda, z_0)$. Denoting the multiplicity of $p(\lambda, z)$ with respect to the decomposition of $\chi(\lambda, A(z))$ into prime factors by m, one obtains, because $n(\lambda_0) = m$:

Corollary 2. *If (z_0, λ_0) is normal, then*

$$\dim P(z; z_0, \lambda_0) = m \tag{3.10}$$

and

$$\chi(\lambda; A(z) \upharpoonright P(z; z_0, \lambda_0) X) = (\lambda - \lambda(z))^m, \tag{3.11}$$

where m denotes the multiplicity of the irreducible polynomial which belongs to (z_0, λ_0) in the decomposition of $\chi(\lambda, A(z))$ into prime factors.

One may consider the group projections defined by (3.9) as the perturbations of the eigenprojections of $A(z_0)$ generated by the perturbation $A(z)$ of $A(z_0)$. In general, the perturbation $A(z)$ produces group projections from eigenprojections. However, the eigenprojections of normal points remain eigenprojections.

3.3.4. Analytic eigenprojection-valued functions on the Riemann surfaces defined by the characteristic polynomial

Let $A(z)$ be meromorphic in G. According to Subsect. 3.2.3, $\chi(\lambda, A(z))$ defines analytic functions $\mathfrak{P}_1, \mathfrak{P}_2, \ldots, \mathfrak{P}_r$ over G corresponding to the irreducible factors $p_1, p_2, \ldots, p_r$ of χ. In the following, let $\mathfrak{P}$ be one of these functions and let $p(\lambda, z)$ be the corresponding irreducible polynomial. We consider $\mathfrak{P}$ only over $G \setminus S_A$. Let $z_0 \in G \setminus S_A$ and let (z_0, λ_0) be normal with respect to $A(z)$, where λ_0 belongs to $\mathfrak{P}(z_0)$. Then there is a one-to-one correspondence between the pair (z_0, λ_0) and the holomorphic function element $(z_0, \lambda(z))$ of $\mathfrak{P}$ with trace point z_0 and $\lambda(z_0) = \lambda_0$. In the following we consider only the family of all normal points (z_0, λ_0) of $\mathfrak{P}$ over $G \setminus S_A$ and the family $\mathfrak{P}'$ of all assigned holomorphic function elements $(z_0, \lambda(z))$ of $\mathfrak{P}$, respectively. Note that all these function elements can be analytically continued into each other over $G \setminus S_A$.

Let (z_0, λ_0) be normal and let $(z_0, \lambda(z))$ be the assigned function element of $\mathfrak{P}$. According to Corollary 1, the point $(z, \lambda(z))$ is normal for all z with $|z - z_0| \leqq r$ and the projection-valued function $P(z; z_0, \lambda_0)$, defined in $|z - z_0| \leqq r$, is the eigenprojection of $\lambda(z)$ with respect to $A(z)$. One can state:

The holomorphic projection-valued function element $(z_0, P(z; z_0, \lambda_0)$ is uniquely assigned to the function element $(z_0, \lambda(z))$. Both elements are holomorphic for $|z - z_0| \leqq r$.

If $(z_0, \lambda(z))$ runs through $\mathfrak{P}'$, then one obtains a family $\mathfrak{Q}'$ of holomorphic projection-valued function elements over $G \setminus S_A$.

In order to show that the elements of $\mathfrak{Q}'$ can be analytically continued into each other over $G \setminus S_A$, and thus define a unique analytic projection-valued function $\mathfrak{Q}$ over G, a theorem is used that has been proved in Subsect. A 6.1 (the proof has been given for scalar-functions $Q(z)$, but the theorem is also true for operator-valued functions).

Theorem 1. *Let $\mathfrak{P}$ define an analytic function over G. Assume that the meromorphic elements of $\mathfrak{P}$ can be analytically continued into each other within $\mathfrak{P}$. Furthermore, there is an assignment from all meromorphic elements $P(z)$ of $\mathfrak{P}$, where z denotes their trace point, to meromorphic elements $Q(z)$ with the same trace point. Assume that the following condition is satisfied: If $P'(z')$ is a direct continuation of $P(z)$ and if $Q'(z')$ denotes the element assigned to $P'(z')$, then $Q'(z')$ is a direct continuation of $Q(z)$. Then the family $\tilde{\mathfrak{Q}}$ of all assigned function elements $Q(z)$ defines a unique analytic function over G.*

For the application of this theorem, let $\mathfrak{P}$ be identified with the family $\mathfrak{P}'$ of all holomorphic elements $(z_0, \lambda(z))$ of $\mathfrak{P}$ over $G \setminus S_A$, where $(z_0, \lambda(z_0))$ is normal. The assignment $P \to Q$ is identified with the assignment $(z_0, \lambda(z)) \to (z_0, P(z; z_0, \lambda_0))$, described above. One obtains

Lemma 5. *The family $\mathfrak{Q}'$ of all holomorphic projection-valued elements $(z_0, P(z; z_0, \lambda_0))$, where $z_0 \in G \setminus S_A$, (z_0, λ_0) is normal and $(z_0, \lambda(z))$ is an element from $\mathfrak{P}$, defines a unique analytic function $\mathfrak{Q}$ over $G \setminus S_A$.*

Proof. According to the preceding remarks and Theorem 1 it is sufficient to prove that the correspondence $(z_0, \lambda(z)) \to (z_0, P(z; z_0, \lambda_0))$ is compatible with the relation between two elements given by the concept "direct continuation". Let the element $(z_0, \lambda(z))$ be given and let $(z, \lambda(z'))$ be a direct continuation of $(z_0, \lambda(z))$:

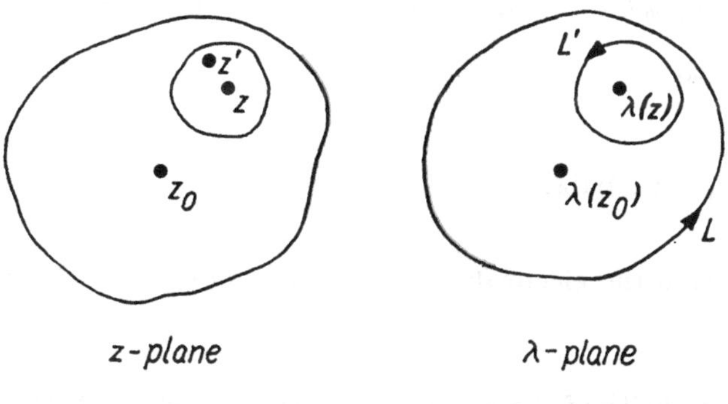

Fig. 3

To the element $(z, \lambda'(z))$ there corresponds the projection-valued element

$$Q(z') = \frac{1}{2\pi i} \int_{L'} R(\lambda, A(z'))\, d\lambda,$$

where L' denotes, for example, a sufficiently small positively oriented sphere around $\lambda(z)$. $Q(z')$ is holomorphic in a certain disc $|z' - z| < \delta$ and represents the eigenprojection of $\lambda(z')$ with respect to $A(z')$. Let $\delta > 0$ be chosen such that the disc $|z' - z| < \delta$ is contained in the disc $|z' - z_0| < r$; note that $|z - z_0| < r$.

Also, the projection

$$P(z'; z_0, \lambda_0) = \frac{1}{2\pi i} \int_{L} R(\lambda, A(z'))\, d\lambda$$

is the eigenprojection of $\lambda(z')$ with respect to $A(z')$ for all z' with $|z' - z_0| < r$ (L and r are defined as before, see Subsect. 3.3.3). Because of the uniqueness of the eigenprojection, one obtains (see Sect. 2.4)

$$P(z'; z_0, \lambda_0) = Q(z'), \qquad |z' - z| < \delta,$$

that is, $(z, Q(z'))$ is a direct continuation of $(z_0, P(z; z_0, \lambda_0))$. Thus, it is shown that the projection-valued function element $(z, Q(z'))$, which is assigned to the direct continuation $(z, \lambda(z'))$ of the element $(z_0, \lambda(z))$, is the direct continuation of the projection-valued element $(z_0; P(z; z_0, \lambda_0))$, which is assigned to the element $(z_0, \lambda(z))$. Hence, the compatibility mentioned before has been proved. ■

According to Lemma 5, a unique analytic function $\mathfrak{Q}$ is defined over $G \setminus S_A$, where the function elements are the projection-valued elements $(z_0, P(z; z_0, \lambda_0))$. The properties of the assignment

$$(z_0, \lambda(z)) \to (z_0, P(z; z_0, \lambda_0))$$

noted in the proof of Lemma 5 can be interpreted such that $\mathfrak{Q}$ *is an analytic function, defined at the normal points* $(z_0, \lambda(z))$ *of* $\mathfrak{P}$ *over* $G \setminus S_A$. *Hence,* $\mathfrak{Q}$ *is an analytic function defined "on* $\mathfrak{P}$". (For the definition of analytic functions on Riemann surfaces see Sect. A 6.)

The points of $\mathfrak{P}$ are denoted by ζ. For normal points whose trace points are in $G \setminus S_A$, i.e., for points $\zeta = (z_0, \lambda(z))$, $z_0 \in G \setminus S_A$, with $(z_0, \lambda(z_0))$ normal, one writes

$$P(\zeta) = P(z_0; z_0, \lambda_0);$$

that is,

$$P(\zeta) = \frac{1}{2\pi i} \int_L R(\lambda, A(z_0))\, d\lambda,$$

where one integrates around the eigenvalue $\lambda(z_0)$. Because of Corollary 2, one obtains

$$\dim P(\zeta) = m, \tag{3.12}$$

where m denotes the multiplicity of the irreducible polynomial $p(\lambda, z))$ belonging to $\mathfrak{P}$, with respect to the decomposition of $\chi(\lambda, A(z))$ into prime factors.

The function $P(\zeta)$ has not yet been defined for all splitting points and for all points whose trace points belong to S_A. The set of all multiple points is denoted by M. Note that $S_A \cup M$ is isolated and closed in G. Hence: the set of all exceptional points on $\mathfrak{P}$ is isolated in $\mathfrak{P}$ and there is no accumulation point of this set in the interior of $\mathfrak{P}$. Therefore, the behavior of the function $P(\zeta)$ in the neighborhood of splitting points and of points whose trace points lie in S_A should be studied.

3.3.5. Eigenprojections and spectral representation of $A(z)$ for simple points

Let $\chi(\lambda, A(z)) = \prod_{\varrho=1}^{r} p_\varrho(\lambda, z)^{m_\varrho}$ be the decomposition of χ into prime factors (cf. Corollary 2.1). The algebroid functions assigned to $p_1, \ldots, p_r$ are denoted by $\mathfrak{P}_1, \mathfrak{P}_2, \ldots, \mathfrak{P}_r$. Let

$z_0 \in G \setminus S_A$ and let z_0 be simple. In particular, for all $\mathfrak{P}_1, \ldots, \mathfrak{P}_r$, only normal points with corresponding holomorphic function elements lie over z_0. According to Subsect. 3.2.5, these elements are denoted by $(z_0, \lambda_{\alpha\sigma}(z))$, where the elements $(z_0, \lambda_{\alpha 1}(z))$, $(z_0, \lambda_{\alpha 2}(z))$, $\ldots$, $(z_0, \lambda_{\alpha g_\alpha}(z))$ represent the branches of $\mathfrak{P}_\alpha$ lying over the neighborhood of z_0 (note $g_\alpha = \deg p_\alpha$). The projection-valued function element which is uniquely assigned to the element $(z_0, \lambda_{\alpha\sigma}(z))$ according to Subsect. 3.3.4 (see Corollary 1), is denoted by $(z_0, P_{\alpha\sigma}(z))$. Thus one has

$$P_{\alpha\sigma}(z) = \frac{1}{2\pi i} \int\limits_L R(\lambda, A(z))\, d\lambda, \qquad \alpha = 1, \ldots, r; \qquad \sigma = 1, \ldots, g_\alpha,$$

where one integrates around the eigenvalue $\lambda_{\alpha\sigma}(z_0)$ (hence also around the eigenvalue $\lambda_{\alpha\sigma}(z)$). The elements $(z_0, P_{\alpha 1}(z)), \ldots, (z_0, P_{\alpha g_\alpha}(z))$ represent the g_α branches of the projection-valued function $P(\zeta_\alpha)$, $\zeta_\alpha \in \mathfrak{P}_\alpha$, with ζ_α normal, given on $\mathfrak{P}_\alpha$ over the neighborhood of z_0. From Corollary 1 and Lemma 2.4.1 one obtains the relations

$$P_{\alpha\sigma}(z)\, P_{\alpha'\sigma'}(z) = \delta_{\alpha\alpha'}\delta_{\sigma\sigma'} P_{\alpha\sigma}(z), \tag{3.13}$$

$$\sum_{\alpha=1}^{r} \sum_{\lambda=1}^{g_\alpha} P_{\alpha\sigma}(z) = 1, \tag{3.14}$$

$$A(z)\, P_{\alpha\sigma}(z) = P_{\alpha\sigma}(z)\, A(z), \tag{3.15}$$

in a neighborhood of z_0. $P_{\alpha\sigma}(z)$ is the eigenprojection of $A(z)$ belonging to $\lambda_{\alpha\sigma}(z)$; one has $\dim P_{\alpha\sigma}(z) = m_\alpha$ and $\chi(\lambda; A(z) \restriction P_{\alpha\sigma}(z)\, X) = (\lambda - \lambda_{\alpha\sigma}(z))^{m_\alpha}$.

The corresponding eigennilpotents

$$D_{\alpha\sigma}(z) = (A(z) - \lambda_{\alpha\sigma}(z))\, P_{\alpha\sigma}(z) \tag{3.16}$$

are also holomorphic in the neighborhood of z_0. Hence, for the behavior of the spectral representation of $A(z)$ in the neighborhood of simple points one obtains the following:

Theorem 2. *Let $A(z)$ be meromorphic in G, $z_0 \in G \setminus S_A$, where z_0 is simple. Then there is a disc $|z - z_0| < \delta$ such that the spectral representation of $A(z)$ takes the form*

$$A(z) = \sum_{\alpha=1}^{r} \sum_{\sigma=1}^{g_\alpha} \left(\lambda_{\alpha\sigma}(z)\, P_{\alpha\sigma}(z) + D_{\alpha\sigma}(z)\right), \qquad |z - z_0| < \delta, \tag{3.17}$$

where the eigenvalues $\lambda_{\alpha\sigma}(z)$, the eigenprojections $P_{\alpha\sigma}(z)$ and the eigennilpotents $D_{\alpha\sigma}(z)$ are holomorphic in $|z - z_0| < \delta$. The number $s = \sum_{\alpha=1}^{r} g_\alpha$ of different eigenvalues and the dimensions $\dim P_{\alpha\sigma}(z) = m_\alpha$ of the eigenprojections are global objects and do not depend on the special simple point z_0.

This behavior of $A(z)$ in $|z - z_0| < \delta$ can be summarized as follows: $A(z)$ *has an invariant holomorphic spectral structure in* $|z - z_0| < \delta$. In this case, the holomorphic perturbation $A(z)$ of $A(z_0)$ causes only a holomorphic perturbation of the spectral objects occuring in the spectral representation.

Therefore, the multiple points z_0 lead to complications under perturbation. The multiple points z_0 are isolated. Hence one can state: If z_0 is multiple, then there is

$\delta > 0$ such that all points with $0 < |z - z_0| < \delta$ are simple. That is, immediately after "switching on" the perturbation $A(z)$ of $A(z_0)$, the spectral representation has an invariant holomorphic behavior with fixed number of eigenvalues and fixed dimensions of the eigenprojections. However, the question concerning the behavior of the spectral objects at multiple points and at points which are poles for $A(z)$ remains open.

3.3.6. Algebroid behavior of the eigenprojections at splitting points

Let $\mathfrak{P}$ be one of the functions over G assigned to the characteristic polynomial $\chi(\lambda, A(z))$. Let g be the valuedness of $\mathfrak{P}$ and let $P(\zeta)$ be the corresponding eigenprojection-valued function. Furthermore, let $z_0 \in G \setminus S_A$ be multiple and let ζ_0 be a splitting of $\mathfrak{P}$ over z_0. Then there is a $\delta > 0$ such that $0 < |z - z_0| < \delta$ consists of simple points z only. Let $\mathfrak{H}$ be one of the branches of $\mathfrak{P}$ over $0 < |z - z_0| < \delta$ containing ζ_0. Let $\mathfrak{H}$ be k-valued. Except for ζ_0, $\mathfrak{H}$ consists of normal points. Hence $P(\zeta)$ is well-defined over $0 < |z - z_0| < \delta$ on $\mathfrak{H}$. The projection-valued elements of $\mathfrak{H}$ lying over z are denoted by $P_1(z), P_2(z), \ldots, P_k(z)$. These elements are distinct, because, if $P_1(z) = P_i(z)$ for some i with $2 \leqq i \leqq k$, then $P_1(z) = 0$, because $P_1(z)\, P_i(z) = 0$. But $P_1(z) = 0$ is impossible. Hence, ζ_0 is a branching point of $P(\zeta)$ with respect to $\mathfrak{H}$ of order k exactly. The subscripts of the projection-valued elements can typically be chosen such that a cyclic permutation appears by a single positive winding around z_0.

Therefore, *the branching behavior of $P(\zeta)$ on $\mathfrak{P}$ over $G \setminus S_A$ corresponds exactly to the branching behavior of $\mathfrak{P}$ itself.*

The question remains whether a splitting point ζ_0 causes an algebraic or an essential singularity with respect to $P(\zeta)$. For this investigation there are several possibilities. In Sect. 5.3, we will give a global proof that splitting points cause at most algebraic singularities with respect to $P(\zeta)$, by giving a global representation of $P(\zeta)$ on $\mathfrak{P}$. For this purpose some preparations are necessary; these are treated in Ch. 4. But the question can also be answered by a purely local consideration as follows (see T. Kato [5, p. 69 sqq.]).

Let $\zeta_0 = (z_0, \lambda_0)$ be a splitting point of $\mathfrak{P}$. Let $\delta > 0$ be such that the disc $0 < |z - z_0| \leqq \delta$ consists of simple points only. Furthermore, let

$$\chi(\lambda, A(z)) = q(\lambda - \lambda_0, z)^m\, s(\lambda, z), \qquad |z - z_0| \leqq \delta,$$

where the polynomial $q(\lambda - \lambda_0, z)$ corresponds to the (z_0, λ_0)-group and where $s(\lambda, z)$ denotes a rest polynomial with $s(\lambda_0, z_0) \neq 0$ (see Sect. 2.6). One has $q(\lambda - \lambda_0, z) = \prod_{\varrho=1}^{k} (\lambda - \lambda_\varrho(z))$. The numbers $\lambda_1(z), \ldots, \lambda_k(z)$ coincide with the eigenvalues of the (z_0, λ_0)-group. They are distinct for $0 < |z - z_0| \leqq \delta$.

Let L be a disc $|\lambda - \lambda_0| < \varepsilon$ that contains all eigenvalues of the (z_0, λ_0)-group for $|z - z_0| \leqq \delta$; moreover, the closed disc $|\lambda - \lambda_0| \leqq \varepsilon$ does not contain any other eigenvalue for $|z - z_0| \leqq \delta$. Hence, one obtains $s(\lambda, z) \neq 0$ for $|\lambda - \lambda_0| \leqq \varepsilon$ and $|z - z_0| \leqq \delta$. Furthermore, let $\Gamma_\varrho(z)$ denote a disc with center $\lambda_\varrho(z)$ and radius $r_\varrho(z)$, contained in L

and separating $\lambda_\varrho(z)$ from the other eigenvalues of the (z_0, λ_0)-group. Hence, the interior of $\Gamma_\varrho(z)$ contains exactly one eigenvalue, namely $\lambda_\varrho(z)$.

The projection-valued element belonging to $\lambda_\varrho(z)$ is given by

$$P_\varrho(z) = \frac{1}{2\pi i} \int\limits_{\partial\Gamma_\varrho(z)} R(\lambda, A(z))\,d\lambda, \qquad \varrho = 1, 2, \dots, k.$$

One has

$$\|P_\varrho(z)\| \leqq r_\varrho(z) \max_{\lambda\in\partial\Gamma_\varrho(z)} \|R(\lambda, A(z))\|.$$

According to Subsect. 1.8.2, the estimate

$$\|R(\lambda, A(z))\| \leqq \gamma \frac{\|\lambda - A(z)\|^{N-1}}{|\det(\lambda - A(z))|} \leqq \gamma \frac{\|\lambda - A(z)\|^{N-1}}{|q(\lambda - \lambda_0, z)^m| \cdot |s(\lambda, z)|}$$

holds, where $\gamma > 0$ depends only on the norm used. Because $|s(\lambda, z)| \geqq \alpha^{-1}$ for some $\alpha > 0$ and for $|\lambda - \lambda_0| \leqq \varepsilon$, $|z - z_0| \leqq \delta$, the estimate

$$\|R(\lambda, A(z))\| \leqq \alpha\gamma \frac{(|\lambda| + \|A(z)\|)^{N-1}}{\left|\prod\limits_{\varrho=1}^{k} (\lambda - \lambda_\varrho(z))^m\right|},$$

$$\lambda \neq \lambda_\varrho(z), \qquad |z - z_0| \leqq \delta, \qquad |\lambda - \lambda_0| \leqq \varepsilon$$

follows. Hence one obtains

$$\|P_\varrho(z)\| \leqq K r_\varrho(z) \max_{\lambda\in\partial\Gamma_\varrho(z)} \left(1 \Big/ \prod_{\sigma=1}^{k} |\lambda - \lambda_\sigma(z)|^m\right).$$

The eigenvalues $\lambda_\sigma(z)$ of the (z_0, λ_0)-group can be represented by Puiseux series with respect to $z - z_0$ in $|z - z_0| \leqq \delta$ (see Subsect. 3.2.6). Hence there are constants $\beta_1 > 0$ and $\mu > 0$ such that

$$|\lambda_\sigma(z) - \lambda_{\sigma'}(z)| \geqq \beta_1 |z - z_0|^\mu, \qquad \sigma \neq \sigma', \qquad |z - z_0| \leqq \delta.$$

Let $r_\varrho(z) = \beta_2 |z - z_0|^\mu$ with $0 < \beta_2 < \beta_1$. Obviously, $\Gamma_\varrho(z)$ separates the eigenvalue $\lambda_\varrho(z)$ from the remaining eigenvalues of the (z_0, λ_0)-group and one has: $\lambda \in \partial\Gamma_\varrho(z)$ if and only if $\lambda = \lambda_\varrho(z) + w\beta_2 |z - z_0|^\mu$, where $|w| = 1$. One obtains $|\lambda - \lambda_\sigma(z)| = |(\lambda_\varrho(z) - \lambda_\sigma(z)) + w\beta_2 |z - z_0|^\mu|$. Hence,

$$|\lambda - \lambda_\sigma(z)| \geqq \begin{cases} \beta_2 |z - z_0|^\mu, & \sigma = \varrho, \\ (\beta_1 - \beta_2) |z - z_0|^\mu, & \sigma \neq \varrho, \end{cases} \qquad \lambda \in \partial\Gamma_\varrho(z), \qquad |z - z_0| \leqq \delta;$$

that is,

$$|\lambda - \lambda_\sigma(z)| \geqq \beta |z - z_0|^\mu, \qquad \lambda \in \partial\Gamma_\varrho(z), \qquad |z - z_0| \leqq \delta,$$

where $\beta > 0$ is some constant. Therefore one obtains

$$\prod_{\sigma=1}^{k} |\lambda - \lambda_\sigma(z)|^m \geqq \beta^{km} |z - z_0|^{km\mu}, \qquad \lambda \in \partial\Gamma_\varrho(z), \qquad |z - z_0| \leqq \delta,$$

and, finally,

$$\|P_\varrho(z)\| \leqq K\beta_2\beta^{-km} |z - z_0|^{\mu(1-km)}, \qquad 0 < |z - z_0| \leqq \delta. \tag{3.18}$$

The estimate (3.18) shows that the splitting point ζ_0 is at most an algebraic singularity for $P(\zeta)$.

Thus, if $\mathfrak{H}$ is an h-valued branch of $\mathfrak{P}$ over $|z - z_0| \leqq \delta$ containing ζ_0 (of course, ζ_0 can be contained in several branches of $\mathfrak{P}$ over $|z - z_0| \leqq \delta$), then ζ_0 is an algebraic branching point of order h for $P(\zeta)$. According to Subsect. A 1.5, in this case one obtains a representation of $P(\zeta)$ in the neighborhood of ζ_0 by convergent Laurent-Puiseux series.

Theorem 3. *Let (z_0, λ_0) be a splitting point and assume that the dotted disc $0 < |z - z_0| < \delta$ consists of simple points only. Furthermore, let $\mathfrak{H}$ be an h-valued branch of $\mathfrak{P}$ over $|z - z_0| < \delta$, where ζ_0 is a branching point of $\mathfrak{H}$ of order h (in this case $\mathfrak{H}$ defines a fixed cycle of the (z_0, λ_0)-group). Then ζ_0 is also a branching point of order h for $P(\zeta)$, which is algebroid, and the branches of $P(\zeta)$ over $0 < |z - z_0| < \delta$ on $\mathfrak{H}$ can be represented by Laurent-Puiseux series of the form*

$$P_\varrho(z) = \sum_{\sigma=-K}^{\infty} B_\sigma[\xi^\varrho(z - z_0)^{1/h}]^\sigma, \qquad \varrho = 1, \dots, h; \qquad 0 < |z - z_0| < \delta, \tag{3.19}$$

where ξ denotes a primitive h-th root of unity and $(z - z_0)^{1/h}$ denotes a fixed root. The B_σ denote operators from $\mathscr{L}(X)$.

In retrospect, the h-valued branching projection-valued function element, given by (3.19), can be added to the analytic function $P(\zeta)$ on $\mathfrak{P}$. Then this function is well-defined at all normal and all splitting points of $\mathfrak{P}$. The behavior of $P(\zeta)$ at points whose trace points belong to S_A remains to be studied.

3.3.7. Poles of $P(\zeta)$ at branching points

There is a connection between the actual existence of a branching at ζ_0 (i.e., a branching order $\geqq 2$) and the actual existence of a pole at ζ_0 (i.e., $K \leqq -1$) (see J. Butler [1] or T. Kato [5, p. 70sqq.]). For the following theorem see Subsect. A 1.5 and A 1.6.

Theorem 4 (Butler). *Let ζ_0 be an actual branching point of $\mathfrak{P}$ (i.e., ζ_0 is an h-valued element of $\mathfrak{P}$ with $h \geqq 2$). Then there is a pole at $\zeta = \zeta_0$ for $P(\zeta)$ (i.e., $K \leqq -1$ in (3.19)), in particular $\|P(\zeta)\| \to \infty$ as $\zeta \to \zeta_0$.*

Proof. Let $P_1(z), \dots, P_h(z)$ denote the projection-valued elements belonging to $P(\zeta)$ and ζ_0, defined on a neighborhood of the trace point z_0 of ζ_0. Assume that $\zeta = \zeta_0$ is not a pole for $P(\zeta)$. Then the expansion (3.19) takes the form

$$P_\varrho(z) = B_0 + B_1(\xi^\varrho(z - z_0)^{1/h}) + B_2(\xi^\varrho(z - z_0)^{1/h}) + \cdots, \quad \varrho = 1, \dots, h. \tag{3.20}$$

From $P_\varrho(z)^2 = P_\varrho(z)$ for $z \neq z_0$ one obtains $B_0^2 = B_0$. From $P_\varrho(z)\, P'_\varrho(z) = 0$ for $\varrho \neq \varrho'$ and $z \neq z_0$ one obtains $B_0^2 = 0$. Hence, $B_0 = 0$. Therefore, $\lim_{z \to z_0} \|P_\varrho(z)\| = 0$, $\varrho = 1, 2, \ldots, h$. According to Subsect. 3.3.4 (see also Corollary 2), one has $\dim P(\zeta) = m$, where $m \geqq 1$ denotes the multiplicity assigned to $\mathfrak{P}$. Hence, $\|P_\varrho(z)\| \geqq 1$ for all $z \neq z_0$ (see Subsect. 1.9.1). This is a contradiction. ■

If ζ_0 is not a branching point of $\mathfrak{P}$ but yet a splitting point, then $P(\zeta)$ may be holomorphic at $\zeta = \zeta_0$. On the other hand, $\zeta = \zeta_0$ may also be a pole for $P(\zeta)$. This is illustrated by the following two examples. Let $X = \mathbb{C}^2$.

(a) $$A(z) = \begin{pmatrix} z & 0 \\ 0 & -z \end{pmatrix}.$$

One has $\chi(\lambda, A(z)) = (\lambda - z)(\lambda + z)$. $z_0 = 0$ is multiple; all other points $z \neq 0$ are simple. The (0, 0)-group consists of two simple cycles. The eigenprojections assigned to $\lambda_1(z) = z$ and $\lambda_2(z) = -z$ take the form

$$P_1(z) = \begin{pmatrix} 1 & 0 \\ 0 & 0 \end{pmatrix}, \qquad P_2(z) = \begin{pmatrix} 0 & 0 \\ 0 & 1 \end{pmatrix};$$

that is, both eigenprojections are holomorphic at the splitting point (0, 0).

(b) $$A(z) = \begin{pmatrix} z & 0 \\ 1 & -z \end{pmatrix}.$$

Again one obtains: $\chi(\lambda, A(z)) = (\lambda - z)(\lambda + z)$, $z_0 = 0$ multiple, $z \neq 0$ simple. But in this case,

$$P_1(z) = \begin{pmatrix} 1 & 1/2z \\ 0 & 0 \end{pmatrix}, \qquad P_2(z) = \begin{pmatrix} 0 & -1/2z \\ 0 & 0 \end{pmatrix}.$$

Hence, $P_1(z)$ and $P_2(z)$ have a pole of the first order at $z_0 = 0$.

Example (a) also shows that the well-defined projection $P_1(0)$ is not the eigenprojection of $A(0)$ belonging to the eigenvalue $\lambda_1(0) = 0$. The same is true for $P_2(0)$. The eigenprojection of $\lambda_0 = 0$ equals $P_1(0) + P_2(0)$.

3.3.8. Algebroid behavior of the eigenprojections at points whose trace points are poles of $A(z)$

Let $z_0 \in S_A$, i.e., z_0 is a pole of $A(z)$ such that $A(z) = (z - z_0)^{-m} \cdot B(z)$, where m denotes the order of the pole; $B(z)$ is meromorphic in G, holomorphic at $z = z_0$, and $B(z_0) \neq 0$. Let $\delta > 0$ be such that $0 < |z - z_0| < \delta$ consists of simple points with respect to $B(z)$ only.

Let μ_0 be an eigenvalue of $B(z_0)$. Then one can apply the considerations of Subsect. 3.3.6 and 3.3.7 to the eigenprojections of the (z_0, μ_0)-group of $B(z)$ over $0 < |z - z_0| < \delta$. In particular, Theorem 3 is true. Therefore, the eigenprojection-valued functions $P(\zeta)$ have an algebroid behavior at $\zeta = \zeta_0 = (z_0, \mu_0)$.

The eigenprojections of $B(z)$ over $0 < |z - z_0| < \delta$ coincide with the eigenprojections of $A(z)$, because, if $A = \gamma B$ with $\gamma \neq 0$ and if μ_0 is an eigenvalue of B, then $\gamma\mu_0$ is an eigenvalue of A and one has

$$P = \frac{1}{2\pi i} \int\limits_{(\gamma\mu_0)} (\zeta - A)^{-1} d\zeta = \frac{1}{2\pi i} \int\limits_{(\gamma\mu_0)} (\gamma\zeta' - A)^{-1} d(\gamma\zeta') = \frac{1}{2\pi i} \int\limits_{(\mu_0)} (\zeta' - B)^{-1} d\zeta'.$$

Thus, *if $P(\zeta)$ is an eigenprojection-valued function of $A(z)$ and if $\zeta_0 \in \mathfrak{P}$, where the trace point z_0 of ζ_0 belongs to S_A, then ζ_0 is at most a branching pole for $P(\zeta)$. The branching orders of ζ_0 and $P(\zeta)$ at ζ_0 coincide.*

3.3.9. Global representation of the projection-valued functions $P(\zeta)$

The results of Subsect. 3.3.4, 3.3.6 and 3.3.8 with respect to projection-valued functions $P(\zeta)$ are summarized in the following proposition:

Lemma 6. *Let $\mathfrak{P}$ be one of the eigenvalue functions over G assigned to the characteristic polynomial $\chi(\lambda, A(z))$ and let $P(\zeta)$ be the corresponding eigenprojection-valued function. Then $P(\zeta)$ is an algebroid projection-valued function defined on $\mathfrak{P}$.*

Proof. An immediate consequence of the results of Subsect. 3.3.4, 3.3.6 and 3.3.8 is that $P(\zeta)$ is an algebroid projection-valued function over G. Moreover, from the existence of the assignment $(z_0, \lambda(z)) \to (z_0, P(z))$ for meromorphic elements $(z_0, \lambda(z))$ of $\mathfrak{P}$, one obtains that $P(\zeta)$ is even an algebroid function on $\mathfrak{P}$ (see Subsect. 3.3.4 and 3.3.8; with respect to the concept of an algebroid function on Riemann surfaces, see Subsect. A 6.1 and A 6.2). ■

The algebroid functions on $\mathfrak{P}$ have several important properties (see Subsect. A 6.2 and A 6.3). In the present context, the following property is of particular interest (its proof is given in Subsect. A 6.3).

Theorem 5. *Let $\mathfrak{P}$ be algebroid over G. Then the function $\mathfrak{F} = f(\zeta)$ is algebroid on $\mathfrak{P}$ if and only if it can be represented in the form*

$$\mathfrak{F} = a_0(z) + a_1(z)\, \mathfrak{P} + a_2(z)\, \mathfrak{P}^2 + \cdots + a_{g-1}(z)\, \mathfrak{P}^{g-1},$$

where g is the valuedness of $\mathfrak{P}$ and where the coefficients $a_0(z), a_1(z), \ldots, a_{g-1}(z)$ are meromorphic functions on G.

In this theorem $\mathfrak{F}$ can be a scalar-, vector- or operator-valued function. In the latter case, the coefficients $a_0(z), \ldots, a_{g-1}(z)$ are meromorphic operator-valued functions on G. As a consequence of Lemma 6 and Theorem 5 one obtains

Theorem 6. *Let $\mathfrak{P}$ be one of the algebroid functions assigned to the characteristic polynomial $\chi(\lambda, A(z))$ and let $\lambda(\zeta)$, $\zeta \in \mathfrak{P}$ denote the corresponding eigenvalue function. Then the corresponding algebroid eigenprojection-valued function $P(\zeta)$, defined on $\mathfrak{P}$, is uniquely* *represented in the form*

$$P(\zeta) = B_0(z) + B_1(z)\,\lambda(\zeta) + B_2(z)\,\lambda(\zeta)^2 + \cdots + B_{g-1}(z)\,\lambda(\zeta)^{g-1}, \tag{3.21}$$

where g is the valuedness of $\mathfrak{P}$ and where the coefficients $B_0(z), B_1(z), \ldots, B_{g-1}(z)$ are meromorphic operator-valued functions on G.

Note that this representation (3.21) will be proved directly in Sect. 5.3; from this representation the algebroid character of the splitting points with respect to $P(\zeta)$ can be derived.

The following corollary gives details concerning the position of the poles of the coefficients $B_0(z), B_1(z), \ldots, B_{g-1}(z)$

Corollary 3. *The operator-valued coefficients $B_0(z), B_1(z), \ldots, B_{g-1}(z)$ have poles at most at points that are multiple points or poles of $A(z)$; that is, if z_0 is simple, then all operator-valued coefficients are holomorphic at z_0.*

Proof. Let z_0 be simple. Let $\lambda_1(z), \lambda_2(z), \ldots, \lambda_g(z)$ denote the g different branches of $\mathfrak{P}$ over a neighborhood of z_0. Let $P_1(z), \ldots, P_g(z)$ denote the corresponding eigenprojection elements. Then,

$$P_\alpha(z) = B_0(z) + B_1(z)\,\lambda_\alpha(z) + B_2(z)\,\lambda_\alpha(z)^2 + \cdots + B_{g-1}(z)\,\lambda_\alpha(z)^{g-1},$$

$$\alpha = 1, 2, \ldots, g,$$

is obtained. These g equations can be used to determine $B_0(z), B_1(z), \ldots, B_{g-1}(z)$ uniquely by $P_\alpha(z), \lambda_\alpha(z), \alpha = 1, 2, \ldots, g$, according to Cramer's rule. One obtains

$$\det\left(\lambda_\alpha(z)^\sigma\right) = \prod_{\mu>\nu}\left(\lambda_\mu(z) - \lambda_\nu(z)\right),$$

according to Vandermonde's determinant theorem. This function is obviously holomorphic at $z = z_0$ and it does not vanish. Furthermore, $P_1(z), \ldots, P_g(z)$ are holomorphic at $z = z_0$. Hence, according to Cramer's rule, the coefficients $B_0(z)$ are holomorphic at $z = z_0$. ■

3.3.10. Eigennilpotent

Together with the eigenvalue $\lambda(\zeta)$ and the eigenprojection $P(\zeta)$, the eigennilpotent

$$D(\zeta) = \left(A(z) - \lambda(\zeta)\right) P(\zeta) \tag{3.22}$$

is also an algebroid function defined on $\mathfrak{P}$ (which is one of the analytic functions, corresponding to $\lambda(\zeta)$, assigned to the characteristic polynomial); z is the trace point of ζ.

Recall that $\lambda(\zeta)$ and $P(\zeta)$ have the same branching points and branching orders. This need not be true for $D(\zeta)$.

For example: If $A(z)$ does not have any permanent degeneration, then $A(z)$ is a diagonal operator for all simple points z. Hence, $D(\zeta) = 0$ for all ζ with simple trace points. As an immediate consequence, $D(\zeta) \equiv 0$, even though $A(z)$ may have splitting points and $A(z)$ is not necessarily also a diagonal operator for multiple points where nonvanishing eigennilpotents may appear. A concrete example may illustrate this situation:

$$A(z) = \begin{pmatrix} 0 & 1 \\ z & 0 \end{pmatrix}.$$

$z_0 = 0$ is multiple; $z \neq 0$ is simple. Eigenvalues are $\lambda_+(z) = z^{1/2}$, $\lambda_-(z) = -z^{1/2}$. There is no permanent degeneration. The eigenprojections take the form

$$P_+(z) = \begin{pmatrix} \frac{1}{2} & \frac{1}{2} z^{-1/2} \\ \frac{1}{2} z^{1/2} & \frac{1}{2} \end{pmatrix}, \quad P_-(z) = \begin{pmatrix} \frac{1}{2} & -\frac{1}{2} z^{-1/2} \\ -\frac{1}{2} z^{1/2} & \frac{1}{2} \end{pmatrix},$$

and one obtains

$$A(z) = \lambda_+(z) P_+(z) + \lambda_-(z) P_-(z), \qquad z \neq 0.$$

Hence, $D_+(z) = D_-(z) \equiv 0$, even though for $z_0 = 0$ the eigennilpotent $D(0) = A(0) = \begin{pmatrix} 0 & 1 \\ 0 & 0 \end{pmatrix}$ appears.

3.3.11. Holomorphic and polynomial operator-valued functions

If $A(z)$ is holomorphic in G, then S_A is empty (see Subsect. 3.2.4). Hence, the points of the eigenvalue functions $\mathfrak{P}_1, \mathfrak{P}_2, \ldots, \mathfrak{P}_r$ are either normal or splitting points. The eigenprojection-valued functions $P(\zeta_\varrho)$, $\zeta_\varrho \in \mathfrak{P}_\varrho$, are holomorphic at normal points and have at most branching poles at splitting points ζ_ϱ^0, where the branching order is given by ζ_ϱ^0.

If $A(z)$ is polynomial and if G is the closed z-plane, then $z = \infty$ is the only exceptional point of $A(z)$, the eigenvalue functions $\mathfrak{P}_1, \ldots, \mathfrak{P}_r$ are algebraic and poles can occur only over $z = \infty$. The eigenprojection-valued functions $P(\zeta_\varrho)$ are also algebraic; that is, also over $z = \infty$, they have at most an algebraic singularity. The only possible trace points of poles are $z = \infty$ and the finite multiple points. According to example (a) in Subsect. 3.3.7, the eigenprojections can be holomorphic at $z = \infty$ even though the corresponding eigenvalues have poles there.

3.3.12. Spectral representation of $A(z)$ in G

The result of Subsect. 3.3.5 concerning the spectral representation of $A(z)$ in the neighborhood of simple points, together with the results of the foregoing subsections, in particular Subsect. 3.3.9, can be summarized and supplemented as follows:

Let S_A denote the set of poles and let M denote the set of multiple points of $A(z)$. Then the spectral representation of $A(z)$ in $G \setminus (S_A \cup M)$ can be written in the form

$$A(z) = \sum_{\alpha=1}^{r} \sum_{\sigma=1}^{g_\alpha} \left(\lambda_{\alpha\sigma}(z) P_{\alpha\sigma}(z) + D_{\alpha\sigma}(z)\right), \qquad z \in G \setminus (S_A \cup M), \tag{3.23}$$

where analytic continuation in $G \setminus (S_A \cup M)$ leads at most to a permutation of the terms of the inner sum. The exceptional set $S_A \cup M$ is isolated and closed in G. There are Laurent-Puiseux series expansions in the neighborhood of every exceptional point for all spectral objects. The eigenprojections and eigennilpotents of the exceptional points need not coincide with possible limits of the spectral objects $P_{\alpha\sigma}(z)$, $D_{\alpha\sigma}(z)$, respectively. (This is always the case only for the eigenvalues.)

3.3.13. Spectral representation of $A(z)$ in the neighborhood of an arbitrary point

Next, the result of Subsect. 3.3.12 will be formulated locally.

Let $A(z) = A_0 + zA_1 + z^2A_2 + \cdots$ be holomorphic in $|z| < r$. Then there is $\delta > 0$ such that the dotted disc $0 < |z| < \delta$ consists of simple points only and the spectral representation of $A(z)$ in $|z| < \delta$ can be written in the form

$$A(z) = \sum_{\alpha=1}^{p} \sum_{\beta=1}^{s_\alpha} \sum_{\sigma=1}^{g_{\alpha\beta}} \left(\lambda_{\alpha\beta\sigma}(z) P_{\alpha\beta\sigma}(z) + D_{\alpha\beta\sigma}(z)\right), \qquad |z| < \delta, \tag{3.24}$$

where $A_0 = \sum_{\alpha=1}^{p} (\lambda_\alpha P_\alpha + D_\alpha)$ is the spectral representation of A_0. The subscript β enumerates the different cycles of the $(0, \lambda_\alpha)$-groups; the subscript σ the eigenvalues within a fixed cycle. Analytic continuation within $0 < |z| < \delta$ leads at most to a permutation of the terms of the inner sum. The perturbed eigenvalues and eigenprojections can be expanded into series of the form

$$\lambda_{\alpha\beta\sigma}(z) = \lambda_\alpha + \sum_{m=1}^{\infty} a_{\alpha\beta m} z^{m/g_{\alpha\beta}}, \qquad |z| < \delta, \tag{3.25}$$

$$P_{\alpha\beta\sigma}(z) = \sum_{m=-K_{\alpha\beta}} B_{\alpha\beta m} z^{m/g_{\alpha\beta}}, \qquad 0 < |z| < \delta. \tag{3.26}$$

The series (3.25) and (3.26) are called *perturbation series* of the analytic perturbation $A(z)$ of A_0. With the insights just obtained, new problems arise: for example, the problem of determining the splitting and the corresponding cycles of the eigenvalues λ_α; further, the problem of determining the coefficients of the perturbation series and their radius of convergence (estimation of δ). These problems are treated systematically in Ch. 8 and Ch. 9.

The operator $\sum_{\beta,\sigma} P_{\alpha\beta\sigma}(z) = P_\alpha(z)$ is the group-projection belonging to the $(0, \lambda_\alpha)$-group. It is holomorphic at $z = 0$.

Omitting the subscript α, one denotes by λ_0 one of the eigenvalues, by $\lambda_{\beta\sigma}(z)$ and $P_{\beta\sigma}(z)$ the perturbed eigenvalues and eigenprojections, respectively, and by $P(z)$ the corresponding group-projection.

From the eigenprojections $P_{\beta\sigma}(z)$, one obtains reducing projections of $A(z)$ by addition. One may ask for special properties of such reducing projections. For example, to every cycle β of the $(0, \lambda_0)$-group corresponds the projection

$$P_\beta(z) = \sum_{\alpha=1}^{g_\beta} P_{\beta\sigma}(z), \qquad \beta = 1, 2, \ldots, s.$$

$P_\beta(z)$ is a single-valued function in $0 < |z| < \delta$, because $P_\beta(z)$ is a symmetric function in $P_{\beta\sigma}(z)$.

Conversely, if a sum $Q(z) = \sum_{\alpha'} \sum_{\sigma'} P_{\alpha'\sigma'}(z)$ is single-valued in $0 < |z| < \delta$, then $Q(z) = \sum_{\alpha'} P_{\alpha'}(z)$. This can be proved by performing one positively oriented winding around $z = 0$. Then one obtains $Q(z) = \sum_{\alpha'} \sum_{\sigma'} P_{\alpha',\sigma'+1}(z)$, if the subscript σ is chosen as usual (cyclic permutation by positively oriented winding). Hence, the sum $\sum_{\sigma'}$ together with $P_{\alpha'\sigma'}$ also contains $P_{\alpha',\sigma'+1}$; thus, all projections of the cycle considered occur in the sum.

In this sense, the projections $P_\beta(z)$ are the "smallest" single-valued projections *corresponding to the* $(0, \lambda_0)$*-group*, because they are subprojections of arbitrary single-valued reducing projections formed by the eigenprojections $P_{\beta\sigma}$.

However, the projections $P_\beta(z)$ are *not necessarily* the "smallest" single-valued reducing subprojections of $P(z)$ with respect to $A(z)$. The minimality property of the $P_\beta(z)$ as single-valued projections is obtained here only because the eigenprojection $P_{\beta\sigma}(z)$ are used as the "atomic" projections (see in this connection the consideration in Sect. 4.6).

In general, the projections $P_\beta(z)$ have a pole at $z = 0$ (see Example (b) of Subsect. 3.3.7), but a holomorphic behavior at $z = 0$ is also possible (see Example (a) of Subsect. 3.3.7).

Thus the question arises whether there are single-valued reducing projections corresponding to the $(0, \lambda_0)$-group, which are holomorphic at $z = 0$; that is, one asks for sums $\sum_{\alpha'} P_{\alpha'}(z)$ that are holomorphic at $z = 0$.

There are other ways to formulate this question. On the one hand, one can ask for reducing projections, corresponding to the $(0, \lambda_0)$-group, which are bounded in $0 < |z| < \delta$ (see Theorem 4). On the other hand, one can ask for reducing subprojections Q of $P(0; 0, \lambda_0) = P(0)$ with respect to A_0, whichbe have holomorphically under the perturbation $A(z)$; that is, is there a reducing subprojection $Q(z)$ of $P(z)$ with respect to $A(z)$ corresponding to the $(0, \lambda_0)$-group, which is holomorphic at $z = 0$, such that $Q(0) = Q$? Such a projection is necessarily a certain sum of the $P_\alpha(z)$. In any case: The eigenvalues of a fixed cycle over $0 < |z| < \delta$ cannot be separated by a holomorphic decomposition of $P(z)$.

The questions outlined above will be treated systematically in Ch. 7 (relevant results are due to F. Wolf [1] and J. Butler [1]).

3.3.14. Case of a simple eigenvalue λ_0 of A_0

If, in particular, λ_0 is a *simple eigenvalue of* A_0, then $(0, \lambda_0)$ is normal, that is, the $(0, \lambda_0)$-group consists of a single eigenvalue $\lambda(z)$, which is holomorphic in $|z| < \delta$ (see Lemma 2.2). *The group projection* $P(z)$ *coincides with the eigenprojection of* $\lambda(z)$ (see Corollary 3.1) and one obtains $\dim P(z) = 1$ (see Lemma 3.4).

3.3.15. Examples (cf. Subsect. 3.2.10)

(a) $$A(z) = \begin{pmatrix} 1+z & 0 & 0 \\ 0 & 0 & 1 \\ 0 & z & 0 \end{pmatrix}, \qquad \lambda_1(z) = 1 + z; \qquad \lambda_{2,3}(z) = \pm z^{1/2}.$$

One has

$$R\left(\lambda, A(z)\right) = \begin{bmatrix} \dfrac{1}{\lambda - (1+z)} & 0 & 0 \\ 0 & \dfrac{\lambda}{\lambda^2 - z} & \dfrac{1}{\lambda^2 - z} \\ 0 & \dfrac{z}{\lambda^2 - z} & \dfrac{\lambda}{\lambda^2 - z} \end{bmatrix},$$

and obtains

$$P_1(z) = \operatorname*{Res}_{\lambda=1+z} R\left(\lambda, A(z)\right) = \begin{pmatrix} 1 & 0 & 0 \\ 0 & 0 & 0 \\ 0 & 0 & 0 \end{pmatrix};$$

$$P_2(z) = \operatorname*{Res}_{\lambda=z^{1/2}} R\left(\lambda, A(z)\right) = \begin{bmatrix} 0 & 0 & 0 \\ 0 & \dfrac{1}{2} & \dfrac{1}{2} z^{1/2} \\ 0 & \dfrac{1}{2} z^{-1/2} & \dfrac{1}{2} \end{bmatrix}.$$

$P_3(z)$ is obtained by replacing the root $z^{1/2}$ in $P_2(z)$ by $-z^{1/2}$. For $z \neq 0$, one has $D_1(z) = D_2(z) = D_3(z) = 0$. The eigennilpotent D_0 belonging to the eigenvalue $\lambda_0 = 0$ of

$A(0)$ equals $\begin{pmatrix} 0 & 0 & 0 \\ 0 & 0 & 1 \\ 0 & 0 & 0 \end{pmatrix}$.

(b) $$A(z) = \begin{pmatrix} 0 & z & 0 \\ -z & 2z & 0 \\ 0 & 0 & -z \end{pmatrix}, \qquad \lambda_1(z) = z, \qquad \lambda_2(z) = -z.$$

One has

$$R(\lambda, A(z)) = \begin{pmatrix} \frac{\lambda - 2z}{(\lambda - z)^2} & \frac{z}{(\lambda - z)^2} & 0 \\ \frac{z}{(\lambda - z)^2} & \frac{\lambda}{(\lambda - z)^2} & 0 \\ 0 & 0 & \frac{1}{\lambda + z} \end{pmatrix},$$

and obtains

$$P_1(z) = \operatorname*{Res}_{\lambda = z} R(\lambda, A(z)) = \begin{pmatrix} 1 & 0 & 0 \\ 0 & 1 & 0 \\ 0 & 0 & 0 \end{pmatrix};$$

$$P_2(z) = \operatorname*{Res}_{\lambda = -z} R(\lambda, A(z)) = \begin{pmatrix} 0 & 0 & 0 \\ 0 & 0 & 0 \\ 0 & 0 & 1 \end{pmatrix}.$$

Moreover,

$$D_1(z) = \begin{pmatrix} -z & z & 0 \\ -z & z & 0 \\ 0 & 0 & 0 \end{pmatrix}; \qquad D_2(z) = \begin{pmatrix} 0 & 0 & 0 \\ 0 & 0 & 0 \\ 0 & 0 & 0 \end{pmatrix}.$$

The eigenprojection of $\lambda_0 = 0$ with respect to $A(0)$ is $P_1(0) + P_2(0)$. The corresponding eigennilpotent is $D_1(0) = D_2(0)$.

(c) $$A(z) = \begin{pmatrix} 1 & z^3 - z^2 \\ -1 & 1 \end{pmatrix}, \qquad \lambda_\pm(z) = 1 \pm z(z - 1)^{1/2}.$$

One has $\chi(\lambda, A(z)) = \left(\lambda - 1 - z(z - 1)^{1/2}\right)\left(\lambda - 1 + z(z - 1)^{1/2}\right)$ and

$$R(\lambda, A(z)) = \begin{pmatrix} \frac{\lambda - 1}{\chi} & \frac{z^3 - z^2}{\chi} \\ -\frac{1}{\chi} & \frac{\lambda - 1}{\chi} \end{pmatrix}.$$

Hence,

$$P_+(z) = \operatorname*{Res}_{\lambda = 1 + z(z-1)^{1/2}} R(\lambda, A(z)) = \begin{pmatrix} \frac{1}{2} & \frac{1}{2} z(z - 1)^{1/2} \\ -\frac{1}{2} z^{-1}(z - 1)^{-1/2} & \frac{1}{2} \end{pmatrix}.$$

$P_-(z)$ is obtained from $P_+(z)$ by replacing the root $(z - 1)^{1/2}$ by $-(z - 1)^{1/2}$. Over $z = 1$ there is a splitting point $(1, 1)$, which is a branching point of order 2. Correspondingly, the projections $P_{\pm}(z)$ have a branching pole at $z = 1$ (see Subsect. 3.3.7). Over $z = 0$ there is a splitting point $(0, 1)$, which is not a branching point. Nevertheless, the projections $P_{\pm}(z)$ have a pole of first order at $z = 0$. One has $A(0) = A(1) = \begin{pmatrix} 1 & 0 \\ -1 & 0 \end{pmatrix}$, $D(0) = D(1) = \begin{pmatrix} 0 & 0 \\ -1 & 0 \end{pmatrix}$, but $D_{\pm}(z) \equiv 0$.

The calculation of the eigenprojections in the foregoing examples has been done by resolvent integration. That is, the integral (3.9) is considered and calculated by the calculus of residues.

In Ch. 5 a further method for the calculation of eigenprojections will be developed. This new method is more algebraically oriented. It starts from the decomposition of the characteristic polynomial into prime factors with respect to $\mathscr{M}_G[\lambda]$ and does not use the explicit knowledge of the roots of the characteristic polynomial.

3.4. Transformation of complete disjoint systems of holomorphic projections

In the foregoing Sect. 3.3, the spectral representation of meromorphic operator-valued functions was studied. The central point for this investigation was the determination of essential properties of the complete disjoint system of eigenprojections of $A(z)$. Eigenvectors and Jordan vectors of $A(z)$ were not considered. However, one is often interested in perturbed eigenvectors. In this section, properties of complete disjoint systems of holomorphic projections are studied, which give partial information about the behavior of perturbed eigen- and Jordan vectors. In some special cases (for example for normal and selfadjoint perturbations, see Subsect. 3.4.6 to 3.4.8), final results are obtained, if one applies the results derived in this section. These results are due to T. Kato [2] (for the following see, in particular, T. Kato [5, p. 98sqq.]).

3.4.1. Motivation

Let the starting point be the spectral representation

$$A(z) = \sum_{\alpha=1}^{r} \sum_{\sigma=1}^{g_\alpha} \big(\lambda_{\alpha\sigma}(z)\, P_{\alpha\sigma}(z) + D_{\alpha\sigma}(z)\big), \qquad z \in G \setminus (S_A \cup M), \tag{4.1}$$

of a meromorphic operator-valued function $A(z)$. $A(z)$ yields a complete disjoint system of projections, holomorphic in $G \setminus (S_A \cup M)$, namely, the eigenprojections of $A(z)$. One has

$$X = \sum_{\alpha,\sigma} \oplus\, P_{\alpha\sigma}(z)\, X .$$

One is interested in the function-theoretic properties of the eigen- and Jordan vectors in the algebraic eigenspace $P_{\alpha\sigma}(z)\, X$. First, one constructs bases of X, holomorphic in $G \setminus (S_A \cup M)$, which are adapted to the decomposition $1 = \sum_{\alpha,\sigma} P_{\alpha\sigma}(z)$ (see Subsect. 1.9.3). Secondly, at a more detailed level, one constructs holomorphic and adapted bases in X with respect to $A(z)$, i.e., holomorphic Jordan bases in $P_{\alpha\sigma}(z)\, X$ with respect to $A(z)$.

The question can be answered immediately if one restricts the consideration to special cases and the local case.

For example, let λ_0 be a simple eigenvalue of $A(z_0)$. Then, according to Subsect. 3.3.14, the perturbed eigenvalue $\lambda(z)$ and the corresponding eigenprojection $P(z)$ are holomorphic in some neighborhood $|z - z_0| < \delta$, where $\dim P(z) = 1$. In this case, the following is obviously true: let x_0 be an eigenvector of $A(z_0)$ with respect to λ_0, i.e., $A(z_0)\, x_0 = \lambda_0 x_0$, $x_0 \neq 0$. Then, $x(z) = P(z)\, x_0$ belongs to $P(z)\, X$ and one has $A(z)\, x(z) = \lambda(z)\, x(z)$. Moreover, $P(z)\, x_0 \neq 0$ in a sufficiently small neighborhood $|z - z_0| < \varepsilon$. Thus, the question posed above is answered for this case.

Of course, in general complications already appear in the local case. In order to discuss them, let $\lambda(z)$ be one of the eigenvalues $\lambda_{\alpha\sigma}(z)$, let $P(z)$ be the corresponding eigenprojection and assume $\dim P(z) = m > 1$, as well as $z_0 \in G \setminus (S_A \cup M)$. Let x_0 be an eigenvector of $A(z_0)$ belonging to $\lambda(z_0)$, i.e., $A(z_0)\, x_0 = \lambda(z_0)\, x_0$, $x_0 \neq 0$. Then $P(z)\, x_0 \in P(z)\, X$, that is, $x(z) = P(z)\, x_0$ is always a Jordan vector belonging to $\lambda(z)$, but $x(z)$ need not be an eigenvector for $z \neq z_0$; moreover, $x(z) = 0$ is also possible for some $z \neq z_0$.

To illustrate the situation, an example is considered:

$$A(z) = \begin{pmatrix} \dfrac{z^2}{1-z^2} & 1 & -\dfrac{z}{1-z^2} \\ -\dfrac{1}{1-z^2} & 2 & \dfrac{z}{1-z^2} \\ \dfrac{z}{1-z^2} & z & -\dfrac{1}{1-z^2} \end{pmatrix}, \qquad |z| < 1.$$

In particular,

$$A(0) = A_0 = \begin{pmatrix} 0 & 1 & 0 \\ -1 & 2 & 0 \\ 0 & 0 & -1 \end{pmatrix}.$$

One obtains $\chi(\lambda, A(z)) = (\lambda - 1)^2 (\lambda + 1)$, that is, the characteristic polynomial is independent of z. For the eigenprojection corresponding to $\lambda_1 = +1$, one calculates

$$P_1(z) = \begin{pmatrix} \dfrac{1}{1-z^2} & 0 & -\dfrac{z}{1-z^2} \\ 0 & 1 & 0 \\ \dfrac{z}{1-z^2} & 0 & -\dfrac{z^2}{1-z^2} \end{pmatrix}.$$

There is a permanent degeneration. Consider an eigenvector x_0 of A_0 belonging to 1,

$$x_0 = \begin{pmatrix} 1 \\ 1 \\ 0 \end{pmatrix}; \qquad A_0 x_0 = x_0,$$

and write:

$$x(z) = P(z)\, x_0 = \begin{pmatrix} \dfrac{1}{1-z^2} \\ 1 \\ \dfrac{z}{1-z^2} \end{pmatrix}.$$

Then,

$$A(z)\, x(z) = \begin{pmatrix} 1 \\ \dfrac{1-2z^2}{1-z^2} \\ z \end{pmatrix} \neq x(z),$$

that is, for any $z \neq 0$ the vector $x(z)$ is not an eigenvector of $A(z)$ any longer. However, $x(z)$ is contained in the algebraic eigenspace belonging to the eigenvalue 1, hence $x(z)$ is a Jordan vector, more precisely, it is a Jordan vector of degree 2, because of $\dim P_1(z) = 2$.

More generally, the situation can be described as follows: take a Jordan basis, $e_1, e_2, \ldots, e_m$ of $P(z_0)\, X$ with respect to $A(z_0)$. Consider the vectors $P(z)\, e_1, P(z)\, e_2, \ldots, P(z)\, e_m$. Then obviously, $P(z)\, e_i \in P(z)\, X$, i.e., the vectors $P(z)\, e_i$ are Jordan vectors with respect to $\lambda(z)$, but they do not necessarily form a basis of $P(z)\, X$ for $z \neq z_0$. If they do, then this basis need not be a Jordan basis.

Of course, if only the local case is considered, the question for holomorphic bases, adapted to the decomposition $1 = \sum P_{\alpha\sigma}(z)$, can be answered easily (but not the question for Jordan bases!).

Lemma 1. *Let $e_1, e_2, \ldots, e_m$ be a basis of $P(z_0)\, X$. Then there exists a $\delta > 0$ such that the vectors $P(z)\, e_1, P(z)\, e_2, \ldots, P(z)\, e_m$ form a (holomorphic) basis of $P(z)\, X$ for all z with $|z - z_0| < \delta$.*

Proof. Note that $P(z_0)\, x = x$ for all $x \in P(z_0)\, X$. If $x \neq 0$, $x \in P(z_0)\, X$, then $P(z_0)\, x = x \neq 0$. First the existence of a $\delta > 0$ is proved, such that, for all z with $|z - z_0| < \delta$ and for all x from $P(z_0)\, X$ with $x \neq 0$, the relation $P(z)\, x \neq 0$ holds. In the contrary case, there are sequences $z_\varrho \to z_0$ and $x_\varrho \neq 0$ from $P(z_0)\, X$ such that $P(z_\varrho)\, x_\varrho = 0$. Without loss of generality one can assume x_ϱ to be normalized, $\|x_\varrho\| = 1$. Then, according to Subsect. 1.12.2., there is a convergent subsequence, again denoted by x_ϱ, such that $x_\varrho \to x_0$ for some x_0. Thus one obtains $P(z_0)\, x_0 = 0$, where $\|x_0\| = 1$. This is a contradiction. The assertion of the lemma follows immediately. Choosing z from $|z - z_0| < \delta$, the relation $\sum_\varrho \lambda_\varrho P(z)\, e_\varrho = P(z) \sum_\varrho \lambda_\varrho e_\varrho = 0$ implies $\sum_\varrho \lambda_\varrho e_\varrho = 0$ i.e., $\lambda_1 = \lambda_2 = \cdots = \lambda_m = 0$. Therefore, the vectors $P(z)\, e_1, \ldots, P(z)\, e_m$ are linearly independent; hence they form a basis of $P(z)\, X$. ■

In the following subsections we will move away from local considerations and construct holomorphic bases, adapted to the decomposition $1 = \sum P_{\alpha\sigma}(z)$, for sufficiently large subregions of $G \setminus (S_A \cup M)$.

The construction of holomorphic Jordan bases will be dealt with in Sect. 5.7.

3.4.2. Transformation theorem

The following theorem, basic for the further considerations, is due to T. KATO [2].

Theorem 1 (Kato). *Let the projection-valued functions $P_1(z), P_2(z), \ldots, P_r(z)$ be holomorphic in the simply connected region G. Furthermore, for every $z \in G$, let the system $P_1(z), \ldots, P_r(z)$ be disjoint and complete, i.e., the relations*

$$\sum_{\varrho=1}^{r} P_\varrho(z) = 1, \qquad P_\varrho(z)\, P_\sigma(z) = \delta_{\varrho\sigma} P_\varrho(z), \qquad \varrho, \sigma = 1, \ldots, r, \tag{4.2}$$

are assumed to be satisfied. Then there is an operator-valued function $U(z, z_0)$, holomorphic in $G \times G$, $z, z_0 \in G$, such that

I. $\quad U(z, z_0)^{-1} = U(z_0, z)$,

II. $\quad U(z_0, z_0) = 1$,

III. $\quad U(z_3, z_2)\, U(z_2, z_1) = U(z_3, z_1), \qquad z_1, z_2, z_3 \in G$,

IV. $\quad P_\varrho(z) = U(z, z_0)\, P_\varrho(z_0)\, U(z, z_0)^{-1}, \qquad z, z_0 \in G, \qquad \varrho = 1, \ldots, r.$

Proof. First let $r = 2$. Let $P(z) := P_1(z)$. Then, $P_2(z) = 1 - P(z)$. Since $P^2 = P$,

$$P'P + PP' = P', \tag{4.3}$$

and, correspondingly,

$$(1 - P)'\,(1 - P) + (1 - P)\,(1 - P)' = (1 - P)'. \tag{4.4}$$

Because $P' + (1 - P)' = 0$,

$$P'P + PP' + (1 - P)'\,(1 - P) + (1 - P)\,(1 - P)' = 0. \tag{4.5}$$

Multiplying (4.3) by P to the right and to the left, one obtains

$$PP'P = 0. \tag{4.6}$$

Furthermore,

$$\begin{aligned} P'P - PP' &= P(1 - P)' - (1 - P)'\,P \\ &= (1 - P)'\,(1 - P) - (1 - P)\,(1 - P)' \end{aligned} \tag{4.7}$$

and

$$\begin{aligned} P'P + (1 - P)'\,(1 - P) &= P'P - P'(1 - P) \\ &= P'P - P' + P'P = P'P - PP'. \end{aligned} \tag{4.8}$$

Abbreviated,

$$[A, B]_- = AB - BA,$$

$$[A, B]_+ = AB + BA;$$

hence, from (4.5),

$$[P', P]_+ + [(1 - P)', 1 - P]_+ = 0, \tag{4.9}$$

and from (4.7),

$$[P', P]_- = [(1 - P)', 1 - P]_- . \tag{4.10}$$

Hence,

$$[P', P]_- = \frac{1}{2}\left([P', P]_- + [(1 - P)', 1 - P]_-\right). \tag{4.11}$$

Moreover, (4.8) implies

$$P'P + (1 - P)'(1 - P) = [P', P]_- . \tag{4.12}$$

Consider the operator differential equation

$$X'(z) = Q(z)\, X(z), \qquad X(z_0) = 1 \tag{4.13}$$

in G, where

$$Q(z) = [P'(z), P(z)]_- . \tag{4.14}$$

$Q(z)$ is holomorphic in G. According to Subsect. 1.16.2, the equation (4.13) has exactly one solution $U(z, z_0)$, which is holomorphic in $G \times G$ and satisfies the conditions

$$U(z_0, z_0) = 1, \tag{4.15}$$

and

$$U(z_3, z_2)\, U(z_2, z_1) = U(z_3, z_1). \tag{4.16}$$

In particular, from (4.16),

$$U(z, z_0)\, U(z_0, z) = 1 = U(z_0, z)\, U(z, z_0),$$

such that

$$U(z, z_0)^{-1} = U(z_0, z). \tag{4.17}$$

Thus, conditions I, II, III of Theorem 1 have been satisfied.

For the (uniquely determined) solution of the equation $Y' = QY$, with the initial condition $Y(z_0) = A$, the relation $Y(z) = U(z, z_0)\, A$ is obtained. From (4.12) the relation

$$P'P = QP \tag{4.18}$$

follows and the operator-valued function $P(z)\, U(z, z_0)$ satisfies the following differential equation with respect to z:

$$\begin{aligned}(PU)' &= P'U + PU' = (P' + PQ)\, U = (P' + PP'P - PP')\, U \\ &= (P' - PP')\, U = P'PU = QPU.\end{aligned}$$

Moreover,

$$P(z_0)\,U(z_0, z_0) = P(z_0).$$

Since the solution of the differential equation $Y' = QY$ with prescribed initial condition is uniquely determined,

$$P(z)\,U(z, z_0) = U(z, z_0)\,P(z), \tag{4.19}$$

which is property IV for $P(z)$. The same property follows for $1 - P(z)$.

If $r > 2$ then one writes (cf. (4.11))

$$Q(z) = \frac{1}{2}\sum_{\varrho=1}^{r} [P'_\varrho, P_\varrho]_- \tag{4.20}$$

and by analogy with (4.9) and (4.12), one obtains the relations

$$Q(z) = \sum_{\varrho=1}^{r} P'_\varrho P_\varrho = -\sum_{\varrho=1}^{r} P_\varrho P'_\varrho. \tag{4.21}$$

From (4.21),

$$Q(z)\,P_\sigma = \sum_\varrho P'_\varrho P_\varrho P_\sigma = P'_\sigma P_\sigma$$

and

$$P_\sigma Q(z) = -\sum_\varrho P_\sigma P_\varrho P'_\varrho = -P_\sigma P'_\sigma;$$

hence,

$$QP_\sigma - P_\sigma Q = P'_\sigma P_\sigma + P_\sigma P'_\sigma = P'_\sigma. \tag{4.22}$$

Again, consider the operator differential equation (4.13) and its uniquely determined solution $U(z, z_0)$, which, (according to Subsect. 1.16.2), is holomorphic in $G \times G$ and has the properties (4.15), (4.16) and (4.17).

If z_0 is fixed, one obtains the following differential equation for $X_\varrho(z) = P_\varrho(z)\,U(z, z_0)$:

$$X'_\varrho = P'_\varrho U + P_\varrho U' = P'_\varrho U + P_\varrho QU = (P'_\varrho + P_\varrho Q)\,U$$

and, because of (4.22),

$$(P_\varrho U)' = Q(P_\varrho U);$$

that is, $P_\varrho(z)\,U(z, z_0)$ is a solution of equation (4.13). The initial condition in this case is $X_\varrho(z_0) = P_\varrho(z_0)\,U(z_0, z_0) = P_\varrho(z_0)$. Therefore,

$$U(z, z_0)\,P_\varrho(z_0) = P_\varrho(z)\,U(z, z_0), \qquad \varrho = 1, \ldots, r,$$

and this is property IV. ∎

The construction of a single-valued holomorphic operator-valued function $U(z, z_0)$ in $G \times G$, given in the foregoing proof, is based on the solution of the operator differential equation (4.13) (see Subsect. 1.16.2). Therefore, this construction is possible because the basic region G is simply connected. If G is multiply connected, then the construction of $U(z, z_0)$ is still possible (cf. formula (1.16.2)), but $U(z, z_0)$ is multiple-valued (even infinitely many values are possible). In Sect. 4.7, the problem of transforming complete disjoint holomorphic systems of projection-valued functions, i.e., the problem of constructing an operator-valued function $U(z, z_0)$ with property IV of Theorem 1, will be treated once more but by means of another method. In this second approach the result obtained in this section can be supplemented. For example, it will be possible to omit the assumption of the basic region G being simply connected but, nevertheless, to obtain single-valued transformations in G. In general, these transformations are merely meromorphic in G. With the use of Theorem 1, the construction of holomorphic adapted bases (see Subsect. 3.4.1) is straightforward.

Lemma 2. *Let G be simply connected and let $P_1(z), \ldots, P_r(z)$ be a holomorphic complete disjoint system of projection-valued functions in G. Let* $\dim P_\varrho(z) = m_\varrho$. *If $x_1^\varrho, x_2^\varrho, \ldots, x_m^\varrho$ denotes a basis of $P_\varrho(z_0) X$, $z_0 \in G$, then the vectors $x_\sigma^\varrho(z) = U(z, z_0)\, x_\sigma^\varrho$, $\sigma = 1, 2, \ldots, m_\varrho$, form a basis of $P_\varrho(z) X$ for all $z \in G$. The vector-valued functions $x_\sigma^\varrho(z)$, $\sigma = 1, \ldots, m_\varrho$, are holomorphic in G.*

Proof. The equation

$$P_\varrho(z)\, x_\sigma^\varrho(z) = P_\varrho(z)\, U(z, z_0)\, x_\sigma^\varrho = U(z, z_0)\, P_\varrho(z_0)\, x_\sigma^\varrho = U(z, z_0)\, x_\sigma^\varrho = x_\sigma^\varrho(z)$$

holds, that is, the vectors $x_1^\varrho(z), \ldots, x_{m_\varrho}^\varrho(z)$ belong to $P_\varrho(z) X$ for all $z \in G$. Moreover, they are linearly independent, since it follows from

$$\sum \lambda_i x_i^\varrho(z) = \sum \lambda_i U(z, z_0)\, x_i^\varrho = U(z, z_0) \sum \lambda_i x_i^\varrho = 0$$

that $\sum \lambda_i x_i^\varrho = 0$; hence, $\lambda_1 = \lambda_2 = \cdots = \lambda_{m_\varrho} = 0$. Obviously, the functions $x_\sigma^\varrho(z)$ are holomorphic in G. ■

3.4.3. Transformation of the eigenprojections of $A(z)$

The system of the eigenprojection of $A(z)$ in $G \setminus (S_A \cup M)$ is denoted by $\{P_{\alpha\sigma}(z)\}$, according to (4.1). Let H be a simply connected subregion of $G \setminus (S_A \cup M)$. Then the transformation theorem (Theorem 1), hence Lemma 2, is applicable.

Let $z_0 \in H$ be fixed, let $\dim P_{\alpha\sigma}(z) = m_\alpha$ and let $e_1^{\alpha,\sigma}, \ldots, e_{m_\alpha}^{\alpha,\sigma}$ be a basis of $P_{\alpha\sigma}(z_0) X$. Then, according to Lemma 2, the vector-valued functions $x_\varrho^{\alpha\sigma}(z) = U(z, z_0)\, e_\varrho^{\alpha,\sigma}$ form holomorphic bases with respect to $P_{\alpha\sigma}(z) X$ if $U(z, z_0)$ is chosen according to Theorem 1 with respect to the system $\{P_{\alpha\sigma}(z)\}$.

With the use of the transformation operator $U(z, z_0)$, the operator-valued function $A(z)$ can be transformed, such that its algebraic eigenspaces are constant. In perturbation theory such transformations are frequently used (see for example F. Wolf [1]; T. Kato [5, p. 104]). Let

$$B(z) = U(z, z_0)^{-1} A(z)\, U(z, z_0), \qquad z \in H. \tag{4.23}$$

Then $B(z_0) = A(z_0)$ and one obtains

$$\begin{aligned} B(z)\, P_{\alpha\sigma}(z_0) &= U(z, z_0)^{-1} A(z)\, U(z, z_0)\, P_{\alpha\sigma}(z_0) = U(z, z_0)^{-1} A(z)\, P_{\alpha\sigma}(z)\, U(z, z_0) \\ &= U(z, z_0)^{-1} P_{\alpha\sigma}(z)\, A(z)\, U(z, z_0) = P_{\alpha\sigma}(z_0)\, U(z, z_0)^{-1} A(z)\, U(z, z_0) \\ &= P_{\alpha\sigma}(z_0)\, B(z), \end{aligned}$$

that is,

$$B(z)\, P_{\alpha\sigma}(z_0) = P_{\alpha\sigma}(z_0)\, B(z), \qquad z \in H. \tag{4.24}$$

Hence the constant projections $P_{\alpha\sigma}(z_0)$ reduce the operator-valued function $B(z)$. Moreover, it follows from

$$P_{\alpha\sigma}(z_0) = U(z, z_0)^{-1} P_{\alpha\sigma}(z)\, U(z, z_0) \tag{4.25}$$

that the $P_{\alpha\sigma}(z_0)$ are the eigenprojections of $B(z)$. As before, the corresponding eigenvalues are the numbers $\lambda_{\alpha\sigma}(z)$. Thus, the spectral representation of $B(z)$ takes the form

$$B(z) = \sum_{\alpha,\sigma} \big(\lambda_{\alpha\sigma}(z)\, P_{\alpha\sigma}(z_0) + \tilde{D}_{\alpha\sigma}(z)\big), \qquad z \in H, \tag{4.26}$$

where $\tilde{D}_{\alpha\sigma}(z) = \big(B(z) - \lambda_{\alpha\sigma}(z)\big)\, P_{\alpha\sigma}(z_0)$. As noted before, the algebraic eigenspaces $P_{\alpha\sigma}(z_0)\, X$ of $B(z)$ are independent of z.

If z varies in H only, the study of $A(z)$ in $P_{\alpha\sigma}(z)\, X$ is reduced to the study of $B(z)$ in $P_{\alpha\sigma}(z_0)\, X$. There is only one eigenvalue $\lambda_{\alpha\sigma}(z)$, corresponding to $B(z) \restriction P_{\alpha\sigma}(z_0)\, X$, which is holomorphic in H; hence $D(z) = \big(B(z) - \lambda_{\alpha\sigma}(z)\big) \restriction P_{\alpha\sigma}(z_0)\, X$ is nilpotent and holomorphic in H. For every $z \in H$, one can decompose $D(z)$ into direct sums of Jordan operators (see Lemma 2.9.4). Now one can ask for special decompositions equipped with simple function-theoretic properties. Such special decompositions will be studied in Ch. 5. We will prove that one can in fact choose the whole region G as the basic region. It is not necessary to restrict the consideration to simply connected subregions.

3.4.4. Transformation of the group projections of $A(z)$

Let z_0 be multiple and assume that $0 < |z - z_0| < \delta$ consists of simple points only. Note that the region $0 < |z - z_0| < \delta$ is not simply connected. Hence Theorem 1 and Lemma 2 cannot be used to study the eigenprojections $P_{\alpha\sigma}(z)$ in $0 < |z - z_0| < \delta$, unless multiple-valued transformation operators are allowed.

However, Theorem 1 and Lemma 2 can be used to transform the group projections of every (z_0, λ_ϱ)-group of $A(z)$.

For this purpose, let $A(z) = \sum_{\varrho=1}^{r} (\lambda_\varrho P_\varrho + D_\varrho)$ be the spectral representation of $A(z_0)$ and let $\dim P_\varrho = m_\varrho$. The group projections of the (z_0, λ_1)-, (z_0, λ_2)-, ..., (z_0, λ_r)-group are denoted by $P_1(z), P_2(z), \ldots, P_r(z)$, respectively. They form a holomorphic complete disjoint system in $|z - z_0| < \delta$. The set of all eigenvalues of the (z_0, λ_ϱ)-group at the point z is denoted by $K_\varrho(z)$.

According to Theorem 1 there is an operator-valued function $U(z) = U(z, z_0)$, holomorphic in $|z - z_0| < \delta$, with the properties noted there, in particular,

$$P_\varrho(z) = U(z)\, P_\varrho U(z)^{-1}, \qquad |z - z_0| < \delta. \tag{4.27}$$

If $x_1^\varrho, \ldots, x_{m_\varrho}^\varrho$ denotes a basis of $P_\varrho(z_0)\, X = P_\varrho X$, then Lemma 2 shows that the vector-valued functions $x_\sigma^\varrho(z) = U(z)\, x_\sigma^\varrho$, which are holomorphic in $|z - z_0| < \delta$, form bases in the algebraic group eigenspace $P_\varrho(z)\, X$.

In this case, too, the perturbation $A(z)$ of $A(z_0)$ can be transformed by $U(z)$, through the introduction of the operator-valued function $B(z)$,

$$B(z) = U(z)^{-1}\, A(z)\, U(z), \qquad |z - z_0| < \delta. \tag{4.28}$$

Hence, $B(z_0) = A(z_0)$ and from (4.27) one obtains

$$B(z)\, P_\varrho = P_\varrho B(z), \qquad \varrho = 1, \ldots, r. \tag{4.29}$$

Therefore, the transformed perturbation has the constant reducing projections P_ϱ. These projections, in general, are *not* eigenprojections of $B(z)$ for $z \neq z_0$. Since (4.28) and (4.29) hold, the equation $\operatorname{spec}\big(A(z) \restriction P_\varrho(z)\, X\big) = \operatorname{spec}\big(B(z) \restriction P_\varrho X\big)$ follows and, because $\operatorname{spec}\big(A(z) \restriction P_\varrho(z)\, X\big) = K_\varrho(z)$ (see property VI of Lemma 3.4), one obtains

$$\operatorname{spec}\big(B(z) \restriction P_\varrho X\big) = K_\varrho(z), \qquad \varrho = 1, \ldots, r, \qquad |z - z_0| < \delta; \tag{4.30}$$

that is, the eigenvalues of $B(z)$ in $P_\varrho X$ form exactly the (z_0, λ_ϱ)-group of $A(z)$ and

$$\chi\big(\lambda, A(z)\big) = \prod_{\varrho=1}^{r} \chi\big(\lambda, B(z) \restriction P_\varrho X\big), \qquad |z - z_0| < \delta. \tag{4.31}$$

$\chi\big(\lambda, B(z) \restriction P_\varrho X\big)$ is a polynomial of degree m_ϱ and the coefficients of this polynomial are holomorphic in $|z - z_0| < \delta$. Moreover, $\chi\big(\lambda, B(z_0) \restriction P_\varrho X\big) = (\lambda - \lambda_\varrho)^{m_\varrho}$, hence $\chi\big(\lambda, B(z) \restriction P_\varrho X\big) = q_\varrho(\lambda - \lambda_\varrho, z)$, where $q_\varrho(\mu, z)$ is a polynomial from $\mathscr{H}^{z_0}_{|z-z_0|<\delta}[\mu]$; that is, q_ϱ is exactly that factor of $\chi\big(\lambda, A(z)\big)$ that belongs to the eigenvalue λ_ϱ, if the polynomial $\chi\big(\lambda, A(z)\big)$ is related to the point z_0. Therefore in this case Theorem 1 leads to a method to produce the decomposition of $\chi\big(\lambda, A(z)\big)$ over $|z - z_0| < \delta$, according to Theorem 2.5 (Weierstrass preparation theorem for polynomials $p(\lambda, z)$ in λ).

In particular, if one of the eigenvalues λ_ϱ of $A(z_0)$ is simple and if $\lambda_\varrho(z)$ denotes the unique eigenvalue of the (z_0, λ_ϱ)-group in this case, then $\dim P_\varrho(z) = 1$. If $x_\varrho \neq 0$ is an eigenvector belonging to λ_ϱ, i.e., $A(z_0)\, x_\varrho = \lambda_\varrho x_\varrho$, then $x_\varrho(z) = U(z)\, x_\varrho$ is an eigenvector belonging to $\lambda_\varrho(z)$, which is holomorphic in $|z - z_0| < \delta$, i.e., $A(z)\, x_\varrho(z) = \lambda_\varrho(z)\, x_\varrho(z)$.

More specifically, if all eigenvalues $\lambda_1, \lambda_2, \ldots, \lambda_N$ are simple and if one denotes the corresponding perturbed eigenvalues and eigenprojections by $\lambda_1(z), \ldots, \lambda_N(z)$ and $P_1(z), \ldots, P_N(z)$, respectively, then the spectral representation of $A(z)$ takes the form

$$A(z) = \sum_{\varrho=1}^{N} \lambda_\varrho(z)\, P_\varrho(z), \qquad |z - z_0| < \delta;$$

that is, $A(z)$ is a diagonal operator. The vector-valued functions $x_\varrho(z) = U(z)\, x_\varrho$, which are holomorphic in $|z - z_0| < \delta$, form a basis of eigenvectors of $A(z)$ in $|z - z_0| < \delta$; that is, the equations

$$A(z)\, x_\varrho(z) = \lambda_\varrho(z)\, x_\varrho(z), \qquad \varrho = 1, \ldots, r, \qquad |z - z_0| < \delta, \tag{4.32}$$

hold.

3.5. Normal and selfadjoint perturbations

The results of Sect. 3.3 on the spectral representation of meromorphic operator-valued functions have been summarized in Subsect. 3.3.12 and 3.3.13. As mentioned in Subsect. 3.3.13, these results raise new problems (e.g., determining the periods of cycles of eigenvalue groups). The importance of these results is first of all given by the fact that they present a qualitative framework for more special investigations. In this section, the results of Sect. 3.3 are sharpened by means of additional assumptions like normality and selfadjointness. We will now concentrate on local considerations.

The treatment of selfadjoint perturbations of selfadjoint operators is one of the first problems which was posed and solved within the framework of perturbation theory (see Subsect. 3.1.2). The present section shows how this treatment proceeds if it can be based on the general results of Sect. 3.3. One of the main results of this paragraph is the proof of the famous theorems of F. Rellich which justify the postulate of E. Schrödinger mentioned in Subsect. 3.1.2. (See E. Schrödinger [1, 2]. Concerning the results of F. Rellich, see F. Rellich [1—6], in particular [1, pp. 602, 610]; see also T. Kato [5, p. 71 and p. 120sqq.]. Note that the results of F. Rellich are related to selfadjoint perturbations in Hilbert space.)

3.5.1. Normal perturbations

In the following discussion the basic space $X = H$ is assumed to be unitary (see Sect. 1.4). Let $A(z)$ be a holomorphic perturbation of $A_0 = A(0)$ defined in a disc $|z| < r$ (see Subsect. 3.1.3).

Definition 1. The holomorphic perturbation $A(z)$ is said to be *normal*, if there is a sequence $z_n \to 0$ such that $A(z_n)$, $n = 1, 2, 3, \ldots$, is a normal operator (see Subsect. 1.10.2).

If the perturbation $A(z)$ is normal, then $A(0) = A_0$ is also normal. Moreover, there is a sequence $z_n \to 0$ of simple points such that $A(z_n)$ is normal (note that there is always a dotted disc $0 < |z| < \delta$ consisting of simple points only).

For normal perturbations the spectral representation of the perturbed operator is much simpler than in the general case; the perturbed spectral objects are holomorphic at $z = 0$. One has

Theorem 1 (Rellich). *Let $A(z)$ be a normal perturbation; let the spectral representation of A_0 be given by $A_0 = \sum_{\alpha=1}^{r} \lambda_\alpha P_\alpha$; the spectral representation of $A(z)$ takes the form*

$$A(z) = \sum_{\alpha=1}^{r} \sum_{\varrho=1}^{s_\alpha} \sum_{\sigma=1}^{g_{\alpha\varrho}} \left(\lambda_{\alpha\varrho\sigma}(z)\, P_{\alpha\varrho\sigma}(z) + D_{\alpha\varrho\sigma}(z)\right)$$

in the dotted disc $0 < |z| < \delta$ which consists of simple points only (see Subsect. 3.3.13). *Then*

I. $g_{\alpha\varrho} = 1$, $\alpha = 1, \ldots, r$; $\varrho = 1, \ldots, s_\alpha$; *that is, the periods of cycles in all eigenvalue groups are equal to* 1; *i.e., every* $(0, \lambda_\alpha)$*-group consists of* s_α *different eigenvalues*

$$\lambda_{\alpha\varrho}(z) = \lambda_\alpha + z\lambda_{\alpha\varrho}^{(1)} + z^2\lambda_{\alpha\varrho}^{(2)} + \cdots, \qquad |z| < \delta, \tag{5.1}$$

which are holomorphic at $z = 0$.

II. *The eigenprojections* $P_{\alpha\varrho}(z)$ *assigned to* $\lambda_{\alpha\varrho}(z)$ *are also holomorphic at* $z = 0$. *Therefore there are expansions of the form*

$$P_{\alpha\varrho}(z) = P_{\alpha\varrho} + zP_{\alpha\varrho}^{(1)} + z^2P_{\alpha\varrho}^{(2)} + \cdots, \qquad |z| < \delta. \tag{5.2}$$

III. *The corresponding eigennilpotents* $D_{\alpha\varrho}(z)$ *vanish identically,* $D_{\alpha\varrho}(z) \equiv 0$. *Therefore, the spectral representation of* $A(z)$ *reduces to*

$$A(z) = \sum_{\alpha=1}^{r} \sum_{\varrho=1}^{s_\alpha} \lambda_{\alpha\varrho}(z)\, P_{\alpha\varrho}(z), \qquad 0 < |z| < \delta. \tag{5.3}$$

Proof. Let $z_n \to 0$ be a sequence of simple points such that $A(z_n)$ is normal. Then $\|P_{\alpha\varrho\sigma}(z_n)\| = 1$ for all $n = 1, 2, 3, \ldots$ (see Lemma 2.7.1 and Subsect. 1.10.2). This implies that the eigenvalue $\lambda_{\alpha\varrho\sigma}(z)$ cannot have a branching point at $z = 0$, since otherwise $P_{\alpha\varrho\sigma}(z)$ would have a (branching) pole at $z = 0$, according to Theorem 3.4, but this, obviously, cannot be true. Therefore, $\lambda_{\alpha\varrho\sigma}(z)$ is holomorphic at $z = 0$, i.e., $g_{\alpha\varrho} = 1$ (period of the corresponding cycle) and the subscript σ can be omitted. The corresponding eigenprojection $P_{\alpha\varrho}(z)$ is single-valued in $0 < |z| < \delta$. The former argument can be repeated: Since $\|P_{\alpha\varrho}(z_n)\| = 1$, the point $z = 0$ cannot be a pole of $P_{\alpha\varrho}(z)$; hence $P_{\alpha\varrho}(z)$ is holomorphic at $z = 0$. This implies that the eigennilpotents $D_{\alpha\varrho}(z)$ are also holomorphic at $z = 0$. Since $A(z_n)$ is normal in any case, $D_{\alpha\varrho}(z_n) = 0$; hence, $D_{\alpha\varrho}(z) \equiv 0$. ■

For $z \to 0$, it follows from (5.3) that

$$A_0 = \sum_{\alpha=1}^{r} \lambda_\alpha \sum_{\varrho=1}^{s_\alpha} P_{\alpha\varrho}.$$

Since the spectral representation is uniquely determined, this implies

$$P_\alpha = \sum_{\varrho=1}^{s_\alpha} P_{\alpha\varrho}, \qquad \alpha = 1, 2, \ldots, r. \tag{5.4}$$

Of course, Theorem 1 has a global aspect, too. One has

Corollary 1. *Let* $A(z)$ *be meromorphic in* G *and let* $z_0 \in G \setminus S_A$. *Assume* $A(z)$ *to be a normal perturbation in the neighborhood of* z_0. *The spectral representation of* $A(z)$ *for simple points is denoted by* $A(z) = \sum_{\alpha=1}^{r} \sum_{\sigma=1}^{g_\alpha} \big(\lambda_{\alpha\sigma}(z)\, P_{\alpha\sigma}(z) + D_{\alpha\sigma}(z)\big)$ (*see Subsect.* 3.3.12). *Then the functions* $\lambda_{\alpha\sigma}(z), P_{\alpha\sigma}(z)$ *are holomorphic at* $z = z_0$ *and* $D_{\alpha\sigma}(z)$ *vanishes identically; therefore, the spectral representation reduces to* $A(z) = \sum_{\alpha=1}^{r} \sum_{\sigma=1}^{g_\alpha} \lambda_{\alpha\sigma}(z)\, P_{\alpha\sigma}(z)$; *i.e.,* $A(z)$ *is a diagonal operator for all simple points* z.

Proof. The propositions concerning the neighborhood of z_0 are implied by Theorem 1. If $D_{\alpha\sigma}(z) \equiv 0$ in a neighborhood of z_0, then $D_{\alpha\sigma}(z) \equiv 0$ in the whole region $G \setminus (S_A \cup M)$. ■

This result can be compared with Subsect. 3.2.2 (see also Subsect. 3.4.4). In that section it was proved that $A(z)$ is diagonal for all simple points if there is no permanent degeneration. In the present result this assumption is dropped. That is, a permanent degeneration of $A(z)$ may exist in the present case.

According to Theorem 1, the property that $A(z)$ is normal has consequences for the function-theoretic behavior of eigenvalues and eigenprojections. As before, $z = 0$ may be multiple, i.e., splittings of eigenvalues are possible. Splitting criteria must therefore be established; e.g. for the zero-order approximations $P_{\alpha\varrho}$ of the perturbed eigenprojections. These questions will be treated systematically in Ch. 7.

Theorem 1 leads to the question of the existence of a holomorphic basis of eigenvectors. This question can be answered with the results of Sect. 3.4.

Theorem 2 (Rellich). *Let $A(z)$ be a normal perturbation in $|z| < r$ with the perturbed eigenvalues and eigenprojections $\lambda_{\alpha\sigma}(z)$ and $P_{\alpha\sigma}(z)$, respectively (see Theorem 1). Let* $\dim P_{\alpha\sigma}(z) = m_{\alpha\sigma}$ *and assume that the dotted disc $0 < |z| < \delta$ consists of simple points only. Then there is a holomorphic basis of eigenvectors of $A(z)$ in $|z| < \delta$, i.e., a system $x_{\alpha\varrho\sigma}(z)$ of holomorphic vector-valued functions in $|z| < \delta$, which form a basis in H for every z, where*

$$A(z)\, x_{\alpha\varrho\sigma}(z) = \lambda_{\alpha\varrho}(z)\, x_{\alpha\varrho\sigma}(z), \qquad |z| < \delta, \qquad \sigma = 1, 2, \dots, m_{\alpha\varrho}. \tag{5.5}$$

Moreover, the vectors $x_{\alpha\varrho\sigma}(z)$, $\sigma = 1, \dots, m_{\alpha\varrho}$ form a holomorphic basis of $P_{\alpha\varrho}(z)\, X$.

Proof. The system $\{P_{\alpha\varrho}(z)\}$ is disjoint, complete and holomorphic in $|z| < \delta$. According to Theorem 4.1, there is an operator-valued function $U(z) = U(z, 0)$, holomorphic in $|z| < \delta$ such that its inverse $U(z)^{-1}$ is also holomorphic in $|z| < \delta$ and the relation

$$P_{\alpha\varrho}(z) = U(z)\, P_{\alpha\varrho} U(z)^{-1}, \qquad |z| < \delta, \qquad \alpha = 1, \dots, r; \qquad \sigma = 1, \dots, s_\alpha, \tag{5.6}$$

holds. With a basis $x_{\alpha\varrho\sigma}$, $\sigma = 1, 2, \dots, m_{\alpha\varrho}$, in $P_{\alpha\varrho}X$, according to Lemma 4.2, the vector-valued functions $x_{\alpha\varrho\sigma}(z) = U(z)\, x_{\alpha\varrho\sigma}$ form a holomorphic basis of $P_{\alpha\varrho}(z)\, X$. Therefore (5.5) is valid. ∎

It should be emphasized that the holomorphic eigenvectors $x_{\alpha\varrho\sigma}(z)$ are not uniquely determined, not even if the basis $x_{\alpha\varrho\sigma}$ at $z = 0$ is fixed, since the transformation operator $U(z)$ is not uniquely determined by the equation (5.6). Together with $U(z)$, $\alpha(z)\, U(z)$ also satisfies these equations, for example, where $\alpha(z)$ denotes a holomorphic non-vanishing scalar function in $|z| < \delta$. On the other hand, the eigenvalues and eigenprojections are uniquely determined, as is well-known.

In order to fix the eigenvectors uniquely, one needs some normalization condition (see, e.g., Subsect. 3.1.2). Later, such conditions will play a role (see Ch. 7).

3.5.2. Transformation of normal perturbations

This section is based on Subsect. 3.4.3. We start with a normal perturbation $A(z)$, holomorphic in $|z| < \delta$, with the representation

$$A(z) = \sum_{\alpha,\varrho} \lambda_{\alpha\varrho}(z)\, P_{\alpha\varrho}(z), \qquad |z| < \delta, \tag{5.7}$$

which is the spectral representation for $0 < |z| < \delta$. Using the operator-valued function $U(z)$ from the proof of Theorem 2 and defining the transformation

$$B(z) = U(z)^{-1} A(z) U(z), \qquad |z| < \delta, \tag{5.8}$$

one obtains, because of (5.6), the representation

$$B(z) = \sum_{\alpha,\varrho} \lambda_{\alpha\varrho}(z) P_{\alpha\varrho}, \qquad |z| < \delta. \tag{5.9}$$

One has $B(0) = B_0 = A_0 = A(0)$ and

$$B(z_1) B(z_2) = B(z_2) B(z_1), \qquad |z_1| < \delta, \qquad |z_2| < \delta; \tag{5.10}$$

in particular

$$B(z) A_0 = A_0 B(z), \qquad |z| < \delta. \tag{5.11}$$

Therefore, the perturbation $B(z)$ "consists" of mutually commuting operators; in particular, $B(z)$ commutes with the unperturbed operator.

This method of transforming given perturbations into "mutually commuting" perturbations was first applied by K. Friedrichs in the perturbation theory of continuous spectra (see K. Friedrichs [1, 2]; see also T. Kato [5, p. 55sqq.]). Later this Friedrichs method was also used in the perturbation theory of discrete spectra (see, e.g., R. E. L. Turner [1]).

3.5.3. Selfadjoint perturbations

Selfadjoint perturbations are special cases of normal perturbations. Let $A(z)$ be a holomorphic perturbation of $A_0 = A(0)$, defined in $|z| < r$.

Definition 2. The holomorphic perturbation $A(z)$ is said to be *selfadjoint* if $A(\varepsilon) = A(\varepsilon)^*$ holds for all real $z = \varepsilon$, $|\varepsilon| < r$, i.e., if $A(\varepsilon)$ is a selfadjoint operator.

The holomorphic perturbation $A(z) = A_0 + zA_1 + z^2A_2 + \cdots$ in $|z| < r$ is selfadjoint if and only if all operator-valued coefficients $A_0, A_1, A_2, \ldots$ are selfadjoint.

For a selfadjoint perturbation, the relation $A(z)^* = A(\bar{z})$ holds for all z with $|z| < r$, because this equation is valid for all real $z = \varepsilon$, hence everywhere in $|z| < r$.

Let G be a region of the complex plane symmetric with respect to the real axis (this is the case if, together with z, $\bar{z}$ also belongs to G). Furthermore, let the operator-valued function $A(z)$ be meromorphic in G. Then $A(z)$ is said to be *selfadjoint* if $A(z)^* = A(\bar{z})$ in G. If the symmetric region G contains a real interval and if $A(\varepsilon)^* = A(\varepsilon)$ is valid for all real $z = \varepsilon \in G$, then $A(z)$ is selfadjoint in G.

Obviously, a selfadjoint perturbation $A(z)$ in $|z| < r$ is normal. Furthermore, a selfadjoint meromorphic operator-valued function $A(z)$ in G is a normal perturbation with respect to a neighborhood of any point $z_0 \in (G \setminus S_A) \cap \mathbb{R}$. Therefore the results of Subsect. 3.5.1 may be transferred to this situation.

Theorem 3 (Rellich). *Let $A(z)$ be selfadjoint in G and let*

$$A(z) = \sum_{\alpha=1}^{r} \sum_{\sigma=1}^{g_\alpha} \left(\lambda_{\alpha\sigma}(z) P_{\alpha\sigma}(z) + D_{\alpha\sigma}(z)\right)$$

be the spectral representation of $A(z)$ at simple points. Then: the eigenvalue functions $\lambda_{\alpha\sigma}(z)$ are holomorphic and real-valued on $(G \setminus S_A) \cap \mathbb{R}$, the eigenprojections $P_{\alpha\sigma}(z)$ are holomorphic on $(G \setminus S_A) \cap \mathbb{R}$ and selfadjoint in G, i.e., the relation $P_{\alpha\sigma}(z)^* = P_{\alpha\sigma}(\bar{z})$ holds in G; moreover, the eigennilpotents vanish identically, $D_{\alpha\sigma}(z) \equiv 0$. Therefore, the spectral representation takes the form $A(z) = \sum_{\alpha=1}^{r} \sum_{\sigma=1}^{g_\alpha} \lambda_{\alpha\sigma}(z) P_{\alpha\sigma}(z)$ and it is holomorphic on $(G \setminus S_A) \cap \mathbb{R}$.

Proof. The assertion is obvious from Corollary 1 and the remarks preceding this theorem. ■

In particular, one obtains the following result for selfadjoint perturbations.

Theorem 4 (Rellich). *Let $A(z)$ be a holomorphic perturbation in $|z| < r$ and assume that $0 < |z| < \delta$ consists of simple points only. Then $A(z)$ can be represented by*

$$A(z) = \sum_{\alpha=1}^{r} \sum_{\varrho=1}^{s_\alpha} \lambda_{\alpha\varrho}(z) P_{\alpha\varrho}(z), \qquad |z| < \delta. \tag{5.12}$$

This representation is the spectral representation in $0 < |z| < \delta$; the functions $\lambda_{\alpha\varrho}(z)$ and $P_{\alpha\varrho}(z)$ are holomorphic in $|z| < \delta$; the functions $\lambda_{\alpha\varrho}(\varepsilon)$ are real-valued for real $z = \varepsilon$; and the projections $P_{\alpha\varrho}(\varepsilon)$ are orthogonal. One obtains power series expansions of the form

$$\lambda_{\alpha\varrho}(z) = \lambda_\alpha + z\lambda_{\alpha\varrho}^{(1)} + z^2\lambda_{\alpha\varrho}^{(2)} + \cdots, \qquad |z| < \delta, \tag{5.13}$$

$$P_{\alpha\varrho}(z) = P_{\alpha\varrho} + zP_{\alpha\varrho}^{(1)} + z^2P_{\alpha\varrho}^{(2)} + \cdots, \qquad |z| < \delta, \tag{5.14}$$

where $A_0 = \sum_{\alpha=1}^{r} \lambda_\alpha \left(\sum_{\varrho=1}^{s_\alpha} P_{\alpha\varrho}\right) = \sum_{\alpha=1}^{r} \lambda_\alpha P_\alpha$ is the spectral representation of A_0.

3.5.4. Transformation of selfadjoint systems of projections

According to Theorem 3 and Theorem 4, the eigenprojections of selfadjoint operator-valued functions $A(z)$ form complete, disjoint, holomorphic and *selfadjoint* systems of projections. Therefore, in the present subsection the considerations of Sect. 3.4 concerning transformation of systems of projections are supplemented with respect to the selfadjoint case. Let G be simply connected and symmetric with respect to the real axis. Then the following result holds.

Lemma 1. *Let the complete disjoint system of projections $P_1(z), P_2(z), \ldots, P_r(z)$ be holomorphic and selfadjoint in G; i.e., let*

$$P_\varrho(z)^* = P_\varrho(\bar{z}), \qquad z \in G, \qquad \varrho = 1, \ldots, r. \tag{5.15}$$

Furthermore, let $U(z, z_0)$ be the transformation operator of the system constructed according to Theorem 4.1. Then

$$U(\bar{z}, \bar{z}_0)^* = U(z, z_0)^{-1}, \qquad z, z_0 \in G. \tag{5.16}$$

In particular, if $z = \varepsilon$ and $z_0 = \varepsilon_0$ are real and from G then

$$U(\varepsilon, \varepsilon_0)^* = U(\varepsilon, \varepsilon_0)^{-1}; \tag{5.17}$$

that is, the transformation operator $U(\varepsilon, \varepsilon_0)$ is unitary for real arguments.

Proof. $U(z, z_0)$ is a solution of the operator differential equation $X'(z) = Q(z)\,X(z)$, $X(z_0) = 1$ (see (4.13)). From (5.15),

$$[P_\varrho'(z)]^* = P_\varrho'(\bar{z});$$

hence,

$$Q(z)^* = \left(\sum_{\varrho=1}^{r} P'(z)\,P_\varrho(z)\right)^* = \sum_{\varrho=1}^{r} P_\varrho(z)^*\,[P'(z)]^* = \sum_{\varrho=1}^{r} P_\varrho(\bar{z})\,P_\varrho'(\bar{z});$$

that is,

$$Q(z)^* = -Q(\bar{z}), \qquad z \in G. \tag{5.18}$$

Sometimes $Q(z)$ is called ***antisymmetric***. If $\dfrac{\partial}{\partial z}\,U(z, z_0) = Q(z)\,U(z, z_0)$, then $\left(\dfrac{\partial}{\partial z}\,U(z, z_0)\right)^* = U(z, z_0)^*\,Q(z)^*$, and, if z is replaced by $\bar{z}$,

$$\frac{\partial}{\partial z}\,[U(\bar{z}, z_0)]^* = \left(\frac{\partial}{\partial \bar{z}}\,U(\bar{z}, z_0)\right)^* = U(\bar{z}, z_0)^*\,Q(\bar{z})^*. \tag{5.19}$$

Therefore,

$$\frac{\partial}{\partial z}\,[U(\bar{z}, z_0)]^* = -U(\bar{z}, z_0)^*\,Q(z). \tag{5.20}$$

Moreover, if $\bar{z} = z_0$, i.e., $z = \bar{z}_0$, then the operator-valued function $U(\bar{z}, z_0)$ is equal to 1. Hence the operator-valued function

$$V(z, z_0) = U(\bar{z}, \bar{z}_0)^*$$

is a solution of the operator differential equation

$$Y'(z) = -Y(z)\,Q(z), \qquad Y(z_0) = 1. \tag{5.21}$$

According to Subsect. 1.16.2, the function $U(z, z_0)^{-1} = U(z_0, z)$ is a solution of one and the same operator differential equation. Since the solution of the initial value problem is uniquely determined, (5.16) follows, hence also (5.17). ■

3.5.5. Perturbed eigenvectors of a selfadjoint perturbation

The system $\{P_{\alpha\varrho}(z)\}$ of eigenprojections introduced in Theorem 4 is selfadjoint in $|z| < \delta$. Therefore, from Lemma 1 one obtains the following insight into the behavior of the perturbed eigenvectors of a selfadjoint perturbation, which is a strengthening of Theorem 5.1 and Theorem 5.2.

Theorem 5 (Rellich). *Let the assumptions of Theorem 4 be satisfied. Let* $\dim P_{\alpha\varrho} = m_{\alpha\varrho}$. *Then there exists a holomorphic basis* $x_{\alpha\varrho\sigma}(z)$, $\sigma = 1, 2, \ldots, m_{\alpha\varrho}$, *of eigenvectors of* $A(z)$ *in* $|z| < \delta$, *that is, a system of holomorphic vector-valued functions in* $|z| < \delta$, *which form a basis in* H *for every* z *with* $|z| < \delta$ *such that*

$$A(z)\, x_{\alpha\varrho\sigma}(z) = \lambda_{\alpha\varrho}(z)\, x_{\alpha\varrho\sigma}(z), \qquad |z| < \delta, \qquad \sigma = 1, \ldots, m_{\alpha\varrho}, \tag{5.22}$$

and such that the basis $x_{\alpha\varrho\sigma}(\varepsilon)$ *is orthonormal for all real* $z = \varepsilon$, *i.e.,*

$$\big(x_{\alpha\varrho\sigma}(\varepsilon), x_{\alpha'\varrho'\sigma'}(\varepsilon)\big) = \delta_{\alpha\alpha'} \cdot \delta_{\varrho\varrho'} \cdot \delta_{\sigma\sigma'}, \qquad |\varepsilon| < \delta. \tag{5.23}$$

In particular, the vectors $x_{\alpha\varrho\sigma} = x_{\alpha\varrho\sigma}(0)$, $\varrho = 1, \ldots, s_\alpha$, $\sigma = 1, \ldots, m_{\alpha\varrho}$ *form an orthonormal basis of the eigenspace* $P_\alpha H$ *with respect to the unperturbed eigenvalue* λ_α *of* A_0.

Proof. Assume that the vectors $e_{\alpha\varrho 1}, e_{\alpha\varrho 2}, \ldots, e_{\alpha\varrho m_{\alpha\varrho}}$ form an orthonormal basis of $P_{\alpha\varrho}H$. Then the system $x_{\alpha\varrho\sigma}(z) = U(z, 0)\, e_{\alpha\varrho\sigma}$ satisfies all required conditions of the theorem (see Theorem 5.2). In particular, since $U(\varepsilon, 0)$ is unitary for real ε (see Lemma 1), the system $x_{\alpha\varrho\sigma}(\varepsilon)$ is orthonormal. From $x_{\alpha\varrho\sigma}(0) = e_{\alpha\varrho\sigma}$ and $P_\alpha = \sum_{\varrho=1}^{s_\alpha} P_{\alpha\varrho}$, the last part of the assertion follows. ■

Therefore, the perturbed eigenvectors can be expanded in the form

$$x_{\alpha\varrho\sigma}(z) = e_{\alpha\varrho\sigma} + z e^{(1)}_{\alpha\varrho\sigma} + z^2 e^{(2)}_{\alpha\varrho\sigma} + \cdots, \qquad |z| < \delta, \tag{5.24}$$

where the condition of orthonormality is satisfied for all real $z = \varepsilon$. Additionally, for selfadjoint holomorphic operator-valued functions $A(z)$ in G one obtains the following result:

Corollary 2. *Let* $A(z)$ *be holomorphic and selfadjoint in* G. *Then the eigenvalues* $\lambda_{\alpha\sigma}(z)$ *are real-valued and holomorphic on* $G \cap \mathbb{R}$ *and the eigenprojections are selfadjoint and holomorphic on* $G \cap \mathbb{R}$ *(Theorem 3). If* $[\alpha, \beta]$ *denotes a subinterval of* $G \cap \mathbb{R}$ *and if* U *is simply connected, where* $[\alpha, \beta] \subset U \subset G$, *and such that the* $\lambda_{\alpha\sigma}(z)$ *and* $P_{\alpha\sigma}(z)$ *are holomorphic in* U, *then there is a holomorphic basis* $\{x_{\alpha\varrho\sigma}(z)\}$ *of eigenvectors of* $A(z)$ *in* H, *defined in* U, *which is orthonormal for all real* $z = \varepsilon \in [\alpha, \beta]$.

Proof. Apply Lemma 1 to the system $\{P_{\alpha\varrho}(z)\}$ of eigenprojections in U. ■

Corollary 2 means that to a holomorphic and selfadjoint operator-valued function defined on an interval $[\alpha, \beta]$, there always corresponds a holomorphic orthonormal system of eigenvectors defined on $[\alpha, \beta]$, where the eigenvalues are also holomorphic (and obviously real-valued) on $[\alpha, \beta]$.

This result and also the other results obtained in this section represent remarkable results of the analytic perturbation theory. They are due to F. Rellich and mark the beginning of the development of this theory. These results, in particular Theorems 4 and 5, justify the assumptions of Subsect. 3.1.2 (postulates of E. Schrödinger) concerning the dependence of eigenvalues and eigenvectors of selfadjoint perturbations on the perturbation parameter.

The result of Corollary 2, that a holomorphic selfadjoint operator given on an interval $[\alpha, \beta]$ can be transformed holomorphically into diagonal form (or into the main axes form, as it is sometimes said), is also remarkable, because it is false if the assumption of holomorphy is replaced by a weaker condition, e.g., by arbitrary differentiability. In this connection, the following example is instructive (see T. Kato [5, p. 111]; it is a modification of an example given by F. Rellich [1, p. 666]). Let $H = \mathbb{C}^2$ and

$$A(\varepsilon) = e^{-1/\varepsilon^2} \begin{pmatrix} \cos \frac{2}{\varepsilon} & \sin \frac{2}{\varepsilon} \\ \sin \frac{2}{\varepsilon} & -\cos \frac{2}{\varepsilon} \end{pmatrix}, \qquad \varepsilon \neq 0; \qquad A(0) = 0.$$

$A(\varepsilon)$ is selfadjoint for real ε. The eigenvalues are $\lambda_\pm(\varepsilon) = e^{-1/\varepsilon^2}$ for $\varepsilon \neq 0$ and $\lambda_\pm(0) = 0$. The calculation of the eigenprojections $P_\pm(\varepsilon)$ yields

$$P_+(\varepsilon) = \begin{pmatrix} \cos^2 \frac{1}{\varepsilon} & \cos \frac{1}{\varepsilon} \sin \frac{1}{\varepsilon} \\ \cos \frac{1}{\varepsilon} \sin \frac{1}{\varepsilon} & \sin^2 \frac{1}{\varepsilon} \end{pmatrix},$$

$$P_-(\varepsilon) = \begin{pmatrix} \sin^2 \frac{1}{\varepsilon} & -\cos \frac{1}{\varepsilon} \sin \frac{1}{\varepsilon} \\ -\cos \frac{1}{\varepsilon} \sin \frac{1}{\varepsilon} & \cos^2 \frac{1}{\varepsilon} \end{pmatrix}, \qquad \varepsilon \neq 0.$$

Therefore, the eigenprojections are not continuous at $\varepsilon = 0$.

Of course, the result concerning the possibility of holomorphic transformation into the main axes form for holomorphic selfadjoint operators can also be formulated in matrix form:

Corollary 3. *Let $A(\varepsilon)$ be a holomorphic and selfadjoint $N \times N$-matrix, defined on the interval $[\alpha, \beta]$. The eigenvalues of $A(\varepsilon)$ are denoted by $\lambda_1(\varepsilon), \lambda_2(\varepsilon), \ldots, \lambda_N(\varepsilon)$, where their multiplicity is taken into account. These eigenvalues are holomorphic on $[\alpha, \beta]$. Then there exists a unitary matrix $U(\varepsilon)$, holomorphic on $[\alpha, \beta]$ such that*

$$U^{-1}(\varepsilon)\, A(\varepsilon)\, U(\varepsilon) = \{\lambda_1(\varepsilon), \lambda_2(\varepsilon), \ldots, \lambda_N(\varepsilon)\}; \tag{5.25}$$

i.e., $A(\varepsilon)$ can be transformed into diagonal form unitarily and holomorphically.

The transformation problem of holomorphic $N \times N$-matrices into the Jordan normal form will be considered in Ch. 5.

3.5.6. Branching for selfadjoint perturbations

Of course, for non-real arguments, eigenvalue and eigenprojection branching may also exist in the case of selfadjoint perturbations. A simple example is given by

$$A(z) = \begin{pmatrix} 2 & iz \\ -iz & 0 \end{pmatrix}.$$

$A(z)$ is defined in the whole complex plane and it is holomorphic and selfadjoint. Calculation of the eigenvalues yields $\lambda_\pm(z) = 1 \pm (1 + z^2)^{1/2}$. These eigenvalues are obviously holomorphic on the whole real axis, and so are the eigenprojections, but there are branching points at $z = \pm i$.

3.6. Perturbations of product type with selfadjoint factors

In applications nonnormal perturbations of type $C(\varepsilon) = B(\varepsilon)^{-1} A(\varepsilon)$ sometimes appear, where $A(\varepsilon)$ and $B(\varepsilon)$ are selfadjoint perturbations (see the Introduction). Perturbations of this type $C(\varepsilon)$ behave like normal or selfadjoint perturbations under certain additional assumptions (see T. Kato [5, p. 429sqq]).

3.6.1. Definition and simple properties

Let H be unitary and let $A(z)$ and $B(z)$ be holomorphic and selfadjoint perturbations in $|z| < r$. Furthermore, let $B(z)$ be invertible in $|z| < r$. Then $B(z)^{-1}$ is also a holomorphic and selfadjoint perturbation. Let $B(0) = B_0$ be positive (see Subsect. 1.10.1). If the eigenvalues of B_0 are denoted by $\mu_1, \ldots, \mu_N$, then $\min\limits_{\varrho} \mu_\varrho > 0$. According to Sect. 3.5, the eigenvalues of $B(\varepsilon)$ are holomorphic, in particular continuous, at $\varepsilon = 0$. Therefore, there are constants $\delta_1 > 0$, $\alpha > 0$, $\beta > 0$ such that

$$\alpha \|x\|^2 \leqq \big(B(\varepsilon)\, x, x\big) \leqq \beta \|x\|^2, \qquad |\varepsilon| < \delta_1, \qquad x \in H, \tag{6.1}$$

(see formula (2.7.10)). Hence $B(\varepsilon)$ is positive for $|\varepsilon| < \delta_1$. The operator-valued function

$$C(z) = B(z)^{-1} A(z), \qquad |z| < r,$$

is a holomorphic perturbation of $C_0 = B_0^{-1} A_0$. Let $\delta_2 > 0$ be chosen such that $0 < |z| < \delta_2$ consists of simple points only. Let $\delta = \min(\delta_1, \delta_2)$. In the following, $|z| < \delta$ is the basic region.

Let ε be real and $0 < |\varepsilon| < \delta$. $\lambda(\varepsilon)$ denotes one of the perturbed eigenvalues of $C(\varepsilon)$, $P(\varepsilon)$ the corresponding eigenprojection. Some simple properties of $\lambda(\varepsilon)$ and $P(\varepsilon)$ are noted:

I. *$\lambda(\varepsilon)$ is real.* To prove this property, let $x(\varepsilon) \neq 0$ be a corresponding eigenvector. Then $C(\varepsilon)\, x(\varepsilon) = \lambda(\varepsilon)\, x(\varepsilon)$ or, equivalently, $A(\varepsilon)\, x(\varepsilon) = \lambda(\varepsilon)\, B(\varepsilon)\, x(\varepsilon)$. Hence, $\big(A(\varepsilon)\, x(\varepsilon), x(\varepsilon)\big) = \lambda(\varepsilon) \big(B(\varepsilon)\, x(\varepsilon), x(\varepsilon)\big)$ and, because $\big(B(\varepsilon)\, x(\varepsilon), x(\varepsilon)\big) > 0$ (see (6.1))

$$\lambda(\varepsilon) = \frac{\big(A(\varepsilon)\, x(\varepsilon), x(\varepsilon)\big)}{\big(B(\varepsilon)\, x(\varepsilon), x(\varepsilon)\big)}.$$

II. *The relation* $P(\varepsilon)^* = B(\varepsilon)\,P(\varepsilon)\,B(\varepsilon)^{-1}$ *is valid.* To prove this relation, let $\lambda \in \operatorname{res} C(\varepsilon)$. Then $\big(\lambda B(\varepsilon) - A(\varepsilon)\big)^{-1}$ exists because

$$\big(\lambda - C(\varepsilon)\big)^{-1} = \big(\lambda - B(\varepsilon)^{-1} A(\varepsilon)\big)^{-1} = \big(\lambda B(\varepsilon) - A(\varepsilon)\big)^{-1} B(\varepsilon). \tag{6.2}$$

Integrating along a positively oriented contour around $\lambda(\varepsilon)$ one obtains from (6.2)

$$P(\varepsilon) = \frac{1}{2\pi i} \int \big(\lambda B(\varepsilon) - A(\varepsilon)\big)^{-1} \,d\lambda\, B(\varepsilon), \tag{6.3}$$

where the pseudoprojection

$$S(\varepsilon) = \frac{1}{2\pi i} \int \big(\lambda B(\varepsilon) - A(\varepsilon)\big)^{-1} \,d\lambda$$

is selfadjoint, $S(\varepsilon)^* = S(\varepsilon)$. (For the concept pseudoprojection, see M. V. Pattabhiraman and P. Lancaster [1].) Therefore, one has $P(\varepsilon) = S(\varepsilon)\,B(\varepsilon)$ and $P(\varepsilon)^* = B(\varepsilon)\,S(\varepsilon)$. Since $S(\varepsilon) = P(\varepsilon)\,B(\varepsilon)^{-1}$, the assertion follows.

One introduces a new scalar product:

$$[x, y]_\varepsilon = \big(B(\varepsilon)\,x, y\big), \qquad x, y \in H, \tag{6.4}$$

with the corresponding norm

$$[x]_\varepsilon^2 = \big(B(\varepsilon)\,x, x\big), \qquad x \in H \tag{6.5}$$

(note Subsect. 1.10.1). From (6.1) follows

$$\sqrt{\alpha}\,\|x\| \leqq [x]_\varepsilon \leqq \sqrt{\beta}\,\|x\|, \qquad x \in H. \tag{6.6}$$

III. *The eigenprojection* $P(\varepsilon)$ *is selfadjoint with respect to* $[x, y]_\varepsilon$. This property is obvious, because

$$\begin{aligned}[P(\varepsilon)\,x, y]_\varepsilon &= \big(B(\varepsilon)\,P(\varepsilon)\,x, y\big) = \big(P(\varepsilon)^*\,B(\varepsilon)\,x, y\big)\\ &= \big(B(\varepsilon)\,x, P(\varepsilon)\,y\big) = [x, P(\varepsilon)\,y]_\varepsilon.\end{aligned} \tag{6.7}$$

The properties noted above lead to an estimate of $\|P(\varepsilon)\|$ which is independent of ε.

Lemma 1. *For all real* ε, $0 < |\varepsilon| < \delta$, *the estimate*

$$1 \leqq \|P(\varepsilon)\| \leqq (\beta/\alpha)^{1/2} \tag{6.8}$$

holds.

Proof. According to Subsect. 1.8.1 and 1.10.2,

$$[P(\varepsilon)]_\varepsilon = \sup_{x \neq 0} \frac{[P(\varepsilon)\,x]_\varepsilon}{[x]_\varepsilon} = 1.$$

From (6.6) one obtains the estimate

$$\sqrt{\frac{\alpha}{\beta}}\,\frac{[P(\varepsilon)\,x]_\varepsilon}{[x]_\varepsilon} \leqq \frac{\|P(\varepsilon)\,x\|}{\|x\|} \leqq \sqrt{\frac{\beta}{\alpha}}\,\frac{[P(\varepsilon)\,x]_\varepsilon}{[x]_\varepsilon}, \qquad 0 \neq x \in H. \tag{6.9}$$

From this estimate follows

$$\sqrt{\frac{\alpha}{\beta}} \leqq \|P(\varepsilon)\| \leqq \sqrt{\frac{\beta}{\alpha}}$$

and because $\|P(\varepsilon)\| \geqq 1$, assertion (6.8) holds. ∎

IV. *The operator $C(\varepsilon)$ is selfadjoint with respect to $[x, y]_\varepsilon$.* This property is obvious, because

$$\begin{aligned}[C(\varepsilon)\,x, y]_\varepsilon &= \big(B(\varepsilon)\,C(\varepsilon)\,x, y\big) = \big(A(\varepsilon)\,x, y\big) = \big(x, A(\varepsilon)\,y\big)\\ &= \big(x, B(\varepsilon)\,B(\varepsilon)^{-1}\,A(\varepsilon)\,y\big) = \big(B(\varepsilon)\,x, C(\varepsilon)\,y\big) = [x, C(\varepsilon)\,y]_\varepsilon.\end{aligned}$$

3.6.2. Perturbation theorem

The properties of the perturbation $C(z)$ mentioned in Subsect. 3.6.1 are sufficient to prove some propositions concerning the behavior of the perturbed eigenvalues, eigenprojections and eigenvectors.

Theorem 1. *Let $A(z)$ and $B(z)$ be holomorphic and selfadjoint in $|z| < r$. Let $B(z)$ be invertible in $|z| < r$, i.e., $B(z)^{-1}$ is also holomorphic and selfadjoint in $|z| < r$. Furthermore, let $B(0) = B_0$ be positive. Let $\delta > 0$ be chosen according to Subsect. 3.6.1. Then the holomorphic perturbation $C(z) = B(z)^{-1}\,A(z)$ can be represented in the form*

$$C(z) = \sum_{\alpha=1}^{r} \sum_{\varrho=1}^{s_\alpha} \lambda_{\alpha\varrho}(z)\,P_{\alpha\varrho}(z), \qquad |z| < \delta, \tag{6.10}$$

which is its spectral representation for $0 < |z| < \delta$. The perturbed eigenvalues $\lambda_{\alpha\varrho}(z)$ and eigenprojections $P_{\alpha\varrho}(z)$ are holomorphic in $|z| < \delta$; i.e., they have power series expansions of the form

$$\lambda_{\alpha\varrho}(z) = \lambda_\alpha + z\lambda_{\alpha\varrho}^{(1)} + z^2\lambda_{\alpha\varrho}^{(2)} + \cdots, \qquad |z| < \delta, \tag{6.11}$$

$$P_{\alpha\varrho}(z) = P_{\alpha\varrho} + zP_{\alpha\varrho}^{(1)} + z^2P_{\alpha\varrho}^{(2)} + \cdots, \qquad |z| < \delta. \tag{6.12}$$

The spectral representation of C_0 is given by $C_0 = \sum_{\alpha=1}^{r} \lambda_\alpha P_\alpha$, where $P_\alpha = \sum_{\varrho=1}^{s_\alpha} P_{\alpha\varrho}$.

To every perturbed eigenvalue $\lambda_{\alpha\varrho}(z)$ correspond exactly $m_{\alpha\varrho} = \dim P_{\alpha\varrho}$ linearly independent eigenvectors $x_{\alpha\varrho\sigma}(z)$ which are holomorphic in $|z| < \delta$.

Proof. Let $P(z)$ be one of the perturbed eigenprojections of $C(z)$, the corresponding eigenvalue is denoted by $\lambda(z)$. According to Lemma 1, the estimate $\|P(\varepsilon)\| \leqq (\beta/\alpha)^{1/2}$ holds. Therefore, the eigenvalue $\lambda(z)$ cannot possess a branching point at $z = 0$. Otherwise, according to Theorem 3.4, the function $P(z)$ should possess a branching pole at $z = 0$ leading to $\|P(z)\| \to \infty$ as $z \to 0$ and this is contradictory to our estimate. Therefore, $\lambda(z)$ is holomorphic at $z = 0$. Hence, $P(z)$ is single-valued in $0 < |z| < \delta$ and, again because $z = 0$ cannot be a pole for $P(z)$, also holomorphic at $z = 0$. Now the perturbed eigenvalues and eigenprojections, holomorphic in $|z| < \delta$, are denoted by $\lambda_{\alpha\varrho}(z)$, $P_{\alpha\varrho}(z)$ as usual (corresponding to different $(0, \lambda_\alpha)$-groups). The system $\{P_{\alpha\varrho}(z)\}$ is complete, disjoint and holomorphic in $|z| < \delta$. According to Lemma 4.2, there is a basis $x_{\alpha\varrho\sigma}(z)$ of H, holomorphic in $|z| < \delta$ and adapted to the decomposition $1 = \sum_{\alpha,\varrho} P_{\alpha\varrho}(z)$. According to IV, for real ε with $0 < |\varepsilon| < \delta$, the algebraic eigenspace $P_{\alpha\varrho}(\varepsilon)\, H$ consists of eigenvectors only (see Subsect. 2.7.2); i.e., the relation

$$C(\varepsilon)\, x_{\alpha\varrho\sigma}(\varepsilon) = \lambda_{\alpha\varrho}(\varepsilon)\, x_{\alpha\varrho\sigma}(\varepsilon), \qquad \sigma = 1, 2, \ldots, m_{\alpha\varrho},$$

is valid for all real ε with $|\varepsilon| < \delta$. Then these equations must hold in the whole region $|z| < \delta$, that is, $D_{\alpha\varrho}(z) \equiv 0$ and the system $\{x_{\alpha\varrho\sigma}(z)\}$ is a holomorphic basis of eigenvectors of $C(z)$ in $|z| < \delta$. ■

In general, the perturbed eigenvectors $x_{\alpha\varrho\sigma}(z)$ are not orthogonal for different eigenvalues $\lambda_{\alpha\varrho}(z)$ not even for real $z = \varepsilon$, as can be seen in the following example. Let

$$A(z) = A_0 + zA_1, \qquad B(z)^{-1} = 1 + zB_1,$$

where

$$A_0 = \begin{pmatrix} 0 & 0 \\ 0 & 0 \end{pmatrix}, \qquad A_1 = \begin{pmatrix} 0 & 1 \\ 1 & 0 \end{pmatrix}, \qquad B_1 = \begin{pmatrix} 0 & 0 \\ 0 & 1 \end{pmatrix}$$

and $H = \mathbb{C}^2$ with $(x, y) = x_1\bar{y}_1 + x_2\bar{y}_2$. Then

$$C(z) = (1 + zB_1)\,(A_0 + zA_1) = \begin{pmatrix} 0 & z \\ z + z^2 & 0 \end{pmatrix},$$

and the eigenvalues are $\lambda_\pm(z) = \pm z(1 + z)^{1/2}$. Eigenvectors are given, for example, by

$$x_\pm(z) = \begin{pmatrix} 1 \\ \pm(1 + z)^{1/2} \end{pmatrix},$$

and, for real $z = \varepsilon \neq 0$,

$$\big(x_+(\varepsilon), x_-(\varepsilon)\big) = 1 - (1 + \varepsilon) = -\varepsilon \neq 0.$$

3.7. Perturbation series

In this section the coefficients of the perturbation series (Taylor series) are determined for the group projection $P(z) = P(z; 0, \lambda_0)$ of a $(0, \lambda_0)$-group introduced in Subsect. 3.3.3 (see Lemma 3.4). Moreover, the calculation of the coefficients for the perturbation series of the mean value of the $(0, \lambda_0)$-group is added (see Subsect. 3.7.1; this section relies heavily on T. Kato [5, p. 74sqq.]).

3.7.1. Group projection and mean value

Let $A(z)$ be a holomorphic perturbation of A_0 in $|z| < r$ and let λ_0 be an eigenvalue of A_0. The group projection of the $(0, \lambda_0)$-group, which is holomorphic in a neighborhood $|z| < \delta$, is denoted by $P(z)$.

According to Corollary 3.3.1, the group projection $P(z)$ for normal points $(0, \lambda_0)$ is the eigenprojection of the (holomorphic) perturbed eigenvalue $\lambda(z)$. In particular, this is true if λ_0 is a simple eigenvalue of A_0 (see Subsect. 3.2.8 and 3.3.14).

If $(0, \lambda_0)$ is normal, then the relation $\lambda(z) = (1/n) \operatorname{tr} A(z) P(z)$ holds, where n denotes the algebraic multiplicity of λ_0 and, hence, of $\lambda(z)$. According to Subsect. 2.1.1, one obtains $\chi(\lambda; A(z) \upharpoonright P(z) X) = (\lambda - \lambda(z))^n = \lambda^n - n\lambda(z) \lambda^{n-1} \pm \cdots = \lambda^n - \operatorname{tr}(A(z) P(z)) \times \lambda^{n-1} \pm \cdots$ and the relation mentioned above follows by comparing the coefficients. Since the trace is additive, the relation $\lambda(z) = \alpha + (1/n) \operatorname{tr}(A(z) - \alpha) P(z)$ also holds; in particular,

$$\lambda(z) = \lambda_0 + \frac{1}{n} \operatorname{tr}(A(z) - \lambda_0) P(z). \tag{7.1}$$

If λ_0 is simple, then $n = 1$ in formula (7.1).

If $(0, \lambda_0)$ is not normal, then $\operatorname{tr}(A(z) P(z)) = \sum_{\varrho=1}^{r} m_\varrho \lambda_\varrho(z)$, where $\lambda_1(z), \ldots, \lambda_r(z)$ denote the perturbed eigenvalues and $m_1, \ldots, m_r$ their algebraic multiplicities, respectively. This follows immediately from

$$\chi(\lambda A(z) \upharpoonright P(z) X) = \prod_{\varrho=1}^{r} (\lambda - \lambda_\varrho(z))^{m_\varrho} = \lambda^n - \left(\sum_{\varrho=1}^{r} m_\varrho \lambda_\varrho(z)\right) \lambda^{n-1} \pm \cdots.$$

The expression

$$\lambda(z) = \frac{1}{n} \sum_{\varrho=1}^{r} m_\varrho \lambda_\varrho(z) \tag{7.2}$$

is called the *mean value* of the $(0, \lambda_0)$-group. Thus, for the mean value one obtains

$$\lambda(z) = \frac{1}{n} \operatorname{tr} A(z) P(z) = \lambda_0 + \frac{1}{n} \operatorname{tr}(A(z) - \lambda_0) P(z) \tag{7.3}$$

and $\lambda(z)$ is holomorphic in $|z| < \delta$. Therefore, together with $P(z)$ one can explicitly calculate the coefficients of the perturbation series for $\lambda(z)$. This will be done in the following section.

In this section, the splitting case will not be treated. That is, there are no calculations in this section for perturbation series of eigenvalues and eigenprojections in the splitting case, i.e., if $(0, \lambda_0)$ is not normal. This problem will be treated systematically in Ch. 7 (reduction theory).

3.7.2. Perturbation series of the group projection

Let

$$A(z) = A_0 + zA_1 + z^2A_2 + \cdots, \qquad |z| < r, \tag{7.4}$$

$$P(z) = P_0 + zP_1 + z^2P_2 + \cdots, \qquad |z| < \delta. \tag{7.5}$$

Then one obtains, according to proposition VIII of Lemma 3.4 and (3.5), the formula

$$P_\varrho = \frac{1}{2\pi i} \sum_{\substack{\mu_1+\mu_2+\cdots+\mu_k=\varrho \\ \mu_1,\ldots,\mu_k \geqq 1 \\ k\geqq 1}} \int_L R_0(\lambda)\, A_{\mu_1} R_0(\lambda)\, A_{\mu_2} R_0(\lambda) \cdots A_{\mu_k} R_0(\lambda)\, d\lambda, \tag{7.6}$$

where L denotes a sufficiently small positively oriented sphere with center λ_0.

For the calculation of the integral (7.6) the Laurent expansion of the resolvent $R_0(\lambda) = R(\lambda, A_0)$ should be used. It takes the form (see (2.6.8))

$$R(\lambda, A_0) = \frac{P_0}{\lambda - \lambda_0} + \sum_{\sigma=2}^{\nu} \frac{D_0^{\sigma-1}}{(\lambda - \lambda_0)^\sigma} + \sum_{\sigma=0}^{\infty} (-1)^\sigma S_0^{\sigma+1} (\lambda - \lambda_0)^\sigma,$$

where ν denotes the index of λ_0, $D_0 = (A_0 - \lambda_0) P_0$ and S_0 denotes the reduced resolvent at $\lambda = \lambda_0$. This Laurent expansion is written in the form

$$R(\lambda, A_0) = \sum_{\sigma=-\nu}^{\infty} S_0^{(\sigma+1)} (\lambda - \lambda_0)^\sigma, \tag{7.7}$$

where

$$S_0^{(0)} = P_0, \qquad S_0^{(-\sigma)} = D_0^\sigma, \qquad S_0^{(\sigma)} = (-1)^{\sigma-1} S_0^\sigma, \qquad \sigma \geqq 1.$$

Substituting (7.7) into (7.6), one obtains for the term

$$R_0(\lambda)\, A_{\mu_1} R_0(\lambda)\, A_{\mu_2} \cdots A_{\mu_k} R_0(\lambda)$$

the expression

$$\sum_{\varrho_1,\varrho_2,\ldots,\varrho_{k+1}} (\lambda - \lambda_0)^{\varrho_1+\varrho_2+\cdots+\varrho_{k+1}} S_0^{(\varrho_1+1)} A_{\mu_1} S_0^{(\varrho_2+1)} A_{\mu_2} \cdots A_{\mu_k} S_0^{(\varrho_{k+1}+1)},$$

and its residue is

$$\sum_{\varrho_1+\varrho_2+\cdots+\varrho_{k+1}=-1} S_0^{(\varrho_1+1)} A_{\mu_1} S_0^{(\varrho_2+1)} A_{\mu_2} \cdots A_{\mu_k} S_0^{(\varrho_{k+1}+1)}$$

or

$$\sum_{\sigma_1+\sigma_2+\cdots+\sigma_{k+1}=k} S_0^{(\sigma_1)} A_{\mu_1} S_0^{(\sigma_2)} A_{\mu_2} \cdots A_{\mu_k} S_0^{(\sigma_{k+1})}, \qquad \sigma_i \geqq -\nu + 1.$$

Therefore,

$$P_\varrho = \sum_{k=1}^{\varrho} \sum_{\substack{\mu_1+\mu_2+\cdots+\mu_k=\varrho \\ \sigma_1+\cdots+\sigma_{k+1}=k \\ \mu_i\geqq 1,\sigma_i\geqq -\nu+1}} S_0^{(\sigma_1)} A_{\mu_1} S_0^{(\sigma_2)} A_{\mu_2} \cdots A_{\mu_k} S_0^{(\sigma_{k+1})}, \tag{7.8}$$

for $\varrho \geqq 1$. For example, the first-order approximation P_1 of the group projection takes the form

$$\begin{aligned} P_1 &= \sum_{\substack{\sigma_1+\sigma_2=1 \\ \sigma_i\geqq -\nu,+1}} S_0^{(\sigma_1)} A_1 S_0^{(\sigma_2)} = S_0^{(-\nu+1)} A_1 S_0^{(\nu)} + S_0^{(-\nu+2)} A_1 S_0^{(\nu-1)} + \cdots + S_0^{(\nu)} A_1 S_0^{(-\nu+1)} \\ &= D_0^{\nu-1} A_1 (-1)^{\nu-1} S_0^\nu + D_0^{\nu-2} A_1 (-1)^{\nu-2} S_0^{\nu-1} + \cdots + D_0 A_1 (-1) S_0^2 \\ &\quad + P_0 A_1 S_0 + S_0 A_1 P_0 + (-1) S_0^2 A_1 D_0 + \cdots + (-1)_0^{\nu-1} S_0^\nu A_1 D_0^{\nu-1}. \end{aligned} \tag{7.9}$$

These formulas can be simplified if λ_0 is *semisimple*. (The fact that λ_0 is semisimple does *not* imply that $(0, \lambda_0)$ is normal, see Subsect. 3.2.8; conversely, $(0, \lambda_0)$ is normal does *not* imply that λ_0 is semisimple, as is shown by the example (c) of Subsect. 3.2.10.) In this case, $\nu = 1$ and in (7.8) one only accounts for those terms where $\sigma_i \geqq 0$. In particular, the first-order approximation takes the form

$$P_1 = P_0 A_1 S_0 + S_0 A_1 P_0. \tag{7.10}$$

In particular, (7.10) holds if λ_0 is *simple*. This implies that $(0, \lambda_0)$ is normal; hence, P_1 is the first-order approximation of the perturbed eigenprojection.

3.7.3. Perturbation series of the mean value

According to (7.3), for the calculation of the perturbation series of the mean value $\lambda(z)$ of the $(0, \lambda_0)$-group, one needs the expression $\big(A(z) - \lambda_0\big) P(z)$. Since

$$\big(A(z) - \lambda_0\big) R\big(\lambda, A(z)\big) = -1 + (\lambda - \lambda_0) R\big(\lambda, A(z)\big),$$

one obtains

$$\big(A(z) - \lambda_0\big) P(z) = \frac{1}{2\pi i} \int_L (\lambda - \lambda_0) R\big(\lambda, A(z)\big) \, d\lambda .$$

Because $(A_0 - \lambda_0) P_0 = D_0$,

$$\big(A(z) - \lambda_0\big) P(z) = D_0 + \sum_{k=1}^{\infty} z^k B_k \tag{7.11}$$

and one obtains, in analogy to (7.6),

$$B_\varrho = \frac{1}{2\pi i} \sum_{\substack{\mu_1+\mu_2+\cdots+\mu_k=\varrho \\ \mu_1,\ldots,\mu_k \geqq 1 \\ k \geqq 1}} \int_L R_0(\lambda) A_{\mu_1} R_0(\lambda) A_{\mu_2} \cdots A_{\mu_k} R_0(\lambda) (\lambda - \lambda_0) \, d\lambda \tag{7.12}$$

for $\varrho \geqq 1$. Hence, by substituting the Laurent expansion of the resolvent, one obtains

$$B_\varrho = \sum_{k=1}^{\varrho} \sum_{\substack{\mu_1+\mu_2+\cdots+\mu_k=\varrho \\ \sigma_1+\cdots+\sigma_{k+1}=k-1 \\ \mu_j \geqq 1, \sigma_j \geqq -\nu+1}} S_0^{(\sigma_1)} A_{\mu_1} S_0^{(\sigma_2)} \cdots A_{\mu_k} S_0^{(\sigma_{k+1})}. \tag{7.13}$$

In particular, for $\varrho = 1$,

$$\begin{aligned} B_1 &= \sum_{\substack{\sigma_1+\sigma_2=0 \\ \sigma_j \geqq -\nu+1}} S_0^{(\sigma_1)} A_1 S_0^{(\sigma_2)} \\ &= S_0^{(-\nu+1)} A_1 S_0^{(\nu-1)} + S_0^{(-\nu+2)} A_1 S_0^{(\nu-2)} + \cdots + S_0^{(\nu-1)} A_1 S_0^{(-\nu+1)} \\ &= D_0^{\nu-1} A_1 (-1)^{\nu-2} S_0^{\nu-1} + \cdots + D_0 A_1 S_0 + P_0 A_1 P_0 \\ &\quad + S_0 A_1 D_0 + \cdots + (-1)^{\nu-2} S_0^{\nu-1} A_1 D_0^{\nu-1}. \end{aligned} \tag{7.14}$$

The formulas for B_ϱ can be simplified if λ_0 is semisimple. In this case,

$$B_1 = P_0 A_1 P_0. \tag{7.15}$$

According to (7.3), the mean value $\lambda(z)$ is a scalar power series

$$\lambda(z) = \lambda_0 + \sum_{\varrho=1}^{\infty} \lambda_\varrho z^\varrho. \tag{7.16}$$

From (7.3) and (7.11),

$$\lambda_\varrho = \frac{1}{n} \operatorname{tr} B_\varrho, \qquad \varrho = 1, 2, 3, \ldots, \tag{7.17}$$

follows such that, with the use of (7.13), the coefficients λ_ϱ can be calculated explicitly.
Thus, because $\operatorname{tr} AB = \operatorname{tr} BA$, (7.15), and $S_0 P_0 = P_0 S_0 = 0$,

$$\lambda_1 = \frac{1}{n} \operatorname{tr} (P_0 A_1 P_0) = \frac{1}{n} \operatorname{tr} (A_1 P_0). \tag{7.18}$$

3.7.4. Second method for the calculation of the perturbation series of the mean value

A second method leads to simpler formulas for the λ_ϱ in (7.16) than those in (7.17). For this purpose, one starts from

$$\lambda(z) - \lambda_0 = \frac{1}{n} \operatorname{tr} \big(A(z) - \lambda_0\big) P(z)$$

and

$$\big(A(z) - \lambda_0\big) P(z) = \frac{1}{2\pi i} \int_L (\lambda - \lambda_0) R\big(\lambda, A(z)\big) \, d\lambda.$$

From these equations, the relation

$$\lambda(z) - \lambda_0 = \frac{1}{2\pi i n} \operatorname{tr} \int_L (\lambda - \lambda_0) R\big(\lambda, A(z)\big) \, d\lambda$$

follows. Replacing $R\big(\lambda, A(z)\big)$ by its second Neumann series (see (3.1)), one obtains

$$\lambda(z) - \lambda_0 = \frac{1}{2\pi i n} \operatorname{tr} \int_L (\lambda - \lambda_0) R_0(\lambda) \sum_{p=1}^{\infty} \big\{\big(A(z) - A_0\big) R_0(\lambda)\big\}^p \, d\lambda,$$

where the term for $p = 0$ can be omitted, because

$$\int_L (\lambda - \lambda_0) R_0(\lambda) \, d\lambda = (A_0 - \lambda_0) P_0 = D_0$$

and $\operatorname{tr} D_0 = 0$.

Furthermore, with $B = A(z) - A_0$ for short

$$\frac{\mathrm{d}}{\mathrm{d}\lambda}\left(BR_0(\lambda)\right)^p = \frac{\mathrm{d}}{\mathrm{d}\lambda}\left(BR_0BR_0 \cdots BR_0\right)$$

$$= BR_0'BR_0 \cdots BR_0 + BR_0BR_0'BR_0 \cdots BR_0 + \cdots + BR_0BR_0 \cdots BR_0'$$

$$= -(BR_0^2BR_0 \cdots BR_0 + BR_0BR_0^2BR_0 \cdots BR_0 + \cdots + BR_0BR_0 \cdots BR_0^2),$$

and, since $\operatorname{tr} AB = \operatorname{tr} BA$,

$$\operatorname{tr}\left[\frac{\mathrm{d}}{\mathrm{d}\lambda}\left(BR_0(\lambda)\right)^p\right] = -p \operatorname{tr}\left[R_0(BR_0)^p\right].$$

Finally, because the trace and integration operations are interchangeable, one obtains

$$\lambda(z) - \lambda_0 = -\frac{1}{2\pi i n} \operatorname{tr} \int_L (\lambda - \lambda_0) \sum_{p=1}^{\infty} \frac{1}{p} \frac{\mathrm{d}}{\mathrm{d}\lambda} (BR_0)^p \,\mathrm{d}\lambda.$$

By partial integration,

$$\int_L (\lambda - \lambda_0) \frac{\mathrm{d}}{\mathrm{d}\lambda} (BR_0)^p \,\mathrm{d}\lambda = [(\lambda - \lambda_0)(BR_0)^p]_L - \int_L (BR_0)^p \,\mathrm{d}\lambda.$$

Hence,

$$\lambda(z) - \lambda_0 = \frac{1}{2\pi i n} \operatorname{tr} \int_L \sum_{p=1}^{\infty} \frac{1}{p} \left\{\left(A(z) - A_0\right) R_0(\lambda)\right\}^p \,\mathrm{d}\lambda. \tag{7.19}$$

This expression can be simplified formally, if one notes that, for example for $\|C\| < 1$,

$$\log(1 - C) = -\sum_{p=1}^{\infty} \frac{1}{p} C^p$$

(see Theorem 2.3.1). Therefore,

$$\lambda(z) - \lambda_0 = -\frac{1}{2\pi i n} \operatorname{tr} \int_L \log\left[1 - \left(A(z) - A_0\right) R_0(\lambda)\right] \mathrm{d}\lambda. \tag{7.20}$$

From (7.19), by substituting the power series $A(z) - A_0 = \sum_{\varrho=1}^{\infty} z^\varrho A_\varrho$, the relation

$$\lambda_\varrho = \frac{1}{2\pi i n} \operatorname{tr} \sum_{\substack{\mu_1 + \cdots + \mu_p = p \\ p \geq 1, \mu_i \geq 1}} \frac{1}{p} \int_L A_{\mu_1} R_0(\lambda) A_{\mu_2} R_0(\lambda) \cdots A_{\mu_p} R_0(\lambda) \,\mathrm{d}\lambda \tag{7.21}$$

follows. Substituting the Laurent series (7.7) for $R_0(\lambda)$ into (7.21), one obtains

$$\lambda_\varrho = \frac{1}{n} \sum_{p=1}^{\varrho} \frac{1}{p} \sum_{\substack{\mu_1 + \cdots + \mu_p = p \\ k_1 + \cdots + k_p = p - 1}} \operatorname{tr} \{A_{\mu_1} S_0^{(k_1)} A_{\mu_2} S_0^{(k_2)} \cdots A_{\mu_p} S_0^{(k_p)}\}. \tag{7.22}$$

For $\varrho = 1$, (7.18) is reproduced, and for the second-order approximation λ_2 one obtains

$$\lambda_2 = \frac{1}{n}\left[\operatorname{tr}(A_2P_0) + \frac{1}{2}\sum_{k_1+k_2=1} \operatorname{tr}(A_1S_0^{(k_1)}A_2S_0^{(k_2)})\right]$$

$$= \frac{1}{n}\left[\operatorname{tr}(A_2P_0) + \operatorname{tr}\left(A_1P_0A_1S_0 - A_1D_0A_1S_0^2 \pm \cdots + (-1)^{\nu-1}A_1D_0^{\nu-1}A_0S_0^{\nu}\right)\right].$$

The formulas (7.22) are simplified if λ_0 is *semisimple*. For λ_1

$$\lambda_1 = \frac{1}{n}\operatorname{tr}(A_1P_0) \tag{7.23}$$

is reproduced, and for λ_2 one obtains

$$\lambda_2 = \frac{1}{n}\operatorname{tr}(A_2P_0 + A_1P_0A_1S_0). \tag{7.24}$$

In particular, (7.23) and (7.24) hold if λ_0 is *simple* ($n = 1$). In this case, λ_1 and λ_2 represent the first- and the second-order approximations, respectively, of the perturbed eigenvalue $\lambda(z)$.

3.7.5. Linear perturbations

A special case, important for applications, is that of linear perturbations

$$A(z) = A_0 + zA_1.$$

Then the formulas (7.8) and (7.22) for the coefficients of the group projection and the mean value can be simplified. From (7.6) and (7.21) one obtains

$$P_\varrho = \frac{1}{2\pi i}\int_L R_0(\lambda)\,A_1R_0(\lambda)\,A_1 \cdots R_0(\lambda)\,A_1R_0(\lambda)\,d\lambda,$$

$$= \frac{1}{2\pi i}\int_L R_0(\lambda)\,[A_1R_0(\lambda)]^\varrho\,d\lambda, \qquad \varrho = 1, 2, 3, \ldots,$$

and

$$\lambda_\varrho = \frac{1}{2\pi i n\varrho}\operatorname{tr}\int_L [A_1R_0(\lambda)]^\varrho\,d\lambda, \qquad \varrho = 1, 2, 3, \ldots$$

Hence, in particular,

$$\lambda_\varrho = \frac{1}{n\varrho}\operatorname{tr}(A_1P_{\varrho-1}), \qquad \varrho = 1, 2, 3, \ldots$$

3.8. Some coefficients of perturbation series represented with respect to a basis

3.8.1. λ_0 semisimple

In this case, the simplified formulas (7.8) and (7.22) for the coefficients P_ϱ and λ_ϱ hold. In this subsection, the expressions for λ_1 and λ_2 will be calculated with the use of suitable bases.

Let $e_1, e_2, \ldots, e_n$ be a basis in P_0X and $f_1, f_2, \ldots, f_n$ the corresponding adjoint basis in $P_0^*X^*$ (see Subsect. 1.2.2.). With respect to the basis $e_1, e_2, \ldots, e_n$, the operator $P_0A_1P_0 \restriction P_0X$ has the matrix elements (A_1e_i, f_j), $i, j = 1, \ldots, n$. Therefore,

$$\operatorname{tr}(P_0A_1P_0) = \sum_{i=1}^{n} (A_1e_i, f_i),$$

and one obtains

$$\lambda_1 = \frac{1}{n} \sum_{i=1}^{n} (A_1e_i, f_i). \tag{8.1}$$

Furthermore, let $x_1, x_2, \ldots, x_r$ a basis of $(1 - P_0)X$ and $y_1, y_2, \ldots, y_r$ the corresponding adjoint basis of $(1 - P_0^*)X^*$. Then

$$\operatorname{tr}(A_1P_0A_1S_0) = \operatorname{tr}(A_1S_0A_1P_0) = \operatorname{tr}(P_0A_1S_0A_1P_0);$$

hence,

$$\operatorname{tr}(P_0A_1S_0A_1P_0) = \sum_{i=1}^{n} (A_1S_0A_1e_i, f_i) = \sum_{i=1}^{n} (S_0A_1e_i, A_1^*f_i).$$

For $x \in (1 - P_0)X$ one has $S_0x = (\lambda_0 - A_0)^{-1}x$. This operator may be assumed known. Let

$$(\lambda_0 - A_0)^{-1} x_j = \sum_{k=1}^{r} c_{kj}x_k, \qquad c_{kj} = (S_0x_j, y_k).$$

If $x = \sum_{j=1}^{r} (x, y_j)\, x_j$, which implies $S_0x = \sum_{j=1}^{r} (x, y_j)\, S_0x_j$, and $y = \sum_{j=1}^{r} (y, x_j)\, y_j$, then

$$(S_0x, y) = \sum_{i,j=1}^{r} (x, y_j)\,(x_i, y)\, c_{ij}.$$

Hence,

$$\operatorname{tr}(P_0A_1S_0A_1P_0) = \sum_{i=1}^{n} \sum_{j,k=1}^{r} c_{kj}(A_1e_i, y_j)\,(x_k, A_1^*f_i),$$

and finally

$$\lambda_2 = \frac{1}{n} \sum_{i=1}^{n} \left[(A_2e_i, f_i) + \sum_{j,k=1}^{r} c_{kj}(A_1e_i, y_j)\,(A_1x_k, f_i)\right]. \tag{8.2}$$

If more is known about the action of the reduced resolvent in $(1 - P_0) X$, then the last formula can be further simplified.

For example, let A_0 be a diagonal operator in $(1 - P_0) X$ with eigenvalues $\beta_1, \beta_2, \ldots, \beta_r$ (where the multiplicities are taken into account). Let $x_1, x_2, \ldots, x_r$ denote a corresponding basis of eigenvectors. Then

$$A_0 x_\varrho = \beta_\varrho x_\varrho, \qquad (\lambda_0 - A_0)^{-1} x_\varrho = \frac{1}{\lambda_0 - \beta_\varrho} x_\varrho, \qquad \varrho = 1, \ldots, r.$$

In this case,

$$c_{kj} = \frac{1}{\lambda_0 - \beta_k} \delta_{kj}, \qquad k, j = 1, \ldots, r,$$

and the formula (8.2) takes the form

$$\lambda_2 = \frac{1}{n} \sum_{i=1}^{n} \left[(A_2 e_i, f_i) + \sum_{j=1}^{r} \frac{1}{\lambda_0 - \beta_j} (A_1 e_i, y_j) (A_1 x_j, f_i) \right]. \tag{8.3}$$

3.8.2. λ_0 simple

In this case, $n = 1$, $e_1 = e$ is an eigenvector belonging to λ_0, $f_1 = f$ is an eigenvector of A_0^* belonging to the eigenvalue $\bar{\lambda}_0$ and one obtains

$$\lambda_1 = (A_1 e, f), \tag{8.4}$$

$$\lambda_2 = (A_2 e, f) + \sum_{j,k=1}^{N-1} c_{kj} (A_1 e, y_j) (A_1 x_k, f). \tag{8.5}$$

If, additionally, A_0 is a diagonal operator in $(1 - P_0) X$, then the formula for λ_2 follows from (8.5):

$$\lambda_2 = (A_2 e, f) + \sum_{j=1}^{N-1} \frac{1}{\lambda_0 - \beta_j} (A_1 e, y_j) (A_1 x_j, f). \tag{8.6}$$

3.8.3. Calculation of P_1; λ_0 semisimple

The expression (7.10) for P_1 will be evaluated for a semisimple λ_0 with the use of suitable basis. In other words, the matrix elements of P_1 will be calculated. According to (7.10), $P_1 = P_0 A_1 S_0 + S_0 A_1 P_0$.

The bases $e_1, \ldots, e_n$; $f_1, \ldots, f_n$ and $x_1, \ldots, x_r$; $y_1, \ldots, y_r$ are chosen according to Subsect. 3.8.1. One calculates $(P_1 e_i, f_j)$, $(P_1 e_i, y_j)$, $(P_1 x_i, f_j)$ and $(P_1 x_i, y_j)$. Since

$$P_0 P_1 P_0 = 0, \qquad (1 - P_0) P_1 (1 - P_0) = 0$$

(see Subsect. 3.4.2), one obtains

$$(P_1 e_i, f_j) = 0; \qquad (P_1 x_i, y_j) = 0.$$

The remaining matrix elements are

$$(P_1 e_i, y_j) = (P_0 A_1 S_0 e_i, y_j) + (S_0 A_1 P_0 e_i, y_j) = (S_0 A_1 e_i, y_j) = (A_1 e_i, S_0^* y_j),$$

$$(P_1 x_i, f_j) = (P_0 A_1 S_0 x_i, f_j) + (S_0 A_1 P_0 x_i, f_j) = (A_1 S_0 x_i, f_j) = (S_0 x_i, A_1^* f_j).$$

Assuming that A_0 is a diagonal operator in $(1 - P_0) X$, as in Subsect. 3.8.1, these matrix elements can be calculated explicitly.

3.8.4. Calculation of P_1; λ_0 simple

In this case, $n = 1$, $(0, \lambda_0)$ is normal. If $e_1 = e$ (eigenvector of A_0 belonging to λ_0), and $f_1 = f$ (eigenvector of A_0^* belonging to $\bar{\lambda}_0$), one obtains

$$\begin{pmatrix} 0 & (S_0 x_1, A_1^* f) & (S_0 x_2, A_1^* f) & \cdots & (S_0 x_{N-1}, A_1^* f) \\ (A_1 e, S_0^* y_1) & 0 & 0 & \cdots & 0 \\ (A_1 e, S_0^* y_2) & 0 & 0 & \cdots & 0 \\ \vdots & \vdots & \vdots & \cdots & \vdots \\ (A_1 e, S_0^* y_{N-1}) & 0 & 0 & \cdots & 0 \end{pmatrix} \tag{8.7}$$

for the matrix of P_1. A perturbed eigenvector with "zero-order approximation" e is given by $P(z)\, e = e + zP_1 e + O(z^2)$. Hence,

$$P_1 e = \sum_{j=1}^{N-1} (A_1 e, S_0^* y_j)\, x_j .$$

Therefore, the approximation $P_1 e$ has a vanishing component with respect to e (cf. Subsect. 3.1.2).

In particular, if A_0 is a diagonal operator in $(1 - P_0) X$ with eigenvalues $\beta_1, \beta_2, \ldots, \beta_{N-1}$ and eigenvectors $x_1, x_2, \ldots, x_{N-1}$, then one obtains for the matrix elements (8.7) the expressions

$$(A_1 e, S_0^* y_j) = \frac{1}{\lambda_0 - \beta_j} (A_1 e, y_j), \qquad j = 1, 2, \ldots, N - 1,$$

$$(S_0 x_i, A_1^* f) = \frac{1}{\lambda_0 - \beta_i} (x_i, A_1^* f), \qquad i = 1, 2, \ldots, N - 1.$$

In particular,

$$P_1 e = \sum_{j=1}^{N-1} \frac{1}{\lambda_0 - \beta_j} (A_1 e, y_j)\, x_j . \tag{8.8}$$

3.8.5. Selfadjoint operators A_0, A_1, A_2; λ_0 simple

If A_0, A_1 and A_2 are selfadjoint and λ_0 is semisimple, then the formulas for λ_1, λ_2 and P_1e can be further simplified. First one has $e = f$; hence,

$$\lambda_1 = (A_1e, e). \tag{8.9}$$

Moreover, in this case A_0 is necessarily a diagonal operator in $(1 - P_0)\,X$ and one can choose $x_j = y_j$ (an orthonormal basis is chosen). Since

$$(A_1e, x_j) = (e, A_1x_j) = \overline{(A_1x_j, e)},$$

from (8.6) the expression

$$\lambda_2 = (A_2e, e) + \sum_{j=1}^{N-1} \frac{1}{\lambda_0 - \beta_j} |(A_1e, x_j)|^2 \tag{8.10}$$

follows. Finally, one obtains

$$P_1e = \sum_{j=1}^{N-1} \frac{1}{\lambda_0 - \beta_j} (A_1e, x_j)\, x_j. \tag{8.11}$$

The formulas (8.9) and (8.10) coincide with the expressions obtained in Subsect. 3.1.2 (note that $A_2 = 0$ in Subsect. 3.1.2). Formula (8.11) for the correction of the perturbed eigenvector in the first-order approximation also coincides with the corresponding correction in Subsect. 3.2.1, even though different "normalizations" have been used for the perturbed eigenvector.

3.9. Applications

3.9.1. Determination of perturbed groups of eigenvalues

The results of Sect. 3.3 concerning the behavior of perturbed eigenvalues suggest the development of methods for the explicit determination of perturbed groups of eigenvalues (cycle periods etc.). One of these methods is the treatment of the characteristic polynomial by means of the Puiseux-Newton diagram. In this connection the following remarks are useful.

Let λ_0 be an unperturbed eigenvalue of algebraic multiplicity n of the linear perturbation $A_0 + zA_1$ and let $P(z) = P_0 + zP_1 + \cdots$ be the group projection of the $(0, \lambda_0)$-group (λ_0-group for short). Then the power series expansion of the transformation operator $U(z)$ used in Subsect. 3.4.4 takes the form

$$U(z) = 1 + z(P_1P_0 - P_0P_1) + \cdots,$$

$$U(z)^{-1} = 1 - z(P_1P_0 - P_0P_1) + \cdots.$$

Recall that $B(z) = U(z)^{-1} A(z) U(z)$ has a constant reducing projection, namely P_0; hence,

$$P_0 B(z) P_0 = P_0 A_0 P_0 + z P_0 A_1 P_0 + \cdots.$$

Therefore, the behavior of the λ_0-group with respect to the first-order approximation depends only on $P_0 A_1 P_0$; that is, without loss of generality, for the study of the first-order approximation of the λ_0-group, it can be assumed that A_0 has only one eigenvalue: λ_0.

For example, let $X = \mathbb{C}^n$ and $A_0 = \lambda_0 E_n + \Gamma_n(0)$ (see Subsect. 2.9.5); that is, let A_0 be a Jordan operator in $\mathscr{L}(\mathbb{C}^n)$. Furthermore, let A_1 be any $n \times n$-matrix. If $a_{n1} \neq 0$, then the λ_0-group of the perturbation $A_0 + zA_1$ consists of only one cycle of period n.

One has $\chi(\lambda, A(z)) = (\lambda - \lambda_0)^n + \cdots - a_{n1} z$ and this polynomial is necessarily irreducible in $\mathscr{H}_{|z|<\delta}[\lambda - \lambda_0]$, where $\delta > 0$ can be chosen arbitrarily small. In the other case, i.e., if $\chi = \chi_1 \cdot \chi_2$ with $\chi_1 = (\lambda - \lambda_0)^p + \cdots$, $\chi_2 = (\lambda - \lambda_0)^q + \cdots$, then at least one of the polynomials χ_1, χ_2 must have a coefficient belonging to $(\lambda - \lambda_0)^0 = 1$, nonvanishing at $z = 0$. This is impossible, because $(\lambda - \lambda_0)^n = \chi_1 \cdot \chi_2$ at $z = 0$. Hence,

$$\lambda_\varrho(z) = \lambda_0 + \sum_{\alpha=1}^{\infty} \gamma_\alpha z^{\alpha/n}.$$

Comparing the coefficients (of the eigenvalue equation), one obtains $\gamma_1 = a_{n1}^{1/n}$.

In the general case (i.e., if the unperturbed operator A_0 also has other eigenvalues different from λ_0), this result takes the following form: Let λ_0 be prime, where the algebraic multiplicity is n, and let e_0 be the only linearly independent unperturbed eigenvector; furthermore, let f_0 be the only linearly independent eigenvector of A_0^* belonging to $\bar{\lambda}_0$. If $(A_1 e_0, f_0) \neq 0$, then the λ_0-group consists of only one cycle of period n and one has

$$\lambda_\varrho(z) = \lambda_0 + (A_1 e_0, f_0)^{1/n} z^{1/n} + \sum_{\alpha=2}^{\infty} \gamma_\alpha z^{\alpha/n}.$$

In Ch. 7, a systematic determination will be made of the first-order approximations of the cycles and the perturbed eigenvalues by means of methods which do not use the characteristic polynomial and the Puiseux-Newton diagram. There, the result noted above will be strengthened and strongly generalized (see M. I. Višik and L. A. Ljusternik [2], V. B. Lidskii [1], W. S. Loud [1]).

3.9.2. Finite selfadjoint perturbations of selfadjoint operators

This subsection is based on T. Kato [5, pp. 125sqq.]. Let A and B be selfadjoint, $C = B - A$. Let $\alpha_n, \beta_n, \gamma_n$, $n = 1, \ldots, N$, denote the eigenvalues of A, B, C, respectively, where the multiplicity is taken into account. Furthermore, they are ordered according to $\alpha_1 \leqq \alpha_2 \leqq \cdots \leqq \alpha_N$. Consider the selfadjoint perturbation $A(\varepsilon) = A + \varepsilon C$ on $[0, 1]$. One has $A(0) = A$, $A(1) = B$. $A(\varepsilon)$ is holomorphic on $[0, 1]$. According to Corollary 5.2, the perturbed eigenvalues $\lambda_n(\varepsilon)$ are holomorphic on $[0, 1]$. There are at most finitely many multiple points $\varepsilon_1, \varepsilon_2, \ldots, \varepsilon_r$ on $[0, 1]$. Numbering the eigenvalues

$\lambda_n(\varepsilon)$ in their natural order, $\lambda_1(\varepsilon) \leqq \lambda_2(\varepsilon) \leqq \cdots \leqq \lambda_N(\varepsilon)$, we find that the points ε_1, $\varepsilon_2, \ldots, \varepsilon_r$ are at most singularities, namely points where the derivation $\lambda_\varrho'(\varepsilon)$ is discontinuous. That is, the functions $\lambda_\varrho(\varepsilon)$ are piecewise holomorphic. Likewise according to Corollary 5.2, there are corresponding orthonormal eigenvectors $x_n(\varepsilon)$ which are also piecewise holomorphic on $[0, 1]$, where these functions may be discontinuous at the points $\varepsilon_1, \ldots, \varepsilon_r$. The equations

$$\big(A(\varepsilon) - \lambda_\varrho(\varepsilon)\big)\, x_\varrho(\varepsilon) = 0, \qquad \varrho = 1, \ldots, N,$$

hold. Hence,

$$\big(C - \lambda_\varrho'(\varepsilon)\big)\, x_\varrho(\varepsilon) + \big(A(\varepsilon) - \lambda_\varrho(\varepsilon)\big)\, x_\varrho'(\varepsilon) = 0, \qquad \varepsilon \neq \varepsilon_i,$$

and by scalar multiplication by $x_\varrho(\varepsilon)$,

$$\lambda_\varrho'(\varepsilon) = \big(Cx_\varrho(\varepsilon), x_\varrho(\varepsilon)\big), \qquad \varepsilon \neq \varepsilon_i, \qquad \varrho = 1, \ldots, N.$$

Integrating this equation over $[0, 1]$, one obtains

$$\beta_\varrho - \alpha_\varrho = \lambda_\varrho(1) - \lambda_\varrho(0) = \int_0^1 \big(Cx_\varrho(\varepsilon), x_\varrho(\varepsilon)\big)\, d\varepsilon. \tag{9.1}$$

Let $e_1, e_2, \ldots, e_N$ be an orthonormal basis consisting of eigenvectors of C: $Ce_\varrho = \gamma_\varrho e_\varrho$, $\varrho = 1, \ldots, N$. Then

$$\big(Cx_\varrho(\varepsilon), x_\varrho(\varepsilon)\big) = \sum_{j=1}^N \big(Cx_\varrho(\varepsilon), e_j\big)\big(e_j, x_\varrho(\varepsilon)\big) = \sum_{j=1}^N \gamma_j \left|\big(x_\varrho(\varepsilon), e_j\big)\right|^2.$$

If

$$\sigma_{\varrho j} = \int_0^1 \left|\big(x_\varrho(\varepsilon), e_j\big)\right|^2 d\varepsilon,$$

then, since the systems $x_\varrho(\varepsilon)$ and e_j are orthonormal,

$$\sum_{\varrho=1}^N \sigma_{\varrho j} = 1, \qquad \sum_{j=1}^N \sigma_{\varrho j} = 1, \qquad \sigma_{\varrho j} \geqq 0. \tag{9.2}$$

Thus equation (9.1) takes the form

$$\beta_\varrho - \alpha_\varrho = \sum_{j=1}^N \gamma_j \sigma_{\varrho j}, \qquad \varrho = 1, \ldots, N. \tag{9.3}$$

Now one uses a theorem due to G. Birkhoff. (See G. Birkhoff [1]; in L. Mirsky [1] one finds a simple proof of this theorem.)

Theorem 1. *Let $A = (\sigma_{ij})$ be an $N \times N$-matrix with the properties* (9.2). *Then A can be represented in the form*

$$A = \sum_\pi w_\pi P_\pi, \qquad w_\pi \geqq 0, \qquad \sum_\pi w_\pi = 1, \tag{9.4}$$

where $\{P_\pi\}$ denotes the system of all permutation matrices of order N.

Note that a matrix $P = (p_{ij})$ is called a *permutation matrix* if there is a permutation π such that $p_{ij} = 1$ if $\pi(i) = j$, and $p_{ij} = 0$ otherwise.

Thus the matrices of the form (9.4) form exactly the convex hull of all permutation matrices. If γ denotes the column formed by the coordinates $\gamma_1, \ldots, \gamma_N$, and if $\gamma_\pi = P_\pi \gamma$ (i.e., γ_π is obtained from γ by application of the permutation π to the coordinates), then $A\gamma = \sum_\pi \omega_\pi \gamma_\pi$ and one obtains the following result proved by V. B. Lidskii [2].

Theorem 2 (Lidskii). *Let $A, B, C; \alpha_n, \beta_n, \gamma_n$ be chosen according to Subsect.* 3.9.2. *Then the vector $(\beta_1 - \alpha_1, \beta_2 - \alpha_2, \ldots, \beta_N - \alpha_N)$ lies in the convex hull of all vectors obtained from $(\gamma_1, \gamma_2, \ldots, \gamma_N)$ by permutation of the coordinates.*

In T. Kato [5, p. 126] one finds further consequences of (9.3).

4. Linear operators on linear spaces over commutative fields; first applications

In Ch. 3 the spectral representation of meromorphic operator-valued functions $A(z)$ was investigated. As pointed out in Ch. 2, the spectral representation of an operator appears as an intermediate step in the determination of its Jordan structure.

If one starts from the spectral representation of $A(z)$ (see Subsect. 3.3.12), the determination of its Jordan structure, i.e., the determination of decompositions of $A(z)$ into a direct sum of Jordan operators, requires the investigation of the internal structure (or refined structure) of the eigennilpotents $D(\zeta_a)$ of $A(z)$. Recall that the $D(\zeta_a)$ are defined on the analytic functions $\mathfrak{P}_a$ given by the characteristic polynomial of $A(z)$.

Tools necessary for the investigation of the eigennilpotents and also for further problems mentioned in Ch. 3 have been collected in the present chapter. (For example, in Subsect. 2.3.6, a direct proof for the representation of the projector-valued function $P(\zeta_a)$ given in Theorem 3.3.6 was announced; references in Ch. 3 to problems to be treated later are given in Subsect. 3.3.15, 3.4.1, 3.4.2, 3.4.3, 3.5.5.)

The main topic in this connection is the investigation of the structure of linear operators on linear spaces over an arbitrary commutative field K.

In the case $K = \mathbb{C}$, this investigation is contained in Ch. 2 (see Subsect. 2.9.5). For arbitrary fields K, the situation is much more complicated. In general, a decomposition into a direct sum of Jordan operators does not exist. Therefore, the problem is, on the one hand, characterizing the irreducible operators and, on the other hand, constructing a decomposition of a given operator A into a direct sum of irreducible operators.

Furthermore, it is important to select properties of A which guarantee that the irreducible parts of A are Jordan operators, i.e., that A is a direct sum of Jordan operators. Conditions of this kind are needed for the applications mentioned.

The theory of linear operators on linear spaces over K is well-known, see, for example, the monographs of B. L. van der Waerden [2], F. R. Gantmacher [1] and especially W. Graeub [1] (in particular Ch. XI); see also R. Kochendörffer [1]. In this chapter only those parts of the theory will be given which are important for the following analysis. This presentation continues the theory of Ch. 2.

As in Ch. 2, the problem of decomposing A is studied in two steps. Even though the characteristic polynomial does not necessarily possess roots in K, it is the starting point for the investigation (note that, for example, for $K = \mathbb{R}$ and $A = \begin{pmatrix} 0 & 1 \\ -1 & 0 \end{pmatrix}$, one obtains $\chi(\lambda, A) = \lambda^2 + 1$). As in Ch. 2, a certain analog of the spectral representation of A is derived by means of $\chi(\lambda, A)$; this representation is also uniquely determined in a sense to be specified. In the second step an analog of the Jordan decomposition of Ch. 2 is constructed. This second step is essentially taken from W. GRAEUB [1, Ch. XI], whereas the first step, which differs from the presentation given there, is adapted for later applications.

The application of the general theory to the problems mentioned before, for example to the problem of the "refined structure" of the eigennilpotents $D(\zeta_a)$ of $A(z)$, requires the identification of the field K with suitable function fields, namely, with the fields of algebroid functions defined on $\mathfrak{P}_1, \mathfrak{P}_2, \ldots, \mathfrak{P}_r$. This very idea leads to a sufficient insight into the function-theoretic properties of the Jordan structure of $A(z)$.

In order to prepare for these applications, which will be presented in Ch. 5, at the end of this chapter two problems have been treated which deal with the determination of the "smallest" meromorphic reducing projections of $A(z)$ in G and the transformation of meromorphic systems of projections. These problems are a continuation of the considerations in Ch. 3 and can be solved by applying the tools collected in this chapter.

The reader who is familiar with this theory can skip this chapter but should consider both applications at the end.

4.1. Basic material

4.1.1. Algebra of linear operators

Let K be a commutative field. A linear space X of finite dimension $\dim X = N > 0$ over the field K is defined as in Subsect. 1.1.1, where the complex numbers are replaced by the elements of K throughout. All definitions, concepts and theorems of Sect. 1.1 remain valid. A linear operator acting on X is defined as in Subsect. 1.5.1. All considerations of Sect. 1.5 and 1.6 remain valid. Again, the algebra of linear operators on X is denoted by $\mathscr{L}(X)$. The concept of a projection P is defined as in Sect. 1.9. The propositions of Subsect. 1.9.2, 1.9.4 and 2.2.1 to 2.2.4 remain valid.

4.1.2. Polynomial operators

The ring of all polynomials over K is denoted by $K[\lambda]$. In $K[\lambda]$ the Euclidean algorithm can be established (see Subsect. A 2.4). A polynomial $p(\lambda)$ from $K[\lambda]$ is called *normalized* if the "highest coefficient" equals 1, i.e., if $p(\lambda) = \lambda^n + a_1\lambda^{n-1} + \cdots + a_n$.

In the following analysis one considers polynomials $A(\lambda)$ with operator-valued coefficients, i.e., polynomial operator-valued functions,

$$A(\lambda) = A_0\lambda^m + A_1\lambda^{m-1} + \cdots + A_{m-1}\lambda + A_m, \qquad A_i \in \mathscr{L}(X).$$

$A(\lambda)$ is called a *polynomial operator*. If one replaces λ by an operator $B \in \mathscr{L}(X)$ then one has to distinguish whether B and its powers are a left factor or a right factor. One writes

$$A(B) = A_0B^m + A_1B^{m-1} + \cdots + A_{m-1}B + A_m$$

and

$$(B)\,A = B^m A_0 + B^{m-1} A_1 + \cdots + B A_{m-1} + A_m,$$

and one speaks of the ***right value*** $A(B)$ and of the ***left value*** $(B)\,A$ of $A(\lambda)$ with respect to the substitution $\lambda \to B$.

4.1.3. Division algorithm

Let

$$A(\lambda) = A_0\lambda^m + A_1\lambda^{m-1} + \cdots + A_m, \qquad A_0 \neq 0,$$

$$B(\lambda) = B_0\lambda^p + B_1\lambda^{p-1} + \cdots + B_p, \qquad \det B_0 \neq 0,$$

be two polynomial operators such that $\deg A(\lambda) = m$, $\deg B(\lambda) = p$ and B_0 is regular. For such polynomials the right and left division algorithms hold. In the right-division algorithm, an equation of the form

$$A(\lambda) = Q(\lambda)\,B(\lambda) + R(\lambda), \qquad \deg R < \deg B, \tag{1.1}$$

is valid, where Q and R are uniquely determined polynomial operators. Q is called the ***right quotient*** and R is called the ***right remainder***. The left-division algorithm is formulated analogously.

The existence of a representation (1.1) is proved as in Subsect. A 2.3. For $m < p$ the proposition is obvious. Let $m \geqq p$. Then $A(\lambda) - A_0 B_0^{-1} \lambda^{m-p} B(\lambda)$ is a polynomial of degree $g \leqq m - 1$. By successive continuation of this procedure the degree of the remainder can be made less than p.

To prove the uniqueness of the representation, let $A(\lambda) = Q^*B + R^*$ be a second representation. Then,

$$(Q - Q^*)\,B = R^* - R.$$

Since $\det B_0 \neq 0$,

$$\deg\,(Q - Q^*)\,B = \deg\,(Q - Q^*) + \deg B \geqq p,$$

if $Q - Q^* \neq 0$. On the other hand, $\deg\,(R^* - R) < p$. Hence, $Q \equiv Q^*$, which implies $R \equiv R^*$.

Let

$$F(\lambda) = F_0\lambda^m + F_1\lambda^{m-1} + \cdots + F_m, \qquad F_0 \neq 0,$$

be an arbitrary polynomial operator of degree m. Then

$$F(\lambda) = Q(\lambda)\,(\lambda - A) + R, \tag{1.2}$$

where the right remainder R is a constant operator and where $Q(\lambda)$ denotes a polynomial operator of degree $m - 1$,

$$Q(\lambda) = Q_0\lambda^{m-1} + Q_1\lambda^{m-2} + \cdots + Q_{m-1}.$$

One obtains

$$F(\lambda) = Q_0\lambda^m + (Q_1 - Q_0A)\,\lambda^{m-1} + (Q_2 - Q_1A)\,\lambda^{m-1} + \cdots$$
$$+ (Q_{m-1} - Q_{m-2}A)\,\lambda - Q_{m-1}A + R.$$

From this equation $R = F(A)$ follows. Correspondingly, in the second case one obtains $F(\lambda) = (\lambda - A)\,Q(\lambda) + \hat{R}$; analogously $\hat{R} = (A)\,F$. Therefore, the following proposition holds (Generalized Bezout Theorem):

Lemma 1 (Bezout). *Let $F(\lambda)$ be a polynomial operator of degree $m \geqq 1$. Then there exist polynomial operators $Q_L(\lambda)$, $Q_R(\lambda)$ of degree $m - 1$ such that*

$$F(\lambda) = Q_R(\lambda)\,(\lambda - A) + F(A),$$
$$F(\lambda) = (\lambda - A)\,Q_L(\lambda) + (A)\,F.$$

4.1.4. Theorem of Cayley-Hamilton

Let $\chi(\lambda, A) = \det(\lambda - A)$ denote the characteristic polynomial of the operator A. According to Cramer's rule, for the operator $(\lambda - A)^{-1}$ one obtains a representation of the form

$$(\lambda - A)^{-1} = \frac{1}{\chi(\lambda, A)}\,B(\lambda),$$

where $B(\lambda)$ is a polynomial operator of degree $N - 1$. One has

$$B(\lambda) = \lambda^{N-1}\,1 + \cdots. \tag{1.3}$$

Obviously, $B(\lambda)$ satisfies the equation

$$(\lambda - A)\,\frac{B(\lambda)}{\chi(\lambda, A)} = \frac{B(\lambda)}{\chi(\lambda, A)}\,(\lambda - A) = 1$$

or

$$(\lambda - A)\,B(\lambda) = B(\lambda)\,(\lambda - A) = \chi(\lambda, A)\,1.$$

That is, the polynomial operator $\chi(\lambda, A)\,1$ of degree N can be right-divided and left-divided by $\lambda - A$ without remainder. Thus, according to Lemma 1, one obtains

Lemma 2 (Cayley-Hamilton). *The characteristic polynomial $\chi(\lambda, A)$ annihilates the operator A, i.e., $\chi(A, A) = 0$.*

Note that $\chi(A, A) = A^N + \alpha_1A^{N-1} + \cdots + \alpha_{N-1}A + \alpha_N$ is an "operator polynomial" in A with coefficients from K.

In the following discussion $\chi(\lambda, A)$ is abbreviated to $\chi(\lambda)$. Note that $\chi(\lambda)$ is a normalized polynomial from $K[\lambda]$ of degree N.

4.1.5. Minimal polynomials

Let $\mu(\lambda)$ be a normalized polynomial from $K[\lambda]$. Then $\mu(\lambda)$ is said to be a *minimal polynomial* with respect to A if $\mu(\lambda)$ annihilates the operator A, i.e., if $\mu(A) = 0$, and if $\deg \mu$ is minimal.

The minimal polynomial is uniquely determined. Let $\alpha(\lambda)$ be an annihilating polynomial with respect to A. Then, according to the division algorithm,

$$\alpha(\lambda) = q(\lambda)\,\mu(\lambda) + r(\lambda), \qquad \deg r < \deg \mu,$$

where $\mu(\lambda)$ denotes a minimal polynomial. Since $\alpha(A) = \mu(A) = 0$, $r(A) = 0$. This implies $r(\lambda) \equiv 0$ and therefore, $\alpha(\lambda) = q(\lambda)\,\mu(\lambda)$, i.e., every annihilating polynomial is a multiple of a minimal polynomial.

Assume that $\mu^*(\lambda)$ is a second minimal polynomial with respect to A. That is, one assumes $\mu^*(A) = 0$ and $\deg \mu = \deg \mu^*$. Then $\mu^*(\lambda) = q_1(\lambda)\,\mu(\lambda)$ and $\mu(\lambda) = q_2(\lambda)\,\mu^*(\lambda)$, which implies $\mu = \mu^*$ because μ and μ^* are normalized.

The minimal polynomial $\mu(\lambda)$ has at least the degree 1 (provided that $\dim X = N > 0$). If $\deg \mu = 0$, i.e., $\mu(\lambda) \equiv 1$, then $\mu(A) \equiv 1 = 0$, that is, the zero and the unit operator coincide. This is the case only for $N = 0$.

4.1.6. Determination of the minimal polynomial

First an auxiliary remark: Let $C(\lambda)$ be a polynomial operator of degree m with $m \geqq 0$, $C(\lambda) \neq 0$. Consider scalar divisors of $C(\lambda)$, i.e., polynomials $\alpha(\lambda)$ from $K[\lambda]$ satisfying $C(\lambda) = \alpha(\lambda)\,\tilde{C}(\lambda)$, where $\tilde{C}(\lambda)$ is a polynomial operator again. Let a representation of the form

$$C(\lambda) = \alpha(\lambda)\,\tilde{C}(\lambda)$$

be given, where $\alpha(\lambda)$ is normalized and $\deg \alpha$ is maximal. Then $\alpha(\lambda)$ is uniquely determined. If $\deg \alpha$ is maximal, then $\tilde{C}(\lambda)$ does not possess any scalar divisor of positive degree. Now let two representations $C = \alpha\tilde{C} = \beta\tilde{D}$ be given which satisfy the properties mentioned. If π is a prime divisor of α, then π is a divisor of β. If on the contrary, π and β have no common divisor, then $1 = \nu\pi + \mu\beta$ follows with certain polynomials ν and μ (see Subsect. A 2.2). Therefore one would obtain $\tilde{D} = \nu\pi\tilde{D} + \mu\beta\tilde{D}$ $= \pi\left(\nu\tilde{D} + \dfrac{\alpha}{\pi}\tilde{C}\right)$, which contradicts the properties of $\tilde{D}$. Hence, $\alpha = \beta$ and finally, $\tilde{C} = \tilde{D}$.

Now let $B(\lambda) = \chi(\lambda)\,(\lambda - A)^{-1}$. $B(\lambda)$ is a polynomial operator of degree $N - 1$ and (1.3) holds. Let

$$B(\lambda) = \gamma(\lambda)\,C(\lambda),$$

where $\gamma(\lambda)$ is scalar, normalized and of maximal degree. According to the auxiliary remark, $\gamma(\lambda)$ and $C(\lambda)$ are uniquely determined.

Now one obtains the proposition *$\gamma(\lambda)$ is a divisor of $\chi(\lambda)$.* To prove this proposition, note that

$$(\lambda - A)\,B(\lambda) = B(\lambda)\,(\lambda - A) = \chi(\lambda)$$

or

$$\gamma(\lambda)\,(\lambda - A)\,C(\lambda) = \gamma(\lambda)\,C(\lambda)\,(\lambda - A) = \chi(\lambda).$$

The last equation already implies the assertion. Let

$$\delta(\lambda) = \frac{\chi(\lambda)}{\gamma(\lambda)}.$$

Then

$$(\lambda - A)\,C(\lambda) = C(\lambda)\,(\lambda - A) = \delta(\lambda). \tag{1.4}$$

In any case therefore, the polynomial $\delta(\lambda)$ annihilates the operator A. Moreover, $\delta(\lambda)$ is normalized. Now one asserts: $\delta(\lambda) = \mu(\lambda)$. To prove this assertion, note that

$$\delta(\lambda) = \varrho(\lambda)\,\mu(\lambda)$$

with some polynomial $\varrho(\lambda)$, because $\delta(\lambda)$ is an annihilating polynomial. Since δ and μ are normalized the same is true for ϱ. Because $\mu(A) = 0$,

$$\mu(\lambda) = (\lambda - A)\,D(\lambda) = D(\lambda)\,(\lambda - A),$$

i.e., $\mu(\lambda)\,1$ can be divided by $\lambda - A$ from the left and from the right without remainder. Then,

$$\delta(\lambda) = (\lambda - A)\,\varrho(\lambda)\,D(\lambda).$$

Moreover, (1.4) is valid. Hence,

$$\varrho(\lambda)\,D(\lambda) = C(\lambda),$$

since the division algorithm is uniquely determined. That is, $\varrho(\lambda)$ is a scalar and normalized divisor of $C(\lambda)$. According to its definition, there are no proper scalar divisors for $C(\lambda)$, hence $\varrho(\lambda) \equiv 1$. Therefore, $\delta(\lambda) = \mu(\lambda)$, which proves the assertion. This equation yields a characterization of the minimal polynomial.

4.1.7. Consequence

According to the previous subsection,

$$\mu(\lambda) = (\lambda - A)\,C(\lambda) = C(\lambda)\,(\lambda - A).$$

Forming the determinant, one obtains

$$\big(\mu(\lambda)\big)^N = \det\,(\lambda - A)\,\det C(\lambda);$$

hence,

$$\big(\mu(\lambda)\big)^N = \chi(\lambda)\,\det C(\lambda).$$

Note that $\chi(\lambda) = \mu(\lambda)\,\gamma(\lambda)$. Then,

$$\big(\mu(\lambda)\big)^{N-1} = \gamma(\lambda)\,\det C(\lambda).$$

That is, $\gamma(\lambda)$ is a divisor of $(\mu(\lambda))^{N-1}$. In the case $N = 1$, this means $\gamma(\lambda) \equiv 1$. In all other cases one obtains: *If $\pi(\lambda)$ is a prime factor of $\gamma(\lambda)$, then $\pi(\lambda)$ is also a prime factor of $\mu(\lambda)$; in other words, all prime factors of $\chi(\lambda)$ are already contained in $\mu(\lambda)$.*

4.2. Spectral representation of an operator from $\mathscr{L}(X)$

In this section a first decomposition of an operator A from $\mathscr{L}(X)$ into a direct sum of parts of A will be constructed. This decomposition is called the *spectral representation.* These parts of A and the corresponding reducing projections are characterized uniquely by the fact that the corresponding characteristic polynomials are powers of a certain uniquely determined prime polynomial. Moreover, the corresponding reducing projections are polynomials in A. This construction starts from the characteristic polynomial and its decomposition into prime polynomials.

4.2.1. Characteristic polynomial

The characteristic polynomial $\chi(\lambda) = \det(\lambda - A)$ can be decomposed into a product of prime (irreducible) polynomials (see Subsect. A 2.6. Subsect. A 2.1 to A 2.6 also hold for an arbitrary commutative field). Let

$$\chi(\lambda) = \prod_{\varrho=1}^{r} \pi_\varrho(\lambda)^{\alpha_\varrho} \tag{2.1}$$

be this decomposition, where $\pi_1(\lambda), \pi_2(\lambda), \ldots, \pi_r(\lambda)$ denote the different prime factors of $\chi(\lambda)$. They are assumed to be normalized. The α_ϱ are natural numbers (multiplicities). Let

$$\varphi_\varrho(\lambda) = \pi_\varrho(\lambda)^{\alpha_\varrho}, \qquad \varrho = 1, \ldots, r. \tag{2.2}$$

The polynomials $\varphi_1, \ldots, \varphi_r$ are normalized and mutually relatively prime. Furthermore, one defines

$$\psi_\varrho(\lambda) = \prod_{\substack{\sigma=1 \\ \sigma \neq \varrho}}^{r} \varphi_\sigma(\lambda), \qquad \varrho = 1, \ldots, r. \tag{2.3}$$

The polynomials $\psi_1(\lambda), \ldots, \psi_r(\lambda)$ are normalized and relatively prime. Then there exist polynomials $q_1(\lambda), q_2(\lambda), \ldots, q_r(\lambda)$ such that

$$1 = \sum_{\varrho=1}^{r} q_\varrho(\lambda)\, \psi_\varrho(\lambda). \tag{2.4}$$

(The proof of (2.4) can be given according to Subsect. A 2.5).

4.2.2. Reducing projections

One defines operators $P_1, \ldots, P_r$ by

$$P_\varrho = q_\varrho(A)\, \psi_\varrho(A), \qquad \varrho = 1, \ldots, r. \tag{2.5}$$

These operators possess some remarkable properties, summarized in the following proposition.

Lemma 1. *The operators $P_1, P_2, \ldots, P_r$ are polynomials in A, which form a complete disjoint system of reducing projections with respect to A; all these operators are different from zero, i.e., the properties*

I. $P_\varrho P_\sigma = \delta_{\varrho\sigma} P_\varrho, \qquad \varrho, \sigma = 1, \ldots, r,$

II. $\sum_{\varrho=1}^{r} P_\varrho = 1,$

III. $P_\varrho \neq 0, \qquad \varrho = 1, \ldots, r,$

IV. $AP_\varrho = P_\varrho A, \qquad \varrho = 1, \ldots, r,$

hold. If $A_\varrho = A \upharpoonright P_\varrho X$, then

$$A = A_1 \oplus A_2 \oplus \cdots \oplus A_r, \tag{2.6}$$

and

V. $\chi(\lambda, A_\varrho) = \varphi_\varrho(\lambda), \qquad \varrho = 1, \ldots, r.$

Proof. First, $P_\varrho P_\sigma = 0$ if $\varrho \neq \sigma$. To prove this result, note that

$$P_\varrho P_\sigma = q_\varrho(A)\, \psi_\varrho(A)\, q_\sigma(A)\, \psi_\sigma(A) = q_\varrho(A)\, q_\sigma(A)\, \psi_\varrho(A)\, \psi_\sigma(A) = 0,$$

because $\chi(\lambda)$ is a divisor of $\psi_\varrho(\lambda)\, \psi_\sigma(\lambda)$ and because $\chi(A) = 0$. Since (2.4) holds, the equation $\sum_{\varrho=1}^{r} P_\varrho = \sum_{\varrho=1}^{r} q_\varrho(A)\, \psi_\varrho(A) = 1$ follows. Moreover, $P_\varrho^2 = P_\varrho$, $\varrho = 1, \ldots, r$, because

$$P_\varrho^2 = P_\varrho \Big(1 - \sum_{\sigma \neq \varrho} P_\sigma\Big) = P_\varrho - \sum_{\sigma \neq \varrho} P_\varrho P_\sigma = P_\varrho.$$

Furthermore, $AP_\varrho = P_\varrho A$, $\varrho = 1, \ldots, r$, because P_ϱ, according to its definition, is a polynomial in A. Thus, I, II and IV have been proved. Next,

$$\varphi_\varrho(A)\, P_\varrho = q_\varrho(A)\, \varphi_\varrho(A)\, \psi_\varrho(A) = q_\varrho(A)\, \chi(A) = 0.$$

Since the relation $\varphi_\varrho(A)\, P_\varrho = \varphi_\varrho(AP_\varrho)\, P_\varrho$ is valid, $P_\varrho \varphi_\varrho(AP_\varrho) = 0$; hence $\varphi_\varrho(AP_\varrho) \upharpoonright P_\varrho X = 0$. Because $AP_\varrho = 0 \oplus 0 \oplus \cdots \oplus A_\varrho \oplus \cdots \oplus 0$ and $\varphi_\varrho(AP_\varrho) = 0 \oplus \cdots \oplus \varphi_\varrho(A_\varrho) \oplus \cdots \oplus 0$, one finally obtains $\varphi_\varrho(A_\varrho) = 0$, $\varrho = 1, \ldots, r$, that is, $\varphi_\varrho(\lambda)$ is an annihilating polynomial for A_ϱ.

Let $\mu_\varrho(\lambda)$ be the minimal polynomial of A_ϱ. Then $\mu_\varrho(\lambda)$ is certainly a divisor of $\varphi_\varrho(\lambda)$, and because of (2.2),

$$\mu_\varrho(\lambda) = \pi_\varrho(\lambda)^{\beta_\varrho}, \qquad 0 \leqq \beta_\varrho \leqq \alpha_\varrho, \qquad \varrho = 1, \ldots, r.$$

For example, $\beta_\varrho = 0$ implies $\dim P_\varrho = 0$, i.e., $P_\varrho X = \{0\}$. Let $m_\varrho = \dim P_\varrho$, $\varrho = 1, \ldots, r$. Then $m_\varrho > 0$ if and only if $P_\varrho \neq 0$. Furthermore, let $\chi(\lambda, A_\varrho) = \det(\lambda - A_\varrho)$ be the characteristic polynomial of A_ϱ, where $\deg \chi(\lambda, A_\varrho) = m_\varrho$. According to Subsect. 4.1.7, every proper prime divisor of

$$\frac{\chi(\lambda, A_\varrho)}{\mu_\varrho(\lambda)}$$

is simultaneously a prime divisor of $\mu_\varrho(\lambda)$. That is, such a proper prime divisor can only be equal to $\pi_\varrho(\lambda)$. Thus, an equation of the form

$$\chi(\lambda, A_\varrho) = \pi_\varrho(\lambda)^{\gamma_\varrho}, \qquad \beta_\varrho \leqq \gamma_\varrho, \qquad \varrho = 1, \ldots, r, \tag{2.7}$$

follows. Now

$$\chi(\lambda, A) = \prod_{\varrho=1}^{r} \chi(\lambda, A_\varrho).$$

Therefore,

$$\chi(\lambda, A) = \prod_{\varrho=1}^{r} \pi_\varrho(\lambda)^{\gamma_\varrho}.$$

Since the decomposition of $\chi(\lambda, A)$ into prime factors is uniquely determined, the identity

$$\gamma_\varrho = \alpha_\varrho, \qquad \varrho = 1, \ldots, r,$$

follows, that is,

$$\chi(\lambda, A_\varrho) = \pi_\varrho(\lambda)^{\alpha_\varrho} = \varphi_\varrho(\lambda), \qquad \varrho = 1, \ldots, r.$$

Thus V is true. This implies, in particular, that all projections $P_1, \ldots, P_r$ are different from zero, because $\dim P_\varrho = m_\varrho = \deg \chi(\lambda, A_\varrho) = \alpha_\varrho \deg \pi_\varrho > 0$. ■

4.2.3. Spectral representation

According to Lemma 1, the decomposition (2.6) of the operator A is a proper decomposition into a direct sum of r operators, which are parts of A and act on subspaces of nonvanishing dimensions, where the characteristic polynomial in this subspace is given by V.

This representation of A is called a *spectral representation.* Note that for $K = \mathbb{C}$, one obtains the "usual" spectral representation of Ch. 2. The notation "spectral representation" for this decomposition is also justified by the fact that it is uniquely determined, as in the usual case.

In other words, the projections P_ϱ are uniquely determined solely by the fact that the characteristic polynomial of the part of A with respect to $P_\varrho X$ is given by $\pi_\varrho(\lambda)^{\alpha_\varrho}$.

Lemma 2. *The reducing projections $P_1, P_2, \ldots, P_r$ of A, assigned to A via* (2.5), *are uniquely determined by their characteristic polynomials $\pi_1(\lambda)^{\alpha_1}, \pi_2(\lambda)^{\alpha_2}, \ldots, \pi_r(\lambda)^{\alpha_r}$, respectively. That means: if Q is any reducing projection of A, $AQ = QA$, such that*

$$\chi(\lambda, AQ \restriction QX) = \pi_\varrho(\lambda)^{\alpha_\varrho},$$

then necessarily $Q = P_\varrho$.

Proof. Recall that $P_\varrho = \sigma_\varrho(A)$, where the polynomial $\sigma_\varrho(\lambda)$ is defined by $\sigma_\varrho(\lambda) = q_\varrho(\lambda) \psi_\varrho(\lambda)$. Because $AQ = QA$, $P_\varrho Q = QP_\varrho$, i.e., the operator $R_\varrho = P_\varrho Q = QP_\varrho$ is a projection reducing A, $R_\varrho A = AR_\varrho$. R_ϱ is a subprojection of Q as well as of P_ϱ; therefore,

$$Q = R_\varrho + (Q - R_\varrho),$$
$$P_\varrho = R_\varrho + (P_\varrho - R_\varrho).$$

Let $S = P_\varrho - R_\varrho$. Then

$$QS_\varrho = QP_\varrho - QR_\varrho = QP_\varrho - Q^2P_\varrho = 0$$

and correspondingly $S_\varrho Q = 0$. Hence, $Q + S_\varrho$ is a projection reducing A. S_ϱ is a subprojection of P_ϱ; hence, $\chi(\lambda, AS_\varrho \restriction S_\varrho X) = \pi_\varrho(\lambda)^\gamma, \gamma \geqq 0$. Then one obtains $\chi\big(\lambda, A(Q + S_\varrho) \restriction (Q + S_\varrho) X\big) = \pi_\varrho(\lambda)^{\alpha_\varrho + \gamma}$. $\chi(\lambda, A)$ contains only the factor $\pi_\varrho(\lambda)^{\alpha_\varrho}$; i.e., $\chi(\lambda, A)$ contains the prime factor $\pi_\varrho(\lambda)$ exactly α_ϱ times but no more. Hence $\gamma = 0$, that is, $\chi(\lambda, AS_\varrho \restriction S_\varrho X) = 1$ or $\dim S_\varrho X = 0$ or $S_\varrho X = \{0\}$ or $S_\varrho = 0$. Therefore one obtains $P_\varrho = R_\varrho \leqq Q$. $\dim Q = \dim P_\varrho$ because the characteristic polynomials of A in $P_\varrho X$ and QX coincide. Therefore $P_\varrho = Q$. ■

This uniqueness proposition will be used later as an essential step (see Subsect. 5.2.5). According to Lemma 2, the following conclusion is possible: Assume that by some method one is able to construct a reducing projection Q equipped with the characteristic polynomial $\pi_\varrho(\lambda)^{\alpha_\varrho}$. Then this projection Q necessarily coincides with the projection P_ϱ; in particular it must be a polynomial in A.

4.2.4. Consequence

Consider the question of decomposing an arbitrary operator A into a direct sum of irreducible operators. According to the results of Subsect. 4.2.2 and 4.2.3, one can consider, without loss of generality, an operator A for which the corresponding characteristic polynomial is a certain power of a prime polynomial; that is, one can assume that $\chi(\lambda, A) = \pi(\lambda)^m$, where $\pi(\lambda)$ denotes a prime polynomial. If $\deg \chi = \dim X = N$ and $\deg \pi = g$, then $N = mg$.

In the case just described, the minimal polynomial is also a power of $\pi(\lambda)$, $\mu(\lambda) = \pi(\lambda)^r$, $1 \leqq r \leqq m$. Therefore, the operator A satisfies the equation $\pi(A)^r = 0$, and there is no polynomial of lesser degree annihilating A.

Of course, in the case $K = \mathbb{C}$, the operators A just described correspond to the operators whose spectrum consist of exactly one eigenvalue.

4.3. Decomposition of operators whose characteristic polynomials are powers of a prime polynomial into a direct sum of irreducible operators

The following section depends strongly on the decomposition presented by W. Graeub [1, p. 202 sqq.].

4.3.1. Preliminaries

Let $\chi(\lambda, A) = \pi(\lambda)^m$ and $\mu(\lambda) = \pi(\lambda)^r$, where $\pi(\lambda)$ is a prime polynomial. Then $\mu(A)^r = 0$, and there is no polynomial of lesser degree annihilating A. Let K_{r-1} denote the kernel of $\pi(A)^{r-1}$ and let Y be a subspace. Assume that the sum $Y + K_{r-1}$ is a direct sum, $Y \oplus K_{r-1}$ (i.e., $Y \cap K_{r-1} = \{0\}$). Then the following propositions are equivalent:

I. *$Y \oplus K_{r-1}$ is invariant with respect to A,*
II. *$\pi(A)^{r-1} Y$ is invariant with respect to A,*
III. *Y is invariant* mod K_{r-1} *with respect to A.*

Assume I to be true. Let $x = \pi(A)^{r-1} y$, $y \in Y$. Then $Ax = \pi(A)^{r-1} Ay$ and $Ay = z + w$, $z \in Y$, $w \in K_{r-1}$. Hence, $Ax = \pi(A)^{r-1} z$; i.e., one obtains $Ax \in \pi(A)^{r-1} Y$, and II has been proved. Assume II to be true. Let $y \in Y$. Then, $A\pi(A)^{r-1} y = \pi(A)^{r-1} \times Ay = \pi(A)^{r-1} y'$, where $y' \in Y$. Therefore, $\pi(A)^{r-1} (Ay - y') = 0$, that is, $Ay \equiv y'$ (mod K_{r-1}), or $Ay \in Y$ mod K_{r-1}. Thus III has been proved. Assume III to be true. Let $z = y + w$, $z \in Y \oplus K_{r-1}$, $y \in Y$, $w \in K_{r-1}$. Then $Az = Ay + Aw$ and $Ay \equiv y'$ (mod K_{r-1}), i.e., $Ay = y' + w'$, where $y' \in Y$, $w' \in K_{r-1}$. Then $Az = y' + w' + Aw \in Y \oplus K_{r-1}$, because $w' + Aw \in K_{r-1}$. Thus I has been proved.

The desired decomposition of A can be founded on the following proposition (see W. Graeub [1, p. 202]).

Lemma 3. *Let $\pi(\lambda)^r$, $r \geqq 1$, denote the minimal polynomial of A in X. Furthermore, let K_{r-1} be the kernel of $\pi(A)^{r-1}$ and let Y be a subspace. Assume that the sum $Y + K_{r-1}$ is direct: $Y \oplus K_{r-1}$, and that this subspace $Y \oplus K_{r-1}$ is invariant with respect to A. Then there exists a direct decomposition of X of the form*

$$X = K_{r-1} \oplus Y \oplus U \oplus AU \oplus A^2U \oplus \cdots \oplus A^{g-1}U, \qquad g = \deg \pi, \tag{2.8}$$

where U denotes a suitable subspace of X.

Proof. Consider A as an operator in X mod $(Y \oplus K_{r-1})$. Then it is obvious that A is uniquely defined, because $x \equiv y (\mathrm{mod}\ Y \oplus K_{r-1})$, i.e., $x - y = k \in Y \oplus K_{r-1}$ implies $Ax - Ay = Ak \in Y \oplus K_{r-1}$; hence $Ax \equiv Ay (\mathrm{mod}\ Y \oplus K_{r-1})$.

$\pi(\lambda)$ annihilates A in X mod $(Y \oplus K_{r-1})$, because $\pi(A)^r = 0$ or $\pi(A)^{r-1} \pi(A) x = 0$ for all $x \in X$ or, equivalently, $\pi(A) x \in K_{r-1} \subseteqq K_{r-1} \oplus Y$, that is, $\pi(A) x \equiv 0 (\mathrm{mod}\ K_{r-1} \oplus Y)$ for all $x \in X$. In other words, $\pi(A)$ equals the zero operator in X mod $(Y \oplus K_{r-1})$. Since $\pi(\lambda)$ is irreducible (prime), the minimal polynomial of A in X mod $(Y \oplus K_{r-1})$ equals either 1 or $\pi(\lambda)$. If the minimal polynomial is equal to 1, then, according to Subsect. 4.1.5, the dimension of X mod $(Y \oplus K_{r-1})$ is equal to 0; that is, $X = Y \oplus K_{r-1}$, and it is sufficient to take $U = \{0\}$.

Let the minimal polynomial be equal to $\pi(\lambda)$. Then, certainly, $X \supset Y \oplus K_{r-1}$. Let $x_1 \in X \setminus (Y \oplus K_{r-1})$ and $x_1 \neq 0$. Consider the vectors $x_1, Ax_1, A^2x_1, \ldots, A^{g-1}x_1$. These g vectors are linearly independent mod $(Y \oplus K_{r-1})$.

To prove this assertion, let $f(\lambda)$ be a polynomial with $\deg f < g$ and assume $f(A) x_1 \equiv 0 (\mathrm{mod}\ Y \oplus K_{r-1})$. Then $f(A) A^\varrho x_1 \equiv 0 (\mathrm{mod}\ Y \oplus K_{r-1})$ for $\varrho = 1, 2, \ldots, g-1$; that is, $f(A) x \equiv 0 (\mathrm{mod}\ Y \oplus K_{r-1})$ for all $x \in \mathrm{spa}\,\{x_1, Ax_1, \ldots, A^{\varrho-1}x_1\} + (Y \oplus K_{r-1})$ or, equivalently, $f(\lambda)$ annihilates A in spa $\{x_1, \ldots, A^{g-1}x_1\}$ mod $(Y \oplus K_{r-1})$. Now $\pi(\lambda)$ is an annihilating polynomial in X mod $(Y \oplus K_{r-1})$, in particular in spa $\{x_1, \ldots, A^{\varrho-1}x_1\}$

$\times \bmod (Y \oplus K_{r-1})$; it is the minimal polynomial because dim spa $\{x_1, \ldots, A^{g-1}x_1\}$ mod $\times (Y \oplus K_{r-1}) > 0$. Then $\pi(\lambda)$ must necessarily be a divisor of $f(\lambda)$. Because $\deg f < g$, $f(\lambda) \equiv 0$, i.e., $f(\lambda)$ is the zero polynomial. In other words, it has been proved that the vectors $x_1, Ax_1, \ldots, A^{g-1}x_1$, are linearly independent mod $Y \oplus K_{r-1}$. If $H_1 = \text{spa}\,\{x_1, \ldots, A^{g-1}x_1\}$, the sum $H_1 + (Y \oplus K_{r-1})$ is direct; i.e., one can write $H_1 \oplus Y \oplus K_{r-1}$.

The subspace $H_1 \oplus Y \oplus K_{r-1}$ is invariant with respect to A. To prove this assertion, it is sufficient to consider the action of A on $A^{g-1}x_1$. One obtains

$$A^g = \pi(A) - \sum_{\varrho=0}^{g-1} \alpha_\varrho A^\varrho.$$

Hence,

$$A^g x_1 \equiv \pi(A)\, x_1 (\text{mod } H_1).$$

Now

$$\pi(A)\, x_1 \equiv 0 (\text{mod } Y \oplus K_{r-1}).$$

Therefore, $A^g x_1 \in H_1 \oplus Y \oplus K_{r-1}$.

The present situation is almost the same as at the beginning of the proof. One only has to replace the subspace Y by the subspace $H_1 \oplus Y$. A is uniquely defined as an operator in X mod $(K_{r-1} \oplus Y \oplus H_1)$; $\pi(\lambda)$ annihilates A in this space; hence the minimal polynomial equals either 1 or $\pi(\lambda)$. If the first case is true, then $X = K_{r-1} \oplus Y \oplus H_1$, and the proof is finished. If the second case is true, then, repeating the conclusion one is led to a subspace $H_2 = \text{spa}\,\{x_2, Ax_2, \ldots, A^{g-1}x_2\}$ such that $K_{r-1} \oplus Y \oplus H_1 \oplus H_2$ (in other words: the sum is direct).

After finitely many steps the first case must necessarily occur. Then

$$X = K_{r-1} \oplus Y \oplus H_1 \oplus H_2 \oplus \cdots \oplus H_m,$$

and the assertion of Lemma 3 is obtained with $U = \text{spa}\,\{x_1, x_2, \ldots, x_m\}$. ■

4.3.2. Construction of a direct decomposition of X

Let $K_\lambda := \ker\left(\pi(A)^\lambda\right)$, $\lambda = 0, 1, 2, \ldots, r$. Then one obtains the chain of inclusions

$$X = K_r \supset K_{r-1} \supset K_{r-2} \supset \cdots \supset K_1 \supset K_0 = \{0\}. \tag{2.9}$$

Now Lemma 3 will be applied with $Y = \{0\}$. In this case all assumptions of this lemma are satisfied and one obtains a direct decomposition of X in the form

$$X = K_{r-1} \oplus \sum_{\varrho=0}^{g-1} \oplus A^\varrho U_r, \qquad g = \deg \pi, \tag{2.10}$$

where U_r denotes some subspace of X.

Consider the subspace $\pi(A) \sum_{\varrho=0}^{g-1} \oplus A^\varrho U_r$. Since $\pi(A)^r = 0$, it is a subspace of K_{r-1}. Moreover, K_{r-2} is a subspace of K_{r-1}. Next, one shows that the sum

$$K_{r-2} + \sum_{\varrho=0}^{g-1} \pi(A)\, A^\varrho U_r$$

is a direct sum. Let $y \in K_{r-2}, x_0, x_1, \ldots, x_{g-1} \in U_r$ and

$$y + \pi(A)\,(x_0 + Ax_1 + \cdots + A^{g-1}x_{g-1}) = 0.$$

Then

$$\pi(A)^{r-2}\, y + \pi(A)^{r-1}\,(x_0 + Ax_1 + \cdots + A^{g-1}x_{g-1}) = 0.$$

Because $\pi(A)^{r-2}\, y = 0$, one obtains

$$x_0 + Ax_1 + \cdots + A^{g-1}x_{g-1} \in K_{r-1}.$$

From this equation and from (2.10) the relation $x_0 = Ax_1 = \cdots = A^{g-1}x_{g-1} = 0$ follows. Hence $y = 0$.

Furthermore, the subspace $K_{r-2} \oplus \sum_{\varrho=0}^{g-1} \oplus\, \pi(A)\, A^\varrho U_r$ is invariant with respect to A. According to the preliminary remarks to Lemma 3, this is equivalent to the proposition: $\pi(A)^{r-2} \sum_{\varrho=0}^{g-1} \oplus\, \pi(A)\, A^\varrho U_r$ is invariant with respect to A. Note that in this connection A is to be considered in K_{r-1}, where $\pi(A)^{r-1} = 0$ and where no polynomial of lesser degree annihilating A exists. In this case K_{r-2} plays the role of K_{r-1}. Obviously, $\pi(A)^{r-2} \sum_{\varrho=0}^{g-1} \oplus\, \pi(A)\, A^\varrho U_r = \pi(A)^{r-1} \sum_{\varrho=0}^{g-1} \oplus\, A^\varrho U_r$. Therefore, one has to show that the images of the subspaces $\pi(A)^{r-1}\, A^\varrho U_r$ under A, are contained in $\pi(A)^{r-1}\, A^\varrho U_r$ again. This is obvious for $\varrho = 0, 1, 2, \ldots, g-2$. Consider $\varrho = g-1$. In this case, one has to show

$$\pi(A)^{r-1}\, A^g U_r \subseteqq \pi(A)^{r-1} \sum_{\varrho=0}^{g-1} \oplus\, A^\varrho U_r.$$

Let $\pi(\lambda) = \lambda^g + \alpha_{g-1}\lambda^{g-1} + \alpha_{g-2}\lambda^{g-2} + \cdots + \alpha_1\lambda + \alpha_0$. Then,

$$\pi(A)^{r-1}\, A^g = \pi(A)^{r-1} \left(\pi(A) - \sum_{\varrho=1}^{g-1} \alpha_\varrho A^\varrho\right)$$

$$= \pi(A)^r - \pi(A)^{r-1} \sum_{\varrho=0}^{g-1} \alpha_\varrho A^\varrho = -\pi(A)^{r-1} \sum_{\varrho=0}^{g-1} \alpha_\varrho A^\varrho,$$

and this implies the assertion.

Therefore, Lemma 3 is applicable to the following case: X is replaced by K_{r-1}, A remains A. Then the minimal polynomial is $\pi(\lambda)^{r-1}$, K_{r-2} is the kernel of $\pi(A)^{r-2}$, and Y is replaced by $\sum_{\varrho=0}^{g-1} \oplus\, \pi(A)\, A^\varrho U_r$. The assumptions of Lemma 3 are satisfied and one obtains the following direct decomposition of K_{r-1}:

$$K_{r-1} = K_{r-2} \oplus \sum_{\varrho=0}^{g-1} \oplus\, \pi(A)\, A^\varrho U_r \oplus \sum_{\varrho=0}^{g=1} \oplus\, A^\varrho U_{r-1}, \tag{2.11}$$

where U_{r-1} denotes a subspace of K_{r-1}.

For the third step, consider the subspace

$$\pi(A)^2 \sum_{\varrho=0}^{g-1} A^\varrho U_r + \pi(A) \sum_{\varrho=0}^{g-1} A^\varrho U_{r-1}.$$

It is a subspace of K_{r-2}. K_{r-3} is also a subspace of K_{r-2}. Again, one shows that the sum

$$K_{r-3} + \sum_{\varrho=1}^{g-1} \pi(A)^2 A^\varrho U_r + \sum_{\varrho=1}^{g-1} \pi(A) A^\varrho U_{r-1}$$

is a direct sum. To prove this, let $y \in K_{r-3}$, $x_0, \ldots, x_{g-1} \in U_r$, $z_0, \ldots, z_{g-1} \in U_{r-1}$, and let

$$y + \pi(A)^2 \sum_{\varrho=0}^{g-1} A^\varrho x_\varrho + \pi(A) \sum_{\varrho=0}^{g-1} A^\varrho z_\varrho = 0.$$

Applying $\pi(A)^{r-3}$, one obtains

$$\pi(A)^{r-1} \sum_{\varrho=1}^{g-1} A^\varrho x_\varrho + \pi(A)^{r-2} \sum_{\varrho=1}^{g-1} A^\varrho z_\varrho = 0$$

or

$$\pi(A)^{r-2} \left[\pi(A) \sum A^\varrho x_\varrho + \sum A^\varrho z_\varrho\right] = 0.$$

Hence,

$$\pi(A) \sum A^\varrho x_\varrho + \sum A^\varrho z_\varrho \in K_{r-2}.$$

From this equation and because of (2.11), it follows that all terms vanish. Finally, $y = 0$. Furthermore, the subspace

$$K_{r-3} \oplus \sum_{\varrho=0}^{g-1} \oplus\, \pi(A)^2 A^\varrho U_r \oplus \sum_{\varrho=0}^{g-1} \oplus\, \pi(A) A^\varrho U_{r-1}$$

is invariant with respect to A. According to the remarks preceding Lemma 3, this result is equivalent to the proposition: $\pi(A)^{r-1} \sum_{\varrho=0}^{g-1} A^\varrho U_r + \pi(A)^{r-2} \sum_{\varrho=0}^{g-1} A^\varrho U_{r-1}$ is invariant with respect to A. The invariance of the first term of this sum has been already shown. Therefore, it is sufficient to show that, by application of A, the images of the subspaces $\pi(A)^{r-2} A^\varrho U_{r-1}$ are contained in $\pi(A)^{r-1} \sum A^\varrho U_r + \pi(A)^{r-2} \sum A^\varrho U_{r-1}$. This property is obvious for $\varrho = 0, 1, \ldots, g - 2$. Let $\varrho = g - 1$. Then $\pi(A)^{r-2} A^g = \pi(A)^{r-1} - \pi(A)^{r-2} \sum_{\varrho=0}^{g-1} \alpha_\varrho A^\varrho$ and, since $U_{r-1} \subset K_{r-1}$ is valid, one obtains

$$\pi(A)^{r-2} A^g U_{r-1} \subseteq \pi(A)^{r-2} \sum_{\varrho=0}^{g-1} A^\varrho U_{r-1}.$$

Therefore, Lemma 3 is applicable once more; X is to be replaced by K_{r-2}, A remains A. Then the minimal polynomial is $\pi(\lambda)^{r-2}$, K_{r-3} is the kernel of $\pi(A)^{r-3}$, and Y is to be replaced by $\sum_{\varrho=0}^{g-1} \oplus\, \pi(A)^2 A^\varrho U_r \oplus \sum_{\varrho=0}^{g-1} \oplus\, \pi(A) A^\varrho U_{r-1}$. Then the assumptions of Lemma 3 are satisfied and one obtains the following direct decomposition of K_{r-2}:

$$K_{r-2} = K_{r-3} \oplus \sum \oplus \pi(A)^2 A^\varrho U_r \oplus \sum \oplus \pi(A) A^\varrho U_{r-1} \oplus \sum \oplus A^\varrho U_{r-2}, \quad (2.12)$$

where U_{r-2} denotes a subspace of K_{r-2}.

These arguments can be continued. Thus one obtains, for every kernel K_λ of the chain of inclusions (2.9), a corresponding decomposition, namely

$$K_{r-\lambda} = K_{r-\lambda-1} \oplus \sum \oplus \pi(A)^\lambda A^\varrho U_r \oplus \sum \oplus \pi(A)^{\lambda-1} A^\varrho U_{r-1}$$
$$\oplus \cdots \oplus \sum \oplus A^\varrho U_{r-\lambda}, \quad (2.13)$$

where $\lambda = 0, 1, 2, \ldots, r-1$, and where U_σ, $\sigma = 1, \ldots, r$, denotes a subspace $U_\sigma \subset K_\sigma$. For K_1 one obtains

$$K_1 = \sum \oplus \pi(A)^{r-1} A^\varrho U_r \oplus \sum \oplus \pi(A)^{r-2} A^\varrho U_{r-1}$$
$$\oplus \cdots \oplus \sum \oplus A^\varrho U_1. \quad (2.14)$$

Finally, taking into account all equations (2.13), one obtains a direct decomposition of X of the following form:

$$X = \sum_{\varrho=0}^{g-1} \oplus A^\varrho U_r$$
$$\oplus \sum_{\varrho=0}^{g-1} \oplus \pi(A) A^\varrho U_r \oplus \sum_{\varrho=0}^{g-1} \oplus A^\varrho U_{r-1}$$
$$\oplus \sum_{\varrho=0}^{g-1} \oplus \pi(A)^2 A^\varrho U_r \oplus \sum_{\varrho=0}^{g-1} \oplus \pi(A) A^\varrho U_{r-1} \oplus \sum_{\varrho=0}^{g-1} \oplus A^\varrho U_{r-2}$$
$$\cdots$$
$$\oplus \sum \oplus \pi(A)^{r-1} A^\varrho U_r \oplus \sum \oplus \pi(A)^{r-2} A^\varrho U_{r-1}$$
$$\oplus \sum \oplus \pi(A)^{r-3} A^\varrho U_{r-2} \oplus \cdots \oplus \sum \oplus A^\varrho U_1. \quad (2.15)$$

4.3.3. Direct decomposition of A. Structure of the irreducible parts

From the direct decomposition of X obtained in the foregoing subsection, one derives a decomposition of A into a direct sum of irreducible operators together with an insight into the structure of the action of A with respect to the single irreducible parts. Let $\dim U_\varrho = n_\varrho$, $\varrho = 1, 2, \ldots, r$. One chooses a basis $x_\varrho^1, x_\varrho^2, \ldots, x_\varrho^{n_\varrho}$ in every subspace U_ϱ. To every vector x_ϱ^μ, there corresponds a subspace M_ϱ^μ spanned by the $g\varrho$ linearly independent vectors

$$x_\varrho^\mu, Ax_\varrho^\mu x, A^{2\mu}_\varrho, \ldots, A^{g-1}x_\varrho^\mu,$$
$$\pi(A)\, x_\varrho^\mu, \pi(A)\, Ax_\varrho^\mu, \pi(A)\, A^2x_\varrho^\mu, \ldots, \pi(A)\, A^{g-1}x_\varrho^\mu, \quad (2.16)$$
$$\pi(A)^{\varrho-1}\, x_\varrho^\mu, \pi(A)^{\varrho-1}\, Ax_\varrho^\mu, \pi(A)^{\varrho-1}\, A_2x_\varrho^\mu, \ldots, \pi(A)^{\varrho-1}\, A^{g-1}x_\mu^\varrho.$$

Some properties of these subspaces are summarized in

Lemma 4. *The subspaces M_ϱ^μ satisfy the following properties:*

I. $X = \sum_{\varrho=1}^{r} \sum_{\mu=1}^{n_\varrho} \oplus M_\varrho^\mu$,

II. $\dim M_\varrho^\mu = g\varrho$,

III. M_ϱ^μ *reduces* A,

IV. M_ϱ^μ *is irreducible with respect to* A.

Proof. II. This relation is obvious, because the $g\varrho$ vectors generating M_ϱ^μ are linearly independent, according to the direct decomposition of X obtained in Subsect. 4.3.2.

I. This relation also follows from the direct decomposition of X obtained in Subsect. 4.3.2, because the family of all vectors generating all M_ϱ^μ forms a basis of X.

To prove III, one has to verify that the images under A of the vectors $\pi(A)^\varkappa A^\sigma x_\varrho^\mu$, which form a basis of M_ϱ^μ, again lie in M_ϱ^μ. This is obvious for $\varrho = 0, 1, 2, \ldots, g-2$. Let $\sigma = g-1$. Then

$$\pi(A)^\varkappa A^g x_\varrho^\mu = \left(\pi(A)\right)^{\varkappa+1} x_\varrho^\mu - \sum_{\sigma=0}^{g-1} \alpha_\sigma \pi(A)^\varkappa A^\sigma x_\varrho^\mu, \qquad \varkappa = 0, 1, \ldots, \varrho - 1.$$

From this equation the assertion follows for $\varkappa = 0, 1, \ldots, \varrho - 2$. If $\varkappa = \varrho - 1$, then $\pi(A)^{\varkappa+1} x_\varrho^\mu = \pi(A)^\varrho x_\varrho^\mu = 0$, because $x_\varrho^\mu \in U_\varrho \subset K_\varrho$. Therefore, the vectors $\pi(A)^{\varrho-1} A^g x_\varrho^\mu$ are also contained in M_ϱ^μ. Hence, M_ϱ^μ is invariant with respect to A. Using I, one obtains III.

To prove IV, first note that the minimal polynomial of A with respect to M_ϱ^μ equals $\pi(\lambda)^\varrho$. On the one hand, $\pi(A)^\varrho x = 0$ for all $x \in M_\varrho^\mu$, because $x_\varrho^\mu \in K_\varrho$; on the other hand, if $f(\lambda)$ is a polynomial with $\deg f < g\varrho$ and $f(A)\, x = 0$ for all $x \in M_\varrho^\mu$, in particular $f(A)\, x_\varrho^\mu = 0$, then, since the vectors $\pi(A)^\varkappa A^\sigma x_\varrho^\mu$, $\varkappa = 0, 1, \ldots, \varrho - 1$, $\sigma = 0, 1, \ldots, g-1$ are linearly independent, this polynomial is necessarily the zero polynomial. Assuming $M_\varrho^\mu = H_1 \oplus H_2$ to be a proper decomposition of M_ϱ^μ into a direct sum of invariant subspaces with respect to A, one gets minimal polynomials $\pi(\lambda)^{\varrho_1}$ and $\pi(\lambda)^{\varrho_2}$ for the subspaces H_1 and H_2, respectively. This implies that the subspaces $\pi(A)^{\varrho_1-1} H_1$ and $\pi(A)^{\varrho_2-1} H_2$ are both different from the zero space. Therefore, the minimal polynomial $\pi(\lambda)$ corresponds to both subspaces. This implies $\dim \pi(A)^{\varrho_1-1} H_1 \geqq g$ and $\dim \pi(A)^{\varrho_2-1} H_2 \geqq g$. The sum $\pi(A)^{\varrho_1-1} H_1 \oplus \pi(A)^{\varrho_2-1} H_2$ is direct; hence $\dim \left(\pi(A)^{\varrho_1-1} \times H_1 \oplus \pi(A)^{\varrho_2-1} H_2\right) \geqq 2g$. However, the dimension of the kernel of $\pi(A)$ in M_ϱ^μ is equal to g, as can be seen from the basis of table (2.16) for M_ϱ^μ. Because $\pi(A)^{\varrho_1-1} H_1 \oplus \pi(A)^{\varrho_2-1} \times H_2$ is necessarily a subspace of this kernel, one obtains a contradiction and IV has been proved. ■

Lemma 4 yields a decomposition of A into a direct sum of irreducible parts. This decomposition consists of n_1 parts of dimension g, of n_2 parts of dimension $2g$, etc., and of n_r parts of dimension rg, where necessarily $n_r \geqq 1$.

The structure of the action of A in the single irreducible parts can be obtained from the basis of Table (2.16). With respect to this basis and to M_ϱ^μ, the matrix of A takes a form which can best be expressed by an example:

Let $g = 4$ and $\varrho = 3$. Then $\dim M_3^\mu = 12$. If $\pi(\lambda) = \lambda^4 - \alpha - \beta\lambda - \gamma\lambda^2 - \delta\lambda^3$, the matrix of A in M_3^μ with respect to the basis (2.16) takes the form

$$\left(\begin{array}{cccc:cccc:cccc}
0 & 1 & 0 & 0 & & & & & & & & \\
0 & 0 & 1 & 0 & & & & & & & & \\
0 & 0 & 0 & 1 & 0 & 0 & & & & \multicolumn{3}{c}{0} \\
\alpha & \beta & \gamma & \delta & 1 & 0 & & & & & & \\
\hdashline
 & & & & 0 & 1 & 0 & 0 & & & & \\
 & 0 & & & 0 & 0 & 1 & 0 & & & & \\
 & & & & 0 & 0 & 0 & 1 & 0 & 0 & & \\
 & & & & \alpha & \beta & \gamma & \delta & 1 & 0 & & \\
\hdashline
 & & & & & & & & 0 & 1 & 0 & 0 \\
 & 0 & & & & 0 & & & 0 & 0 & 1 & 0 \\
 & & & & & & & & 0 & 0 & 0 & 1 \\
 & & & & & & & & \alpha & \beta & \gamma & \delta
\end{array}\right) = \Gamma_3(\pi).$$

In general, the matrix of A with respect to the basis (2.16) of $M_{\varrho_1}^\mu$ is denoted by $\Gamma_\varrho(\pi)$. Thus one obtains the following corresponding matrix of A in X:

$$A \leftrightarrow \{n_1\Gamma_1(\pi), n_2\Gamma_2(\pi), \ldots, n_r\Gamma_r(\pi)\}, \tag{2.17}$$

where the factors n_ϱ of $\Gamma_\varrho(\pi)$ indicate that the corresponding matrix $\Gamma_\varrho(\pi)$ appears exactly n_ϱ times.

The matrices $\Gamma_\varrho(\pi)$ are called *generalized Jordan blocks*. They are uniquely determined by an irreducible polynomial and by a natural number ϱ.

An operator A on X is called a *generalized Jordan operator* if there is a basis in X such that the corresponding matrix of A with respect to this basis is a generalized Jordan block $\Gamma_m(\pi)$.

If A is an arbitrary operator on X, a reducing projection P of A is called a *generalized Jordan projection* if $A \restriction PX$ is a generalized Jordan operator.

Again one obtains (cf. Subsect. 2.9.3): *A is a generalized Jordan operator in X if and only if A is irreducible in X.*

To prove this result, let A be a generalized Jordan operator. Then A is irreducible (see the proof of IV in Lemma 4). If, conversely, A is irreducible in X, then $\chi(\lambda, A) = \pi(\lambda)^m$, where $\pi(\lambda)$ denotes some irreducible polynomial. If further prime factors of χ different from π existed, then the spectral decomposition of A would yield a nontrivial decomposition of A. Furthermore, one can construct the direct decomposition of A according to Lemma 4. Since A is irreducible, this decomposition necessarily contains exactly one term. Hence, A is a generalized Jordan operator.

Simultaneously one obtains the proposition: *A reducing projection P of A is irreducible if and only if it is a generalized Jordan projection.*

With the use of the concepts just introduced, Lemma 4 can be formulated as follows.

Lemma 5. *Let A be an operator on X with $\chi(\lambda, A) = \pi(\lambda)^m$, where π is irreducible. Then there exists a complete disjoint system of reducing generalized Jordan projections P_ϱ^σ $\varrho = 1, 2, \ldots, r$; $\sigma = 1, 2, \ldots, n_\varrho$, with respect to A, where r is defined by the fact that*

$\pi(\lambda)^r$ is the minimal polynomial of A. One has dim $P_\varrho^\sigma = g\varrho$, *where* $g = \deg \pi$, *and* $A \upharpoonright P_\varrho^\sigma X$ *is a generalized Jordan operator with the corresponding generalized Jordan block* $\Gamma_\varrho(\pi)$. *Furthermore,* $n_r \geqq 1$.

The matrix of A on X with respect to a suitable basis takes the form (2.17) in this case. It is called a *generalized Jordan normal form.*

4.3.4. Special case, $g = 1$

The case $\deg \pi = g = 1$, where $\pi(\lambda) = \lambda - \alpha$, is considered separately. In this case, a generalized Jordan block $\Gamma_\varrho(\pi)$ takes the following form

$$\Gamma_\varrho(\pi) = \Gamma_\varrho(\lambda - \alpha) = \begin{pmatrix} \alpha & 1 & 0 & 0 & \cdots & 0 & 0 \\ 0 & \alpha & 1 & 0 & \cdots & 0 & 0 \\ 0 & 0 & \alpha & 1 & \cdots & 0 & 0 \\ & & & \cdots & & & \\ 0 & 0 & 0 & 0 & \cdots & \alpha & 1 \\ 0 & 0 & 0 & 0 & \cdots & 0 & \alpha \end{pmatrix}; \qquad \varrho \times \varrho\text{-matrix};$$

that is, in this case one obtains a usual Jordan block. This fact justifies the notations introduced in Subsect. 4.3.3: A generalized Jordan operator with $\deg \pi = 1$ is a (standard) Jordan operator; a generalized Jordan projection P_0 with $\deg \pi = 1$ (where π denotes the prime polynomial corresponding to $A \upharpoonright P_0 X$) is a (standard) Jordan projection; a generalized Jordan normal form, where $\deg \pi = 1$ for all π, is a (standard) Jordan normal form.

Note that *a Jordan operator A on X is irreducible* in any case, but now the converse is not true in general. A Jordan projection P_0 of A is always irreducible. One obtains

Corollary 1. *Let A be an operator on X with $\chi(\lambda, A) = (\lambda - \alpha)^m$. Then there exists a complete disjoint system of reducing Jordan projections P_ϱ^σ, $\varrho = 1, \ldots, r$; $\sigma = 1, 2, \ldots, n_\varrho$, where r is defined by the fact that $(\lambda - \alpha)^r$ is the minimal polynomial of A. Furthermore,* dim $P_\varrho^\sigma = \varrho$, *and $A \upharpoonright P_\varrho^\sigma X$ is a Jordan operator with the Jordan block $\Gamma_\varrho(\alpha)$.*

4.3.5. Uniqueness of the generalized Jordan structure

The basis (2.16), which was essential for the construction in Subsect. 4.3.3, is not uniquely determined. However, the generalized Jordan normal form (2.17) of A is uniquely determined up to the order of the relevant generalized Jordan blocks. In other words, the following proposition holds.

Lemma 6. *The dimensions and the corresponding numbers of generalized Jordan projections of A on X are uniquely determined.*

Proof. The assertion is that the numbers $n_1, n_2, \ldots, n_r$ are uniquely determined. Consider the operator $\pi(A)$. It is nilpotent because $\pi(A)^r = 0$. The minimal polynomial of $\pi(A)$ is λ^r. That is, for the operator $\pi(A)$ one obtains the case described in Corollary 1.

The decomposition of $\pi(A)$ into a direct sum of Jordan operators and the corresponding decomposition of X into a direct sum of irreducible subspaces can be taken from Table (2.15). Denoting the number of irreducible subspaces of dimension ϱ by p_ϱ, $\varrho = 1, 2, \ldots, r$, one obtains the relation

$$p_\varrho = g n_\varrho, \qquad \varrho = 1, 2, \ldots, r, \tag{2.18}$$

which expresses the fact that an irreducible subspace with respect to A is decomposed into g irreducible subspaces with respect to $\pi(A)$. The numbers $p_1, p_2, \ldots, p_r$ are uniquely determined (see Subsect. 2.9.4). According to the considerations presented there, one obtains

$$p_\varrho = 2 \dim K_\varrho - \dim K_{\varrho-1} - \dim K_{\varrho+1}, \qquad \varrho = 1, 2, \ldots, r,$$

where $K_\varrho = \ker \pi(A)^\varrho$, $\varrho = 1, 2, \ldots, r$, and $K_0 = \{0\}$. Because of the relations $\dim K_\varrho = \operatorname{def} \pi(A)^\varrho$ and (2.18),

$$n_\varrho = \frac{1}{g} \left(2 \operatorname{def} \pi(A)^\varrho - \operatorname{def} \pi(A)^{\varrho-1} - \operatorname{def} \pi(A)^{\varrho+1}\right). \tag{2.19}$$

This proves the uniqueness of those numbers. ■

4.4. Generalized Jordan structure of arbitrary operators

In this section the results of Sect. 4.2 and 4.3 are summarized. The result shows the decomposition of an arbitrary operator A into a direct sum of generalized Jordan operators.

4.4.1. Decomposition of an arbitrary operator

Let A be an arbitrary operator with the characteristic polynomial $\chi(\lambda, A) = \pi_1(\lambda)^{m_1} \times \pi_2(\lambda)^{m_2} \cdots \pi_r(\lambda)^{m_r}$. The spectral representation is denoted by $A = A_1 \oplus A_2 \oplus \cdots \oplus A_r$. The corresponding resolution of the unit operator 1 is given by $1 = P_1 + P_2 + \cdots + P_r$. Then $\chi(\lambda, A_\varrho) = \chi(\lambda, A \upharpoonright P_\varrho X) = \pi_\varrho(\lambda)^{m_\varrho}$, where $\dim P_\varrho = g_\varrho m_\varrho$, $g_\varrho = \deg \pi_\varrho$.

To every subspace $P_\varrho X$, the considerations of Sect. 4.3 can be applied; that is, one can choose a basis in $P_\varrho X$ according to Table (2.16). In this way one obtains a basis in the whole space X called a *generalized Jordan basis* of A in X. The corresponding matrix of A is called a *generalized Jordan normal form* of A. Because the spectral representation is uniquely determined, and because of Lemma 3.6, the generalized Jordan normal form is uniquely determined up to the order of the relevant generalized Jordan blocks. It consists of generalized Jordan blocks $\Gamma_p(\pi_\varrho)$, $p = 1, 2, \ldots, \nu_\varrho$, where $\pi_\varrho(\lambda)^{\nu_\varrho}$ is the minimal polynomial of A_ϱ. The number of identical blocks $\Gamma_p(\pi_\varrho)$ is denoted by $\varkappa(\varrho, p)$. Then the generalized Jordan normal form takes the form

$$\{\varkappa(1,1)\, \Gamma_1(\pi_1), \ldots, \varkappa(1, \nu_1)\, \Gamma_{\nu_1}(\pi_1); \varkappa(2,1)\, \Gamma_1(\pi_2), \ldots, \varkappa(2, \nu_2)\, \Gamma_{\nu_2}(\pi_2), \ldots, \varkappa(r, \nu_r)\, \Gamma_{\nu_r}(\pi_r)\},$$

where the notation $\varkappa(\varrho, p)\, \Gamma_p(\pi_\varrho)$ indicates that the generalized Jordan block $\Gamma_p(\pi_\varrho)$ appears exactly $\varkappa(\varrho, p)$ times. The numbers $\varkappa(\varrho, p)$ are uniquely determined. The result can also be formulated as follows:

Theorem 1 (generalized Jordan structure). *Let A be an arbitrary operator on X, where $1 = P_1 + P_2 + \cdots + P_r$ denotes the resolution of the unit operator corresponding to the spectral representation. Then P_ϱ can be decomposed in the form*

$$P_\varrho = \sum_{\sigma=1}^{\nu_\varrho} \sum_{\varkappa=1}^{\varkappa(\varrho,\sigma)} P_\varrho^{\sigma\varkappa},$$

where the projections $P_\varrho^{\varkappa\sigma}$ form a disjoint system of reducing generalized Jordan projections with respect to A; that is, $A \restriction P_\varrho^{\sigma\varkappa} X$ is a generalized Jordan operator assigned to the irreducible polynomial $\pi_\varrho(\lambda)$. One has $\dim P_\varrho^{\sigma\varkappa} = g_\varrho \sigma$ for all $\varkappa = 1, 2, \ldots, \varkappa(\varrho, \sigma)$ and the numbers $\varkappa(\varrho, \sigma)$ of generalized Jordan projections assigned to π_ϱ and of fixed dimension $g_\varrho \sigma$ are uniquely determined. The generalized Jordan block belonging to $P_\varrho^{\sigma\varkappa}$ takes the form $\Gamma_\sigma(\pi_\varrho)$.

According to Subsect. 4.3.5, the numbers $\varkappa(\varrho, \sigma)$ are determined by the formulas (2.19).

4.4.2. Complete system of invariants of a class of similar matrices; elementary divisors

The result obtained in the foregoing subsection can be interpreted as follows. Recall that an operator A can be represented uniquely by a class of similar matrices. As shown above, there is exactly one generalized Jordan normal form belonging to a given class $\{A\}$ (up to the order of the generalized Jordan blocks). Since the generalized Jordan normal form is uniquely determined by the numbers $\varkappa(\pi, p)$, where π runs through all irreducible polynomials occurring in $\chi(\lambda, A)$ and where p is a natural number, one obtains: *The system of the numbers $\varkappa(\pi, p)$ (together with their arguments) forms a complete system of invariants with respect to the class of similar matrices;* that is, such a class can be represented uniquely by this system.

Let π be an irreducible polynomial occurring in $\chi(\lambda, A)$, where $\deg \pi = g$, and let P_0 be a generalized Jordan projection belonging to π, where $\dim P_0 = g\sigma$. Then the polynomial $\chi(\lambda, A \restriction P_0 X) = \pi(\lambda)^\sigma$ is called the corresponding *elementary divisor*. Of course, the system formed by all elementary divisors of A is also a complete system of invariants with respect to A.

Finally, the special case of the (standard) Jordan normal form is considered. The considerations of Subsect. 4.3.4 lead to the following proposition.

Corollary 1. *A class of similar matrices corresponding to an operator A contains a Jordan normal form if and only if all prime factors $\pi(\lambda)$ of $\chi(\lambda, A)$ are linear, i.e., of degree 1; that is, they take the form $\pi(\lambda) = \lambda - \alpha$. In this case (and only in this case) all irreducible parts of A are Jordan operators.*

4.4.3. Decomposition with respect to a given reducing projection

Let A be an operator and let Q be a corresponding reducing projection. Then $AQ = QA$; hence $P_\varrho Q = QP_\varrho$, $\varrho = 1, \ldots, r$, for all projections of the spectral representation with respect to A. That is, a given reducing projection Q can be "distributed" among the single reducing projections P_ϱ. If $\chi(\lambda, A \upharpoonright P_\varrho X) = \pi_\varrho(\lambda)^{m_\varrho}$, then

$$\chi(\lambda, A \upharpoonright QP_\varrho X) = \pi_\varrho(\lambda)^{m'_\varrho}; \qquad \chi(\lambda, A \upharpoonright (1 - Q) P_\varrho X) = \pi_\varrho(\lambda)^{m''_\varrho},$$

where $m_\varrho = m'_\varrho + m''_\varrho$.

Then the construction of a complete disjoint system of (irreducible) generalized Jordan projections in $P_\varrho X$ with respect to $A \upharpoonright P_\varrho X$ described in Sect. 4.3 can be performed in the spaces $QP_\varrho X$ and $(1 - Q) P_\varrho X$ separately.

Therefore, the projection Q appears as a certain sum of generalized Jordan projections of the decomposition of A in X.

In other words: *The decomposition of A can be chosen such that it is compatible* (*i.e., commuting*) *with a given reducing projection of A.*

Applications

The results of the foregoing sections 4.2 to 4.4 will be applied in solving a problem (Sect. 4.6) which has been already touched on in Subsect. 3.3.13, namely the determination of the "smallest" meromorphic reducing projections of a meromorphic operator-valued function $A(z)$ in a given region G. This application requires some preliminaries (Sect. 4.5). With the help of these preliminaries, a problem will be solved (Sect. 4.7) which has been mentioned in Subsect. 3.4.2. This problem is connected with the transformation of a meromorphic complete disjoint system of projections in a given region G.

4.5. Linear spaces of meromorphic vector-valued functions and their linear operators

4.5.1. Meromorphic vector-valued functions

Let X be a complex linear space of dimension N, $N = \dim X$. Let G be a region of the complex plane. The field of the meromorphic (scalar) functions in G is denoted by $\mathscr{M}$ (see Subsect. A 2.1).

Let $a(z)$ be a meromorphic vector-valued function from G into X (see Subsect. 1.13.5); i.e., $a(z)$ is a vector-valued function such that all scalar products $(a(z), f)$, $f \in X^*$, are meromorphic functions from $\mathscr{M}$. The set of all meromorphic vector-valued functions $a(z)$ from G into X is denoted by $X(G_{\mathscr{M}}) \equiv X(G)$.

$X(G_{\mathscr{M}})$ is a complex linear space, i.e., a linear space over the field $\mathbb{C}$, because $\alpha a(z) + \beta b(z)$, where $\alpha, \beta \in \mathbb{C}$, is a meromorphic vector-valued function, hence from $X(G_{\mathscr{M}})$, whenever $a(z)$ and $b(z)$ are; furthermore, the corresponding rules (see Subsect. 1.1.1) are satisfied.

4.5.2. Embedding of X into $X(G_{\mathcal{M}})$

Let $\alpha \in \mathbb{C}$. The meromorphic function $\alpha(z)$ from $\mathcal{M}$ which is constant and equal to α can be identified with the complex number α, $\alpha(z) \equiv \alpha$. In this sense $\mathbb{C} \subset \mathcal{M}$ holds; that is, $\mathbb{C}$ is the subfield consisting of all constant meromorphic functions, the so-called field of constants with respect to $\mathcal{M}$.

Similarly, X can be considered as a subspace of $X(G_{\mathcal{M}})$. The meromorphic vector-valued function $a(z)$ from $X(G_{\mathcal{M}})$ which is constant and equal to $a \in X$ can be identified with the vector a. In this sense, $X \subset X(G_{\mathcal{M}})$ holds; that is, X is a subset of $X(G_{\mathcal{M}})$, the subset of all constant meromorphic vector-valued functions. Obviously, this subset X is a subspace of $X(G_{\mathcal{M}})$ over $\mathbb{C}$.

4.5.3. $X(G_{\mathcal{M}})$ as a linear space over $\mathcal{M}$

The space $X(G_{\mathcal{M}})$ can also be considered as a linear space over the field $\mathcal{M}$. If $a(z)$ and $b(z)$ are meromorphic vector-valued functions and if $\alpha(z)$ and $\beta(z)$ denote meromorphic scalar functions, then $\alpha(z)\, a(z) + \beta(z)\, b(z)$ is a meromorphic vector-valued function again. The corresponding rules (see Subsect. 1.1.1) are obviously satisfied.

To determine the dimension of $X(G_{\mathcal{M}})$ over $\mathcal{M}$, first let $e_1, e_2, \ldots, e_N$ be a basis of X over $\mathbb{C}$. The corresponding adjoint basis of X^* is denoted by $f_1, \ldots, f_N$. Secondly, let $a(z)$ be an arbitrary meromorphic vector-valued function. Then

$$a(z) = \big(a(z), f_1\big)\, e_1 + \big(a(z), f_2\big)\, e_2 + \cdots + \big(a(z), f_N\big)\, e_N, \tag{5.1}$$

where the scalar functions $\big(a(z), f_\varrho\big)$ are meromorphic, i.e., $\big(a(z), f_\varrho\big) \in \mathcal{M}$. Thus the N vectors $e_1, e_2, \ldots, e_N$ of $X(G_{\mathcal{M}})$ form a system generating $X(G_{\mathcal{M}})$ over $\mathcal{M}$. Let

$$\alpha_1(z)\, e_1 + \alpha_2(z)\, e_2 + \cdots + \alpha_N(z)\, e_N = 0$$

in the sense of an equation in the space $X(G_{\mathcal{M}})$ over $\mathcal{M}$. The functions $\alpha_1(z), \ldots, \alpha_N(z)$ are meromorphic scalar functions. If z_0 denotes a holomorphic point for all functions $\alpha_\varrho(z)$, then

$$\alpha_1(z_0)\, e_1 + \alpha_2(z_0)\, e_2 + \cdots + \alpha_N(z_0)\, e_N = 0$$

and, since the vectors $e_1, \ldots, e_N$ are linearly independent in X over $\mathbb{C}$, one obtains $\alpha_1(z_0) = \alpha_2(z_0) = \cdots = \alpha_N(z_0)$ for all holomorphic points z_0. From the identity theorem for holomorphic functions the identities $\alpha_1(z) \equiv \alpha_2(z) \equiv \cdots \equiv \alpha_N(z) \equiv 0$ follow. Hence, the vectors $e_1, \ldots, e_N$ also form a basis of $X(G_{\mathcal{M}})$ over $\mathcal{M}$, that is

$$\dim \big(X(G_{\mathcal{M}}) : \mathcal{M}\big) = \dim (X : \mathbb{C}), \tag{5.2}$$

where the notation $X : \mathbb{C}$ indicates explicitly that the linear space X is to be considered over the field $\mathbb{C}$. Simultaneously, it has been shown that every basis of X over $\mathbb{C}$ is also a basis of $X(G_{\mathcal{M}})$ over $\mathcal{M}$; that is, in $X(G_{\mathcal{M}})$ there are always bases formed by constant vector-valued functions.

For example, in the case $X = \mathbb{C}^N$ the space $X(G_{\mathcal{M}})$ is the space of all N-tuples $\big(a_1(z), a_2(z), \ldots, a_N(z)\big)$ of meromorphic scalar functions.

4.5.4. Linear operators on $X(G_{\mathscr{M}})$

The linear operators of $X(G):\mathscr{M}$ can be identified with the meromorphic operator-valued functions from G into $\mathscr{L}(X)$.

This can be seen as follows: Let $A(z)$ be a meromorphic operator-valued function from G into $\mathscr{L}(X)$ (see Subsect. 1.5.4) and let $x(z)$ be a meromorphic vector-valued function, i.e., an element from $X(G)$. Then $A(z)\,x(z)$ is also a meromorphic vector-valued function (see Subsect. 1.13.5) and

$$A(z)\,\big(\alpha(z)\,x(z)+\beta(z)\,y(z)\big)=\alpha(z)\,A(z)\,x(z)+\beta(z)\,A(z)\,y(z),$$

where $\alpha(z), \beta(z)$ are meromorphic scalar functions and $x(z), y(z)$ are meromorphic vector-valued functions.

Thus $A(z)$ defines in a unique way a linear operator A from $\mathscr{L}\big(X(G):\mathscr{M}\big)$ via $A\big(x(z)\big) = A(z)\,x(z)$.

Conversely, let $A\in\mathscr{L}\big(X(G):\mathscr{M}\big)$. If $x(z)$ is a meromorphic vector-valued function, then $A\big(x(z)\big)=y\big(z;x(z)\big)$ is also a meromorphic vector-valued function. In particular, for all constant vector-valued functions x the expression $Ax = y(z;x)$ is a meromorphic vector-valued function. Therefore, A defines an operator-valued function $A(z)$ from G into $\mathscr{L}(X)$ by $A(z)\,x=y(z;x)$. This function is meromorphic, because the scalar products $\big(A(z)\,x,f\big)=\big(y(z;x),f\big)$ are meromorphic scalar functions for all $f\in X^*$.

Hence, A defines a unique meromorphic operator-valued function $A(z)$ from G into $\mathscr{L}(X)$ satisfying the equation $A\big(x(z)\big)=A(z)\,x(z)$.

Therefore, the identification mentioned above is possible without ambiguity. Obviously it is compatible with the linear relations.

For example, let $X=\mathbb{C}^N$, such that $\mathscr{L}(X)$ coincides with the set of all complex $N\times N$-matrices. Then the identification described above is reflected in the following simple fact: The set $\mathscr{L}\big(X(G):\mathscr{M}\big)$ of all operators on $X(G):\mathscr{M}$ coincides with the set of all $N\times N$-matrices, where the coefficients are taken from $\mathscr{M}$. On the other hand, these matrices can be considered as meromorphic functions from G into the set of all $N\times N$-matrices with complex coefficients.

4.6. Meromorphic reducing projections of meromorphic operator-valued functions

4.6.1. Construction of meromorphic-irreducible reducing projections

Let $A(z)$ be a meromorphic operator-valued function from G into $\mathscr{L}(X)$. The next problem is constructing a meromorphic reducing projection-valued function in G with respect to $A(z)$. In particular, one is interested in projection-valued functions which are meromorphic-irreducible, where a meromorphic projection-valued function is called meromorphic-irreducible if there is no proper decomposition into meromorphic reducing projections. That is, one asks for the "smallest" meromorphic reducing projections with respect to $A(z)$.

Referring to the results in Subsect. 3.3.12 and 3.3.13, one has the spectral representation $A(z)=\sum\limits_{\alpha,\varrho}\big(\lambda_{\alpha\varrho}(z)\,P_{\alpha\varrho}(z)+D_{\alpha\varrho}(z)\big)$ for simple points z. But, in general, the eigenprojections $P_{\alpha\varrho}(z)$ fail to be meromorphic in G. Meromorphic reducing projections can be obtained by symmetrization $P_\alpha(z) = \sum_{\varrho=1}^{g_\alpha} P_{\alpha\varrho}(z)$. In general, however, the projections $P_\alpha(z)$ fail to be meromorphic-irreducible, as will be shown shortly. Moreover, the results of Sect. 4.2 lead to the construction of the projections $P_\alpha(z)$ without the use of the eigenprojections, namely, as operator polynomials of $A(z)$ with meromorphic coefficients.

4.6.2. Spectral representation of $A(z)$

Note that "spectral representation" in the following discussion means spectral representation in the sense of Sect. 4.2, that is, a decomposition over a certain field K, in the present case the field $\mathscr{M}$.

According to Subsect. 4.5.4, the operator-valued function $A(z)$ can be considered as an operator from $\mathscr{L}(X(G):\mathscr{M})$. This will be indicated here by $A \equiv A(z)$; i.e., the argument z will be omitted. Then the results of Sect. 4.2 to 4.4 are applicable to A as an operator on $X(G):\mathscr{M}$ if one takes $K = \mathscr{M}$ and replaces the linear space X used there by $X(G):\mathscr{M}$.

The decomposition of $\chi(\lambda, A)$ into prime factors with respect to $\mathscr{M}[\lambda]$ takes the form

$$\chi(\lambda, A) = \prod_{\varrho=1}^{r} p_\varrho(\lambda, z)^{m_\varrho}, \qquad \deg p_\varrho = g_\varrho, \qquad \varrho = 1, \ldots, r. \tag{6.1}$$

The irreducible polynomials $p_1(\lambda, z), \ldots, p_r(\lambda, z)$ are assumed to be normalized. Then, according to Lemma 2.1 and Lemma 2.2, one obtains the following results:

To every factor $p_\varrho(\lambda, z)^{m_\varrho}$ (a power of a fixed prime polynomial) corresponds a unique reducing projection $P_\varrho \in \mathscr{L}(X(G):\mathscr{M})$; i.e.,

$$AP_\varrho = P_\varrho A, \qquad \varrho = 1, \ldots, r. \tag{6.2}$$

The projections $P_1, \ldots, P_r$ are polynomials in A; that is, there exist polynomials $q_1(\lambda, z)$, $q_2(\lambda, z), \ldots, q_r(\lambda, z)$ from $\mathscr{M}[\lambda]$ such that

$$P_\varrho = q_\varrho(A, z), \qquad \varrho = 1, \ldots, r. \tag{6.3}$$

(In Sect. 4.2 it has been shown how these polynomials can be determined.) Let

$$A_\varrho = A \upharpoonright P_\varrho X(G), \qquad \varrho = 1, \ldots, r, \tag{6.4}$$

then

$$\chi(\lambda, A_\varrho) = p_\varrho(\lambda, z)^{m_\varrho}, \qquad \varrho = 1, \ldots, r. \tag{6.5}$$

The projection P_ϱ is already uniquely determined by the property (6.5).

According to (6.3), the projection P_ϱ is a meromorphic projection-valued function $P_\varrho \equiv P_\varrho(z)$ in G. For simple points z and for the eigenprojection $P_{\varrho\sigma}(z)$ corresponding to the eigenvalue $\lambda_{\varrho\sigma}(z)$ (see Subsect. 3.3.5), the equation

$$\chi\big(\lambda, A(z) \upharpoonright P_{\varrho\sigma}(z)\, X\big) = \big(\lambda - \lambda_{\varrho\sigma}(z)\big)^{m_\varrho}$$

holds. For the projection $\sum_{\sigma=1}^{g_\varrho} P_{\varrho\sigma}(z)$ one obtains the relation

$$\chi\left(\lambda, A(z) \upharpoonright \sum_{\sigma=1}^{g_\varrho} P_{\varrho\sigma}(z)\, X\right) = \prod_{\sigma=1}^{g_\varrho} \left(\lambda - \lambda_{\varrho\sigma}(z)\right)^{m_\varrho} = p_\varrho(\lambda, z)^{m_\varrho}.$$

According to Lemma 2.2, this implies

$$P_\varrho(z) = \sum_{\sigma=1}^{g_\varrho} P_{\varrho\sigma}(z). \tag{6.6}$$

This equation explains the connection between $P_\varrho(z)$ and the eigenprojections.

4.6.3. Decomposition of $P_\varrho(z)$ into meromorphic-irreducible projections

In general, the projections $P_\varrho(z)$ fail to be irreducible with respect to $A(z)$, even if one aims for a meromorphic decomposition. A decomposition of $P_\varrho(z)$ into meromorphic-irreducible reducing projections with respect to $A(z)$ can be obtained from the considerations of Sect. 4.3 and 4.4, in particular from Theorem 4.1.

According to this theorem, there exist uniquely determined numbers

$$m_\varrho^1 \geqq m_\varrho^2 \geqq \cdots \geqq m_\sigma^{l_\varrho}, \qquad \varrho = 1, \ldots, r, \tag{6.7}$$

and meromorphic projections $P_\varrho^1(z), \ldots, P_\varrho^{l_\varrho}(z)$, in general not uniquely determined, which form a disjoint system such that

$$\dim P_\nu^\sigma(z) = m_\varrho^\sigma, \qquad \sigma = 1, 2, \ldots, l_\varrho,$$

and

$$\sum_{\sigma=1}^{l_\varrho} P_\varrho^\sigma(z) = P_\varrho(z), \qquad \varrho = 1, \ldots, r.$$

The projections $P_\varrho^\sigma(z)$ reduce $A(z)$ and are irreducible with respect to $\mathcal{L}\left(X(G) : \mathcal{M}\right)$; i.e., the projections $P_\varrho^\sigma(z)$ do not possess proper reducing meromorphic subprojections.

The dimensions $m_\varrho^1, \ldots, m_\varrho^{l_\varrho}$ are entire multiples of g_ϱ. They can be determined in accordance with Subsect. 4.3.5.

The projections $P_\varrho^\sigma(z)$ are generalized Jordan projections. If $m_\varrho^\sigma = \nu_\varrho^\sigma g_\varrho$, then the corresponding generalized Jordan block is $\Gamma_{\nu_\varrho^\sigma}\left(p_\varrho(\lambda, z)\right)$.

4.7. Transformation of complete disjoint systems of meromorphic projections

As already mentioned, the following consideration, which uses the concepts of Sect. 4.5, supplements Subsect. 3.4.2. Recall that in that subsection a holomorphic transformation operator $U(z)$ with holomorphic inverse $U(z)^{-1}$ was constructed for a complete disjoint system of holomorphic projections $P_1(z), \ldots, P_r(z)$ in a simply connected region G, according to a method of T. Kato. This transformation operator satisfies the property $P_\varrho(z) = U(z)\, P_\varrho(z_0)\, U(z)^{-1}$, $U(z_0) = 1$, where z_0 is arbitrary from G. Furthermore, recall that this method can also be applied to the case of an arbitrary region, but leads to a multi-valued holomorphic transformation operator in G.

Let $P_1(z), \ldots, P_r(z)$ be a complete disjoint system of meromorphic projections, given in an arbitrary region G. Does there exist a single-valued operator-valued function $T(z)$, such that $T(z)$ and $T(z)^{-1}$ both are meromorphic in G and such that the property $P_\varrho(z) = T(z)\, P_\varrho(z_0)\, T(z)^{-1}$, $\varrho = 1, \ldots, r$, holds, where z_0 is an arbitrarily chosen holomorphic point for all $P_\varrho(z)$ from G?

This question can be answered affirmatively, with the use of the concepts of Sect. 4.5. Note once more that one obtains a single-valued and, together with its inverse, meromorphic transformation operator $T(z)$ with $T(z_0) = 1$ in G, whereas the method of T. Kato yields a multi-valued and, together with its inverse, holomorphic transformation operator $U(z)$ with $U(z_0) = 1$.

Basically, the answer to this question, which is given here, is already contained in Lemma 3.4.1.

4.7.1. Transformation of systems of projections from $\mathscr{L}(X(G):\mathscr{M})$

Let $P_1(z), \ldots, P_r(z)$ be a complete disjoint system of meromorphic projection-valued functions given in a region G. Let $\dim P_\varrho(z) = m_\varrho$, $\varrho = 1, \ldots, r$. According to Subsect. 4.5.4, the meromorphic projection-valued functions $P_\varrho(z) \equiv P_\varrho$ are operators from $\mathscr{L}\big(X(G):\mathscr{M}\big)$ and form a complete system of disjoint projections.

Furthermore, let $Q_1, \ldots, Q_r$ be a second complete system of disjoint projections from $\mathscr{L}\big(X(G):\mathscr{M}\big)$ such that $\dim Q_\varrho = m_\varrho$, $\varrho = 1, \ldots, r$.

Let $e_\varrho^\sigma, \tilde{e}_\varrho^\sigma$, $\sigma = 1, \ldots, m_\varrho$, $\varrho = 1, \ldots, r$, be bases of $X(G):\mathscr{M}$, adapted to the system $P_1, \ldots, P_r$. Correspondingly, let $f_\varrho^\sigma, \tilde{f}_\varrho^\sigma$ be bases, adapted to the system $Q_1, \ldots, Q_r$. In particular,

$$P_\varrho e_\varrho^\sigma = e_\varrho^\sigma, \qquad Q_\varrho f_\varrho^\sigma = f_\varrho^\sigma, \qquad \varrho = 1, \ldots, r; \qquad \sigma = 1, \ldots, m_\varrho; \tag{7.1}$$

similar for the bases $\tilde{e}_\varrho^\sigma, \tilde{f}_\varrho^\sigma$.

Defining the operator $S \in \mathscr{L}\big(X(G):\mathscr{M}\big)$ by

$$S e_\varrho^\sigma = \tilde{e}_\varrho^\sigma, \qquad \varrho = 1, \ldots, r; \qquad \sigma = 1, 2, \ldots, m_\varrho, \tag{7.2}$$

and, correspondingly, T by

$$T f_\varrho^\sigma = \tilde{f}_\varrho^\sigma, \tag{7.3}$$

one obtains

$$S P_\varrho = P_\varrho S, \qquad \varrho = 1, \ldots, r, \tag{7.4}$$

and

$$T Q_\varrho = Q_\varrho T, \qquad \varrho = 1, \ldots, r. \tag{7.5}$$

Moreover, S and T are invertible.

Conversely, an invertible operator S with the property (7.4) transforms a basis adapted to the system $P_1, \ldots, P_r$ into a basis which is again adapted to the system $P_1, \ldots, P_r$. The corresponding proposition for T is also true.

Defining $U \in \mathscr{L}(X(G):\mathscr{M})$ by

$$Uf_\varrho^\sigma = e_\varrho^\sigma, \qquad \varrho = 1, \ldots, r; \qquad \sigma = 1, \ldots, m_\varrho, \tag{7.6}$$

one obtains

$$P_\varrho = UQ_\varrho U^{-1}, \qquad \varrho = 1, \ldots, r; \tag{7.7}$$

that is, every two arbitrary adapted bases define a transformation operator.

If $V = SUT$, where (7.4) and (7.5) hold and where S and T are invertible, then $P_\varrho = VQ_\varrho V^{-1}$, $\varrho = 1, \ldots, r$. Conversely, if V is invertible and if $P_\varrho = VQ_\varrho V^{-1}$, then $V = SUT$, where S and T are suitable operators, both invertible and such that (7.4) and (7.5) hold.

4.7.2. Application

Let $P_\varrho = P_\varrho(z)$, $\varrho = 1, \ldots, r$, be the given complete system of disjoint and meromorphic projections. Let z_0 be a holomorphic point for all these projections. One writes

$$Q_\varrho = P_\varrho(z_0), \qquad \varrho = 1, \ldots, r; \tag{7.8}$$

i.e., the system $Q_1, \ldots, Q_r$ consists of constant projection-valued functions of the form (7.8). Then, according to Subsect. 4.7.1, one obtains the existence of a meromorphic and invertible operator-valued function $T(z)$ such that $P_\varrho(z) = T(z)\, P_\varrho(z_0)\, T(z)^{-1}$, $\varrho = 1, \ldots, r$. Every two adapted bases — adapted to $P_\varrho(z_0)$ and $P_\varrho(z)$, respectively — define a transformation operator $T(z)$.

In particular, let f_ϱ^σ be a constant basis of X adapted to the system $P_\varrho(z_0)$. One writes

$$P_\varrho(z)\, f_\varrho^\sigma = e_\varrho^\sigma(z), \qquad \varrho = 1, \ldots, r, \qquad \sigma = 1, \ldots, m_\varrho.$$

Then $e_\varrho^\sigma(z_0) = f_\varrho^\sigma$ and $P_\varrho(z)\, e_\varrho^\sigma(z) = e_\varrho^\sigma(z)$. The vectors $e_\varrho^\sigma(z)$ are from $X(G):\mathscr{M}$. They form a basis of $X(G):\mathscr{M}$ adapted to the system $P_\varrho(z)$. Let $\sum_{\varrho,\sigma} \alpha_\varrho^\sigma(z)\, e_\varrho^\sigma(z) \equiv 0$ for every z. According to Subsect. 3.4.1, the system $e_\varrho^\sigma(z)$ forms a basis of X for every z from a sufficiently small neighborhood $|z - z_0| < \delta$. Therefore, $\alpha_\varrho^\sigma(z) \equiv 0$ in $|z - z_0| < \delta$; hence $\alpha_\varrho^\sigma(z) \equiv 0$ everywhere.

In particular, if one chooses these adapted bases f_ϱ^σ and $e_\varrho^\sigma(z)$, the corresponding transformation operator $T(z)$ is defined by $T(z)\, f_\varrho^\sigma = e_\varrho^\sigma(z)$. Next, $P_\beta(z_0)\, f_\varrho^\sigma = 0$ for $\beta \neq \varrho$ and $P_\varrho(z)\, P_\varrho(z_0)\, f_\varrho^\sigma = e_\varrho^\sigma(z)$, therefore $\left(\sum_{\alpha=1}^{r} P_\alpha(z)\, P_\alpha(z_0)\right) f_\varrho^\sigma = e_\varrho^\sigma(z)$ and one obtains the equation

$$T(z) = \sum_{\alpha=1}^{r} P_\alpha(z)\, P_\alpha(z_0). \tag{7.9}$$

Additionally, this transformation operator satisfies $T(z_0) = 1$.

The connection of (7.9) with the method of T. Kato is obtained by differentiation of (7.9) at the point z_0. Then

$$T'(z_0) = \sum_{\alpha=1}^{r} P'_\alpha(z_0) P_\alpha(z_0),$$

and this operator-valued function is the factor in the operator differential equation (3.4.13), which is satisfied by the transformation operator U. Hence this operator differential equation can be written in the form $U'(z) = T'(z) U(z)$ at the point z_0. Therefore, Kato constructs his transformation operator from "infinitesimal" operators $T(z_0 + dz)$ by means of an integration process.

5. Jordan structure of meromorphic operator-valued functions

This chapter is a central part of the book. The objective is deriving the theorems formulated in Sect. 5.5 to 5.7. These theorems contain the results on the function-theoretic behavior of the Jordan structure (Jordan normal form) for meromorphic operator-valued functions.

The (local) consequences of these theorems for perturbation theory are presented in Ch. 6.

Let $A(z)$ be a meromorphic operator-valued function defined on a region G. According to Ch. 2, every single operator has a certain Jordan structure. Therefore, the problem is investigating the analytic properties of the Jordan structure of $A(z)$. This means finding connections and simple function-theoretic properties with respect to the individual decompositions

$$A(z) = A_1(z) \oplus A_2(z) \oplus \cdots \oplus A_{p(z)}(z) \tag{0.1}$$

of the single operators $A(z)$, $z \in G \setminus S$, into direct sums of Jordan operators; in particular, establishing internal connection of the individual Jordan normal forms $N(z)$ of $A(z)$.

In Ch. 3 similar questions were answered for the spectral representation

$$A(z) = \sum_{\alpha,\varrho} \left(\lambda_{\alpha\varrho}(z) P_{\alpha\varrho}(z) + D_{\alpha\varrho}(z)\right) \tag{0.2}$$

of $A(z)$ (see Subsect. 3.3.12 and 3.3.13). Recall that the spectral representation is an intermediate step for obtaining the Jordan structure. If one starts from (0.2), the determination of the Jordan structure of $A(z)$ requires the study of the internal structure (refined structure) of the eigennilpotents $D_{\alpha\varrho}(z)$.

Because the decomposition (0.1) is not unique, answering this question becomes more difficult. Recall, for example, that the behavior of the spectral representation is complicated first of all at multiple points. In the present case, "singularities" can appear even at simple points, that is, at points where the spectral representation is given by (0.2) and behaves completely "regularly". This can be shown already by the following simple example (example (c) of Subsect. 3.2.10): Let $X = \mathbb{C}^2$ and let

$$A(z) = \begin{pmatrix} 0 & z \\ -z & 2z \end{pmatrix}.$$

The only eigenvalue is $\lambda(z) = z$. The corresponding eigennilpotent is

$$A(z) - \lambda(z) = \begin{pmatrix} -z & z \\ -z & z \end{pmatrix} = zD, \qquad D = \begin{pmatrix} -1 & 1 \\ -1 & 1 \end{pmatrix}.$$

All points z are simple; the spectral representation is given by $A(z) = z1 + zD$. When one looks for the Jordan normal form, a singular behavior at the point $z = 0$ appears. For $z = 0$, the Jordan normal form is

$$N(0) = \begin{pmatrix} 0 & 0 \\ 0 & 0 \end{pmatrix};$$

that is, there are two Jordan operators N_1 and N_2 of dimension 1 such that $A(0) = N_1 \oplus N_2$. For $z \neq 0$ the Jordan normal form is given by

$$N(z) = \begin{pmatrix} z & 1 \\ 0 & z \end{pmatrix}, \qquad z \neq 0;$$

that is, for $z \neq 0$ the operator $A(z)$ itself is already a Jordan operator. Therefore, the Jordan normal form of $A(z)$ is "discontinuous" at $z = 0$. This discontinuity is essential (not removable) for $A(z)$. It appears even though all the spectral objects $\lambda(z) \equiv z$, $P(z) \equiv 1$ and $D(z) = zD$ are holomorphic at the point $z = 0$.

Therefore, the decompositions (0.1) should be constructed such that the discontinuities mentioned can be handled. Moreover, the individual decompositions of the single operators $A(z)$ should be suitably chosen to obtain simple connections between the corresponding (irreducible) Jordan projections and Jordan operators.

The key for the solution of this problem is given by the analytic functions $\mathfrak{P}_1, \mathfrak{P}_2, \ldots, \mathfrak{P}_r$, defined by the characteristic polynomial of $A(z)$, and by the corresponding fields of functions $\mathcal{M}(\mathfrak{P}_\varrho)$, $\varrho = 1, \ldots, r$. The eigennilpotents $D(\zeta_\alpha)$, $\zeta_\alpha \in \mathfrak{P}_\alpha$, can be considered as operators from $\mathcal{L}(X(\mathfrak{P}_\alpha) : \mathcal{M}(\mathfrak{P}_\alpha))$, where $X(\mathfrak{P}_\alpha) : \mathcal{M}(\mathfrak{P}_\alpha)$ denotes the linear space of all algebroid vector-valued functions defined on $\mathfrak{P}_\alpha$, considered as a linear space over the field $\mathcal{M}(\mathfrak{P}_\alpha)$. Then the results of Sect. 4.2 to 4.4 can be applied in an obvious manner. These applications represent an essential aspect of the proof of Theorem 7.1.

As already indicated by the above example a new kind of exceptional points appears, which will be called *weakly splitting points*. (Recall that the exceptional points known so far are the splitting points and their trace points, the multiple points connected with the spectral representation, and the poles of $A(z)$.) At weakly splitting points the Jordan normal form becomes singular. Weakly splitting points do not cause new branchings, but only poles of the objects to be decomposed. For example, in the example mentioned above, $z = 0$ is a weakly splitting point. If $T(z)$ denotes any matrix which transforms $A(z)$ into the Jordan normal form $N(z)$, $A(z) = T(z)\,N(z)\,T(z)^{-1}$, then at least one of the matrices $T(z)$, $T(z)^{-1}$ has a pole at $z = 0$; for example,

$$\begin{pmatrix} 0 & z \\ -z & 2z \end{pmatrix} = \begin{pmatrix} 1 & 1 \\ 1 & 1 + \dfrac{1}{z} \end{pmatrix} \begin{pmatrix} z & 1 \\ 0 & z \end{pmatrix} \begin{pmatrix} z+1 & -z \\ -z & z \end{pmatrix}.$$

Sect. 5.1 presents some preliminaries for the application of the results of Ch. 4, which are related to Sect. 4.5 and 4.7.

In Sect. 5.2 the results of Sect. 4.2 will be applied. This consideration is related to Subsect. 4.6.2, where the spectral representation (as in Sect. 4.2) with respect to $\mathcal{M}$ led to the group projections of the G-cycles of eigenvalues. Here, the spectral representation of $A(z)$ with respect to $\mathcal{M}(\mathfrak{P}_\alpha)$ yields the eigenprojections of the G-cycle belonging to $\mathfrak{P}_\alpha$. This fact has two interesting consequences: On the one hand, one obtains a purely algebraic calculation of eigenprojections without using integration theory, residue theory or "explicit" knowledge of the eigenvalues. On the other hand, one obtains a direct proof of the representation of the eigenprojections derived in Theorem 3.3.6 on the basis of local considerations. From this representation the local properties can be deduced again.

In Sect. 5.3, the results of Sect. 4.3 and 4.4, are applied in the investigation of the eigennilpotents $D(\zeta_\alpha)$, $\zeta_\alpha \in \mathfrak{P}_\alpha$. One obtains, among other things, a decomposition of the eigenprojection

$$P(\zeta_\alpha) = P_1(\zeta_\alpha) + P_2(\zeta_\alpha) + \cdots + P_{l_\alpha}(\zeta_\alpha) \tag{0.3}$$

into a sum of disjoint (and irreducible with respect to $D(\zeta_\alpha)$) Jordan projections of $X(\mathfrak{P}_\alpha)\colon \mathcal{M}(\mathfrak{P}_\alpha)$. $D(\zeta_\alpha)$ acts on the irreducible subspace given by $P_\varrho(\zeta_\alpha)$ via a Jordan block $\Gamma_{m_\varrho^\alpha}(0)$, where $m_\varrho^\alpha = \dim P_\varrho(\zeta_\alpha)$. Finally, one also obtains transformation operators $T(\zeta_\alpha)$, $T(\zeta_\alpha)^{-1} \in \mathcal{L}\big(X(\mathfrak{P}_\alpha)\colon \mathcal{M}(\mathfrak{P}_\alpha)\big)$, where the relations

$$P_\varrho(\zeta_\alpha) = T(\zeta_\alpha)\, P_\varrho(\zeta_\alpha^0)\, T(\zeta_\alpha)^{-1}, \tag{0.4}$$

$$D(\zeta_\alpha) = T(\zeta_\alpha)\, D(\zeta_\alpha^0)\, T(\zeta_\alpha)^{-1}, \tag{0.5}$$

$$T(\zeta_\alpha^0) = 1 \tag{0.6}$$

hold and where ζ_α^0 denotes a suitable holomorphic point of the system $P_1(\zeta_\alpha), \ldots, P_{l_\alpha}(\zeta_\alpha)$. In particular, ζ_α^0 is normal. Then $D(\zeta_\alpha) \sim D(\zeta_\alpha^0)$, i.e., the eigennilpotent $D(\zeta_\alpha)$ is similar to a constant eigennilpotent $D(\zeta_\alpha^0)$, for all ζ_α such that $T(\zeta_\alpha)$ and $T(\zeta_\alpha)^{-1}$ are both well-defined as individual operators.

In Sect. 5.4 the set of exceptional points ζ_α^0 will be characterized. In this connection, a point ζ_α^0 is called exceptional if no transformation operator $T(\zeta_\alpha)$ exists with the properties (0.4), (0.5) and (0.6). The set of exceptional points coincides with the set of all normal points ζ_α^0 from $\mathfrak{P}_\alpha$ such that the Jordan normal form of $D(\zeta_\alpha^0)$ is different from the Jordan normal form of $D(\zeta_\alpha) \in \mathcal{L}\big(X(\mathfrak{P}_\alpha)\colon \mathcal{M}(\mathfrak{P}_\alpha)\big)$. In other words, the Jordan normal form is singular at those points (weakly splitting points).

Sect. 5.5 contains the study of the (irreducible) Jordan projections and Sect. 5.6 that of Jordan vectors and Jordan bases with respect to $D(\zeta_\alpha)$. Finally, in Sect. 5.7 one obtains the following result on the Jordan structure of $A(z)$:

Let $A(z)$ be meromorphic in G. Then there is a set A, isolated and closed in G (containing the poles of $A(z)$ and the multiple points), such that for all points $z \in G \setminus A$

$$N(z) = A(z) - \sum_{\alpha,\sigma} \lambda_{\alpha\sigma}(z)\, P_{\alpha\sigma}(z) = \sum_{\alpha,\sigma} D_{\alpha\sigma}(z) \sim \sum_{\alpha,\sigma} D_{\alpha\sigma}(z_0) = N(z_0) \tag{0.7}$$

and, in particular, $D_{\alpha\sigma}(z) \sim D_{\alpha 1}(z_0)$ holds, where $z_0 \in G \setminus A$. Moreover, there is an operator-valued function $U(z)$ such that $U(z)$ and $U(z)^{-1}$ are holomorphic in a suitable neigh-

borhood $|z - z_0| < \delta$ *and*

$$N(z) = U(z)\, N(z_0)\, U(z)^{-1}. \tag{0.8}$$

In the neighborhood of exceptional points $z_0 \in A$ *all spectral objects can be expanded into Laurent-Puiseux series.*

$A(z)$ is said to possess an invariant Jordan structure in $G \setminus A$.

The results of this chapter, in particular the theorem just quoted, are due to the author (see H. BAUMGÄRTEL [1, 2 and 4]). Of course, in the literature there are results connected with the results of this chapter (see, for example, I. C. GOCHBERG and M. G. KREIN [1, pp. 64 and 2, p. 37sqq.]). Note that in the earlier literature ideas can be found related to the theorem quoted above, which proved mistaken. (See, for example, the remarks of F. GANTMACHER in [2, p. 128] which correct a conclusion of V. Volterra; see V. VOLTERRA and B. HOSTINSKY [1, pp. 117—120] and V. VOLTERRA [1].)

5.1. Algebroid vector-valued and operator-valued functions

5.1.1. Algebroid vector-valued functions

Let $p(\lambda, z)$ be a normalized irreducible polynomial from $\mathcal{M}[\lambda]$ and $\mathfrak{P}$ the algebroid function over the original region G defined by the equation $p(\lambda, z) = 0$ (see Subsect. A 5.3). Let $\mathcal{M}(\mathfrak{P})$ denote the field of all scalar functions which are algebroid on $\mathfrak{P}$. Let $\deg p = g$. Then the elements of $\mathcal{M}(\mathfrak{P})$ take the form

$$\alpha_0(z) + \alpha_1(z)\, \lambda(\zeta) + \alpha_2(z)\, \lambda(\zeta)^2 + \cdots + \alpha_{g-1}(z)\, \lambda(\zeta)^{g-1}, \qquad \zeta \in \mathfrak{P},$$

where the coefficients $\alpha_0(z), \alpha_1(z), \ldots, \alpha_{g-1}(z)$ are meromorphic scalar functions on G and where $\lambda(\zeta)$ represents the function $\mathfrak{P}$. z is the trace point of ζ (see Subsect. A 6.2 and A 6.3).

Let X be a linear space over $\mathbb{C}$ with $\dim X = N$. Then, as in Sect. A 5, one introduces the linear space of all algebroid vector-valued functions from $\mathfrak{P}$ into X (concerning the concept of an algebroid vector-valued function, see Subsect. A 6.4). This linear space is denoted by $X(\mathfrak{P})$ or, if one wishes to emphasize that $X(\mathfrak{P})$ is considered over the field $\mathbb{C}$, by $X(\mathfrak{P})\colon \mathbb{C}$. The elements of $X(\mathfrak{P})$ take the form

$$a_0(z) + a_1(z)\, \lambda(\zeta) + a_2(z)\, \lambda(\zeta)^2 + \cdots + a_{s-1}(z)\, \lambda(\zeta)^{g-1},$$

where the coefficients $a_0(z), a_1(z), \ldots, a_{s-1}(z)$ are meromorphic vector-valued functions from G into X.

As in Subsect. 4.5.2, one can embed X into $X(\mathfrak{P})$. X is a subspace of $X(\mathfrak{P})\colon \mathbb{C}$, namely, the subspace of all constant vector-valued functions from $\mathfrak{P}$ into X.

Of course, $X(\mathfrak{P})$ can also be considered as a linear space over $\mathcal{M}(\mathfrak{P})$. In this case one writes $X(\mathfrak{P})\colon \mathcal{M}(\mathfrak{P})$. As in Subsect. 4.5.3, one obtains

$$\dim \big(X(\mathfrak{P})\colon \mathcal{M}(\mathfrak{P})\big) = \dim (X\colon \mathbb{C}).$$

That is, every basis of X over $\mathbb{C}$ is simultaneously a basis of $X(\mathfrak{P})$ over $\mathcal{M}(\mathfrak{P})$. In other words, there are bases consisting of constant vector-valued functions.

5.1.2. Algebroid operator-valued functions

The operators of $\mathscr{L}(X(\mathfrak{P}): \mathscr{M}(\mathfrak{P}))$ can be identified with the algebroid operator-valued functions from $\mathfrak{P}$ into $\mathscr{L}(X: \mathbb{C})$, as in Subsect. 4.5.4. This property implies a representation

$$A(\zeta) = A_0(z) + A_1(z)\,\lambda(\zeta) + A_2(z)\,\lambda(\zeta)^2 + \cdots + A_{g-1}(z)\,\lambda(\zeta)^{g-1}, \tag{1.1}$$

where the coefficients $A_0(z), A_1(z), \ldots, A_{g-1}(z)$ are meromorphic operator-valued functions from G into $\mathscr{L}(X)$.

First one proves some auxiliary propositions, which will be used in Sect. 5.5.

Lemma 1. *Let $A \equiv A(\zeta)$ be an algebroid operator-valued function on $\mathfrak{P}$ and let* ran $A = r$. *$A(\zeta)$ is well-defined as an operator from $\mathscr{L}(X)$ except on an isolated set which is closed in $\mathfrak{P}$. Then* ran $A(\zeta) = r$ *for all points $\zeta \in \mathfrak{P}$ except on an isolated set which is closed in $\mathfrak{P}$;* ran $A(\zeta) < r$ *at exceptional points.*

Proof. Let $e_1, \ldots, e_N$ be a basis of X and let $f_1, \ldots, f_N$ denote the adjoint basis. Let $a_{ij}(\zeta) = (A(\zeta)\,e_i, f_j)$, $i, j = 1, 2, \ldots, N$. Then ran $(a_{ij}(\zeta)) = r$. Without loss of generality, let det $(a_{\nu\mu}(\zeta)) = d(\zeta) \not\equiv 0$, $\mu, \nu = 1, 2, \ldots, r$. One has $d(\zeta) \in \mathscr{M}(\mathfrak{P})$. The roots of $d(\zeta)$ are isolated and closed in $\mathfrak{P}$. Every exceptional point is necessarily a root of $d(\zeta)$. Moreover, at exceptional points one necessarily has ran $A(\zeta) < r$, because otherwise ran $A(\zeta) > r$ in a full neighborhood in contradiction to the assumption ran $A = r$. ■

Correspondingly, one obtains

Lemma 2. *Let $A \equiv A(\zeta)$ be defined as in Lemma* 1 *and let* def $A = d$. *Then* def $A(\zeta) = d$ *holds for all ζ from $\mathfrak{P}$ except on an isolated set which is closed in $\mathfrak{P}$. Moreover,* def $A(\zeta) > d$ *at exceptional points. If the operator $A(\zeta_0)$ is well-defined at the point ζ_0 and satisfies* def $A(\zeta_0) = d$, *then the equation* def $A(\zeta) = d$ *holds even in a full neighborhood of ζ_0.*

Lemma 3. *Let $A \equiv A(\zeta)$ belong to $\mathscr{L}(X(\mathfrak{P}): \mathscr{M}(\mathfrak{P}))$. Let* dim ker $A = m$. *Furthermore, let ζ_0 be a holomorphic point from $\mathfrak{P}$, let $A(\zeta_0)$ be well-defined and let* dim ker $A(\zeta_0) = m$. *Then there is a basis $x_1(\zeta), x_2(\zeta), \ldots, x_m(\zeta)$ of* ker A, *consisting of functions which are holomorphic in a neighborhood U of ζ_0 and such that the values $x_j(\zeta)$, $j = 1, \ldots, m$, form a basis of* ker $A(\zeta)$ *for every $\zeta \in U$.*

Proof. Let $e_1, \ldots, e_N$ be a basis of X and let $(a_{ij}(\zeta))$ denote the matrix of $A(\zeta)$ corresponding to this basis. Let det $(a_{\mu\nu}(\zeta)) \neq 0$, $\mu, \nu = m + 1, \ldots, N$, $\zeta \in U$, where U is suitably chosen. Then the equation $A(\zeta)\,x = 0$ is equivalent to $\sum_{j=1}^{N} a_{ij}(\zeta)\,x_j = 0$, $i = 1, \ldots, N$, where $x = \sum_{j=1}^{N} x_j e_j$. Solving this equation, one obtains $x_j = \sum_{\varrho=1}^{m} c_{\varrho j}(\zeta)\,x_\varrho$, $j = m + 1, \ldots, N$, where the $c_{\varrho j}(\zeta)$ denote certain algebroid scalar functions which are holomorphic in U. Then a basis of ker A is obtained by setting $x_\varrho = \delta_{\varrho k}$, $\varrho = 1, \ldots, m$, k fixed, $k = 1, \ldots, m$, and by determining the remaining coordinates from the equations just obtained. This basis $n_1(\zeta), n_2(\zeta), \ldots, n_m(\zeta)$ of ker A consists of algebroid vector-valued functions, holomorphic in U, which form a basis of ker $A(\zeta)$ for every $\zeta \in U$. ■

Lemma 4. *Let* $A \equiv A(\zeta)$ *belong to* $\mathscr{L}(X(\mathfrak{P}): \mathscr{M}(\mathfrak{P}))$. *Let* $\dim \ker A = m$. *Furthermore, let* $x_1(\zeta), \ldots, x_p(\zeta)$ *be algebroid vector-valued functions from* $\ker A$ *which are linearly independent in* $X(\mathfrak{P}): \mathscr{M}(\mathfrak{P})$, $p < m$. *At the holomorphic point* ζ_0 *from* $\mathfrak{P}$, *let* $A(\zeta_0)$ *be well-defined, where* $\dim \ker A(\zeta_0) = m$; *let* $x_1(\zeta), \ldots, x_p(\zeta)$ *be holomorphic in a neighborhood* U *of* ζ_0 *and linearly independent in* $X: \mathbb{C}$ *for every* $\zeta \in U$. *Then there exist algebroid vector-valued functions* $y_{p+1}(\zeta), \ldots, y_m(\zeta)$ *in* $\ker A$ *such that they, together with* $x_1(\zeta), \ldots, x_p(\zeta)$, *form a basis of* $\ker A$ *over* $\mathscr{M}(\mathfrak{P})$; *moreover, there is a suitable neighborhood* V *of* ζ_0 *such that the functions* $y_{p+1}(\zeta), \ldots, y_m(\zeta)$ *are holomorphic in* V *and the values at* $\zeta \in V$, *together with the values of* $x_1(\zeta), \ldots, x_p(\zeta)$ *at* ζ, *form a basis of* $\ker A(\zeta)$ *over* $\mathbb{C}$.

Proof. According to Lemma 3, there is a basis of $\ker A$, consisting of functions holomorphic in some neighborhood V' of ζ_0 and such that the values form a basis of $\ker A(\zeta)$ for every $\zeta \in V'$. Let $v_1(\zeta), v_2(\zeta), \ldots, v_m(\zeta)$ denote such a basis. Then

$$x_i(\zeta) = \sum_{\varrho=1}^{m} \gamma_{i\varrho}(\zeta)\, v_\varrho(\zeta), \qquad i = 1, \ldots, p, \tag{1.2}$$

where the $\gamma_{i\varrho}(\zeta)$ denote algebroid scalar functions which are holomorphic in $U \cap V'$. Moreover, the rank of the $p \times m$-matrix $(\gamma_{i\varrho}(\zeta))$ is equal to p for all $\zeta \in U \cap V'$. Without loss of generality, one can assume that

$$\det\left(\gamma_{ij}(\zeta)\right) \neq 0, \qquad i, j = 1, \ldots, p,$$

for all $\zeta \in V$, where V denotes a suitable neighborhood of ζ_0. Solving (1.2) explicitly with respect to $v_1(\zeta), \ldots, v_p(\zeta)$, one obtains

$$v_i(\zeta) = \sum_{j=1}^{p} \alpha_{ij}(\zeta)\, x_j(\zeta) + \sum_{\varrho=p+1}^{m} \beta_{i\varrho}(\zeta)\, v_\varrho(\zeta), \qquad i = 1, \ldots, p,$$

where the $\alpha_{ij}(\zeta)$, $\beta_{i\varrho}(\zeta)$ denote algebroid scalar functions which are holomorphic in V. Therefore, the vectors $x_1(\zeta), \ldots, x_p(\zeta)$, $v_{p+1}(\zeta), \ldots, v_m(\zeta)$ form a basis of $\ker A$ over $\mathscr{M}(\mathfrak{P})$. The supplementary vectors are holomorphic in V and the values $x_1(\zeta), \ldots, x_p(\zeta)$, $v_{p+1}(\zeta), \ldots, v_m(\zeta)$ form a basis of $\ker A(\zeta)$ over $\mathbb{C}$ for every $\zeta \in V$. ∎

5.1.3. Transformation of complete disjoint systems of algebroid projection-valued functions

Let $P_1(\zeta), P_2(\zeta), \ldots, P_r(\zeta)$ denote a complete disjoint system of projections from $\mathscr{L}(X(\mathfrak{P}): \mathscr{M}(\mathfrak{P}))$. Then the considerations of Sect. 4.7 can be transferred to this case.

Let ζ_0 be a holomorphic point with respect to $\mathfrak{P}$ and the system. To every two adapted bases of $P_\varrho(\zeta_0)$ and $P_\varrho(\zeta)$ corresponds a transformation operator $T(\zeta)$ with

$$P_\varrho(\zeta) = T(\zeta)\, P_\varrho(\zeta_0)\, T(\zeta)^{-1}, \qquad \varrho = 1, \ldots, r. \tag{1.3}$$

In particular, if f_ϱ^a denotes a constant basis adapted to $P_\varrho(\zeta_0)$ and if one uses $P_\varrho(\zeta)\, f_\varrho^a$ as the basis adapted to $P_\varrho(\zeta)$, then one obtains the following expression for the assigned transformation operator $T(\zeta)$:

$$T(\zeta) = \sum_{\alpha=1}^{r} P_\alpha(\zeta)\, P_\alpha(\zeta_0), \qquad T(\zeta_0) = 1.$$

Let $e_\varrho^\sigma(\zeta)$ denote an basis adapted to $P_\varrho(\zeta)$ and let ζ_0 be a holomorphic point with respect to $\mathfrak{P}$, the system and the basis $e_\varrho^\sigma(\zeta)$. If $e_\varrho^\sigma(\zeta_0)$ is a basis (hence an adapted one) with respect to $P_\varrho(\zeta_0)$, then one obtains (1.3) for the transformation operator $T(\zeta)$ assigned to $e_\varrho^\sigma(\zeta_0)$ and $e_\varrho^\sigma(\zeta)$; moreover, $T(\zeta_0) = 1$.

5.2. Algebraic determination of the eigenprojections of meromorphic operator-valued functions

Let $A(z)$ be a meromorphic operator-valued function and let $\mathfrak{P}$ be one of the algebroid functions defined by $\chi(\lambda, A(z))$. According to Sect. 4.2, the spectral representation of $A(z)$ with respect to $\mathscr{M}(\mathfrak{P})$ leads to the eigenprojections of the G-cycle belonging to $\mathfrak{P}$. Therefore, the determination of the eigenprojections is reduced to the calculation of certain polynomials in $\mathscr{M}(\mathfrak{P})[\lambda]$.

5.2.1. Characteristic polynomial

Let $p(\lambda, z)$ be a fixed normalized prime factor of $\chi(\lambda, A(z))$ of degree g with respect to $\mathscr{M}[\lambda]$. The g-valued algebroid function over G defined by the equation $p = 0$ is denoted by $\mathfrak{P}$. The elements of $\mathfrak{P}$ are denoted by ζ, $\zeta \in \mathfrak{P}$. The corresponding "eigenvalue function", defined on $\mathfrak{P}$, is denoted by $\lambda(\zeta)$. The eigenprojection belonging to $\lambda(\zeta)$ is denoted by $P(\zeta)$.

One has $A \equiv A(z) \in \mathscr{L}(X(\mathfrak{P}) : \mathscr{M}(\mathfrak{P}))$. The divisor $p(\lambda, z)$ of $\chi(\lambda, A(z))$, which is irreducible with respect to $\mathscr{M}[\lambda]$, can also be considered as a polynomial of $\mathscr{M}(\mathfrak{P})[\lambda]$. Then $p(\lambda, z)$ is no longer irreducible, but

$$p(\lambda, z) = (\lambda - \lambda(\zeta))\, q(\lambda, z, \lambda(\zeta)), \tag{2.1}$$

where $q(\lambda; z, \lambda(\zeta))$ is a polynomial of $\mathscr{M}(\mathfrak{P})[\lambda]$. If one denotes the number of factors $p(\lambda, z)$ with respect to χ by m, i.e., if

$$\chi(\lambda, A(z)) = p(\lambda, z)^m\, r(\lambda, z), \qquad r(\lambda, z) \in \mathscr{M}[\lambda], \tag{2.2}$$

where r and p are relatively prime, then one obtains the following decomposition of χ in $\mathscr{M}(\mathfrak{P})[\lambda]$:

$$\chi(\lambda, A(z)) = (\lambda - \lambda(\zeta))^m\, s(\lambda; z, \lambda(\zeta)), \tag{2.3}$$

where $s(\lambda; z, \lambda(\zeta))$ denotes a polynomial from $\mathscr{M}(\mathfrak{P})[\lambda]$, which no longer contains the factor $\lambda - \lambda(\zeta)$. So, $(\lambda - \lambda(\zeta))^m$ is the exact power of the prime polynomial $\lambda - \lambda(\zeta)$ with respect to $\chi(\lambda, A(z))$ as a polynomial of $\mathscr{M}(\mathfrak{P})[\lambda]$.

5.2.2. Construction of the projections $Q(\zeta, \mathfrak{P})$

According to Sect. 4.2, to every (maximal) power of a prime polynomial of the characteristic polynomial corresponds a unique reducing projection of $A(z)$. This projection is equipped with some remarkable properties which have been proved in Lemma 4.2.1 and Lemma 4.2.2. This assignment is used here for the factor $(\lambda - \lambda(\zeta))^m$ of $\chi(\lambda, A(z))$

with respect to $\mathscr{M}(\mathfrak{P})[\lambda]$. The assigned projection is denoted by $Q(\zeta, \mathfrak{P})$. From Sect. 4.2, these properties follow:

I. *$Q(\zeta, \mathfrak{P})$ is an algebroid operator-valued function defined on $\mathfrak{P}$, $Q(\zeta, \mathfrak{P}) \in \mathscr{L}(X(\mathfrak{P}): \mathscr{M}(\mathfrak{P}))$. This implies a unique representation of $Q(\zeta, \mathfrak{P})$ of the form*

$$Q(\zeta, \mathfrak{P}) = B_0(z) + B_1(z)\,\lambda(\zeta) + B_2(z)\,\lambda(\zeta)^2 + \cdots + B_{g-1}(z)\,\lambda(\zeta)^{g-1}, \tag{2.4}$$

where the coefficients $B_0(z), B_1(z), \ldots, B_{g-1}(z)$ are meromorphic operator-valued functions from G into $\mathscr{L}(X)$; z is the trace point of ζ.

II. *$Q(\zeta, \mathfrak{P})$ is a projection; the function $Q(\zeta, \mathfrak{P})$ is projection-valued.*

III. *$Q(\zeta, \mathfrak{P})$ reduces $A(z)$, i.e.,*

$$Q(\zeta, \mathfrak{P})\,A(z) = A(z)\,Q(\zeta, \mathfrak{P}). \tag{2.5}$$

IV. *The relation*

$$\chi(\lambda, A(z) \restriction Q(\zeta, \mathfrak{P})\,X(\mathfrak{P})) = (\lambda - \lambda(\zeta))^m \tag{2.6}$$

is valid.

V. *$Q(\zeta, \mathfrak{P})$ is an operator polynomial of $A(z)$ with coefficients from $\mathscr{M}(\mathfrak{P})$; i.e., there exists a polynomial $q(\lambda; z, \lambda(\zeta)) \in \mathscr{M}(\mathfrak{P})[\lambda]$ with the property*

$$Q(\zeta, \mathfrak{P}) = \varrho(A(z); z, \lambda(\zeta)). \tag{2.7}$$

It has been shown in Subsect. 4.2.1 and 4.2.2, how this polynomial q can be constructed. We will come back to this procedure when we discuss some examples.

5.2.3. Fixing of the argument ζ

The operators $B_0(z), \ldots, B_{g-1}(z)$ are meromorphic in G. Let z_0 be a simple point of $A(z)$ and let z_0 be holomorphic for all $B_\varrho(z)$. Then the operators $B_0(z_0), \ldots, B_{g-1}(z_0)$ are well-defined and from $\mathscr{L}(X)$. The exceptional set A, which consists of all poles of $A(z)$, of all multiple points of $A(z)$, and of all poles of the operators $B_0(z), \ldots, B_{g-1}(z)$, is isolated and closed in G. Then, for all points $\zeta \in \mathfrak{P}$ whose trace points z belong to $G \setminus A$, the projection $Q(\zeta, \mathfrak{P})$ is well-defined and from $\mathscr{L}(X)$; it reduces $A(z)$ and

$$\chi(\lambda, A(z) \restriction Q(\zeta, \mathfrak{P})\,X) = (\lambda - \lambda(\zeta))^m. \tag{2.8}$$

5.2.4. Identity of $P(\zeta)$ and $Q(\zeta, \mathfrak{P})$

The eigenprojection-valued function $P(\zeta)$ is well-defined and holomorphic at all normal points $\zeta \in \mathfrak{P}$. According to Corollary 3.3.2,

$$\chi(\lambda, A(z) \restriction P(\zeta)\,X) = (\lambda - \lambda(\zeta))^m. \tag{2.9}$$

Consider $P(\zeta)$ at points $\zeta \in \mathfrak{P}$ where $Q(\zeta, \mathfrak{P})$ is also well-defined, i.e., at points whose trace points belong to $G \setminus A$.

For all points $z \in G \setminus A$, the polynomial $\chi(\lambda, A(z))$ has a representation of the form

$$\chi(\lambda, A(z)) = (\lambda - \lambda(\zeta))^m r(\lambda)$$

as a polynomial from $\mathbb{C}[\lambda]$, where the polynomials $\lambda - \lambda(\zeta)$ and $r(\lambda)$ are relatively prime; i.e., $r(\lambda)$ does not contain the linear factor $\lambda - \lambda(\zeta)$. This is implied by the simplicity of z (see Subsect. 3.2.2.).

Consider $A(z)$ for an arbitrary but fixed $z \in G \setminus A$ as an operator from $\mathscr{L}(X)$. Then the situation is as follows: $P(\zeta)$ and $Q(\zeta, \mathfrak{P})$ are reducing projections with respect to $A(z)$. (Note that, together with z, ζ and therefore $\lambda(\zeta)$ are fixed.) According to (2.8) and (2.9), the operator $A(z)$ has one and the same characteristic polynomial $(\lambda - \lambda(\zeta))^m$ with respect to both corresponding reducing subspaces $P(\zeta) X$ and $Q(\zeta, \mathfrak{P}) X$; this is the exact power of the prime polynomial $\lambda - \lambda(\zeta)$ occurring in the decomposition of $\chi(\lambda, A(z))$ into a product of prime factors with respect to $\mathbb{C}[\lambda]$. One can apply the uniqueness theorem (Lemma 4.2.2), where $K = \mathbb{C}$, $A = A(z)$, and where the corresponding power of the prime polynomial is given by $(\lambda - \lambda(\zeta))^m$. According to this theorem, two reducing projections coincide if they yield the same characteristic polynomial $(\lambda - \lambda(\zeta))^m$. Therefore, $P(\zeta) = Q(\zeta, \mathfrak{P})$. Recall that ζ is arbitrary; the only restriction is that the trace point z of ζ does not belong to A. Therefore, the identity of both functions, $P(\zeta)$ and $Q(\zeta, \mathfrak{P})$, follows,

$$P(\zeta) \equiv Q(\zeta, \mathfrak{P}). \tag{2.10}$$

Hence, one obtains a representation for $P(\zeta)$ of the form

$$P(\zeta) = B_0(z) + B_1(z)\, \lambda(\zeta) + B_2(z)\, \lambda(\zeta)^2 + \cdots + B_{g-1}(z)\, \lambda(\zeta)^{g-1}, \tag{2.11}$$

with meromorphic operator-valued coefficients $B_\varrho(z)$, which implies the algebroid behavior of $P(\zeta)$ at splitting points. Earlier (see Subsect. 3.3.6 to 3.3.9), the representation (2.11) was derived by a local investigation of $P(\zeta)$ in the neighborhood of splitting points. The method developed here proves that the converse conclusion is also possible.

The result (2.10), together with property V of Subsect. 5.2.2, solves the problem of a purely algebraic determination of the eigenprojections. In Subsect. 5.2.6 this calculation will be illustrated by some examples.

5.2.5. Further properties of $P(\zeta)$

Recall that the operator-valued coefficients $B_0(z), \ldots, B_{g-1}(z)$ in the representation (2.11) of $P(\zeta)$ are uniquely determined and that all these coefficients are holomorphic at simple points of $A(z)$ (see Corollary 3.3.3).

If z_0 is multiple and if all operators $B_0(z), \ldots, B_{g-1}(z)$ are holomorphic at $z = z_0$, then $P(\zeta)$ is holomorphic at every point $\zeta = \zeta_0$ whose trace point is z_0.

In this case, the projection $P(\zeta)$ must be continuous at all points ζ_0 lying over z_0, because of representation (2.11). Hence, according to Theorem 3.3.4, $P(\zeta)$ is even holomorphic at ζ_0.

In other words, in this case there is no branching point over z_0. Conversely, if there is a branching point ζ_0 over z_0, then at least one of the operator-valued coefficients

$B_\varrho(z)$ has a pole at $z = z_0$. If, in the latter case, there is also a normal point ζ_1 over z_0, then $P(\zeta)$ is holomorphic at $\zeta = \zeta_1$, despite the fact that some of the operator-valued coefficients have poles at $z = z_0$. This behavior will be demonstrated by an example (Subsect. 5.2.6).

With use of the local Puiseux series for $\lambda(\zeta)$ in the neighborhood of splitting points ζ_0 and the Laurent series for the operators $B_0(z), \dots, B_{g-1}(z)$ in the neighborhood of the corresponding trace point z_0, the expansion (2.11) for $P(\zeta)$ can be rewritten as a local expansion in the neighborhood of ζ_0 (cf. Theorem 3.3.3). If the branching order is denoted by h and if M denotes the maximal order of poles of the coefficients $B_0(z)$, $\dots, B_{g-1}(z)$, then one obtains for the h branches $P_1(z), P_2(z), \dots, P_h(z)$ of $P(\zeta)$ in the neighborhood of ζ_0 expansions of the form

$$P_\sigma(z) = \sum_{\varrho=0}^{\infty} C_\varrho (z - z_0)^{-M+\frac{\varrho}{h}}, \qquad \sigma = 1, 2, \dots, h, \tag{2.12}$$

where the h different values of $(z - z_0)^{1/h}$ yield exactly the h branches $P_1(z), \dots, P_h(z)$. One has $C_\varrho \in \mathscr{L}(X)$; $C_0 \neq 0$ need not be true.

5.2.6. Examples

Three examples are considered. Let G be the whole complex plane and $X = \mathbb{C}^2$.

(a) $$A(z) = \begin{pmatrix} 0 & 1 \\ 0 & z \end{pmatrix}.$$

One has $\chi(\lambda, A(z)) = \lambda(\lambda - z)$. This is already the decomposition into irreducible factors. The eigenvalue functions are given by $\lambda_1(z) \equiv 0$, $\lambda_2(z) \equiv z$. These functions belong to $\mathscr{M}$, the field of all rational functions. One obtains

$$\frac{1}{z}\lambda - \frac{1}{z}(\lambda - z) = 1 \tag{2.13}$$

as the representation of 1 by the irreducible polynomials λ and $\lambda - z$ (see Subsect. 4.2.1). Denoting the eigenprojections belonging to $\lambda_1(z)$ and $\lambda_2(z)$ by $P_1(z)$ and $P_2(z)$, respectively, one obtains

$$P_1(z) = -\frac{1}{z}\begin{pmatrix} 0 & 1 \\ 0 & z \end{pmatrix} + \begin{pmatrix} 1 & 0 \\ 0 & 1 \end{pmatrix} = \begin{pmatrix} 1 & -\frac{1}{z} \\ 0 & 0 \end{pmatrix}$$

and

$$P_2(z) = \frac{1}{z}\begin{pmatrix} 0 & 1 \\ 0 & z \end{pmatrix} = \begin{pmatrix} 0 & \frac{1}{z} \\ 0 & 1 \end{pmatrix}.$$

(b) $$A(z) = \begin{pmatrix} 0 & 0 \\ 0 & z \end{pmatrix}.$$

Obviously, $\chi(\lambda, A(z)) = \lambda(\lambda - z)$, as before. Again the relation (2.13) holds and one obtains

$$P_1(z) = -\frac{1}{z}\begin{pmatrix}0 & 0\\ 0 & z\end{pmatrix} + \begin{pmatrix}1 & 0\\ 0 & 1\end{pmatrix} = \begin{pmatrix}1 & 0\\ 0 & 0\end{pmatrix},$$

$$P_2(z) = \frac{1}{z}\begin{pmatrix}0 & 0\\ 0 & z\end{pmatrix} = \begin{pmatrix}0 & 0\\ 0 & 1\end{pmatrix}.$$

The examples (a) and (b) stay within the considerations of Subsect. 4.6.2. The following example is a proper one. Let $X = \mathbb{C}^3$.

(c) $$A(z) = \begin{pmatrix}0 & 1 & -1\\ 0 & 0 & -1\\ z & 0 & 1\end{pmatrix}.$$

Here, $\chi(\lambda, A(z)) = \lambda^3 - \lambda^2 + z\lambda + z$. This polynomial is already irreducible in $\mathcal{M}[\lambda]$, where, as usual, $\mathcal{M}$ denotes the field of the rational functions. From $\chi = (\lambda + \alpha)(\lambda^2 + \beta\lambda + \gamma) = \lambda^3 + \lambda^2(\alpha + \beta) + \lambda(\gamma + \alpha\beta) + \alpha\gamma$ the relations $\alpha + \beta = -1$, $\gamma + \alpha\beta = z$, $\alpha\gamma = z$ follow; hence $\gamma + \alpha(-1 - \alpha) = z$, i.e., $\alpha^3 + \alpha^2 + \alpha z = z$. On the other hand, α must be constant or a multiple of z.

To determine the set of all multiple points, one forms the derivative $\chi'_\lambda = 3\lambda^2 - 2\lambda + z$ and one calculates all points z where χ and χ'_λ have common roots. After a short calculation one obtains the equation $z^3 + 11z^2 - z = 0$ for these points; i.e., multiple points are $z = 0$ and $z = 2^{-1}(-11 \pm 5\sqrt{5})$.

The eigenvalue expansions in the neighborhood of $z = 0$ can be determined with the use of the Puiseux-Newton diagram (see Sect. A 7). One obtains

$$\lambda_{1/2}(z) = z^{1/2} + \frac{1}{2}(z^{1/2})^2 + \cdots, \qquad \lambda_3(z) = 1 - 2z + \cdots.$$

To the equation $\chi = 0$ corresponds a unique algebroid 3-valued function over the complex z-plane, which is even algebraic in this case. Therefore, there is a unique eigenvalue function, denoted by $\mu = \lambda(\zeta)$. First one obtains

$$\chi = (\lambda - \mu)\left(\lambda^2 + (\mu - 1)\lambda + \mu^2 - \mu + z\right).$$

Now one can apply the method of Subsect. 4.2.1 to determine the eigenprojection-valued function belonging to μ. One forms

$$\begin{aligned}&(\alpha\lambda + \beta)(\lambda - \mu) + \gamma\left(\lambda^2 + (\mu - 1)\lambda + \mu^2 - \mu + z\right)\\ &= \lambda^2(\alpha + \gamma) + \lambda\left(\beta - \alpha\mu + \gamma(\mu - 1)\right) + \gamma(\mu^2 - \mu + z) - \beta\mu.\end{aligned}$$

The coefficients of λ^2 and λ should vanish; hence,

$$\alpha + \gamma = 0, \qquad \beta - \alpha\mu + \gamma(\mu - 1) = 0.$$

Let $\gamma = 1$, then $\alpha = -1$ and $\beta = 1 - 2\mu$, and one obtains

$$\left(-\lambda + (1 - 2\mu)\right)(\lambda - \mu) + \left(\lambda^2 + (\mu - 1)\lambda + \mu^2 - \mu + z\right) = 3\mu^2 - 2\mu + z.$$

Therefore,

$$\frac{-\lambda + (1 - 2\mu)}{3\mu^2 - 2\mu + z} (\lambda - \mu) + \frac{\lambda^2 + (\mu - 1)\lambda + \mu^2 - \mu + z}{3\mu^2 - 2\mu + z} = 1.$$

The function $1/(3\mu^2 - 2\mu + z)$ belongs to $\mathcal{M}(\mu)$. One obtains

$$\frac{1}{3\mu^2 - 2\mu + z} = \frac{3z - 1}{D} \mu^2 + \frac{-8z + 1}{2D} \mu + \frac{2z^2 + z}{2D},$$

where

$$D = z^3 + 11z^2 - z.$$

To determine the projection $P(\zeta)$, one needs the polynomial

$$p(\lambda) = \frac{\lambda^2 + (\mu - 1)\lambda + \mu^2 - \mu + z}{3\mu^2 - 2\mu + z}.$$

Hence, this polynomial takes the form

$$\begin{aligned} p(\lambda) &= \lambda^2 \left(\frac{3z - 1}{2D} \mu^2 + \frac{1 - 8z}{2D} \mu + \frac{2z^2 + z}{2D} \right) \\ &\quad + \lambda \left(\frac{1 - 8z}{2D} \mu^2 - \frac{z^2 - 10z + 1}{2D} \mu - \frac{5z^2}{2D} \right) \\ &\quad + \left(\frac{2z^2 + z}{2D} \mu^2 - \frac{5z^2}{2D} \mu + \frac{2z^3 + 9z^2 - z}{2D} \right). \end{aligned}$$

Therefore,

$$P(\zeta) = B_0(z) + B_1(z)\mu + B_2(z)\mu^2,$$

where the coefficients B_0, B_1, B_2 are given by

$$B_0(z) = \frac{2z^2 + z}{2D} A(z)^2 - \frac{5z^2}{2D} A(z) + \frac{2z^3 + 9z^2 - z}{2D} 1,$$

$$B_1(z) = \frac{1 - 8z}{2D} A(z)^2 - \frac{z^2 - 10z + 1}{2D} A(z) - \frac{5z^2}{2D} 1,$$

$$B_2(z) = \frac{3z - 1}{2D} A(z)^2 + \frac{1 - 8z}{2D} A(z) + \frac{2z^2 + z}{2D} 1.$$

Note that the operator-valued functions $B_0(z), B_1(z), B_2(z)$ are rational functions, where poles can occur at most at the points $z = 0$ and $z = 2^{-1}(-11 \pm 5\sqrt{5})$.

There is also a normal point over $z = 0$, namely $(0, \lambda_3(z))$. Hence, $P(\zeta)$ is holomorphic at that point despite the fact that some of the operator-valued coefficients possess poles at $z = 0$, because a branching point also lies over $z = 0$.

5.3. Eigennilpotents

This section is a continuation of Subsect. 3.3.10. It contains the decomposition of $P(\zeta)$ into a sum of disjoint projections, which are irreducible with respect to $D(\zeta)$ and even Jordan projections. Furthermore, it is proved that the eigennilpotent $D(\zeta)$, considered as an operator of $\mathscr{L}(X(\mathfrak{P}) : \mathscr{M}(\mathfrak{P}))$, is similar to a constant operator $D(\zeta_0)$, where ζ_0 is suitably chosen from $\mathfrak{P}$.

5.3.1. Simple properties

As in Sect. 5.2, let $A(z)$ be a meromorphic operator-valued function in G; let $p(\lambda, z)$ be one of the prime factors of $\chi(\lambda, A(z))$ with respect to $\mathscr{M}[\lambda]$ of degree g, and let $\mathfrak{P}$ denote the corresponding algebroid function. The elements of $\mathfrak{P}$ are denoted by ζ; the corresponding eigenvalue function is denoted by $\lambda(\zeta)$ and the eigenprojection by $P(\zeta)$. Then the eigennilpotent is defined by

$$D(\zeta) = \big(A(z) - \lambda(\zeta)\big)\, P(\zeta), \tag{3.1}$$

where z denotes the trace point of ζ (see (3.3.22)). $D(\zeta)$ is an algebroid operator-valued function defined on $\mathfrak{P}$, $D(\zeta) \in \mathscr{L}(X(\mathfrak{P}) : \mathscr{M}(\mathfrak{P}))$. Hence, there exists a unique representation

$$D(\zeta) = C_0(z) + C_1(z)\, \lambda(\zeta) + \cdots + C_{g-1}(z)\, \lambda(\zeta)^{g-1}, \tag{3.2}$$

where the operator-valued coefficients $C_\varrho(z)$ are meromorphic. This representation is obtained if (2.11) is substituted into (3.1). Moreover, $D(\zeta)$ is an operator polynomial in $A(z)$ with coefficients from $\mathscr{M}(\mathfrak{P})$ (see Subsect. 5.2.2). $D(\zeta)$ is holomorphic at all normal points $\zeta \in \mathfrak{P}$. One has

$$D(\zeta)\, P(\zeta) = P(\zeta)\, D(\zeta) = D(\zeta). \tag{3.3}$$

Hence $D(\zeta)$, restricted to the subspace $\big(1 - P(\zeta)\big)\, X(\mathfrak{P})$, equals the zero operator. If ζ_0 is normal, then $D(\zeta_0) \in \mathscr{L}(X)$, restricted to the subspace $\big(1 - P(\zeta_0)\big)\, X$, equals the zero operator.

If $\dim P(\zeta) = m$, then

$$D(\zeta)^m \equiv 0. \tag{3.4}$$

If ζ_0 is normal, then to $D(\zeta_0)$, restricted to $P(\zeta_0)\, X$, corresponds the only eigenvalue 0. Hence $\chi\big(\lambda, D(\zeta_0) \restriction P(\zeta_0)\, X\big) = \lambda^m$ and the relation $\big(D(\zeta_0)\, P(\zeta_0)\big)^m = D(\zeta_0)^m = 0$ holds Since this is true for all normal points ζ_0, (3.4) follows.

A supplementary remark is added with respect to the behavior of $D(\zeta)$ at splitting points (see Subsect. 3.3.10):

If $D(\zeta) \not\equiv 0$ and if ζ_0 is a branching point of order g with respect to $\lambda(\zeta)$, then ζ_0 is also a branching point of (exact) order g with respect to $D(\zeta)$.

Let $\lambda_1(z), \lambda_2(z), \ldots, \lambda_g(z)$ denote the g branches of $\lambda(\zeta)$ in the neighborhood of ζ_0. The corresponding eigenprojections are denoted by $P_1(z), \ldots, P_g(z)$. Then one obviously obtains g branches for the eigennilpotent $D(\zeta)$ in the neighborhood of ζ_0,

$$D_\varrho(z) = \big(A(z) - \lambda_\varrho(z)\big) P_\varrho(z), \qquad \varrho = 1, \dots, g.$$

Assume that two of these branches coincide; i.e., let the branching order for $D(\zeta)$ be less than g:

$$D_\nu(z) = D_\mu(z), \qquad \nu \neq \mu. \tag{3.5}$$

Multiplying (3.5) by $P_\nu(z)$, one obtains $D_\nu(z) \equiv 0$ because $P_\nu(z)\, P_\mu(z) = 0$, first in a neighborhood of a fixed point z and finally $D(\zeta) \equiv 0$ in contradiction to the assumption.

5.3.2. Jordan structure of the eigennilpotents

Consider $D(\zeta)$ as an operator from $\mathscr{L}\big(X(\mathfrak{P}) : \mathscr{M}(\mathfrak{P})\big)$ and apply the results of Sect. 4.3 and 4.4 to $D(\zeta)$. For this application X is to be replaced by $X(\mathfrak{P})$, the linear space of the algebroid vector-valued functions from $\mathfrak{P}$ into X, and K is to be replaced by the field $\mathscr{M}(\mathfrak{P})$ of the algebroid scalar functions. Because of (3.4), one obtains

$$\chi\big(\lambda, D(\zeta)\big) = \lambda^N. \tag{3.6}$$

Therefore the decomposition of χ into prime factors with respect to $\mathscr{M}(\mathfrak{P})\,[\lambda]$ is trivial. There is only one prime factor, namely λ, and this factor is a linear one. These facts already imply the following proposition (see Corollary 4.4.1):

All irreducible parts of $D(\zeta)$ are necessarily Jordan operators.

Furthermore, one obtains from Subsect. 4.4.4:

The decomposition of $D(\zeta)$ into a direct sum of irreducible operators can be formed compatibly with $P(\zeta)$.

The decomposition of $D(\zeta)$ is given according to Theorem 4.4.1. The result can be formulated as follows.

There are uniquely determined numbers $m_1 \geq m_2 \geq \dots \geq m_l$ with $\sum_{\varrho=1}^{l} m_\varrho = m$ and Jordan projections $P_1(\zeta), P_2(\zeta), \dots, P_l(\zeta)$ (not uniquely determined) with $\dim P_\varrho = m_\varrho$, such that

$$P(\zeta) = P_1(\zeta) + P_2(\zeta) + \dots + P_l(\zeta), \tag{3.7}$$

$$P_\varrho(\zeta)\, P_\sigma(\zeta) = \delta_{\varrho\sigma} P_\varrho(\zeta), \qquad \varrho, \sigma = 1, \dots, l, \tag{3.8}$$

$$P_\varrho(\zeta)\, D(\zeta) = D(\zeta)\, P_\varrho(\zeta), \qquad \varrho = 1, \dots, l. \tag{3.9}$$

Restricted to the subspace $P_\varrho(\zeta)\, X(\mathfrak{P})$, the operator $D(\zeta)$ acts by a Jordan block $\Gamma_{m_\varrho}(0)$ with respect to a suitable (Jordan) basis $e_\varrho^\sigma(\zeta)$, $\sigma = 1, 2, \dots, m_\varrho$, in $P_\varrho(\zeta)\, X(\mathfrak{P})$.

Then the relations

$$D(\zeta)\, e_\varrho^1(\zeta) = 0, \qquad D(\zeta)\, e_\varrho^2(\zeta) = e_\varrho^1(\zeta), \dots, D(\zeta)\, e_\varrho^{m_\varrho}(\zeta) = e_\varrho^{m_\varrho - 1}(\zeta) \tag{3.10}$$

are valid. The dimensions $m_1, m_2, \ldots, m_l$ can be calculated by the Frobenius formulas (see Subsect. 2.9.4 and 4.3.5). The starting point for this calculation is given by the defects def $D(\zeta)^\varrho$, $\varrho = 1, 2, \ldots, m$. These defects satisfy the inequalities

$$N - m + 1 \leqq \operatorname{def} D(\zeta) \leqq \operatorname{def} D(\zeta)^2 \leqq \cdots \leqq \operatorname{def} D(\zeta)^m = N. \tag{3.11}$$

5.3.3. Transformation of the system $P_1(\zeta), P_2(\zeta), \ldots, P_l(\zeta)$ and of the eigennilpotents

Now one applies the results of Subsect. 5.1.3 (see also Sect. 4.7) to the complete disjoint system $P_1(\zeta), P_2(\zeta), \ldots, P_l(\zeta)$, $1 - P(\zeta)$, of projections. The following condition, to be verified after its formulation, is the starting point for this purpose:

Let the system $P_\varrho(\zeta)$ and the Jordan bases $e_\varrho^\sigma(\zeta)$ be fixed. These objects are algebroid operator-valued and vector-valued functions, respectively, from $\mathfrak{P}$ into $\mathscr{L}(X)$ and X, respectively.

The assertion is that in any case *there are normal points ζ_0 from $\mathfrak{P}$ such that ζ_0 is a holomorphic point for the system $P_\varrho(\zeta)$ and for the Jordan bases $e_\varrho^\sigma(\zeta)$ and, moreover, the vectors $e_\varrho^\sigma(\zeta_0)$ form adapted Jordan bases of $P_\varrho(\zeta_0)\, X$ with respect to $D(\zeta_0)$.*

To prove this assertion first note that the set of all normal points ζ of $\mathfrak{P}$, which are poles of $P_\varrho(\zeta)$ or of $e_\varrho^\sigma(\zeta)$, $\varrho = 1, 2, \ldots, l$; $\sigma = 1, \ldots, m_\varrho$, is isolated and closed in $\mathfrak{P}$. Obviously there are normal points ζ_0 which are holomorphic with respect to all functions $P_\varrho(\zeta)$ and $e_\varrho^\sigma(\zeta)$. Let ζ_0 denote such a fixed point and let U denote a neighborhood of ζ_0 such that all these functions are holomorphic in U.

Let $Q(\zeta)$ denote one of the functions $P_\varrho(\zeta)$ and let $f_1(\zeta), \ldots, f_p(\zeta)$ denote the corresponding Jordan basis, where $p = \dim Q(\zeta)$. Furthermore, let $e_1, e_2, \ldots, e_N$ be a basis of X. Then $f_\alpha(\zeta) = \sum_{\varrho=1}^{N} \gamma_{\alpha\varrho}(\zeta)\, e_\varrho$ with algebroid scalar functions $\gamma_{\alpha\varrho}(\zeta)$ which are holomorphic in U. One has ran $(\gamma_{\alpha\varrho}(\zeta)) = p$. This implies the existence of a $p \times p$-submatrix, which without loss of generality may be taken as the matrix $(\gamma_{\alpha\varrho}(\zeta))$, $\alpha, \varrho = 1, 2, \ldots, p$, such that $\det (\gamma_{\alpha\varrho}(\zeta)) \not\equiv 0$. Hence, there is a subneighborhood V of U such that $\det (\gamma_{\alpha\varrho}(\zeta)) \neq 0$ for all $\zeta \in V \setminus \{\zeta_0\}$; i.e., the vectors $f_1(\zeta), f_2(\zeta), \ldots, f_p(\zeta)$ are linearly independent in $X : \mathbb{C}$ for all $\zeta \in V \setminus \{\zeta_0\}$. Hence these vectors form a basis in $Q(\zeta)\, X$ and this basis is obviously a Jordan basis with respect to $D(\zeta)$.

Applying this conclusion to all $P_\varrho(\zeta)$, $e_\varrho^\sigma(\zeta)$, one finally obtains a neighborhood V_0 of ζ_0 such that all $P_\varrho(\zeta)$, $e_\varrho^\sigma(\zeta)$ are holomorphic for all $\zeta \in V_0 \setminus \{\zeta_0\}$, and, moreover, the vectors $e_\varrho^\sigma(\zeta)$, $\sigma = 1, 2, \ldots, m_\varrho$, form a basis, and therefore a Jordan basis with respect to $D(\zeta)$, of $P_\varrho(\zeta)\, X$. Thus the assertion has been verified.

The basis $e_\varrho^\sigma(\zeta)$, adapted to the system $P_\varrho(\zeta)$, can be supplemented by a basis of $(1 - P(\zeta))\, X(\mathfrak{P})$ with the additional property that this basis is holomorphic at $\zeta = \zeta_0$ and that the values of the basis vectors at ζ_0 form a basis of $(1 - P(\zeta_0))\, X$. Such a basis is denoted by $f_0^\sigma(\zeta)$.

Then the vectors $e_\varrho^\sigma(\zeta)$, $f_0^\sigma(\zeta)$ form a basis of $X(\mathfrak{P})$ adapted to the system $P_1(\zeta), \ldots, P_l(\zeta)$, $1 - P(\zeta)$, and, furthermore, the vectors $e_\varrho^\sigma(\zeta_0)$, $f_0^\sigma(\zeta_0)$ form a basis of X adapted to the system $P_1(\zeta_0), \ldots, P_l(\zeta_0)$, $1 - P(\zeta_0)$. That is, the latter basis is a "constant" basis in $X(\mathfrak{P})$ adapted to the "constant" system $P_1(\zeta_0), \ldots, P_l(\zeta_0)$, $1 - P(\zeta_0)$.

Now one can apply the considerations of Subsect. 5.1.3. Therefore, there is a transformation operator $T(\zeta) \in \mathcal{L}(X(\mathfrak{P}) : \mathcal{M}(\mathfrak{P}))$, defined by $T(\zeta)\, e_\varrho^\sigma(\zeta_0) = e_\varrho^\sigma(\zeta)$, $T(\zeta)\, f_0^\sigma(\zeta_0) = f_i^\sigma(\zeta)$, such that the properties

$$P_\varrho(\zeta) = T(\zeta)\, P_\varrho(\zeta_0)\, T(\zeta)^{-1}, \qquad \varrho = 1, 2, \ldots, l, \tag{3.12}$$

$$T(\zeta_0) = 1 \tag{3.13}$$

hold. A further important property of $T(\zeta)$ is given by

$$D(\zeta) = T(\zeta)\, D(\zeta_0)\, T(\zeta)^{-1}. \tag{3.14}$$

To prove this relation, form

$$T(\zeta)\, D(\zeta_0)\, e_\varrho^\sigma(\zeta_0) = T(\zeta)\, e_\varrho^{\sigma-1}(\zeta_0) = e_\varrho^{\sigma-1}(\zeta),$$

where $e_\varrho^0(\zeta_0) = 0$. Moreover,

$$T(\zeta)\, D(\zeta_0)\, f_0^\sigma(\zeta_0) = 0.$$

On the other hand,

$$D(\zeta)\, T(\zeta)\, e_\varrho^\sigma(\zeta_0) = D(\zeta)\, e_\varrho^\sigma(\zeta) = e_\varrho^{\sigma-1}(\zeta),$$

and

$$D(\zeta)\, T(\zeta)\, f_0^\sigma(\zeta_0) = D(\zeta)\, f_0^\sigma(\zeta) = 0.$$

Hence (3.14) follows. Therefore, one obtains

Lemma 1. *The eigennilpotent $D(\zeta)$, as an operator from $\mathcal{L}(X(\mathfrak{P}) : \mathcal{M}(\mathfrak{P}))$, is similar to a "constant" operator $D(\zeta_0)$, where the normal point $\zeta_0 \in \mathfrak{P}$ is suitably chosen; moreover, there exists an operator $T(\zeta) \in \mathcal{L}(X(\mathfrak{P}) : \mathcal{M}(\mathfrak{P}))$ with $T(\zeta_0) = 1$ such that the properties* (3.12) *and* (3.14) *are satisfied.*

Since the transformation operator $T(\zeta)$ belongs to $\mathcal{L}(X(\mathfrak{P}) : \mathcal{M}(\mathfrak{P}))$, it can be uniquely represented in the form

$$T(\zeta) = S_0(z) + S_1(z)\, \lambda(\zeta) + \cdots + S_{g-1}(z)\, \lambda(\zeta)^{g-1}, \tag{3.15}$$

where the operator-valued coefficients are meromorphic in G. Analogously,

$$T(\zeta)^{-1} = R_0(z) + R_1(z)\, \lambda(\zeta) + \cdots + R_{g-1}(z)\, \lambda(\zeta)^{g-1}. \tag{3.16}$$

The operator-valued functions $S_0(z), \ldots, S_{g-1}(z)$, $R_0(z), \ldots, R_{g-1}(z)$ may have poles at poles of $A(z)$, at multiple points but also at simple points. The set A_T of all poles of $A(z)$, of all multiple points, and of all poles of the coefficients $S_0(z), \ldots, S_{g-1}(z)$, $R_0(z), \ldots, R_{g-1}(z)$ is isolated and closed in G.

Hence, $T(\zeta)$ and $T(\zeta)^{-1}$ are both holomorphic at all points $\zeta \in \mathfrak{P}$ whose trace points belong to $G \setminus A_T$. Therefore, at these points the "individual" operator $T(\zeta)$ is well-defined, from $\mathcal{L}(X)$, and invertible. For all these points

$$D(\zeta) = T(\zeta)\, D(\zeta_0)\, T(\zeta)^{-1};$$

that is, the operators $D(\zeta)$ and $D(\zeta_0)$ are similar; i.e., $D(\zeta)$ has an invariant Jordan normal form for all these points which coincides with the Jordan normal form of $D(\zeta_0)$. Therefore, the following proposition is true.

Lemma 2. *Except for a set from $\mathfrak{P}$ whose trace set is isolated and closed in G, the operators $D(\zeta) \in \mathscr{L}(X)$ are mutually similar; i.e., they have a common (constant) Jordan normal form coinciding with the Jordan normal form of $D(\zeta)$ as an operator of $\mathscr{L}\big(X(\mathfrak{P})\colon \mathscr{M}(\mathfrak{P})\big)$.*

5.4. Weakly splitting points

The results of Subsect. 5.3.3 were implied mainly by the fact that there exist special normal points $\zeta_0 \in \mathfrak{P}$ with corresponding transformation operators $T(\zeta) \in \mathscr{L}\big(X(\mathfrak{P})\colon \mathscr{M}(\mathfrak{P})\big)$ such that the properties (3.12), (3.13) and (3.14) hold.

In this section the properties of such special normal points will be investigated in detail. The exceptional set of normal points from $\mathfrak{P}$ such that there does not exist an operator $T(\zeta)$ with the properties mentioned will be characterized.

5.4.1. Strongly normal points

Definition 1. A normal point $\zeta_0 \in \mathfrak{P}$ is called *strongly normal* if there exists an operator-valued function $T(\zeta) \in \mathscr{L}\big(X(\mathfrak{P})\colon \mathscr{M}(\mathfrak{P})\big)$ such that the properties

I. $T(\zeta_0) = 1,$

II. $P(\zeta) = T(\zeta)\, P(\zeta_0)\, T(\zeta)^{-1},$

III. $D(\zeta) = T(\zeta)\, D(\zeta_0)\, T(\zeta)^{-1}$

are satisfied.

By means of this definition, an essential part of the considerations in Subsect. 5.3.3 can be formulated as follows.

Lemma 1. *There exist strongly normal points.*

Moreover, one obtains

Lemma 2. *Let ζ_0 be strongly normal and let $T(\zeta)$ be a corresponding transformation operator. Furthermore, let $\zeta_1 \in \mathfrak{P}$ be normal and assume that $T(\zeta)$ and $T(\zeta)^{-1}$ are holomorphic at the point $\zeta = \zeta_1$. Then ζ_1 is also strongly normal and a corresponding transformation operator is given by $S(\zeta) = T(\zeta)\, T(\zeta_1)^{-1}$.*

Proof. Obviously, $S(\zeta_1) = 1$. Furthermore,

$$T(\zeta_1)^{-1}\, D(\zeta_1)\, T(\zeta_1) = D(\zeta_0)$$

and

$$T(\zeta_1)^{-1}\, P(\zeta_1)\, T(\zeta_1) = P(\zeta_0).$$

Hence,

$$S(\zeta)\, D(\zeta_1)\, S(\zeta)^{-1} = T(\zeta)\, T(\zeta_1)^{-1}\, D(\zeta_1)\, T(\zeta_1)\, T(\zeta)^{-1}$$
$$= T(\zeta)\, D(\zeta_0)\, T(\zeta)^{-1} = D(\zeta)$$

and

$$S(\zeta)\, P(\zeta_1)\, S(\zeta)^{-1} = T(\zeta)\, T(\zeta_1)^{-1}\, P(\zeta_1)\, T(\zeta_1)\, T(\zeta)^{-1}$$
$$= T(\zeta)\, P(\zeta_0)\, T(\zeta)^{-1} = P(\zeta). \blacksquare$$

As noted in Subsect. 5.3.3, the operators $T(\zeta)$, $T(\zeta)^{-1}$ are both holomorphic at normal points, except for a set whose trace set is isolated and closed in G. Hence,

Lemma 3. *All normal points of $\mathfrak{P}$ are strongly normal, except for a set of normal points which is isolated and closed in $\mathfrak{P}$ (i.e. its trace set is isolated and closed in G).*

This exceptional set can be characterized by the set of all normal points ζ_0 of $\mathfrak{P}$ where no transformation operator $T(\zeta) \in \mathscr{L}(X(\mathfrak{P}) : \mathscr{M}(\mathfrak{P}))$ exists with the properties II and III of Definition 1 and such that $T(\zeta)$ and $T(\zeta)^{-1}$ are holomorphic at the point $\zeta = \zeta_0$. If such an operator $T(\zeta)$ exists, and if $S(\zeta) = T(\zeta)\, T(\zeta_0)^{-1}$, one obtains an operator satisfying all properties of Definition 1; hence, ζ_0 would be strongly normal. The converse is obvious.

5.4.2. Weakly splitting points

The next subsection contains a further characterization of strongly normal points (and of the corresponding exceptional set). Simultaneously, the existence of such exceptional points will be discussed. The present subsection deals with some preliminary remarks for that purpose.

Consider $D(\zeta)$ as an operator from $\mathscr{L}(X(\mathfrak{P}) : \mathscr{M}(\mathfrak{P}))$. If

$$d_\varrho = \operatorname{def} D(\zeta)^\varrho, \qquad \varrho = 1, 2, \ldots, m, \tag{4.1}$$

one obtains (see (3.11))

$$N - m + 1 \leqq d_1 \leqq d_2 \leqq \cdots \leqq d_m = N.$$

With use of the numbers $d_1, d_2, \ldots, d_m$, the dimensions of the irreducible subprojections of $P(\zeta)$ with respect to $D(\zeta)$ can be calculated uniquely (Frobenius formulas).

Let ζ_0 be a fixed normal point of $\mathfrak{P}$. Then, according to Lemma 1.2., one obtains the inequalities

$$\operatorname{def} D(\zeta_0)^\varrho \geqq d_\varrho, \qquad \varrho = 1, 2, \ldots, m. \tag{4.2}$$

If all these m inequalities are equalities, i.e. if $\operatorname{def} D(\zeta_0)^\varrho = d_\varrho$, $\varrho = 1, \ldots, m$, then the Jordan normal form of $D(\zeta_0)$, as an operator from $\mathscr{L}(X : \mathbb{C})$, coincides with the Jordan normal form of $D(\zeta)$, as an operator from $\mathscr{L}(X(\mathfrak{P}) : \mathscr{M}(\mathfrak{P}))$.

Definition 2. A normal point ζ_0 from $\mathfrak{P}$ is called a *weakly splitting point* if there is a number ϱ, $1 \leqq \varrho \leqq m$, such that $\operatorname{def} D(\zeta_0)^\varrho > d_\varrho$.

If ζ_0 is a weakly splitting point then the Jordan normal form of $D(\zeta_0) \in \mathscr{L}(X:\mathbb{C})$ is different from the Jordan normal form of $D(\zeta) \in \mathscr{L}\big(X(\mathfrak{P}):\mathscr{M}(\mathfrak{P})\big)$, and conversely.

Lemma 1.2 implies

Lemma 4. *The set of all weakly splitting points is isolated and closed in* $\mathfrak{P}$.

5.4.3. Connection between strongly normal and weakly splitting points

In this section strongly normal points (and the corresponding exceptional set) are characterized by weakly splitting points.

Theorem 1. *A normal point* $\zeta_0 \in \mathfrak{P}$ *is strongly normal if and only if it is not a weakly splitting point.*

Proof. Let ζ_0 be strongly normal. Then there is an operator $T(\zeta) \in \mathscr{L}\big(X(\mathfrak{P}):\mathscr{M}(\mathfrak{P})\big)$ with the properties $T(\zeta_0) = 1$, $D(\zeta) = T(\zeta)\, D(\zeta_0)\, T(\zeta)^{-1}$. Hence, the Jordan normal form of $D(\zeta) \in \mathscr{L}\big(X(\mathfrak{P}):\mathscr{M}(\mathfrak{P})\big)$ coincides with that of $D(\zeta) \in \mathscr{L}(X:\mathbb{C})$ for all ζ from a suitable neighborhood of ζ_0, in particular with that of $D(\zeta_0) \in \mathscr{L}(X:\mathbb{C})$. Thus ζ_0 is not a weakly splitting point.

Conversely, let ζ_0 not be a weakly splitting point, i.e., let $\operatorname{def} D(\zeta_0)^\varrho = d_\varrho$, $\varrho = 1, 2, \ldots, m$. Then there is a full neighborhood $U = U(\zeta_0)$ such that

$$\operatorname{def} D(\zeta)^\varrho = d_\varrho, \qquad \varrho = 1, 2, \ldots, m, \qquad \zeta \in U, \tag{4.3}$$

(see Lemma 1.2). Note that $D(\zeta)^m \equiv 0$. There exists a number p, $1 \leqq p \leqq m$, such that

$$D(\zeta)^p \equiv 0, \qquad D(\zeta)^{p-1} \not\equiv 0 \tag{4.4}$$

or, equivalently,

$$\operatorname{def} D(\zeta)^p = N, \qquad \operatorname{def} D(\zeta)^{p-1} < N. \tag{4.5}$$

In particular,

$$D(\zeta)^{p-1} \neq 0 \tag{4.6}$$

for all $\zeta \in U$. For $\zeta \in U$ the kernels of $D(\zeta)^\varrho$ in $P(\zeta)\, X$ are denoted by $\ker_P \big(D(\zeta)^\varrho\big) = K_\varrho(\zeta) := \ker \big(D(\zeta)^\varrho\big) \cap P(\zeta)\, X$. Then one obtains

$$\{0\} \subset K_1(\zeta) \subset K_2(\zeta) \subset \cdots \subset K_{p-1}(\zeta) \subset K_p(\zeta) = P(\zeta)\, X, \tag{4.7}$$

and the dimensions of $K_\varrho(\zeta)$, $\varrho = 1, 2, \ldots, p$, are constant for $\zeta \in U$.

Now the proof proceeds with the step by step construction of a Jordan basis for $D(\zeta) \in \mathscr{L}\big(X(\mathfrak{P}):\mathscr{M}(\mathfrak{P})\big)$ in $P(\zeta)\, X(\mathfrak{P})$, where all vectors of this basis are holomorphic in a suitable neighborhood V of ζ_0 and where they form a Jordan basis for $D(\zeta)$ in $P(\zeta)\, X$ if $\zeta \in V$ is fixed.

Note that in the following the symbol V is always used for neighborhoods of ζ_0. This does not mean that these neighborhoods necessarily coincide.

1. Let $x_1, x_2, \ldots, x_s$ be a relative basis of $P(\zeta_0) X$ modulo $K_{p-1}(\zeta_0)$. In particular, these vectors are linearly independent and from $P(\zeta_0) X$. Consider the vectors $P(\zeta) x_1, P(\zeta) x_2, \ldots, P(\zeta) x_s$. These vectors are algebroid vector-valued functions which are holomorphic and linearly independent for all ζ from a suitable neighborhood V of ζ_0. Moreover, they belong to $P(\zeta) X$.

The first assertion is that the vectors $P(\zeta) x_1, \ldots, P(\zeta) x_s$ form a relative basis of $P(\zeta) X$ modulo $K_{p-1}(\zeta)$ for all ζ from a certain neighborhood V of ζ_0.

Assume that this is not the case. Then there is a sequence $\zeta_\varrho \to \zeta_0$ and there are numbers $\lambda_1^{(\varrho)}, \lambda_2^{(\varrho)}, \ldots, \lambda_s^{(\varrho)}$ such that for all ϱ at least one of these numbers is different from zero and, furthermore, such that

$$\sum_{\alpha=1}^{s} \lambda_\alpha^{(\varrho)} P(\zeta_\varrho) x_\alpha \in K_{p-1}(\zeta_\varrho).$$

In other words, the vectors $P(\zeta_\varrho) x_1, \ldots, P(\zeta_\varrho) x_s$ are linearly dependent modulo $K_{p-1}(\zeta_\varrho)$ for every point ζ_ϱ of the sequence. Without loss of generality one can assume that

$$\sum_{\alpha=1}^{s} |\lambda_\alpha^{(\varrho)}|^2 = 1, \qquad \varrho = 1, 2, 3, \ldots$$

From the sequence $\Lambda_\varrho = (\lambda_1^{(\varrho)}, \lambda_2^{(\varrho)}, \ldots, \lambda_s^{(\varrho)})$ one can select a convergent subsequence which is again denoted by Λ_ϱ such that a convergence relation

$$\Lambda_\varrho \to \Lambda_0 = (\lambda_1^{(0)}, \lambda_2^{(0)}, \ldots, \lambda_s^{(0)})$$

holds. This implies

$$\sum_{\alpha=1}^{s} |\lambda_\alpha^{(0)}|^2 = 1. \tag{4.8}$$

Taking the limit $\varrho \to \infty$ in the relation

$$D(\zeta_\varrho)^{p-1} \left[\sum_{\alpha=1}^{s} \lambda_\alpha^{(\varrho)} P(\zeta_\varrho) x_\alpha\right] = 0, \qquad \varrho = 1, 2, 3, \ldots,$$

one obtains

$$D(\zeta_0)^{p-1} \left[\sum_{\alpha=1}^{s} \lambda_\alpha^{(0)} P(\zeta_0) x_\alpha\right] = 0$$

or

$$\sum_{\alpha=1}^{s} \lambda_\alpha^{(0)} P(\zeta_0) x_\alpha \in K_{p-1}(\zeta_0),$$

which is impossible, because of (4.8).

2. Form the vectors $D(\zeta) x_1, D(\zeta) x_2, \ldots, D(\zeta) x_s$. These vectors are algebroid vector-valued functions and $D(\zeta) x_\varrho \in K_{p-1}(\zeta)$ for $\varrho = 1, 2, \ldots, s$. Moreover, these vectors are linearly independent modulo $K_{p-2}(\zeta)$ for every $\zeta \in V$ (see Lemma 2.9.1). In particular, these vectors are linearly independent for every $\zeta \in V$.

Then the vectors $D(\zeta) x_1, \ldots, D(\zeta) x_s$ are also linearly independent in $X(\mathfrak{P}) : \mathcal{M}(\mathfrak{P})$. Assuming their linear dependence, one obtains a relation of the form

$$\alpha_1(\zeta) D(\zeta) x_1 + \cdots + \alpha_s(\zeta) D(\zeta) x_s \equiv 0,$$

where one of the algebroid scalar functions, for example $\alpha_1(\zeta)$, does not vanish identically. Hence, there are points ζ_1 from V such that all functions $\alpha_1, \ldots, \alpha_s$ are holomorphic at ζ_1 and such that $\alpha_1(\zeta_1) \neq 0$. Then the relation

$$\alpha_1(\zeta_1)\, D(\zeta_1)\, x_1 + \cdots + \alpha_s(\zeta_1)\, D(\zeta_1)\, x_s = 0, \qquad \alpha_1(\zeta_1) \neq 0,$$

follows, in contradiction to the linear independence of the vectors $D(\zeta_1)\, x_1, \ldots, D(\zeta_1)\, x_s$.

Therefore, the vectors $D(\zeta)\, x_1, \ldots, D(\zeta)\, x_s$ form a system of vectors in $K_{p-1}(\zeta)$ which are linearly independent with respect to $\mathcal{M}(\mathfrak{P})$, holomorphic in V and linearly independent with respect to $\mathbb{C}$ for every $\zeta \in V$.

3. Apply Lemma 1.4 to this situation. According to this lemma, there exist algebroid vector-valued functions $y_{s+1}(\zeta), \ldots, y_r(\zeta)$ $\left(r = d_{p-1} = \dim K_{p-1}(\zeta) = \operatorname{def} D(\zeta)^{p-1}\right)$ from $K_{p-1}(\zeta)$ with the following properties: First, they are holomorphic in a suitable neighborhood V of ζ_0; secondly, together with the vector-valued functions $D(\zeta)\, x_1, \ldots, D(\zeta)\, x_s$, they form a basis of $K_{p-1}(\zeta)$ as a subspace of $X(\mathfrak{P}) : \mathcal{M}(\mathfrak{P})$ and, for every $\zeta \in V$, they form a basis of $K_{p-1}(\zeta)$ as a subspace of $X : \mathbb{C}$.

4. In particular, if $\zeta = \zeta_0$, one obtains a basis of $K_{p-1}(\zeta_0)$ composed of the vectors

$$D(\zeta_0)\, x_1, D(\zeta_0)\, x_2, \ldots, D(\zeta_0)\, x_s, \qquad y_{s+1}(\zeta_0), \ldots, y_r(\zeta_0). \tag{4.9}$$

Hence, the vectors (4.9) also form a system generating the linear space $K_{p-1}(\zeta_0) \bmod K_{p-2}(\zeta_0)$; i.e., their linear span modulo $K_{p-2}(\zeta_0)$ coincides with $K_{p-1}(\zeta_0) \bmod K_{p-2}(\zeta_0)$. The first s vectors of (4.9) are linearly independent modulo $K_{p-2}(\zeta_0)$. The system of these s vectors can be enlarged to a relative basis of $K_{p-1}(\zeta_0) \bmod K_{p-2}(\zeta_0)$ by addition of some of the remaining $r - s$ vectors of (4.9) (see Subsect. 1.1.7). These supplementary vectors are denoted by $y_{s+1}(\zeta_0), \ldots, y_t(\zeta_0)$. Hence, the vectors

$$D(\zeta_0)\, x_1, D(\zeta_0)\, x_2, \ldots, D(\zeta_0)\, x_s, \qquad y_{s+1}(\zeta_0), \ldots, y_t(\zeta_0) \tag{4.10}$$

form a relative basis of $K_{p-1}(\zeta_0) \bmod K_{p-2}(\zeta_0)$ (where $t = d_{p-1} - d_{p-2}$).

5. Consider the algebroid vector-valued functions

$$D(\zeta)\, x_1, D(\zeta)\, x_2, \ldots, D(\zeta)\, x_s, \qquad y_{s+1}(\zeta), \ldots, y_t(\zeta). \tag{4.11}$$

These functions belong to $K_{p-1}(\zeta)$. The assertion is that there is a neighborhood V of ζ_0 such that the vectors (4.11) form a relative basis of $K_{p-1}(\zeta) \bmod K_{p-2}(\zeta)$ for every $\zeta \in V$.

Assume that this assertion is not true. Then there is a sequence $\zeta_\varrho \to \zeta_0$ and a sequence $\Lambda_\varrho = (\lambda_1^{(\varrho)}, \lambda_2^{(\varrho)}, \ldots, \lambda_t^{(\varrho)})$ of t-tuples of complex numbers with $\Lambda_\varrho \neq (0, 0, \ldots, 0)$ such that the relation

$$\sum_{\alpha=1}^{s} \lambda_\alpha^{(\varrho)} D(\zeta_\varrho)\, x_\alpha + \sum_{\alpha=s+1}^{t} \lambda_\alpha^{(\varrho)} y_\alpha(\zeta_\varrho) \in K_{p-2}(\zeta_\varrho), \qquad \varrho = 1, 2, \ldots,$$

is satisfied. Hence,

$$D(\zeta_\varrho)^{p-2} \left[\sum_{\alpha=1}^{s} \lambda_\alpha^{(\varrho)} D(\zeta_\varrho)\, x_\alpha + \sum_{\alpha=s+1}^{t} \lambda_\alpha^{(\varrho)} y_\alpha(\zeta_\varrho)\right] = 0, \qquad \varrho = 1, 2, 3, \ldots \tag{4.12}$$

Without loss of generality one can assume that $\sum_{\alpha=1}^{t} |\lambda_\alpha^{(\varrho)}|^2 = 1$, $\varrho = 1, 2, 3, \ldots$ From the sequence Λ_ϱ one can select a convergent subsequence, again denoted by Λ_ϱ, such that $\Lambda_\varrho \to \Lambda_0 = (\lambda_1^{(0)}, \lambda_2^{(0)}, \ldots, \lambda_t^{(0)})$, where $\sum_{\alpha=1}^{t} |\lambda_\alpha^{(0)}|^2 = 1$. Taking the limit $\zeta_\varrho \to \zeta_0$ in (4.12), one obtains

$$D(\zeta_0)^{p-2} \left[\sum_{\alpha=1}^{s} \lambda_\alpha^{(0)} D(\zeta_0)\, x_\alpha + \sum_{\alpha=s+1}^{t} \lambda_\alpha^{(0)} y_\alpha(\zeta_0)\right] = 0;$$

that is,

$$\sum_{\alpha=1}^{s} \lambda_\alpha^{(0)} D(\zeta_0)\, x_\alpha + \sum_{\alpha=s+1}^{t} \lambda_\alpha^{(0)} y_\alpha(\zeta_0) \in K_{p-2}(\zeta_0).$$

This result is a contradiction, because the vectors $D(\zeta_0)\, x_1, \ldots, D(\zeta_0)\, x_s$, $y_{s+1}(\zeta_0), \ldots, y_t(\zeta_0)$ are linearly independent modulo $K_{p-2}(\zeta_0)$.

6. By the foregoing steps a system of algebroid vector-valued functions

$$\begin{array}{ll} P(\zeta)\, x_1, \ldots, P(\zeta)\, x_s, & \\ D(\zeta)\, x_1, \ldots, D(\zeta)\, x_s, & y_{s+1}(\zeta), \ldots, y_t(\zeta) \end{array}$$

has been constructed with the following properties: All these vector-valued functions are holomorphic in V; the vectors of the first line form a relative basis of $P(\zeta)\, X$ mod $K_{p-1}(\zeta)$ for every $\zeta \in V$ and the vectors of the second line form a relative basis of $K_{p-1}(\zeta)$ mod $K_{p-2}(\zeta)$ for every $\zeta \in V$.

7. This constructive procedure can be continued successively. For the next step, one considers the vectors

$$D(\zeta)^2\, x_1, \ldots, D(\zeta)^2\, x_s, \qquad D(\zeta)\, y_{s+1}(\zeta), \ldots, D(\zeta)\, y_t(\zeta).$$

These functions belong to $K_{p-2}(\zeta)$; they are holomorphic in V and, moreover, linearly independent modulo $K_{p-3}(\zeta)$ for every $\zeta \in V$. According to the foregoing steps 2 to 5, this system can be enlarged to a relative basis modulo $K_{p-3}(\zeta)$ for every ζ from a suitable neighborhood V by adding suitable algebroid vector-valued functions which are holomorphic in V.

8. Finally, one obtains the following table analogous to the table in Subsect. 2.9.2, which is made up of algebroid vector-valued functions holomorphic in a suitable neighborhood V of ζ_0. For every $\zeta \in V$, each row of this table forms a relative basis of the factor space given in the last column. (The notation is adapted to that of Table (2.9.4).):

$$\left.\begin{array}{ll} P(\zeta)\, y_p^{(\sigma_p)}, & \ldots P(\zeta)\, X/K_{p-1}(\zeta), \\ D(\zeta)\, y_p^{(\sigma_p)}, y_{p-1}^{(\sigma_{p-1})}(\zeta), & \ldots K_{p-1}(\zeta)/K_{p-2}(\zeta), \\ D(\zeta)^2\, y_p^{(\sigma_p)}, D(\zeta)\, y_{p-1}^{(\sigma_{p-1})}(\zeta), y_{p-2}^{(\sigma_{p-2})}(\zeta), & \ldots K_{p-2}(\zeta)/K_{p-3}(\zeta), \\ \cdots & \\ D(\zeta)^{p-1}\, y_p^{(\sigma_p)}, D(\zeta)^{p-2}\, y_{p-1}^{(\sigma_{p-1})}(\zeta), & \\ D(\zeta)^{p-3}\, y_{p-2}^{(\sigma_{p-2})}(\zeta), \ldots, D(\zeta)\, y_2^{(\sigma_2)}(\zeta), y_1^{(\sigma)_1}(\zeta) & \ldots K_1(\zeta)/\{0\}. \end{array}\right\} \qquad (4.13)$$

The vectors in this table yield a basis of $P(\zeta) X$ for every $\zeta \in V$. They are algebroid vector-valued functions, which also form a basis of $P(\zeta) X(\mathfrak{P})$ over $\mathcal{M}(\mathfrak{P})$, more precisely, a Jordan basis of $P(\zeta) X(\mathfrak{P})$ with respect to $D(\zeta)$.

As in Sect. 2.9, another notation for this basis is introduced. This new notation groups the vectors which belong to the same irreducible subspace and, moreover, expresses explicitly the degree of these vectors (as Jordan vectors). For example, let

$$P(\zeta)\, D(\zeta)^\varrho\, y_p^{(1)} = e_1^{p-\varrho}(\zeta), \qquad \varrho = 0, 1, 2, \ldots, p-1.$$

Then the basis can be written in the form $e_\varrho^\sigma(\zeta)$, $\varrho = 1, 2, \ldots, l$ (l denotes the number of different irreducible subspaces), $\sigma = 1, 2, \ldots, m_\varrho$ (the numbers m_ϱ denote the dimensions of the irreducible subspaces).

Moreover, one chooses a basis $e_0^\sigma(\zeta)$ of $(1 - P(\zeta))\, X(\mathfrak{P})$ such that the vectors $e_0^\sigma(\zeta)$ are holomorphic in V and such that they form a basis of $(1 - P(\zeta))\, X$ for every $\zeta \in V$. Then one can define a transformation operator $T(\zeta)$ by

$$T(\zeta)\, e_\varrho^\sigma(\zeta_0) = e_\varrho^\sigma(\zeta). \tag{4.14}$$

Obviously, $T(\zeta) \in \mathcal{L}(X(\mathfrak{P}){:}\mathcal{M}(\mathfrak{P}))$ and $T(\zeta_0) = 1$. Hence, $T(\zeta)$ and $T(\zeta)^{-1}$ are both holomorphic in a neighborhood of $\zeta = \zeta_0$. Because of the construction of $T(\zeta)$, one obtains

$$P(\zeta) = T(\zeta)\, P(\zeta_0)\, T(\zeta)^{-1} \tag{4.15}$$

and

$$D(\zeta) = T(\zeta)\, D(\zeta_0)\, T(\zeta)^{-1}. \tag{4.16}$$

Therefore, the point ζ_0 is strongly normal. ■

5.4.4. Examples

Theorem 1 also answers the question of the existence of exceptional (normal) points from $\mathfrak{P}$, which are not strongly normal. This theorem expresses the fact that these exceptional points coincide with the weakly splitting points. However, it is very easy to construct examples where weakly splitting points occur. Once more, consider the example from the introduction to this chapter.

(a) $$A(z) = \begin{pmatrix} 0 & z \\ -z & 2z \end{pmatrix}.$$

There are no splitting points in this case; $\lambda(z) \equiv z$ is the only eigenvalue. Hence,

$$D(z) = A(z) - z = \begin{pmatrix} -z & z \\ -z & z \end{pmatrix}$$

and one obtains

$$\operatorname{def} D(0) = 2; \qquad \operatorname{def} D(z) = 1, \qquad z \neq 0.$$

Therefore, $z = 0$ is a weakly splitting point. All other points are strongly normal. The Jordan normal form of $D(0)$ is $\begin{pmatrix} 0 & 0 \\ 0 & 0 \end{pmatrix}$, the Jordan normal form of $D(z)$ for $z \neq 0$ is $\begin{pmatrix} 0 & 1 \\ 0 & 0 \end{pmatrix}$.

This example shows that simple examples can be constructed, where weakly splitting points occur, if there are no splitting points. Moreover, one can assume for simplicity that only one eigenvalue appears and that this eigenvalue vanishes. That is, one considers meromorphic and nilpotent operator-valued functions $D(z)$ in a region G, $D(z)^N \equiv 0$, and looks for weakly splitting points.

(b) $D(z) = z\Gamma_N(0) = z\Gamma, \qquad X = \mathbb{C}^N.$

One obtains $D(z)^2 = z^2\Gamma^2, \ldots, D(z)^{N-1} = z^{N-1}\Gamma^{N-1}$, $D(z)^N \equiv 0$. Obviously, $z = 0$ is a weakly splitting point and all other points are strongly normal, for $\operatorname{def} D(0) = N$; however, $\operatorname{def} D(z) = 1$ if $z \neq 0$. For example,

$$D(z) = T(z)\, \Gamma T(z)^{-1}, \tag{4.17}$$

where

$$T(z) = \{z^{N-1}, z^{N-2}, \ldots, z^2, z, 1\}$$

and

$$T(z)^{-1} = \left\{\frac{1}{z^{N-1}}, \frac{1}{z^{N-2}}, \ldots, \frac{1}{z^2}, \frac{1}{z}, 1\right\}.$$

Note that all other transformations $T(z)$ with the property (4.17) also have a structure such that $z = 0$ is a pole either of $T(z)$ or of $T(z)^{-1}$.

This example can be generalized without difficulties.

(c) $D(z) = a_1(z)\, \Gamma + a_2(z)\, \Gamma^2 + \cdots + a_{N-1}(z)\, \Gamma^{N-1}, \qquad X = \mathbb{C}^N.$

The functions $a_1(z), a_2(z), \ldots, a_{N-1}(z)$ denote, for example, holomorphic scalar functions in a region G. Let $z_0 \in G$ and $a_1(z_0) \neq 0$. Then $D(z_0) \sim \Gamma$. If $a_1(z_0) = 0$, but $a_2(z_0) \neq 0$, then $D(z_0) \sim \Gamma^2$. In general, if $a_1(z_0) = a_2(z_0) = \cdots = a_{k-1}(z_0) = 0$, but $a_k(z_0) \neq 0$, then $D(z_0) \sim \Gamma^k$, $k = 1, 2, \ldots, N-1$. If $a_1(z) \not\equiv 0$ in G, then all zeros of $a_1(z)$ are weakly splitting points, for these points are exactly those points where the Jordan normal form of $D(z)$ is different from Γ. The form of the Jordan normal form at weakly splitting points depends on the vanishing or nonvanishing of further coefficients.

5.4.5. Index and geometric multiplicity for weakly splitting points

Let $\zeta_0 \in \mathfrak{P}$ be normal with respect to $D(\zeta)$. $\lambda(\zeta)$ denotes the corresponding eigenvalue function. Recall that $\operatorname{def} D(\zeta)$ is the number of different irreducible subspaces for $\lambda(\zeta)$ with respect to $A(z)$, i.e., the geometric multiplicity of $\lambda(\zeta)$. If ζ_0 is a weakly splitting point, then $\operatorname{def} D(\zeta_0) \geq \operatorname{def} D(\zeta)$. Thus one obtains:

The geometric multiplicity of $\lambda(\zeta)$ with respect to $A(z)$ can at most increase at weakly splitting points.

Furthermore, for strongly normal points ζ the index of $\lambda(\zeta)$ is constant, $\nu(\lambda(\zeta)) = p$. If ζ_0 is a weakly splitting point, then there is a number r, $1 \leqq r \leqq p$, such that $D(\zeta_0)^r = 0$, but $D(\zeta_0)^{r-1} \neq 0$. Therefore, $\nu(\lambda(\zeta_0)) = r \leqq p$. That is:

The index of $\lambda(\zeta)$ with respect to $A(z)$ can at most decrease at weakly splitting points.

5.5. Irreducible projections

In Subsect. 5.3.2 and 5.3.3 irreducible subprojections of $P(\zeta)$ with respect to $D(\zeta)$ appeared. In this section, further properties of these projections will be derived. The results obtained in Sect. 5.4 will be used for this purpose.

5.5.1. Construction

According to the previous results, especially those of Sect. 5.4, the decomposition of $P(\zeta)$ into a sum of disjoint (irreducible) algebroid Jordan projections with respect to $D(\zeta)$ can be given as follows: Let $\zeta_0 \in \mathfrak{P}$ be a fixed strongly normal point and let $T(\zeta)$ be a corresponding transformation operator with the properties I to III of Definition 4.1. Furthermore, let

$$P(\zeta_0) = \sum_{i=1}^{l} P_i$$

be a decomposition of $P(\zeta_0)$ into a sum of disjoint Jordan projections with respect to $D(\zeta_0)$. Defining $P_i(\zeta)$ by

$$P_i(\zeta) = T(\zeta)\, P_i T(\zeta)^{-1}, \qquad i = 1, 2, \ldots, l, \tag{5.1}$$

one obtains

$$P(\zeta) = \sum_{i=1}^{l} P_i(\zeta), \tag{5.2}$$

which is a decomposition of $P(\zeta)$ into a sum of disjoint (irreducible) algebroid Jordan projections with respect to $D(\zeta)$. This decomposition has the property that the Jordan projections $P_i(\zeta)$ are holomorphic at the strongly normal point $\zeta = \zeta_0$ and that the relation $P_i(\zeta_0) = P_i$ holds; i.e., the Jordan projections coincide at ζ_0 with the prescribed decomposition operators P_i.

As the Jordan projections $P_i(\zeta)$, $i = 1, \ldots, l$, are algebroid operator-valued functions, there are unique representations of the form

$$P_i(\zeta) = C_{i0}(z) + C_{i1}(z)\, \lambda(\zeta) + \cdots + C_{i,g-1}(z)\, \lambda(\zeta)^{g-1}, \qquad i = 1, 2, \ldots, l, \tag{5.3}$$

where the coefficients $C_{i\alpha}(z)$ are meromorphic operator-valued functions in G.

5.5.2. Behavior at splitting points

As the irreducible projections $P_1(\zeta), \ldots, P_l(\zeta)$ are holomorphic at all normal points of $\mathfrak{P}$, except for a certain set which is isolated and closed in $\mathfrak{P}$, to every splitting point ζ_0 corresponds a neighborhood U of ζ_0 such that the projections $P_1(\zeta), \ldots, P_l(\zeta)$ are holomorphic in $U \setminus \{\zeta_0\}$. (Note that U could be a multi-valued neighborhood, namely in the case that ζ_0 is a branching point of order $a \geq 2$.)

The branching behavior of all irreducible projections at branching points of $\lambda(\zeta)$ coincides with that of $\lambda(\zeta)$ and $P(\zeta)$. More precisely:

Lemma 1. *Let ζ_0 be a branching point or $\lambda(\zeta)$ of order a. Then ζ_0 is also a branching point of $P_1(\zeta), \ldots, P_l(\zeta)$ of exactly the same order a.*

Proof. Let $P_\alpha(\zeta)$ be one of the irreducible projections. Its (for the moment formal) branches in the neighborhood of the branching point ζ_0 are denoted by $P_\alpha^1(z), P_\alpha^2(z), \ldots, P_\alpha^a(z)$. Correspondingly, the branches of $P(\zeta)$ are denoted by $P^1(z), P^2(z), \ldots, P^a(z)$. Then $P^\nu(z)\, P^\mu(z) = \delta_{\nu\mu} P^\nu(z)$. The identity $\sum_{i=1}^{l} P_i^\nu(z) = P^\nu(z)$ is valid, because of (5.2). Hence, $P_\alpha^\nu(z)\, P^\nu(z) = P_\alpha^\nu(z)$; i.e., the irreducible projections $P_\alpha^\nu(z)$ are subprojections of $P^\nu(z)$. Assume that $P_\alpha^\nu(z) = P_\alpha^\mu(z)$ holds, where $\nu \neq \mu$. Then, $P_\alpha^\nu(z)\, P^\nu(z) = P_\alpha^\mu(z)\, P^\mu(z)$ and $P_\alpha^\nu(z)\, P^\nu(z)\, P^\mu(z) = P_\alpha^\mu(z)\, P^\mu(z)$; i.e., $P_\alpha^\mu(z)\, P^\mu(z) = P_\alpha^\mu(z) = 0$; hence, $P_\alpha(\zeta) \equiv 0$. This result is a contradiction. ■

5.5.3. The coefficients $C_{ij}(z)$

Let $\mathfrak{P}$ be g-valued and let $\lambda_1(z), \ldots, \lambda_g(z)$ denote the g branches of $\lambda(\zeta)$ over a given simple point z_0. Correspondingly, the g branches of the irreducible projection-valued function $P_i(\zeta)$, $i = 1, \ldots, l$ are denoted by $P_i^1(z), P_i^2(z), \ldots, P_i^g(z)$. Then from (5.3), the equation

$$P_i^\alpha(z) = \sum_{j=0}^{g-1} C_{ij}(z)\, \lambda_\alpha(z)^j, \qquad \alpha = 1, 2, \ldots, g,$$

follows. From this equation, (in analogy with Subsect. 3.3.9), one obtains a representation of the form

$$C_{ij}(z) = \sum_{\varrho=1}^{g} \alpha_{ij}^{(\varrho)}(z)\, P_i^\varrho(z), \qquad i = 1, 2, \ldots, l, \qquad j = 0, 1, 2, \ldots, g-1,$$

for the meromorphic operator-valued coefficients $C_{ij}(z)$; note that the $\alpha_{ij}^{(\varrho)}(z)$ are meromorphic scalar functions such that poles occur at most at multiple points.

5.5.4. Puiseux-Laurent series

Let ζ_0 be an arbitrary but fixed point from $\mathfrak{P}$ (splitting point, weakly splitting point or strongly normal point). Then the expansion (5.3) for $P_i(\zeta)$ can be rewritten into a local expansion of $P_i(\zeta)$ in the neighborhood of ζ_0 if one substitutes the Puiseux series for $\lambda(\zeta)$ and the Laurent series for the coefficients $C_{ij}(z)$ into (5.3). Thus one obtains

Theorem 1. *Let $\zeta_0 = (z_0, \lambda_0)$ be a point from $\mathfrak{P}$. Let the dotted disc $0 < |z - z_0| < \delta$ consist of simple points only and not contain trace points of poles of $P_1(\zeta), \ldots, P_l(\zeta)$. Furthermore, let $\mathfrak{H}$ be a h-valued cycle of the (z_0, λ_0)-group of $\mathfrak{P}$ $(1 \leq h \leq g)$ over $0 < |z - z_0| < \delta$ (where $\zeta_0 \in \mathfrak{H}$). Then the irreducible projections $P_1(\zeta), \ldots, P_l(\zeta)$ have branches $P_i^1(z), P_i^2(z), \ldots, P_i^h(z)$, $i = 1, \ldots, l$, corresponding to $\mathfrak{H}$, and these branches can be expanded in the neighborhood $0 < |z - z_0| < \delta$ of z_0 into Puiseux-Laurent series of the form*

$$P_i^\alpha(z) = (z - z_0)^{-M_i} \sum_{\mu=0}^{\infty} B_{i\mu}[\xi^\alpha (z - z)^{l/h}]^\mu, \quad i = 1, \ldots, l; \alpha = 1, \ldots, h, \tag{5.4}$$

where $B_{i\mu} \in \mathscr{L}(X)$ (note that not necessarily $B_{i0} \neq 0$); M_i denotes the maximal pole order of $C_{ij}(z)$, $j = 0, 1, \ldots, g - 1$, at $z = z_0$; ξ denotes a primitive h-th root of unity; $(z - z_0)^{1/h}$ denotes a fixed root.

5.5.5. Poles of $P_i(\zeta)$ at branching points

The result of J. Butler on the connection between branching and poles for eigenprojections $P(\zeta)$ (see Subsect. 3.3.7) can be formulated also for the irreducible projections.

Theorem 2. *Let ζ_0 be a branching point of $\lambda(\zeta)$ of order a (≥ 2). Then every irreducible projection-valued function $P_i(\zeta)$, $i = 1, 2, \ldots, l$, has a pole at $\zeta = \zeta_0$.*

Proof. First, according to Lemma 1, every function $P_1(\zeta), \ldots, P_l(\zeta)$ has the same branching order a as $\lambda(\zeta)$ at $\zeta = \zeta_0$. Assume that $P_i(\zeta)$ is continuous at $\zeta = \zeta_0$. This means that the local expansion (5.4) of $P_i(\zeta)$ in the neighborhood of $\zeta = \zeta_0$ takes the form

$$P_i^\alpha(z) = B_{i0} + B_{i1}(z - z_0)^{1/a} + B_{i2}(z - z_0)^{2/a} + \cdots, \qquad \alpha = 1, 2, \ldots, a.$$

Therefore,

$$P_i^\alpha(z)\, P_i^{\alpha'}(z) = P_i^\alpha(z)\, P^\alpha(z)\, P_i^{\alpha'}(z)\, P^{\alpha'}(z) = P_i^\alpha P_i^{\alpha'} P^\alpha P^{\alpha'} = 0, \qquad \alpha \neq \alpha'. \tag{5.5}$$

Hence, by taking the limit $z \to z_0$, one obtains

$$B_{i0}^2 = 0.$$

On the other hand, $P_i^\alpha(z)^2 = P_i^\alpha(z)$. Hence, $B_{i0}^2 = B_{i0}$ and

$$B_{i0} = 0.$$

Thus one obtains

$$\lim_{z \to z_0} \|P_i^\alpha(z)\| = 0,$$

in contradiction to $\|P_i^\alpha(z)\| \geq 1$ for $z \neq z_0$. ■

5.6. Jordan vectors

Jordan vectors and Jordan bases of $D(\zeta)$ in $P(\zeta) X$ have been considered in Subsect. 5.3.2 and 5.3.3. In this section supplementary remarks will be made.

5.6.1. Construction

According to the results of Sect. 5.3 and especially 5.4, a Jordan basis of $D(\zeta)$ in $P(\zeta) X$ can be given as follows: Let ζ_0 be a fixed strongly normal point of $\mathfrak{P}$ with respect to $D(\zeta)$ and let $T(\zeta)$ denote a corresponding transformation operator. Then, choosing a Jordan basis e_ϱ^σ of $D(\zeta_0)$ in $P(\zeta_0) X$, $\varrho = 1, 2, \ldots, l$; $\sigma = 1, 2, \ldots, m_\varrho$ (l denotes the number of irreducible parts, m_ϱ their dimensions, e_ϱ^σ is a Jordan vector of degree σ), one defines $e_\varrho^\sigma(\zeta)$ by

$$e_\varrho^\sigma(\zeta) = T(\zeta)\, e_\varrho^\sigma, \qquad \varrho = 1, \ldots, l; \qquad \sigma = 1, \ldots, m_\varrho. \tag{6.1}$$

The vectors (6.1) are algebroid vector-valued functions from $\mathfrak{P}$ into X, that is, there are unique representations of the form

$$e_\varrho^\sigma(\zeta) = f_{\varrho 0}^\sigma(z) + f_{\varrho 1}^\sigma(z)\, \lambda(\zeta) + \cdots + f_{\varrho, g-1}^\sigma(z)\, \lambda(\zeta)^{g-1}, \tag{6.2}$$

where the coefficients $f_{\varrho\alpha}^\sigma(z)$ are meromorphic vector-valued functions in G.

The vectors (6.1) form a Jordan basis of $D(\zeta) \in \mathscr{L}(X(\mathfrak{P}) : \mathscr{M}(\mathfrak{P}))$ in $P(\zeta)\, X(\mathfrak{P})$. Note that $e_\varrho^\sigma(\zeta)$ is a Jordan vector of degree σ; i.e., the relations $P(\zeta)\, e_\varrho^\sigma(\zeta) = e_\varrho^\sigma(\zeta)$ and $D(\zeta)\, e_\varrho^\sigma(\zeta) = e_\varrho^{\sigma-1}(\zeta)$, $\sigma = 1, 2, \ldots, m_\varrho$ hold, where $e_\varrho^0(\zeta) \equiv 0$.

Moreover, the vectors $e_\varrho^\sigma(\zeta)$ form a Jordan basis of $D(\zeta) \in \mathscr{L}(X : \mathbb{C})$ in $P(\zeta) X$ (see Subsect. 5.3.3 and 5.4.1) for all points ζ of $\mathfrak{P}$, except for a set from $\mathfrak{P}$ which is isolated and closed in $\mathfrak{P}$. This exceptional set contains at least all splitting points and all weakly splitting points.

In particular, to every splitting point and weakly splitting point corresponds a neighborhood U of ζ_0 such that for all $\zeta \in U \setminus \{\zeta_0\}$ the vectors $e_\varrho^\sigma(\zeta)$ form a Jordan basis of $D(\zeta)$ in $P(\zeta) X$.

5.6.2. Behavior at splitting points

The branching order a of a splitting point ζ_0 remains invariant also for all vectors $e_\varrho^\sigma(\zeta)$.

Theorem 1. *Let ζ_0 be a splitting point with branching order a ($\geqq 2$). Then every vector $e_\varrho^\sigma(\zeta)$ has a branching of (exact) order a at $\zeta = \zeta_0$.*

Proof. Let $\lambda_1(z), \ldots, \lambda_a(z)$ denote the a branches of $\lambda(\zeta)$ in the neighborhood of ζ_0; let $P_i^1(z), P_i^2(z), \ldots, P_i^a(z)$ denote the a branches of the irreducible projection-valued function $P_i(\zeta)$, $i = 1, \ldots, l$, which belong to the Jordan basis $e_\varrho^\sigma(\zeta)$. The (formal) branches of $e_i^\sigma(\zeta)$ are denoted by $e_{i1}^\sigma(z), \ldots, e_{ia}^\sigma(z)$. Then

$$P_i^\alpha(z)\, e_{i\alpha}^\sigma(z) = e_{i\alpha}^\sigma(z), \qquad i = 1, \ldots, l, \qquad \sigma = 1, \ldots, m_i, \qquad \alpha = 1, \ldots, a.$$

Assuming $e_{i\alpha}^\sigma(z) = e_{i\beta}^\sigma(z)$ for $\alpha \neq \beta$, one obtains

$$P_i^\alpha(z)\, P_i^\beta(z)\, e_{i\alpha}^\sigma(z) = e_{i\alpha}^\sigma(z).$$

Then from (5.5) the equality $e_{i\alpha}^\sigma(z) \equiv 0$ follows. However, this is not true. ∎

5.6.3. Puiseux-Laurent series

Let ζ_0 be an arbitrary but fixed point from $\mathfrak{P}$. Then the expansion (6.2) for $e_\varrho^\sigma(\zeta)$ can be rewritten as a local expansion of $e_\varrho^\sigma(\zeta)$ in the neighborhood of ζ_0. One only has to substitute the corresponding Puiseux series for $\lambda(\zeta)$ and the Laurent series for the coefficients $f_{\varrho\alpha}^\sigma(z)$ into (6.2). Thus one obtains

Theorem 2. *Let $\zeta_0 = (z_0, \lambda_0)$ be a point from $\mathfrak{P}$. Let the dotted disc $0 < |z - z_0| < \delta$ consist of simple points only, which do not contain trace points of poles of $e_\varrho^\sigma(\zeta)$. Let $\mathfrak{H}$ denote an h-valued cycle of the (z_0, λ_0)-group of $\mathfrak{P}$ over $0 < |z - z_0| < \delta$. Then the vectors $e_\varrho^\sigma(\zeta)$ of the Jordan basis of $D(\zeta)$ in $P(\zeta)\,X$, with the branches $e_{\varrho\alpha}^\sigma(z)$, $\alpha = 1, \ldots, h$ over $\mathfrak{H}$, can be expanded in the neighborhood $0 < |z - z_0| < \delta$ of z_0 into Puiseux-Laurent series of the form*

$$e_{\varrho\alpha}^\sigma(z) = (z - z_0)^{-M_\varrho^\sigma} \sum_{\mu=0}^{\infty} b_{\varrho\mu}^\sigma [\xi^\alpha (z - z_0)^{1/h}]^\mu, \qquad \alpha = 1, 2, \ldots, h, \tag{6.3}$$

where $b_{\varrho\mu}^\sigma \in X$, M_ϱ^σ denotes the maximal order of the poles of the coefficients $f_{\varrho\mu}^\sigma(z)$ at the point $z = z_0$, and where ξ is a primitive h-th root of unity, $(z - z_0)^{1/h}$ a fixed root.

The following consequence of this theorem should be emphasized. Let $b_{\varrho k_\varrho^\sigma}^\sigma = b_\varrho^\sigma$ denote the first non-vanishing coefficient of the expansion (6.3). Then one obtains

$$e_{\varrho\alpha}^\sigma(z) = (z - z_0)^{-M_\varrho^\sigma + \frac{k_\varrho^\sigma}{h}} \left(b_\varrho^\sigma + b_{\varrho, k_\varrho^\sigma + 1}^\sigma (z - z_0)^{1/h} + \cdots\right).$$

If $-M_\varrho^\sigma + h^{-1} k_\varrho^\sigma = h^{-1} c_\varrho^\sigma$, then,

$$f_{\varrho\alpha}^\sigma(z) = (z - z_0)^{-c_\varrho^\sigma/h}\, e_{\varrho\alpha}^\sigma(z) = b_\varrho^\sigma + b_{\varrho, k_\varrho^\sigma+1}^\sigma (z - z_0)^{1/h} + \cdots,$$

that is, the algebroid vector-valued function $f_{\varrho\alpha}^\sigma(z)$ is continuous at $z = z_0$ and different from zero at this point.

Let $D_1(z), \ldots, D_h(z)$ denote the h branches of $D(\zeta)$ in a neighborhood of ζ_0 over $\mathfrak{H}$. The property that $e_{\varrho\alpha}^\sigma(z)$ is a Jordan vector of degree σ with respect to $D_\alpha(z)$ remains unchanged under multiplication with a scalar non-vanishing factor. Hence,

$$D_\alpha(z)^\varrho\, f_{\varrho\alpha}^\sigma(z) \equiv 0, \qquad D_\alpha(z)^{\sigma-1}\, f_{\varrho\alpha}^\sigma(z) \not\equiv 0. \tag{6.4}$$

In other words,

$$\big(A(z) - \lambda_\alpha(z)\big)^\sigma f_{\varrho\alpha}^\sigma(z) \equiv 0, \qquad \big(A(z) - \lambda_\alpha(z)\big)^{\sigma-1} f_{\varrho\alpha}^\sigma(z) \not\equiv 0. \tag{6.5}$$

From (6.5), by taking the limit $z \to z_0$,

$$\big(A(z_0) - \lambda_0\big)^\sigma b_\varrho^\sigma = 0. \tag{6.6}$$

That is, b_ϱ^σ is also a Jordan vector of $A(z_0)$ with respect to λ_0; but b_ϱ^σ is a Jordan vector *at most of degree* σ. In other words, the exact degree of b_ϱ^σ may be smaller than σ; in that case, the relation

$$\big(A(z_0) - \lambda_0\big)^{\sigma-1} b_\varrho^\sigma = 0$$

holds.

5.7. Jordan structure of $A(z)$

The results of the foregoing Sect. 5.3 to 5.6 refer to those spectral objects $P(\zeta)$, $D(\zeta)$ etc. which are defined on a fixed eigenvalue function $\mathfrak{P}$ with respect to $\chi(\lambda, A(z))$.

Summarizing the results for all eigenvalue functions $\mathfrak{P}_1, \mathfrak{P}_2, \ldots, \mathfrak{P}_r$, one obtains detailed information on the Jordan structure of the operator $A(z)$. This summary will be given in this section.

5.7.1. Invariance of the Jordan structure

Let $A(z)$ be a meromorphic operator-valued function in G. By $\mathfrak{P}_1, \mathfrak{P}_2, \ldots, \mathfrak{P}_r$ one denotes the different eigenvalue functions over G belonging to $\chi(\lambda, A(z))$. The corresponding valuednesses are denoted by $g_1, g_2, \ldots, g_r$. As before, let S be the set of all poles of $A(z)$ and M the set of all multiple points (see, for example, Subsect. 3.3.12). To every function $\mathfrak{P}_\alpha$ corresponds a set of weakly splitting points. The assigned trace set is denoted by H_α. Let $H = H_1 \cup H_2 \cup \cdots \cup H_r$ and form the set $A = S \cup M \cup H$. The subset A of G is isolated and closed in G. In the following discussion A serves as the exceptional set.

Let $z_0 \in G \setminus A$. Then, in particular, z_0 is simple. The branches of the eigenvalue function $\mathfrak{P}_\alpha$ lying over z_0 are denoted by $\lambda_{\alpha 1}(z), \lambda_{\alpha 2}(z), \ldots, \lambda_{\alpha g_\alpha}(z)$. Then all points $(z_0, \lambda_{\alpha\sigma}(z))$, $\alpha = 1, \ldots, r$; $\sigma = 1, 2, \ldots, g_\alpha$, are strongly normal.

Furthermore, recall the concept of similarity of two operators (see Subsect. 1.5.2). Note that for different linear spaces X and Y with $\dim X = \dim Y$ the similarity of two operators A and B is defined as in the case $X = Y$, namely by the existence of an invertible operator T from X onto Y with the property $A = TBT^{-1}$.

One can formulate

Lemma 1. *Let $z \in G \setminus A$; let $P_{\alpha\sigma}(z)$ and $D_{\alpha\sigma}(z)$ denote the eigenprojections and the eigennilpotents, respectively, with respect to $\lambda_{\alpha\sigma}(z)$. Then,*

$$D_{\alpha\sigma}(z) \restriction P_{\alpha\sigma}(z)\, X \sim D_{\alpha 1}(z_0) \restriction P_{\alpha 1}(z_0)\, X\,, \quad \alpha = 1, \ldots, r; \quad \sigma = 1, \ldots, g_\alpha; \tag{7.1}$$

hence,

$$D_{\alpha\sigma}(z) \sim D_{\alpha 1}(z_0)\,, \qquad \alpha = 1, \ldots, r; \qquad \sigma = 1, \ldots, g_\alpha. \tag{7.2}$$

Proof. The relation (7.1), which expresses the similarity of $D_{\alpha\sigma}(z) \restriction P_{\alpha\sigma}(z)\, X$ and $D_{\alpha 1}(z_0) \restriction P_{\alpha 1}(z_0)\, X$, and the relation (7.2) follow from Theorem 4.1, and from the fact that $(z, \lambda_{\alpha\sigma}(z))$ and $(z_0, \lambda_{\alpha 1}(z_0))$ are both strongly normal. ∎

For short,

$$D_{\alpha\varrho}(z_0) \restriction P_{\alpha\varrho}(z_0)\, X = N_\alpha^\varrho\,, \qquad \alpha = 1, \ldots, r; \qquad \varrho = 1, \ldots, g_\alpha. \tag{7.3}$$

Then, if α is fixed, the operators N_α^ϱ are mutually similar,

$$N_\alpha^\varrho \sim N_\alpha^\sigma\,, \qquad \varrho, \sigma = 1, \ldots, g_\alpha. \tag{7.4}$$

From (7.1) and (7.2), the relation

$$\sum_{\alpha=1}^{r} \sum_{\sigma=1}^{g_\alpha} D_{\alpha\sigma}(z) \sim \sum_{\alpha=1}^{r} \sum_{\sigma=1}^{g_\alpha} D_{\alpha\sigma}(z_0) \tag{7.5}$$

follows; hence,

$$\sum_{\alpha=1}^{r} \sum_{\sigma=1}^{g_\alpha} D_{\alpha\sigma}(z) \sim \sum_{\alpha=1}^{r} \oplus \sum_{\sigma=1}^{g_\alpha} \oplus N_\alpha^\sigma . \tag{7.6}$$

If

$$A(z) - \sum_{\alpha,\sigma} \lambda_{\alpha\sigma}(z) P_{\alpha\sigma}(z) = N(z), \tag{7.7}$$

then (7.5) takes the form

$$N(z) \sim N(z_0), \qquad z, z_0 \in G - A, \tag{7.8}$$

and

$$N(z) \sim \sum_{\alpha=1}^{r} \oplus \sum_{\sigma=1}^{g_\alpha} \oplus N_\alpha^\sigma, \qquad z \in G - A. \tag{7.9}$$

Summarizing these facts, one obtains

Theorem 1. *Let $A(z)$ be meromorphic in G. Then there exists an exceptional set A from G, where A is denumerable and where its accumulation points can be found at most on ∂G, such that for all $z, z_0 \in G \setminus A$, the relation*

$$N(z) = A(z) - \sum_{\alpha,\sigma} \lambda_{\alpha\sigma}(z) P_{\alpha\sigma}(z) \sim N(z_0) \tag{7.10}$$

holds. One has $D_{\alpha\sigma}(z) \restriction P_{\alpha\sigma}(z) X \sim D_{\alpha 1}(z_0) \restriction P_{\alpha 1}(z_0) X$ for all $\alpha = 1, \ldots, r$, $\sigma = 1, \ldots, g_\alpha$, $z, z_0 \in G \setminus A$ and, if $D_{\alpha\sigma}(z_0) \restriction P_{\alpha\sigma}(z_0) X = N_\alpha^\sigma$, the relation

$$N(z) \sim \sum_{\alpha=1}^{r} \oplus \sum_{\sigma=1}^{g_\alpha} \oplus N_\alpha^\sigma, \qquad z \in G - A,$$

is valid.

The property of $A(z)$ in $G \setminus A$, which is expressed by this theorem, can be called *the invariance of the Jordan structure* of $A(z)$ in $G \setminus A$.

5.7.2. Jordan normal form of $A(z)$

To determine the Jordan normal form of $A(z)$ in $G \setminus A$, let

$$N_1^\alpha = N_\alpha = N_{\alpha 1} \oplus N_{\alpha 2} \oplus \cdots \oplus N_{\alpha l_\alpha} \tag{7.11}$$

be a decomposition of N_α into a direct sum of Jordan operators, where the corresponding dimensions are denoted by $m_\alpha^1 \geqq m_\alpha^2 \geqq \cdots \geqq m_\alpha^{l_\alpha}$. Denote the Jordan normal form of an operator C by $J(C)$. The relation

$$J(N_\alpha) = \left\{\Gamma_{m_\alpha^1}(0), \Gamma_{m_\alpha^2}(0), \ldots, \Gamma_{m_{l_\alpha}^\alpha}(0)\right\} \tag{7.12}$$

holds and one obtains

$$J(N(z)) = J(N(z_0)) = J(\sum \oplus N_\alpha^\sigma) = \{g_1 J(N_1), g_2 J(N_2), \ldots, g_r J(N_r)\}, \qquad (7.13)$$

where the notation $g_\alpha J(N_\alpha)$ means that the diagonal block $J(N_\alpha)$ occurs exactly g_α times. (7.13) together with (7.10) also yields the Jordan normal form of $A(z)$ in $G \setminus A$.

5.7.3. Transformation of $N(z)$

Let $z_0 \in G \setminus A$. Then there are only strongly normal points $\zeta_\alpha^\sigma \in \mathfrak{P}_\alpha$, $\alpha = 1, \ldots, r$; $\sigma = 1, \ldots, g_\alpha$, over z_0. According to Definition 4.1, to every point ζ_α^σ corresponds a transformation operator $T_{\alpha\sigma}(\zeta_\alpha)$. Consider these transformation operators over some neighborhood $|z - z_0| < \delta$. They can be represented as functions of z: $T_{\alpha\sigma} = T_{\alpha\sigma}(z)$. In particular, $T_{\alpha\sigma}(z_0) = 1$. The neighborhood $|z - z_0| < \delta$ is chosen such that $T_{\alpha\sigma}(z)$ and $T_{\alpha\sigma}(z)^{-1}$ are holomorphic in $|z - z_0| < \delta$ for all $\alpha = 1, \ldots, r$; $\sigma = 1, \ldots, g_\alpha$.

One chooses some decomposition of $P_{\alpha\sigma}(z_0)$ into a sum of disjoint Jordan projections $P_{\alpha\sigma}^\varrho(z_0)$, $\varrho = 1, \ldots, l_\alpha$, with respect to $D_{\alpha\sigma}(z_0)$, where $\dim P_{\alpha\sigma}^\varrho(z_0) = m_\alpha^\varrho$. Furthermore, one chooses Jordan bases $e_{\alpha\sigma}^{\varrho\varkappa}$, $\varkappa = 1, \ldots, m_\alpha^\varrho$, with respect to $D_{\alpha\sigma}(z_0)$. One writes

$$e_{\alpha\sigma}^{\varrho\varkappa}(z) = T_{\alpha\sigma}(z)\, e_{\alpha\sigma}^{\varrho\varkappa}, \qquad |z - z_0| < \delta. \qquad (7.14)$$

The vector-valued functions $e_{\alpha\sigma}^{\varrho\varkappa}(z)$ are holomorphic in $|z - z_0| < \delta$ and form a basis in $P_{\alpha\sigma}(z_0)\, X$ for every z from this neighborhood, because from the relation

$$\sum_{\varrho,\varkappa} \lambda_{\varrho\varkappa} e_{\alpha\sigma}^{\varrho\varkappa}(z) = T_{\alpha\sigma}(z) \sum_{\varrho,\varkappa} \lambda_{\varrho\varkappa} e_{\alpha\sigma}^{\varrho\varkappa} = 0$$

the equality $\sum_{\varrho,\varkappa} \lambda_{\varrho\varkappa} e_{\alpha\sigma}^{\varrho\varkappa} = 0$ follows, hence, $\lambda_{\varrho\varkappa} = 0$. Therefore, the vector-valued functions $e_{\alpha\sigma}^{\varrho\varkappa}(z)$ form a Jordan basis in X with respect to $\sum_{\varrho,\sigma} D_{\alpha\sigma}(z)$ for every z from $|z - z_0| < \delta$.

An operator-valued function $T(z)$ can be defined in $|z - z_0| < \delta$ by

$$T(z)\, e_{\alpha\sigma}^{\varrho\varkappa} = e_{\alpha\sigma}^{\varrho\varkappa}(z), \qquad |z - z_0| < \delta. \qquad (7.15)$$

$T(z)$ and $T(z)^{-1}$ are both holomorphic in $|z - z_0| < \delta$, and one has $T(z_0) = 1$. Furthermore,

$$N(z) = T(z)\, N(z_0)\, T(z)^{-1}, \qquad |z - z_0| < \delta. \qquad (7.16)$$

This relation follows immediately from

$$D_{\alpha\sigma}(z) = T_{\alpha\sigma}(z)\, D_{\alpha\sigma}(z_0)\, T_{\alpha\sigma}(z)^{-1}, \qquad (7.17)$$

$$P_{\alpha\sigma}(z) = T_{\alpha\sigma}(z)\, P_{\alpha\sigma}(z_0)\, T_{\alpha\sigma}(z)^{-1} \qquad (7.18)$$

and from the definition of $T(z)$. Recall that the relations (7.17) and (7.18) are consequences of the strong normality of the points lying over z_0. Hence,

Theorem 2. *Let $z_0 \in G \setminus A$. Then there is a neighborhood $|z - z_0| < \delta$ and an operator-valued function $T(z)$ such that $T(z)$ and $T(z)^{-1}$ are both holomorphic in $|z - z_0| < \delta$, $T(z_0) = 1$ and the relation*

$$N(z) = T(z)\, N(z_0)\, T(z)^{-1}, \qquad |z - z_0| < \delta, \qquad (7.19)$$

holds.

6. Analytic perturbations

In this chapter the foregoing results will be applied to analytic perturbations (see Subsect. 3.1.3).

Sect. 6.1 is a continuation of Subsect. 3.3.13, where the spectral representation of an analytic perturbation was derived and the perturbation series for the perturbed eigenvalues and eigenprojections were introduced. These considerations will be continued with the application of the results of Sect. 5.2 to 5.7. The essential results are contained in Theorem 5.7.1, which, formulated locally, expresses the invariance of the perturbed Jordan structure, and in the local expansion theorems for the irreducible projections and the Jordan vectors.

Sect. 6.2 contains preliminary relations between the unperturbed and the perturbed Jordan structure (Gochberg-Krein inequalities) and some applications and implications.

6.1. Invariance of the perturbed Jordan structure and perturbation series

6.1.1. Jordan structure

Let $A(z)$ be an analytic (holomorphic) perturbation of $A(0) = A_0$, given in $|z| < R$. The eigenvalues of A_0 are denoted by $\lambda_1, \lambda_2, \ldots, \lambda_r$ and the perturbed eigenvalues and eigenprojections by $\lambda_{\alpha\varrho\sigma}(z)$ and $P_{\alpha\varrho\sigma}(z)$, respectively (see Subsect. 3.3.13). From Theorem 5.7.1, one obtains

Theorem 1. *There is a number $\gamma > 0$ such that only strongly normal points lie over $0 < |z| < \gamma$ (in particular, all points from $0 < |z| < \gamma$ are simple). For arbitrary points z and z_0 with $0 < |z| < \gamma$ and $0 < |z_0| < \gamma$, the relation*

$$N(z) = A(z) - \sum_{\alpha,\varrho,\sigma} \lambda_{\alpha\varrho\sigma}(z)\, P_{\alpha\varrho\sigma}(z) \sim N(z_0) \tag{1.1}$$

holds, expressing the similarity of $N(z)$ and $N(z_0)$. Furthermore, one has $D_{\alpha\varrho\sigma}(z) \restriction P_{\alpha\varrho\sigma}(z)\, X \sim D_{\alpha\varrho 1}(z_0) \restriction P_{\alpha\varrho 1}(z_0)\, X$ for all $\alpha = 1, \ldots, r$; $\varrho = 1, \ldots, s_\alpha$, $\sigma = 1, \ldots, g_{\alpha\varrho}$, $0 < |z| < \gamma$, $0 < |z_0| < \gamma$. If $D_{\alpha\varrho\sigma}(z_0) \restriction P_{\alpha\varrho\sigma}(z_0)\, X = N_{\alpha\varrho\sigma}$, then the relation

$$N(z) \sim \sum_{\alpha,\varrho,\sigma} \oplus N_{\alpha\varrho}^{\sigma} \tag{1.2}$$

is valid, where the operators $N_{\alpha\varrho 1}, N_{\alpha\varrho 2}, \ldots, N_{\alpha\varrho g_{\alpha\varrho}}$ are mutually similar.

For short, the perturbation $A(z)$ of A_0 has an invariant Jordan structure on $0 < |z| < \gamma$; i.e., immediately after "switching on" the perturbation $A(z)$, $z \neq 0$, the Jordan structure is invariant. Therefore, it makes sense to speak of *the* perturbed Jordan structure in contrast to the unperturbed Jordan structure.

The next problem is finding relations between the unperturbed and the perturbed Jordan structure and estimating the number γ. Sect. 6.2 and Ch. 7 deal with the first problem. The second problem will be considered in Ch. 8.

6.1.2. Perturbation series for eigenvalues and eigenprojections

For the sake of completeness, these series will be recalled here; we consider only a fixed $(0, \lambda_0)$-group (for short: λ_0-group) of the perturbation $A(z)$. Let this λ_0-group consist of r cycles with periods $g_1, g_2, \ldots, g_r$. Then one obtains expansions for the eigenvalues $\lambda_{\alpha\sigma}(z)$ of the λ_0-group of the form

$$\lambda_{\alpha\sigma}(z) = \lambda_0 + \sum_{\mu=1}^{\infty} a_{\alpha\mu} z^{\mu/g_\alpha}, \quad \alpha = 1, \ldots, r; \quad \sigma = 1, \ldots, g_\alpha; \quad |z| < \gamma \tag{1.3}$$

(see Subsect. 3.2.6). The eigenprojections $P_{\alpha\sigma}(z)$ can be uniquely represented in the form

$$P_{\alpha\sigma}(z) = \sum_{\mu=0}^{g_\alpha - 1} A_{\alpha\mu}(z)\, \lambda_{\alpha\sigma}(z)^\mu, \qquad 0 < |z| < \gamma \tag{1.4}$$

(see Subsect. 3.3.9 or 5.2.4). There exist polynomials $p_{\alpha\mu}(\lambda)$ whose coefficients are holomorphic in $0 < |z| < \gamma$ and which have at most a pole at the point $z = 0$ such that

$$A_{\alpha\mu}(z) = p_{\alpha\mu}\big(A(z)\big) \tag{1.5}$$

(see Subsect. 5.2.2). In particular, the operators $A_{\alpha\mu}(z)$ are holomorphic in $0 < |z| < \gamma$ and they have at most a pole at $z = 0$.

Substituting the Puiseux series (1.3) for the eigenvalues and the Laurent series for the coefficients $A_{\alpha\mu}(z)$ into (1.4), one obtains Puiseux-Laurent series for the eigenprojections of the form

$$P_{\alpha\sigma}(z) = z^{-N_\alpha} \sum_{\mu=0}^{\infty} B_{\alpha\mu} z^{\mu/g_\alpha}, \qquad 0 < |z| < \gamma, \tag{1.6}$$

where N_α denotes the maximal order of poles of the coefficients $A_{\alpha 0}(z), A_{\alpha 1}(z), \ldots, A_{\alpha g_{\alpha}-1}(z)$ at $z = 0$.

6.1.3. Perturbation series for the irreducible projections

Let $\dim P_{\alpha\sigma}(z) = m_\alpha$. From Sect. 5.5, in particular from Theorem 5.5.1, one obtains the following propositions on the decomposition of $P_{\alpha\sigma}(z)$ into a sum of irreducible projections and their properties:

There are uniquely determined numbers $m_\alpha^1 \geqq m_\alpha^2 \geqq \cdots \geqq m_\alpha^{l_\alpha} \geqq 1$ with $\sum_{\varrho=1}^{l_\alpha} m_\alpha^\varrho = m_\alpha$, and there is a decomposition of $P_{\alpha\sigma}(z)$ into a sum of disjoint and, with respect to $A(z)$, reducing, even irreducible, projections,

$$P_{\alpha\sigma}(z) = \sum_{\varrho=1}^{l_\alpha} P_{\alpha\sigma}^{\varrho}(z), \qquad \dim P_{\alpha\sigma}^{\varrho}(z) = m_{\alpha}^{\varrho}, \qquad 0 < |z| < \gamma' \leqq \gamma, \tag{1.7}$$

with the following properties: There are unique representations of the form

$$P_{\alpha\sigma}^{\varrho}(z) = \sum_{\mu=0}^{g_\alpha-1} B_{\alpha\mu}^{\varrho}(z)\, \lambda_{\alpha\sigma}(z)^{\mu}, \qquad 0 < |z| < \gamma' \tag{1.8}$$

(see Subsect. 5.5.1), where the operator-valued coefficients $B_{\alpha\mu}^{\varrho}(z)$ are holomorphic in $0 < |z| < \gamma'$, and where they have at most a pole at $z = 0$.

Substituting the Puiseux series (1.3) for the eigenvalues and the Laurent series for the coefficients $B_{\alpha\mu}^{\varrho}(z)$ into (1.8), one obtains Puiseux-Laurent series expansions for the irreducible projections of the form

$$P_{\alpha\sigma}^{\varrho}(z) = z^{-N_{\alpha}^{\varrho}} \sum_{\mu=0}^{\infty} B_{\alpha\mu}^{\varrho} z^{\mu/g_\alpha}, \qquad 0 < |z| < \gamma' \tag{1.9}$$

(see Subsect. 5.5.4), where N_{α}^{ϱ} denotes the maximal order of poles of the coefficients $B_{\alpha 0}^{\varrho}(z), B_{\alpha 1}^{\varrho}(z), \ldots, B_{\alpha, g_\alpha - 1}^{\varrho}(z)$ at $z = 0$.

6.1.4. Perturbation series for Jordan vectors

In this subsection, the corresponding consequences from Subsect. 5.6.1 to 5.6.3 will be derived. In every irreducible subspace $P_{\alpha\sigma}^{\varrho}(z)\, X$, $0 < |z| < \gamma'$, one can choose a Jordan basis $e_{\alpha\sigma}^{\varrho\varkappa}(z)$, $\varkappa = 1, 2, \ldots, m_{\alpha}^{\varrho}$, with the following properties:

$$P_{\alpha\sigma}^{\varrho}(z)\, e_{\alpha\sigma}^{\varrho\varkappa}(z) = e_{\alpha\sigma}^{\varrho\varkappa}(z), \qquad 0 < |z| < \gamma', \tag{1.10}$$

$$\big(A(z) - \lambda_{\alpha\sigma}(z)\big)\, e_{\alpha\sigma}^{\varrho\varkappa}(z) = e_{\alpha\sigma}^{\varrho,\varkappa-1}(z), \qquad \varkappa = 1, 2, \ldots, m_{\alpha}^{\varrho}; \qquad e_{\alpha\sigma}^{\varrho 0}(z) \equiv 0; \tag{1.11}$$

that is,

$$e_{\alpha\sigma}^{\varrho\varkappa}(z) = \big(A(z) - \lambda_{\alpha\sigma}(z)\big)^{m_{\alpha}^{\varrho} - \varkappa}\, e_{\alpha\sigma}^{\varrho, m_{\alpha}^{\varrho}}(z), \qquad \varkappa = 1, 2, \ldots, m_{\alpha}^{\varrho}, \tag{1.12}$$

for all z with $0 < |z| < \gamma'$.

The Jordan vectors $e_{\alpha\sigma}^{\varrho\varkappa}(z)$ of this Jordan basis can be uniquely represented in the form

$$e_{\alpha\sigma}^{\varrho\varkappa}(z) = \sum_{\mu=0}^{g_\alpha-1} f_{\alpha\mu}^{\varrho\varkappa}(z)\, \lambda_{\alpha\sigma}(z)^{\mu} \tag{1.13}$$

(see Subsect. 5.6.1), where the vector-valued functions $f_{\alpha\mu}^{\varrho\varkappa}(z)$ are holomorphic in $0 < |z| < \gamma'$ and where they have at most a pole at $z = 0$.

Substituting the Puiseux series (1.3) for the eigenvalues and the Laurent series for the coefficients $f_{\alpha\mu}^{\varrho\varkappa}(z)$ into (1.13), one obtains Puiseux-Laurent series expansions for the Jordan vectors $e^{\varrho\varkappa}_{\alpha\mu}(z)$ of the form

$$e^{\varrho\varkappa}_{\alpha\sigma}(z) = z^{-N^{\varrho\varkappa}_{\alpha}} \sum_{\mu=0}^{\infty} f^{\varrho\varkappa}_{\alpha\mu} z^{\mu/g_\alpha}, \qquad 0 < |z| < \gamma' \tag{1.14}$$

(see Theorem 5.6.2), where $N^{\varrho\varkappa}_{\alpha}$ denotes the maximal order of poles for the coefficients $f^{\varrho\varkappa}_{\alpha 0}(z), f^{\varrho\varkappa}_{\alpha 1}(z), \ldots, f^{\varrho\varkappa}_{\alpha, g_\alpha - 1}(z)$ at $z = 0$.

6.1.5. Perturbation series for Jordan vectors which are continuous at the point $z = 0$

This section is a continuation of Subsect. 5.6.3. The sequence of vectors $f^{\varrho m^\varrho_\alpha}_{\alpha 0}, f^{\varrho m^\varrho_\alpha}_{\alpha 1}, f^{\varrho m^\varrho_\alpha}_{\alpha 2}, \ldots$ of the expansion coefficients of the cyclic vector $e^{\varrho}_{\alpha\sigma}(z) = e^{\varrho m^\varrho_\alpha}_{\alpha\sigma}(z)$ with respect to the expansion (1.14) contains a first coefficient which is different from zero. Let $f^{\varrho m^\varrho_\alpha}_{\alpha p} \neq 0$ be this coefficient, where $p = p(\alpha, \varrho)$. Then

$$e^{\varrho}_{\alpha\sigma}(z) = z^{-N^{\varrho m^\varrho_\alpha}_{\alpha}} z^{p/g_\alpha} \sum_{\mu=0}^{\infty} f^{\varrho, m^\varrho_\alpha}_{\alpha, p+\mu} z^{\mu/g_\alpha}, \qquad f^{\varrho, m^\varrho_\alpha}_{\alpha p} \neq 0.$$

Consider the vector

$$g^{\varrho}_{\alpha\sigma}(z) = z^{N^{\varrho m^\varrho_\alpha}_{\alpha} - p/g_\alpha} e^{\varrho}_{\alpha\sigma}(z). \tag{1.15}$$

Moreover, let

$$g^{\varrho\varkappa}_{\alpha\sigma}(z) = \big(A(z) - \lambda_{\alpha\sigma}(z)\big)^{m^\varrho_\alpha - \varkappa} g^{\varrho}_{\alpha\sigma}(z), \qquad \varkappa = 1, 2, \ldots, m^\varrho_\alpha. \tag{1.16}$$

The vectors $g^{\varrho\varkappa}_{\alpha\sigma}(z)$ again form a Jordan basis in $P^{\varrho}_{\alpha\sigma}(z)\, X$ for every z with $0 < |z| < \gamma'$. For the Jordan basis $g^{\varrho\varkappa}_{\alpha\sigma}(z)$, the Puiseux-Laurent expansions (1.14) take the following form

$$g^{\varrho\varkappa}_{\alpha\sigma}(z) = g^{\varrho\varkappa}_{\alpha} + \sum_{\mu=1}^{\infty} g^{\varrho\varkappa}_{\alpha\mu} z^{\mu/g_\alpha}, \qquad |z| < \gamma', \tag{1.17}$$

where

$$g^{\varrho m^\varrho_\alpha}_{\alpha} \neq 0, \qquad \alpha = 1, \ldots, r, \qquad \varrho = 1, \ldots, l_\alpha. \tag{1.18}$$

The equations (1.11) imply that, for the Jordan basis $g^{\varrho\varkappa}_{\alpha\sigma}(z)$, the equations

$$\big(A(z) - \lambda_{\alpha\sigma}(z)\big)\, g^{\varrho\varkappa}_{\alpha\sigma}(z) = g^{\varrho,\varkappa-1}_{\alpha\sigma}(z), \qquad \varkappa = 1, 2, \ldots, m^\varrho_\alpha; \qquad g^{\varrho 0}_{\alpha\sigma}(z) \equiv 0, \tag{1.19}$$

hold. Then one obtains

$$\big(A(z) - \lambda_{\alpha\sigma}(z)\big)^{\varkappa} g^{\varrho\varkappa}_{\alpha\sigma}(z) \equiv 0, \qquad \varkappa = 1, 2, \ldots, m^\varrho_\alpha, \tag{1.20}$$

and

$$\big(A(z) - \lambda_{\alpha\sigma}(z)\big)^{\varkappa - 1} g^{\varrho\varkappa}_{\alpha\sigma}(z) \not\equiv 0, \qquad \varkappa = 1, 2, \ldots, m^\varrho_\alpha; \tag{1.21}$$

that is, the vector $g^{\varrho\varkappa}_{\alpha\sigma}(z)$ is a Jordan vector of degree $\varkappa$ for all z with $0 < |z| < \gamma'$. By taking the limit $z \to 0$ in the equations (1.20), one obtains

$$(A_0 - \lambda_0)^\varkappa g_\alpha^{\varrho\varkappa} = 0, \qquad \varkappa = 1, 2, \ldots, m_\alpha^\varrho. \tag{1.22}$$

From (1.19), the relation

$$(A_0 - \lambda_0) g_\alpha^{\varrho\varkappa} = g_\alpha^{\varrho,\varkappa-1}, \qquad \varkappa = 1, 2, \ldots, m_\alpha^\varrho; \qquad g_\alpha^{\varrho 0} = 0, \tag{1.23}$$

follows. Finally,

$$g_\alpha^{\varrho\varkappa} = (A_0 - \lambda_0)^{m_\alpha^\varrho - \varkappa} g_\alpha^{\varrho m_\alpha^\varrho}, \qquad \varkappa = 1, 2, \ldots, m_\alpha^\varrho. \tag{1.24}$$

The equation (1.22) expresses the fact that, with respect to A_0, $g_\alpha^{\varrho\varkappa}$ is a Jordan vector of degree at most $\varkappa$ belonging to λ_0. In particular:

$g_\alpha^{\varrho m_\alpha^\varrho} \neq 0$ is a Jordan vector of degree at most m_α^ϱ of λ_0 with respect to A_0.

Let $\varkappa_1 = \varkappa_1(\alpha, \varrho)$ denote the maximal number such that $g_\alpha^{\varrho\varkappa_1} = 0$. Then,

$$(A_0 - \lambda_0)^{m_\alpha^\varrho - \varkappa_1} g_\alpha^{\varrho m_\alpha^\varrho} = 0, \qquad (A_0 - \lambda_0)^{m_\alpha^\varrho - (\varkappa_1 + 1)} g_\alpha^{\varrho m_\alpha^\varrho} \neq 0.$$

Thus the number $m_\alpha^\varrho - \varkappa_1$ is the exact degree of the Jordan vector $g_\alpha^{\varrho m_\alpha^\varrho}$. One has $0 \leqq \varkappa_1 \leqq m_\alpha^\varrho - 1$.

The vectors $g_\alpha^{\varrho,\varkappa_1+1}, g_\alpha^{\varrho,\varkappa_1+2}, \ldots, g_\alpha^{\varrho,m_\alpha^\varrho}$ span a $(m_\alpha^\varrho - \varkappa_1)$-dimensional invariant irreducible subspace of A_0, contained in P_0X (P_0 is the eigenprojection of λ_0). Hence,

$$1 \leqq m_\alpha^\varrho - \varkappa_1(\alpha, \varrho) \leqq \nu(\lambda_0). \tag{1.25}$$

The procedure just described can be repeated, of course. For this purpose, the Jordan vector $g_{\alpha\sigma}^{\varrho\varkappa_1}(z)$ is written in the form

$$g_{\alpha\sigma}^{\varrho\varkappa_1}(z) = z^{m/g_\alpha} \left(\tilde g_\alpha^{\varrho\varkappa_1} + \sum_{\mu=1}^{\infty} \tilde g_{\alpha\mu}^{\varrho\varkappa_1} z^{\mu/g_\alpha} \right), \qquad \tilde g_\alpha^{\varrho\varkappa_1} \neq 0.$$

If

$$\tilde g_{\alpha\sigma}^{\varrho\varkappa_1}(z) = \tilde g_\alpha^{\varrho\varkappa_1} + \sum_{\mu=1}^{\infty} \tilde g_{\alpha\mu}^{\varrho\varkappa_1} z^{\mu/g_\alpha},$$

the procedure described above is repeated for $\tilde g_{\alpha\sigma}^{\varrho\varkappa_1}(z)$ (instead of $g_{\alpha\sigma}^\varrho(z)$) and one obtains: $\tilde g_\alpha^{\varrho\varkappa_1}$ is a Jordan vector for λ_0 with respect to A_0 of exact degree $\varkappa_1 - \varkappa_2$, where $1 \leqq \varkappa_1 \leqq \varkappa_2 \leqq \nu(\lambda_0)$. In finitely many steps, if the indices α, ϱ are fixed, one thus obtains a finite sequence $m_\alpha^\varrho > \varkappa_1 > \varkappa_2 > \cdots > \varkappa_{p-1} > \varkappa_p = 0$ such that $\nu(\lambda_0) \geqq \varkappa_a - \varkappa_{a+1} \geqq 1$, $a = 0, 1, 2, \ldots, p - 1$, where $\varkappa_0 = m_\alpha^\varrho$. By adding, one obtains

$$p\nu(\lambda_0) \geqq m_\alpha^\varrho \geqq p.$$

For example, if λ_0 is semisimple, i.e., $\nu(\lambda_0) = 1$, then $\varkappa_1 = m_\alpha^\varrho - 1$, $\varkappa_{a+1} = \varkappa_a - 1$, $a = 1, 2, \ldots, p - 1$, and $p = m_\alpha^\varrho$. This is obvious, because in this case every "Jordan vector group" consists of exactly one Jordan vector, since the zero-order approximation of an arbitrary perturbed Jordan vector is necessarily an eigenvector.

6.1.6. Examples

(a) $$A(z) = \begin{pmatrix} 0 & z \\ 0 & 0 \end{pmatrix}.$$

One has $\lambda_0 = 0$, $\nu(\lambda_0) = 1$, $\chi(\lambda, A(z)) = \lambda^2$; therefore, $\lambda(z) \equiv 0$ and $\nu(\lambda(z)) = 2$. A perturbed Jordan basis is given by

$$e_1(z) = \begin{pmatrix} z \\ 0 \end{pmatrix}, \qquad e_2(z) = \begin{pmatrix} 0 \\ 1 \end{pmatrix}.$$

In this case $\varkappa_1 = 1$. Therefore, there exist two "Jordan vector groups", both consisting of one Jordan vector; the first is given by $e_2(z)$, the second by $z^{-1}e_1(z)$. The linearly independent eigenvectors $\begin{pmatrix} 1 \\ 0 \end{pmatrix}$ and $\begin{pmatrix} 0 \\ 1 \end{pmatrix}$ of A_0 appear as zero-order approximations of the perturbed Jordan vectors.

(b) $$A(z) = \begin{pmatrix} 0 & 1 & 0 & 0 \\ z & 0 & 1 & 0 \\ 0 & 0 & 0 & 1 \\ 0 & 0 & z & 0 \end{pmatrix}.$$

One has $\lambda_0 = 0$, $\nu(\lambda_0) = 4$, $\chi(\lambda, A(z)) = (\lambda^2 - z)^2$; hence, $r = 1$ and $m = 2$ (m denotes the dimension of the perturbed eigenprojections). The eigenvalues are given by $\lambda_1(z) = z^{1/2}$, $\lambda_2(z) = -z^{1/2}$. From

$$\frac{1}{4} z^{-3/2}(\lambda + 2z^{1/2})(\lambda - z^{1/2})^2 + \frac{1}{4} z^{-3/2}(-\lambda + 2z^{1/2})(\lambda + z^{1/2})^2 \equiv 1$$

one obtains the eigenprojection $P_1(z)$ corresponding to $\lambda_1(z) = z^{1/2}$,

$$P_1(z) = \frac{1}{4} z^{-3/2}\left(-A(z) + 2z^{1/2}\right)\left(A(z) + z^{1/2}\right)^2$$

$$= \begin{pmatrix} \frac{1}{2} & \frac{1}{2} z^{-1/2} & 0 & -\frac{1}{4} z^{-3/2} \\ \frac{1}{2} z^{1/2} & \frac{1}{2} & \frac{1}{4} z^{-1/2} & 0 \\ 0 & 0 & \frac{1}{2} & \frac{1}{2} z^{-1/2} \\ 0 & 0 & \frac{1}{2} z^{1/2} & \frac{1}{2} \end{pmatrix}.$$

From this representation, one gets for example

$$e_1(z) = \begin{pmatrix} \frac{1}{2} \\ \frac{1}{2}\, z^{1/2} \\ 0 \\ 0 \end{pmatrix}, \qquad e_2(z) = \begin{pmatrix} 0 \\ \frac{1}{4}\, z^{-1/2} \\ \frac{1}{2} \\ \frac{1}{2}\, z^{1/2} \end{pmatrix}.$$

After suitable multiplication a Jordan basis is given by

$$g_1(z) = \begin{pmatrix} 1 \\ z^{1/2} \\ 0 \\ 0 \end{pmatrix}, \qquad g_2(z) = \begin{pmatrix} 0 \\ 1 \\ 2z^{1/2} \\ 2z^{3/2} \end{pmatrix},$$

for example. One has $\left(A(z) - z^{1/2}\right) g_2(z) = g_1(z)$, $\left(A(z) - z^{1/2}\right) g_1(z) = 0$. Therefore, $\nu\left(\lambda_1(z)\right) = \nu\left(\lambda_2(z)\right) = 2$ for $z \neq 0$. In this case $\varkappa_1 \neq 0$; i.e., $g_2(0) = g_2$ is a Jordan vector of A_0 of degree 2. The Jordan vectors of degree 3 and 4 with respect to A_0 cannot appear as zero-order approximations of perturbed Jordan vectors.

6.1.7. Summary

The results obtained in Subsect. 6.1.3 to 6.1.5 supplement Subsect. 3.3.13 and represent extensions and refinements of the results of B. v. Sz.-Nagy, T. Kato and F. Wolf. (See for example T. Kato [5, p. 70sqq.; p. 370]. Recall that these authors dealt with perturbation series expansions for eigenvalues, eigenprojections, and, in some cases, eigenvectors.) In the following, a part of these results will be summarized in a theorem. Note that this theorem represents an exact equivalent of the result of F. Rellich (for selfadjoint and normal perturbations $A(z)$ in $|z| < R$; see Theorems 3.6.1 and 3.6.2 and Theorems 3.6.4 and 3.6.5) with respect to arbitrary analytic perturbations $A(z)$ in $|z| < R$. This is so because the propositions on the perturbation $A(z)$ include propositions on the Jordan structure of $A(z)$, i.e., propositions on perturbation series for perturbed irreducible projections and perturbed Jordan bases.

Theorem 2. *Let $A(z)$ be an arbitrary holomorphic perturbation of A_0 in $|z| < R$ and let λ_0 be an eigenvalue of A_0. Then there is a number $\gamma > 0$ with the following properties:*

I. *The perturbed eigenvalues $\lambda_{\alpha\sigma}(z)$ of the λ_0-group have perturbation series expansions of the form*

$$\lambda_{\alpha\sigma}(z) = \lambda_0 + \sum_{\mu=1}^{\infty} a_{\alpha\mu} z^{\mu/g_\alpha}, \quad \alpha = 1, 2, \ldots, r; \quad \sigma = 1, 2, \ldots, g_\alpha; \quad |z| < \gamma. \quad (1.26)$$

II. *The corresponding eigenprojections* $P_{\alpha\sigma}(z)$, *where* $\dim P_{\alpha\sigma}(z) = m_\alpha$, *have perturbation series expansions of the form*

$$P_{\alpha\sigma}(z) = z^{-N_\alpha} \sum_{\mu=0}^{\infty} P_{\alpha\mu} z^{\mu/g_\alpha}, \quad \alpha = 1, \ldots, r; \quad \sigma = 1, \ldots, g_\alpha; \quad 0 < |z| < \gamma. \tag{1.27}$$

III. *There are uniquely determined numbers* $m_\alpha^1 \geqq m_\alpha^2 \geqq \cdots \geqq m_\alpha^{l_\alpha} \geqq 1$ *with* $\sum_{\varrho=1}^{l_\alpha} m_\alpha^\varrho = m_\alpha$, *and there is a decomposition of* $P_{\alpha\sigma}(z)$ *into a sum of disjoint and, with respect to* $A(z)$, *reducing and even irreducible projections,*

$$P_{\alpha\sigma}(z) = \sum_{\varrho=1}^{l_\alpha} P_{\alpha\sigma}^\varrho(z), \qquad \dim P_{\alpha\sigma}^\varrho = m_\varrho^\alpha, \qquad 0 < |z| < \gamma, \tag{1.28}$$

where the irreducible projections $P_{\alpha\sigma}^\varrho(z)$ *have perturbation series expansions of the form*

$$P_{\alpha\sigma}^\varrho(z) = z^{-N_\alpha^\varrho} \sum_{\mu=0}^{\infty} P_{\alpha\mu}^\varrho z^{\mu/g_\alpha}, \qquad \alpha = 1, \ldots, r; \qquad \sigma = 1, \ldots, g_\alpha;$$
$$\varrho = 1, \ldots, l_\alpha; \qquad 0 < |z| < \gamma. \tag{1.29}$$

IV. *One can choose Jordan bases* $g_{\alpha\sigma}^{\varrho\varkappa}(z)$ *in the irreducible subspaces* $P_{\alpha\sigma}^\varrho(z)\, X$, $0 < |z| < \gamma$, *with respect to* $A(z)$, *such that the Jordan vectors* $g_{\alpha\sigma}^{\varrho\varkappa}(z)$ (*of exact degree* $\varkappa$) *have perturbation series expansions of the form*

$$g_{\alpha\sigma}^{\varrho\varkappa}(z) = g_\alpha^{\varrho\varkappa} + \sum_{\mu=1}^{\infty} g_{\alpha\mu}^{\varrho\varkappa} z^{\mu/g_\alpha}, \qquad \varkappa = 1, 2, \ldots, m_\alpha^\varrho; \qquad |z| < \gamma. \tag{1.30}$$

Moreover,

$$\big(A(z) - \lambda_{\alpha\sigma}(z)\big)^\varkappa g_{\alpha\sigma}^{\varrho\varkappa}(z) \equiv 0, \qquad \big(A(z) - \lambda_{\alpha\sigma}(z)\big)^{\varkappa-1} g_{\alpha\sigma}^{\varrho\varkappa}(z) \not\equiv 0,$$
$$\varkappa = 1, \ldots, m_\alpha^\varrho; \qquad 0 < |z| < \gamma.$$

One has $g_\alpha^\varrho = g_\alpha^{\varrho m_\alpha^\varrho} \neq 0$ *and* g_α^ϱ *is a Jordan vector of* A_0 *with respect to* λ_0, *whose degree equals at most* m_α^ϱ. *The vectors* $g_\alpha^{\varrho\varkappa}$, $\varkappa = 1, \ldots, m_\alpha^\varrho - 1$, *are Jordan vectors of* A_0 *with respect to* λ_0 *of degree at most* $\varkappa$.

6.2. Connection between the unperturbed and the perturbed Jordan structures

Theorem 1.1 leads to the question of the relation between the unperturbed and the perturbed Jordan structures. Theorem 1.2 also touches upon the question, how to determine or characterize the zero-order approximations $g_\alpha^{\varrho\varkappa}$ of perturbed Jordan vectors. Note that this problem refers to that part of Theorem 1.2 where the parallel with F. Rellich's result for normal and selfadjoint perturbations is missing. This problem will be treated systematically but not exhaustively in Ch. 7. In this section some relations in that direction will be derived.

6.2.1. Notations

Let A_0 denote the unperturbed operator, λ_0 an unperturbed eigenvalue, $n(\lambda_0)$ the algebraic multiplicity of λ_0. Let

$$d(\lambda_0, q) = \dim \ker [(A_0 - \lambda_0)^q], \qquad \varrho = 1, 2, \ldots, \nu(\lambda_0). \tag{2.1}$$

Then, $d(\lambda_0, \nu(\lambda_0)) = n(\lambda_0)$, and $d(\lambda_0, 1)$ is the geometric multiplicity of λ_0.

As usual, let $\lambda_{\alpha\sigma}(z)$ denote the perturbed eigenvalues of the analytic perturbation $A(z)$. Then the numbers $m_\alpha = n(\lambda_{\alpha\sigma}(z))$ and $d(\lambda_{\alpha\sigma}(z), q)$ are defined correspondingly. One has $n(\lambda_0) = \sum_{\alpha=1}^{r} g_\alpha m_\alpha$, in particular, $m_\alpha \leqq n(\lambda_0)$.

According to Subsect. 2.9.4,

$$d(\lambda_0, q) = \varkappa(\lambda_0, 1) + 2\varkappa(\lambda_0, 2) + \cdots + q\varkappa(\lambda_0, q)$$
$$+ q\varkappa(\lambda_0, q + 1) + \cdots + \varrho\varkappa(\lambda_0, n(\lambda_0)),$$

where $\varkappa(\lambda_0, \varrho)$ denotes the number of irreducible parts of dimension ϱ belonging to λ_0 with respect to A_0. Correspondingly,

$$d(\lambda_{\alpha\sigma}(z), q) = \varkappa(\lambda_{\alpha\sigma}(z), 1) + \cdots + q\varkappa(\lambda_{\alpha\sigma}(z), q)$$
$$+ q\varkappa(\lambda_{\alpha\sigma}(z), q + 1) + \cdots + q\varkappa(\lambda_{\alpha\sigma}(z), m_\alpha).$$

The numbers $\varkappa(\lambda_{\alpha\sigma}(z), \varrho)$ do not depend on σ (see Theorem 1.2); hence, the numbers $d(\lambda_{\alpha\sigma}(z), \varrho)$ do not depend on σ either. For short,

$$d(\lambda_{\alpha\sigma}(z), q) = d_\alpha(q), \qquad \alpha = 1, \ldots, r.$$

In particular,

$$d_\alpha(1) = l_\alpha, \qquad \alpha = 1, \ldots, r. \tag{2.2}$$

6.2.2. Gochberg-Krein inequalities

The following relation between the numbers $d(\lambda_0, q)$ and $d_\alpha(q)$ exist (see for example I. C. Gochberg and M. G. Krein [1, p. 64, Theorem 3.6]):

Theorem 1 (Gochberg-Krein). *Let $A(z)$ be a holomorphic perturbation of A_0 in $|z| < R$ and let λ_0 be an eigenvalue of A_0. Furthermore, let the numbers $d(\lambda_0, q)$, $d_\alpha(q)$ be defined as in Subsect.* 6.2.1. *Then*

$$d(\lambda_0, q) \geqq d_\alpha(q), \qquad q = 1, 2, \ldots, n(\lambda_0); \qquad \alpha = 1, 2, \ldots, r. \tag{2.3}$$

Proof. The vectors

$$g_{\alpha\sigma}^{\varrho\varkappa}(z) = g_\alpha^{\varrho\varkappa} + \sum_{\mu=1}^{\infty} g_{\alpha\mu}^{\varrho\varkappa} z^{\mu/g_\alpha}, \qquad 0 < |z| < \gamma, \tag{2.4}$$

appearing in Theorem 1.2, form a Jordan basis in $P_{\alpha\sigma}(z)\, X$ of $A(z)$ with respect to

$\lambda_{\alpha\sigma}(z)$ if α and σ are fixed. Furthermore, $g_{\alpha\sigma}^{\varrho\varkappa}(z)$ is a Jordan vector of degree $\varkappa$. Let q be fixed, $1 \leqq q \leqq n(\lambda_0)$. Consider the set of all Jordan vectors $g_{\alpha\sigma}^{\varrho\varkappa}(z)$ such that $1 \leqq \varkappa \leqq q$, $1 \leqq \varrho \leqq l_\alpha$, i.e., the set of all Jordan vectors $g_{\alpha\sigma}^{\varrho\varkappa}(z)$ in $P_{\alpha\sigma}(z) X$ of degree at most q. These (linearly independent) Jordan vectors generate exactly the subspace $\ker \{(A(z) - \lambda_{\alpha\sigma}(z))^q\}$; hence, their number is $d_\alpha(q)$. For short, these vectors are denoted by $g_i(z)$, $1 \leqq i \leqq d_\alpha(q)$. Note that they are linearly independent for every z with $0 < |z| < \gamma$ and that each of them is continuous at $z = 0$.

By the introduction of some scalar product (x, y) in X, the system $g_i(z)$ of vectors can be orthonormalized for every z with $0 < |z| < \gamma$ (see Subsect. 1.4.2). Thus one obtains an orthonormal system $e_i(z)$, $1 \leqq i \leqq d_\alpha(q)$, of vectors such that

$$\operatorname{spa} \{e_i(z)\} = \operatorname{spa} \{g_i(z)\}, \qquad 0 < |z| < \gamma,$$

where all vectors $e_i(z)$ are continuous with respect to $z = x + iy$ in $0 < |z| < \gamma$. Therefore,

$$\big(e_i(z), e_j(z)\big) = \delta_{ij}, \qquad i, j = 1, 2, \ldots, d_\alpha(q); \qquad 0 < |z| < \gamma.$$

In particular, since $\|e_i(z)\| = 1$ for $i = 1, \ldots, d_\alpha(q)$, $0 < |z| < \gamma$, one can choose a sequence $z_\varrho \to 0$ such that all vectors $e_i(z_\varrho)$ are convergent as $\varrho \to \infty$. Let

$$\lim_{\varrho \to \infty} e_i(z_\varrho) = e_i, \qquad i = 1, 2, \ldots, d_\alpha(q).$$

Then,

$$(e_i, e_j) = \delta_{ij}, \qquad i, j = 1, 2, \ldots, d_\alpha(q).$$

Recall

$$\big(A(z) - \lambda_{\alpha\sigma}(z)\big)^q e_i(z) = 0, \qquad i = 1, \ldots, d_\alpha(q),$$

for all z with $0 < |z| < \gamma$. Taking the limit $z_\varrho \to 0$, one obtains

$$(A_0 - \lambda_0)^q e_i = 0, \qquad i = 1, 2, \ldots, d_\alpha(q).$$

That is, to the eigenvalue λ_0 corresponds at least $d_\alpha(q)$ linearly independent Jordan vectors of degree at most q. Therefore,

$$d(\lambda_0, q) \geqq d_\alpha(q). \ \blacksquare$$

Note that the considerations of Gochberg and Krein refer to holomorphic operator-valued functions on Banach spaces and to $\dim \ker A(z)$. (For details, see I. C. Gochberg and M. G. Krein [1, § 3].) In the present connection, the considerations of Gochberg and Krein are applicable only for holomorphic eigenvalues $\lambda_{\alpha\sigma}(z) = \lambda_\alpha(z)$; hence, $\big(A(z) - \lambda_\alpha(z)\big)^q$ is also holomorphic. Then, according to the results of Gochberg and Krein, the inequality (2.3) follows. As Theorem 1 shows, (2.3) is also valid if there are eigenvalue branchings at $z = 0$. (Note that the proof of Theorem 1 can also be founded on Lemma 5.1.2).

In particular, if the point $(0, \lambda_0)$ is only weakly splitting, then the inequalities (2.3) can be written in the form

$$d(\lambda_0, q) \geqq d\big(\lambda(z), q\big)$$

and they follow from the defect inequalities

$$\operatorname{def}(A_0 - \lambda_0)^q \geqq \operatorname{def}\big(A(z) - \lambda(z)\big)^q, \qquad 0 < |z| < \gamma$$

(see Subsect. 5.4.2).

The linear space of Jordan vectors of A_0 with respect to λ_0, which is spanned by the vectors e_i, $1 \leqq i \leqq d_\alpha(q)$, is independent of $\sigma = 1, 2, \ldots, g_\alpha$, if α is fixed. This is the case because the limits $g^{\varrho\varkappa}_{\alpha\sigma}(0) = g^{\varrho\varkappa}_{\alpha}$ are independent of σ and, taking the limit $z_\varrho \to 0$ in $e_i(z_\varrho)$, one only has to form limits of vectors $g^{\varrho\varkappa}_{\alpha\sigma}(z)$ and their scalar products.

In particular, for $q = n(\lambda_0)$, this subspace of Jordan vectors has dimension $d_\alpha\big(n(\lambda_0)\big) = m_\alpha$. This subspace contains all zero-order approximations of perturbed Jordan vectors belonging to the eigenvalues $\lambda_{\alpha 1}(z), \lambda_{\alpha 2}(z), \ldots, \lambda_{\alpha g_\alpha}(z)$. Denoting the dimension of that subspace, which is spanned by all zero-order approximations of perturbed Jordan vectors belonging to arbitrary eigenvalues of the λ_0-group, by d, one obtains the inequality $d \leqq \sum_{\alpha=1}^{r} m_\alpha$. If there is a proper eigenvalue branching, i.e., if at least one of the numbers $g_\alpha > 1$, then

$$d \leqq \sum_{\alpha=1}^{r} m_\alpha < \sum_{\alpha=1}^{r} g_\alpha m_\alpha = n(\lambda_0); \tag{2.5}$$

That is,

$$d < n(\lambda_0). \tag{2.6}$$

Hence the following proposition is true:

If the λ_0-group of perturbed eigenvalues corresponding to the perturbation $A(z)$ of A_0 contains a cycle of a period $g > 1$, then the zero-order approximations of perturbed Jordan vectors of $A(z)$, with respect to all eigenvalues of the λ_0-group, do not span the full algebraic eigenspace of λ_0 with respect to A_0.

Probably, the simplest example for this fact is given by

$$A(z) = \begin{pmatrix} 0 & 1 \\ z & 0 \end{pmatrix}.$$

In this case $\lambda_1(z) = z^{1/2}$, $\lambda_2(z) = -z^{1/2}$; corresponding eigenvectors are $g_1(z) = (1, z^{1/2})$, $g_2(z) = (1, -z^{1/2})$. One has $g_1(0) = g_2(0) = (1, 0)$. The Jordan vector of A_0 of second degree $(0, 1)$ does not appear as a zero-order approximation.

6.2.3. Geometric multiplicity

Because of (2.2), the inequalities (2.3) for $q = 1$ take the form

$$d(\lambda_0, 1) \geqq l_\alpha, \qquad \alpha = 1, 2, \ldots, r. \tag{2.7}$$

That is, the number of irreducible projections of a perturbed eigenvalue does not exceed the number of irreducible projections of the corresponding unperturbed eigenvalue.

In particular, if $d(\lambda_0, 1) = 1$, i.e., if λ_0 is a prime eigenvalue (see Subsect. 2.4.3), then $l_1 = l_2 = \cdots = l_r = 1$. That is, one obtains

Corollary 1. *Let the eigenprojection P_0 of λ_0 be irreducible. Then all perturbed eigenprojections $P_{\alpha\sigma}(z)$, $z \neq 0$, are also irreducible. In other words, if λ_0 is prime, then all eigenvalues of the λ_0-group are prime.*

In particular, once more considering example (b) of Subsect. 6.1.6, one can conclude that $P_1(z)$ is irreducible.

6.2.4. Examples

If $r > 1$, then there are at least two numbers $l_1, l_2, \ldots, l_r$. In this case $(0, \lambda_0)$ is necessarily a splitting point. From (2.7), the inequality $d(\lambda_0, 1) \geqq \max_{\alpha=1,\ldots,r} l_\alpha$ follows. The cases $d(\lambda_0, 1) = \max_\alpha l_\alpha$ and $d(\lambda_0, 1) > \max_\alpha l_\alpha$ can both be illustrated by simple examples.

(a) $$A(z) = \begin{pmatrix} 0 & 1 \\ 0 & z \end{pmatrix}.$$

One has $\lambda_1(z) \equiv 0$, $\lambda_2(z) \equiv z$; hence $r = 2$, $g_1 = g_2 = 1$, $m_1 = m_2 = 1$, $l_1 = l_2 = 1$. Furthermore, $d(0, 1) = 1$; hence $d(0, 1) = \max (l_1, l_2)$.

(b) $$A(z) = \begin{pmatrix} 0 & 0 \\ 0 & z \end{pmatrix}.$$

In this case, $r = 2$, $l_1 = l_2 = 1$, but $d(0, 1) = 2$.

If $r = 1$, then there is only one number $l_1 = l$ and the inequality $d(\lambda_0, 1) \geqq l$ follows. Again, both cases $d(\lambda_0, 1) = l$ as well as $d(\lambda_0, 1) > l$ are possible.

(c) $$A(z) = \begin{pmatrix} 0 & 1 \\ z & 1 \end{pmatrix}.$$

One has $r = 1$ and $d(\lambda_0, 1) = 1$; hence $l = 1$.

(d) $$A(z) = \left\{\begin{pmatrix} 0 & 1 \\ z & 0 \end{pmatrix}, \begin{pmatrix} 0 & 1 \\ z & 0 \end{pmatrix}, \ldots, \begin{pmatrix} 0 & 1 \\ z & 0 \end{pmatrix}\right\},$$

where the diagonal block $\begin{pmatrix} 0 & 1 \\ z & 0 \end{pmatrix}$ appears m times. In this case one has $\chi(\lambda, A(z)) = (\lambda^2 - z)^m$; i.e., $r = 1$ and $d(\lambda_0, 1) = m$. On the other hand, $l = m$.

A typical example for the case $d(\lambda_0, 1) > l$ is given by

(e) $$A(z) = \begin{pmatrix} 0 & z & 0 \\ 0 & 0 & 1 \\ z & 0 & 0 \end{pmatrix}.$$

(See for example V. B. Lidskii [1].) In this case one has $\chi(\lambda, A(z)) = \lambda^3 - z^2$. Let $a = z^{1/3}$ denote a fixed root and let ξ be a primitive third root of unity. Then $\lambda_1(z) = a^2$, $\lambda_2(z) = \xi^2 a^2$, $\lambda_3(z) = \xi a^2$. The corresponding eigenvectors are given by

$$e_1(z) = (a, 1, a^2), \qquad e_2(z) = (\xi a, 1, \xi^2 a^2), \qquad e_3(z) = (\xi^2 a, 1, \xi a^2).$$

One has $l = 1$, but $d(\lambda_0, 1) = 2$. For example, the vectors $(1, 0, 0)$ and $(0, 1, 0)$ are eigenvectors of A_0 with respect to $\lambda_0 = 0$. Only $(0, 1, 0)$ appears as a zero-order approximation of perturbed eigenvectors, in accordance to Subsect. 6.2.2.

Example (e) can easily be generalized slightly. Let $k < n$, where k and n are relatively prime. Let $n = m_1 + m_2 + \cdots + m_k$ and

$$A_0 = \{\Gamma_{m_1}(0), \Gamma_{m_2}(0), \ldots, \Gamma_{m_k}(0)\}.$$

Let $A_1 = (a_{ij})$ be an $n \times n$-matrix, where

$$a_{m_1, m_1+1} = a_{m_1+m_2, m_1+m_2+1} = \cdots = a_{m_1+\cdots+m_{k-1}, m_1+\cdots+m_{k-1}+1} = a_{n,1} = 1,$$

all other a_{ij} vanish. Let

(f) $$A(z) = A_0 + zA_1.$$

One has $\chi(\lambda, A(z)) = \lambda^n - z^k$. If a is a fixed root $z^{k/n}$ and ξ a primitive n-th root of unity, then the perturbed eigenvalues are given by $\lambda_\sigma(z) = \xi^\sigma a$, $\sigma = 1, \ldots, n$. Hence, $l = 1$, but $d(\lambda_0, 1) = k$.

Therefore one can state: *The geometric multiplicity of λ_0 with respect to A_0 may be arbitrarily large; nevertheless, there is a perturbation $A(z)$ of A_0 such that $l = 1$.*

6.2.5. Diagonal perturbations

Some of the results obtained in foregoing subsections can be interpreted from a new perspective given by the results of Sect. 6.1 and Subsect. 6.2.1 to 6.2.3. For instance, this is the case for the results of F. Rellich (see Subsect. 3.6.1 and 3.6.3sqq.).

A holomorphic perturbation $A(z)$ of A_0 in $|z| < R$ is said to be *diagonal* if there is a sequence $z_n \to 0$ such that the operators $A(z_n)$ are diagonal operators for all z_n (A_0 itself need not be diagonal). Then the following theorem holds.

Theorem 2. *Let $A(z)$ be a diagonal perturbation. Then $A(z)$ is a diagonal operator for all z from a suitable dotted disc $0 < |z| < \gamma$; that is, all irreducible projections of $A(z)$ are one-dimensional and the spectral representation takes the form*

$$A(z) = \sum_{\alpha=1}^{r} \sum_{\varrho=1}^{s_\alpha} \sum_{\sigma=1}^{g_{\alpha\varrho}} \lambda_{\alpha\varrho\sigma}(z)\, P_{\alpha\varrho\sigma}(z). \tag{2.8}$$

If, additionally, the eigenprojections $P_{\alpha\varrho\sigma}(z_n)$ of $A(z_n)$ are uniformly bounded with respect to n for every α, ϱ, σ, i.e., if

$$\|P_{\alpha\varrho\sigma}(z_n)\| < K_{\alpha\varrho\sigma}, \tag{2.9}$$

then the eigenvalues and eigenprojections are holomorphic at $z = 0$ (i.e., the subscript σ can be omitted) and the operator

$$A_0 = \sum_{\alpha=1}^{r} \lambda_\alpha \sum_{\varrho=1}^{s_\alpha} P_{\alpha\varrho}, \qquad P_{\alpha\varrho} = P_{\alpha\varrho}(0), \tag{2.10}$$

itself is also diagonal. There is a system $g^{\varkappa}_{\alpha\varrho}(z)$ of eigenvectors of $A(z)$, holomorphic in $|z| < \gamma$ and such that it forms a basis in X for every z from $|z| < \gamma$.

Proof. First note that there is a subsequence $z_{n'}$ of the original sequence z_n such that only strongly normal points of $A(z)$ lie over $z_{n'}$. As all irreducible projections of the operators $A(z_n)$ are one-dimensional, the relation

$$m^1_{\alpha\varrho} = m^2_{\alpha\varrho} = \cdots = m^{l_{\alpha\varrho}}_{\alpha\varrho} = 1; \qquad l_{\alpha\varrho} = m_{\alpha\varrho},$$

follows. This means, any system of perturbed irreducible projections $P^{\varkappa}_{\alpha\varrho\sigma}(z)$ consists of one-dimensional projections only. Hence (2.8) follows. According to Theorem 3.3.4, the inequality (2.9) implies that the eigenvalues and eigenprojections are holomorphic at $z = 0$ and one also obtains (2.10). Furthermore, it follows that any system of irreducible projections $P^{\varkappa}_{\alpha\varrho}(z)$ is holomorphic in a dotted disc $0 < |z| < \gamma'$. That is, the point $z = 0$ can be at most a pole with respect to every projection $P^{\varkappa}_{\alpha\varrho}(z)$. Choose a decomposition $P_{\alpha\varrho}(z) = \sum_{\varkappa} P^{\varkappa}_{\alpha\varrho}(z)$ of $P_{\alpha\varrho}(z)$ into a sum of disjoint one-dimensional projections, holomorphic at $z = 0$. Then one obtains a system $g^{\varkappa}_{\alpha\varrho}(z)$ of eigenvectors of $A(z)$, holomorphic in a disc $|z| < \gamma$. ■

6.2.6. The operator $A(z)^{-1}$

Recall the result of I. C. Gochberg and M. G. Krein (I. C. Gochberg and M. G. Krein [1, p. 64]; also [2, p. 39]; I. C. Gochberg [1]). This result appears in the present context as a consequence of Theorem 2.1 and of Lemma 5.1.2, and its content is the following:

Let $A(z)$ be meromorphic in G. Then there is an isolated set $B \subset G \setminus S_A$ which is closed in G, such that

$$\dim \ker \{A(z)\} = k(z) = k, \qquad z \in G - (S_A \cup B),$$

and

$$\dim \ker \{A(z)\} = k(z) > k, \qquad z \in B.$$

Using Theorem 2.1 to prove this result, one distinguishes two cases: either the function $\lambda(\zeta_\alpha) \equiv 0$ is to be found among the eigenvalue functions $\lambda(\zeta_\alpha)$ of $A(z)$, or this is not true. If $k > 0$, then $A(z)$ is nowhere invertible; if $k = 0$, then $A(z)^{-1}$ is meromorphic because $\det A(z) \not\equiv 0$. Thus we obtain

Lemma 1 (Keldyš). *If $A(z)$ is meromorphic in G, then either $A(z)$ is nowhere invertible or $A(z)^{-1}$ is meromorphic.*

For this result, see M. V. KELDYŠ [1]; note that, in this reference, $A(z)$ is compact, and that the proposition refers to $1 - A(z)$.

The fact that $A(z)^{-1}$ is meromorphic in the second case can also be derived from the spectral representation of $A(z)$. Let $A(z)$ be holomorphic in $|z| < R$ and assume that $\lambda_0 = 0$ is an eigenvalue of A_0 but not an eigenvalue of $A(z)$ in $0 < |z| < \delta$. Then $A(z)^{-1}$ is holomorphic in $0 < |z| < \delta$ and has a pole at $z = 0$. Let

$$A(z) = \sum_{\alpha,\sigma} \big(\lambda_{\alpha\sigma}(z)\, P_{\alpha\sigma}(z) + D_{\alpha\sigma}(z)\big) + A(z)\,\big(1 - P(z)\big).$$

The first term is assumed to correspond to the 0-group; $P(z)$ denotes the corresponding group projection. Obviously, $A(z)^{-1}$ exists in $0 < |z| < \delta$. One has

$$A(z)^{-1} = \sum_{\alpha,\sigma} [\lambda_{\alpha\sigma}(z) + D_{\alpha\sigma}(z)]^{-1}\, P_{\alpha\sigma}(z) + [A(z)]^{-1}\,\big(1 - P(z)\big).$$

The second term is holomorphic in $0 < |z| < I$ and has a finite limit as $z \to 0$, because the eigenvalues of $A_0 \restriction (1 - P_0)\, X$ are different from zero. Hence, the second term is holomorphic in $|z| < \delta$. For the first term one obtains

$$\sum_{\alpha,\sigma} [\lambda_{\alpha\sigma}(z) + D_{\alpha\sigma}(z)]^{-1}\, P_{\alpha\sigma}(z) = \sum_{\alpha,\sigma} \sum_{\mu=0}^{n(\lambda_0)} \frac{(-1)^{\mu}\, D_{\alpha\sigma}(z)^{\mu}}{\lambda_{\alpha\sigma}(z)^{\mu+1}}\, P_{\alpha\sigma}(z).$$

This term is single-valued and holomorphic in $0 < |z| < \delta$; the point $z = 0$ is a pole of this term. (For Subsect. 6.2.6, see also M. RIBARIČ and I. VIDAV [1], S. STEINBERG [1].)

7. Reduction theory

In Ch. 6 (Subsect. 6.1.1 and Sect. 6.2) an essential problem in connection with Theorem 6.1.1 was emphasized, namely the problem of determining the connection between unperturbed and perturbed Jordan structures. Part of this problem was solved in Subsect. 6.2.2.

The present chapter contains a systematic presentation of the methods and results connected with this problem. Note that there are different points of view with respect to the relation between unperturbed and perturbed Jordan structures.

The first approach assumes that a fixed type of the perturbed Jordan structure is given. That is, the Jordan structure of the perturbation $A(z)$ for $z \neq 0$ is prescribed. The question is what Jordan structures are possible for A_0. Note that this question played a role in Sect. 6.2, where the Jordan structures that are possible for A_0 were restricted by the Gochberg-Krein inequalities.

The second approach assumes that the perturbation $A(z)$, hence also the unperturbed Jordan structure, is given. The question is determining the perturbed Jordan structure. The considerations of this chapter are based on this point of view.

Apart from Theorem 6.1.1, a further essential result of Ch. 6 is the existence of convergent perturbation series with respect to fractional powers of the perturbation parameter z for all perturbed spectral objects (see Theorem 6.1.2). For those perturbed objects whose perturbation series are continuous at $z = 0$ (for example eigenvalues, eigenvectors, Jordan vectors), i.e., which do not contain negative powers of z, the question arises how the zero-order approximations can be characterized (this problem was emphasized especially by I. C. Gochberg). Further questions concern the calculation of the zero-order approximations and, moreover, the explicit calculation of the coefficients of the perturbation series.

These problems, calculation of perturbation series that are known qualitatively, calculation of the splitting behavior of unperturbed eigenvalues and calculation of the zero-order approximations of perturbed eigenvectors, mark the beginning of the development of perturbation theory. *These* problems were the interesting ones (cf., for example, E. Schrödinger [1]). They will be treated in this chapter, too.

First, a short introduction will be presented, where the case of a simple unperturbed eigenvalue λ_0 will be described from the point of view of this chapter. It will be shown that the questions mentioned have been answered by the investigations of Ch. 3. After that, some special difficulties are indicated which appear if λ_0 is no longer simple.

Secondly, the problems mentioned will be treated in the case of a normal perturbation. A complete solution with respect to all relevant questions will be presented. The case of normal perturbations will be considered separately and at the beginning.

Thirdly, the reduction theory for semisimple eigenvalues follows, which has been established by F. Wolf [1], T. Kato [5] and other authors (for example P. Lancaster [1, 2]). Of course, this treatment cannot be complete, because the reduction process can be continued only if the assumption of semisimplicity (for the perturbed first-order approximations) is satisfied, which need not be true.

Furthermore, results will be presented which have been obtained by generalizing the reduction process to the case of non-semisimple unperturbed eigenvalues. These results have been obtained by M. I. Višik and I. A. Ljusternik [2], V. B. Lidskii [1] and H. Baumgärtel [3]. Their derivation requires a series of propositions on operators acting on finite-dimensional spaces which will be collected and proved first.

For this more general situation, the problem of calculating the perturbation series will be treated in simple cases.

Finally, it will be pointed out that a part of the results can also be obtained by the method of bifurcation equations (see for example M. M. Vainberg and V. A. Trenogin [1]). This method was originally developed for non-linear problems (E. Schmidt [1]), but it also yields results for linear perturbation theory. If the unperturbed eigenvalue is simple, by means of this method, one obtains an iterative approach to determining the perturbed eigenvalue (and also a perturbed eigenvector), i.e., without referring to the perturbation series (P. Rosenbloom [1]).

7.1. Simple unperturbed eigenvalues

In the following discussion let $A(z)$ be a holomorphic perturbation of $A_0 = A(0)$ in $|z| < R$,

$$A(z) = A_0 + zA_1 + z^2A_2 + \cdots, \tag{1.1}$$

let λ_0 be a simple eigenvalue of A_0 and let P_0 be the corresponding eigenprojection; $\dim P_0 = 1$.

7.1.1. Perturbation of eigenvalues and eigenprojections

In the case of perturbation of eigenvalues and eigenprojections, the results of Subsect. 6.1.7, 3.2.8, 3.3.14 and 3.7.2 to 3.7.4 can be summarized in the following theorem:

Theorem 1. *There is $\delta > 0$ such that the λ_0-group of $A(z)$ consists of only a single eigenvalue $\lambda(z)$ for $|z| < \delta$. The group projection $P(z)$ corresponding to this λ_0-group is the eigenprojection of $\lambda(z)$. $\lambda(z)$ and $P(z)$ are holomorphic in $|z| < \delta$. The coefficients λ_i and P_i, $i = 1, 2, 3, \ldots$, of the power series expansions*

$$\lambda(z) = \lambda_0 + z\lambda_1 + z^2\lambda_2 + \cdots, \qquad |z| < \delta,$$

$$P(z) = P_0 + zP_1 + z^2P_2 + \cdots, \qquad |z| < \delta,$$

can be calculated by the formulas

$$P_i = \sum_{k=1}^{i} \sum_{\substack{\mu_1+\cdots+\mu_k=i \\ \sigma_1+\cdots+\sigma_{k+1}=k \\ \mu_j\geq 1,\ \sigma_j\geq 0}} S_0^{(\sigma_1)} A_{\mu_1} S_0^{(\sigma_2)} A_{\mu_2} \cdots A_{\mu_k} S_0^{(\sigma_{k+1})}, \tag{1.2}$$

$$\lambda_i = \sum_{p=1}^{i} \frac{1}{p} \sum_{\substack{\mu_1+\cdots+\mu_p=i \\ k_1+\cdots+k_p=p-1 \\ \mu_j\geq 1,\ k_j\geq 0}} \operatorname{tr}\left(A_{\mu_1} S_0^{(k_1)} \cdots A_{\mu_p} S_0^{(k_p)}\right). \tag{1.3}$$

For example,

$$P_1 = P_0 A_1 S_0 + S_0 A_1 P_0, \tag{1.4}$$

$$\lambda_1 = \operatorname{tr}(A_1 P_0), \tag{1.5}$$

$$\lambda_2 = \operatorname{tr}(A_2 P_0) + \operatorname{tr}(A_1 P_0 A_1 S_0). \tag{1.6}$$

For a simple unperturbed eigenvalue, this theorem establishes the function-theoretic dependence of the perturbed eigenvalue and eigenprojection for an arbitrary holomorphic perturbation. Furthermore, it establishes a method for the explicit calculation of the perturbation series.

With the use of biorthogonal bases in P_0X and $(1 - P_0)X$ and in $P_0^*X^*$ and $(1 - P_0^*)X^*$, the formulas (1.2) and (1.3) can be written in another form (see the calculations in Subsect. 3.9.2 and 3.9.4, where this transformation is performed with respect to the coefficients (1.4), (1.5) and (1.6)).

Therefore, concerning the eigenvalue and the eigenprojection, the questions asked at the beginning of this chapter have been answered. Note that both objects are spectral objects which are uniquely determined, so the perturbation-theoretic question is appropriate.

7.1.2. Perturbation of eigenvectors

An eigenvector belonging to a simple eigenvalue is uniquely determined up to a scalar factor. Therefore, a normalization condition must be prescribed for the perturbed eigenvector. The following proposition (see Subsect. 3.4.4 and 6.1.7) forms a supplement of Theorem 1.

Theorem 2. *There is an eigenvector $g(z)$, holomorphic in $|z| < \delta$,*

$$g(z) = g_0 + z g_1 + z^2 g_2 + \cdots, \qquad g(z) \neq 0, \qquad |z| < \delta, \tag{1.7}$$

such that

$$A(z)\, g(z) = \lambda(z)\, g(z), \qquad |z| < \delta, \tag{1.8}$$

and

$$P(z)\, g(z) = g(z), \qquad |z| < \delta. \tag{1.9}$$

This theorem can be supplemented by the following remark: Let $\bar{g}(z)$ be a second eigenvector equipped with the properties of Theorem 2,

$$\bar{g}(z) = \bar{g}_0 + z\bar{g}_1 + z^2\bar{g}_2 + \cdots, \qquad |z| < \delta;$$

then

$$\bar{g}(z) = \alpha(z)\, g(z), \qquad |z| < \delta,$$

where $\alpha(z)$ denotes a non-vanishing scalar function holomorphic in $|z| < \delta$ and where, in particular, $\bar{g}_0 = \alpha(0)\, g_0$. Conversely, if $g(z)$ is an eigenvector, then $\alpha(z)\, g(z)$ is also an eigenvector for every non-vanishing scalar function $\alpha(z)$ which is holomorphic in $|z| < \delta$.

Therefore, to determine $g(z)$ uniquely, a normalization condition is necessary. There are different possibilities for normalization.

The most convenient choice of normalization is the linear normalization. Let $g_0 \neq 0$ be a fixed unperturbed eigenvector; i.e., let $A_0 g_0 = \lambda_0 g_0$ and $P_0 g_0 = g_0$. Furthermore, let f_0^* be biorthogonal with respect to g_0; i.e., let $P_0^* f_0^* = f_0^*$, $A_0^* f_0^* = \bar{\lambda}_0 f_0^*$ and $(g_0, f_0^*) = 1$. Then one can choose the normalization rule

$$\left(g(z), f_0^*\right) = 1. \tag{1.10}$$

This means that one selects the eigenvector

$$g(z) = \frac{P(z)\, g_0}{\left(P(z)\, g_0, f_0^*\right)}. \tag{1.11}$$

Another "nonlinear" normalization is used by B. v. Sz.-Nagy in [2]. He selects the eigenvector

$$g(z) = \frac{P(z)\, g_0}{\left(P(z)\, g_0, f_0^*\right)^{1/2}}, \tag{1.12}$$

where the branch of the square root is taken which yields the value $+1$ at $z = 0$. This normalization can be established by the following argument: If

$$f^*(z) = \frac{P(z)^*\, f_0^*}{\left(g_0, P(z)^*\, f_0^*\right)^{1/2}}, \tag{1.13}$$

then,

$$\left(g(z), f^*(z)\right) = 1. \tag{1.14}$$

If $X = H$ is unitary and A_0 selfadjoint and, furthermore, if $A(z)$ is a selfadjoint perturbation, then, for real $z = \varepsilon$, (1.14) takes the form

$$\left(g(\varepsilon), g(\varepsilon)\right) = \|g(\varepsilon)\|^2 = 1.$$

Thus one obtains the usual normalization in H (i.e., the eigenvector $g(\varepsilon)$ is normalized). Using the linear normalization from

$$P(z)\, g_0 = g_0 + zP_1 g_0 + z^2 P_2 g_0 + \cdots,$$

one obtains

$$\left(P(z)\,g_0, f_0^*\right) = 1 + z(P_1 g_0, f_0^*) + z^2(P_2 g_0, f_0^*) + \cdots.$$

According to Subsect. 3.8.4, the relation

$$P_1 g_0 = \sum_{j=1}^{N-1} (A_1 g_0, S_0^* y_j)\, x_j,$$

holds, where $x_1, x_2, \ldots, x_{N-1}$ denotes a basis of $(1 - P_0)\,X$ and $y_1, y_2, \ldots, y_{N-1}$ is the corresponding biorthogonal basis of $(1 - P_0^*)\,X^*$. One obtains

$$(P_1 g_0, f_0^*) = 0;$$

hence,

$$\left(P(z)\,g_0, f_0^*\right) = 1 + z^2(P_2 g_0, f_0^*) + \cdots.$$

Therefore,

$$g(z) = \frac{g_0 + zP_1 g_0 + z^2 P_2 g_0 + \cdots}{1 + z^2(P_2 g_0, f_0^*) + \cdots} = g_0 + zP_1 g_0 + \cdots.$$

Using the normalization of B. v. Sz.-Nagy, one obtains

$$g(z) = \frac{g_0 + zP_1 g_0 + z^2 P_2 g_0 + \cdots}{\left(1 + z^2(P_2 g_0, f_0^*) + \cdots\right)^{1/2}} = g_0 + zP_1 g_0 + \cdots.$$

Therefore, the first-order approximations $zP_1 g_0$ coincide for both kinds of normalization.

7.1.3. Holomorphic diagonalization of perturbations

A simple special case is given if all eigenvalues $\lambda_1, \lambda_2, \ldots, \lambda_N$ of the unperturbed operator A_0 are simple. Let $A(z)$ be an arbitrary holomorphic perturbation of A_0 in $|z| < R$ (this case is discussed in B. L. Livšic [1]). From Subsect. 3.4.4 the following result is obtained.

Theorem 3. *Let $A(z)$ be a holomorphic perturbation of $A(0) = A_0$ in $|z| < R$. Let all eigenvalues $\lambda_1, \lambda_2, \ldots, \lambda_N$ of A_0 be simple. Then there is a δ, $0 < \delta \leqq R$, such that all perturbed eigenvalues $\lambda_1(z), \ldots, \lambda_N(z)$ and all perturbed eigenprojections $P_1(z), \ldots, P_N(z)$ are holomorphic in $|z| < \delta$. One has*

$$A(z) = \sum_{i=1}^{N} \lambda_i(z)\, P_i(z), \qquad |z| < \delta;$$

that is, $A(z)$ is holomorphic diagonalizable in $|z| < \delta$. Furthermore, there is a system $g_1(z), \ldots, g_N(z)$ of eigenvectors with respect to $\lambda_1(z), \ldots, \lambda_N(z)$, which is holomorphic in $|z| < \delta$ and linearly independent for every z.

7.1.4. Non-simple unperturbed eigenvalues

As we have already seen, the situation is much more complicated if the algebraic multiplicity of λ_0 is greater than 1. The perturbed eigenvalued need not be holomorphic (see the examples from Subsect. 3.2.10).

Even in the case that the perturbation is linear, i.e., if it takes the form $A(z) = A_0 + zA_1$, and that there is a neighborhood $|z| < \delta$ where $A(z)$ is diagonal (so each unperturbed eigenvalue λ_0 is necessarily semisimple), the perturbed eigenvalues need not be holomorphic at $z = 0$. This fact can be seen from the following example (see T. Kato [5, p. 86]). Let $N = 3$ and

$$A(z) = \begin{pmatrix} 0 & z & 0 \\ 0 & 0 & z \\ z & 0 & 1 \end{pmatrix}.$$

One has $\chi(\lambda, A(z)) = \lambda^3 - \lambda^2 - z^3$ and $\chi_1'(\lambda, A(z)) = 3\lambda^2 - 2\lambda$. The polynomial χ has two roots at $z = 0$, namely $\lambda = 0$ and $\lambda = 1$. $A(0)$ is a diagonal operator. At $z \neq 0$, the polynomial χ has three different roots, except if $z^3 = -4/27$. Therefore, there is a neighborhood of $z = 0$ such that $A(z)$ is diagonal at every point of this neighborhod. The Puiseux-Newton diagram for χ takes the form shown in Fig. 4 (see Sect. A 7).

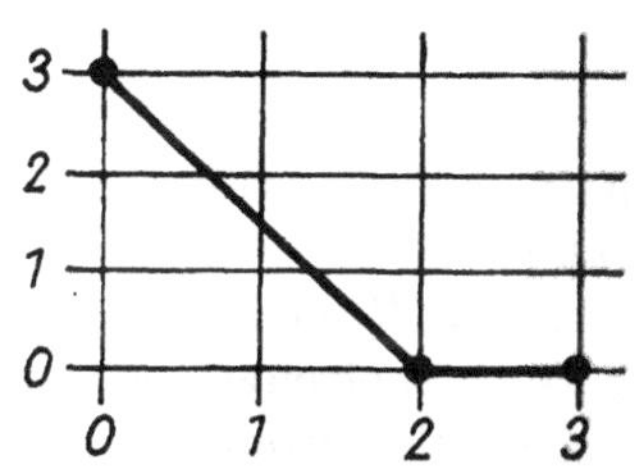

Fig. 4

Hence, possible exponents are 0 and 3/2; that is, one obtains the Puiseux series

$$\lambda_{1,2}(z) = \pm z^{3/2} + \cdots, \qquad \lambda_3(z) = 1 + z^3 + \cdots.$$

Therefore, $\lambda_1(z)$ and $\lambda_2(z)$ are not holomorphic at $z = 0$.

As mentioned above, the question arises what are the criteria of holomorphy for perturbed eigenvalues; more generally, how does one determine the periods of cycles of perturbed eigenvalues of a λ_0-group (see Subsect. 3.2.6).

In addition to the loss of holomorphy, the second important characteristic of the perturbation of non-simple eigenvalues is the phenomenon of splitting. Thus one has to ask what are the criteria for the splitting behavior of an unperturbed eigenvalue λ_0.

7.2. Reduction process for normal and selfadjoint perturbations

The next task is effectively determining the perturbed Jordan structure and the perturbation series for normal perturbations. This is due to the fact that, according to Sect. 3.5, most of the questions involved here have already been answered. The results mentioned give information on the function-theoretic behavior of the spectral objects and on the perturbed spectral structure in general, such that all subsequent qualitative questions are reduced to a *single* question: what is the *number* of perturbed eigenvalues? The remaining question is: how does one calculate the perturbation series effectively and quantitatively? So far, only group projections of λ_0-groups of perturbed eigenvalues and mean values of λ_0-groups have been calculated. Only in the case of a simple unperturbed eigenvalue, these perturbation series coincide with those of eigenvalues and eigenprojections.

In other words, one has to deal with the new problem concerning perturbation series for *splitted* eigenvalues and eigenprojections.

This problem is solved by transforming the original question such that the known results concerning calculation of perturbation series for group projections and group mean values can be applied. This method is called the *reduction process* (see T. Kato [5, p. 81]).

A further reason for treating the case of normal perturbations separately by means of the reduction process, is that the reduction process can be repeated arbitrarily often, as already mentioned, so one can calculate arbitrarily high-order approximations for the perturbation series.

7.2.1. Elementary reduction

In this subsection, the first step of the reduction process will be described in elementary terms. Later, a formulation will be given that is suitable for generalization (see for example I. M. Gelfand [1, p. 278 sqq]; E. Schrödinger [1]).

Let $A(z) = A_0 + zA_1$ be a linear normal perturbation in H, and λ_0 an unperturbed eigenvalue of multiplicity n. Then the λ_0-group of perturbed eigenvalues consists, in a neighborhood $|z| < \delta$, of n functions $\lambda_1(z), \lambda_2(z), \ldots, \lambda_n(z)$, holomorphic in $|z| < \delta$, where the multiplicity of the perturbed eigenvalues is taken into account. Furthermore, there is a system $g_1(z), \ldots, g_n(z)$ of eigenvectors, holomorphic in $|z| < \delta$ and linearly independent for every z from this disc. One obtains

$$A(z)\, g_i(z) = \lambda_i(z)\, g_i(z), \qquad |z| < \delta, \tag{2.1}$$

$$g_i(z) = g_i + z g_i^{(1)} + z^2 g_i^{(2)} + \cdots, \qquad i = 1, \ldots, n, \qquad |z| < \delta, \tag{2.2}$$

$$\lambda_i(z) = \lambda_0 + z\lambda_i^{(1)} + z^2\lambda_i^{(2)} + \cdots, \qquad i = 1, \ldots, n, \qquad |z| < \delta. \tag{2.3}$$

The reduction process practically yields a simultaneous determination of the "first-order approximations" $\lambda_i^{(1)}$ of the perturbed eigenvalues and of the "zero-order approximation" g_i of the perturbed eigenvectors. It is a characteristic feature of the perturbation theory of non-simple eigenvalues that already the zero-order approximation of perturbed eigenvector needs to be determined.

Substituting (2.2) and (2.3) into (2.1) and comparing the coefficients, one obtains the equations

$$A_1 g_i + A_0 g_i^{(1)} = \lambda_0 g_i^{(1)} + \lambda_i^{(1)} g_i, \qquad i = 1, \ldots, n. \tag{2.4}$$

Let $e_1, e_2, \ldots, e_n$ denote any fixed basis of the eigenspace of A_0 with respect to λ_0. Then

$$g_i = \eta_1^{(i)} e_1 + \eta_2^{(i)} e_2 + \cdots + \eta_n^{(i)} e_n. \tag{2.5}$$

Multiplying (2.4) by e_j (scalar multiplication), one obtains

$$(A_1 g_i, e_j) + (A_0 g_i^{(1)}, e_j) = \lambda_0 (g_i^{(1)}, e_j) + \lambda_i^{(1)} (g_i, e_j).$$

Now,

$$(A_0 g_i^{(1)}, e_j) = (g_i^{(1)}, A_0^* e_j) = (g_i^{(1)}, \bar{\lambda}_0 e_j) = \lambda_0 (g_i^{(1)}, e_j),$$

hence,

$$(A_1 g_i, e_j) = \lambda_i^{(1)} (g_i, e_j), \qquad i, j = 1, \ldots, n. \tag{2.6}$$

Substituting (2.5) into (2.6), one obtains the linear system of equations

$$\sum_{\varrho=1}^{n} \eta_\varrho^{(i)} (A_1 e_\varrho, e_j) = \lambda_i^{(1)} \eta_j^{(i)}, \qquad j = 1, \ldots, n, \tag{2.7}$$

where $i = 1, 2, \ldots, n$ is a fixed index. Since $\eta_1^{(i)}, \eta_2^{(i)}, \ldots, \eta_n^{(i)}$ has to be a nontrivial solution of (2.7), one can conclude that $\lambda_i^{(1)}$ is a root of the secular equation

$$\det \left[(A_1 e_\varrho, e_j) - \lambda \delta_{\varrho j} \right] = 0. \tag{2.8}$$

To every root $\lambda^{(1)}$ of (2.8) with multiplicity r are assigned exactly r linearly independent solutions of (2.7), corresponding to the zero-order approximations of those perturbed eigenvectors which belong to the first-order approximation $\lambda^{(1)}$ of the perturbed eigenvalues.

The equation (2.8) for determining the first-order approximations of perturbed eigenvalues was established already by E. Schrödinger [1].

The procedure just sketched can be iterated and leads, in the second step, to the determination of the second-order approximations of the perturbed eigenvalues. After finitely many steps, the "full" splitting of perturbed eigenvalues is reached.

We do not continue the procedure in the way just described, but we intend to give another formulation of the first step. After that the process will be iterated. As mentioned, this second formulation is more appropriate for generalization.

7.2.2. Reduction process

For the following discussion see T. Kato [5, p. 81 sqq.]. Let

$$A(z) = A_0 + z A_1 + z^2 A_2 + \cdots, \qquad |z| < R, \tag{2.9}$$

be a normal perturbation of A_0. Let λ_0 be an unperturbed eigenvalue with the eigenprojection P_0. Let $P(z)$ denote the group projection of the λ_0-group; it is holomorphic in a disc $|z| < \delta$.

Lemma 1. *The operator-valued function*

$$B(z) = \frac{1}{z}\left(A(z) - \lambda_0\right) P(z), \qquad |z| < \delta, \tag{2.10}$$

is holomorphic in $|z| < \delta$, in particular at the point $z = 0$. The relations

$$B(z)\, A(z) = A(z)\, B(z), \qquad |z| < \delta, \tag{2.11}$$

$$B(z)\, P(z) = P(z)\, B(z) = B(z), \qquad |z| < \delta, \tag{2.12}$$

and

$$B_0 = B(0) = P_0 A_1 P_0 \tag{2.13}$$

are valid. $B(z)$ is a normal perturbation of B_0 in $|z| < \delta$.

Proof. Obviously, $B(z)$ is holomorphic in $0 < |z| < \delta$. Since $A_0 P_0 = \lambda_0 P_0$, the function $\left(A(z) - \lambda_0\right) P(z)$ vanishes at the point $z = 0$. This implies that $B(z)$ is also holomorphic at $z = 0$. (2.11) and (2.12) follow because of $A(z)\, P(z) = P(z)\, A(z)$ and $P(z)^2 = P(z)$. The coefficients of the power series

$$B(z) = B_0 + zB_1 + z^2 B_2 + \cdots$$

can be calculated according to Subsect. 3.7.3. In particular,

$$\begin{aligned} B_0 &= \frac{\mathrm{d}}{\mathrm{d}z}\left[\left(A(z) - \lambda_0\right) P(z)\right]_{z=0} = \left(A'(z)\, P(z) + \left(A(z) - \lambda_0\right) P'(z)\right)_{z=0} \\ &= A_1 P_0 + (A_0 - \lambda_0)\, P_1, \end{aligned}$$

and (2.13) follows because of (2.12).

$A(z)$ is a normal perturbation. Hence, there is a sequence $z_n \to 0$ such that $A(z_n)$ is normal; i.e., $A(z_n)\, A(z_n)^* = A(z_n)^*\, A(z_n)$. Then $P(z_n) = P(z_n)^*$ and one obtains $B(z_n)\, B(z_n)^* = B(z_n)^*\, B(z_n)$. ∎

Without loss of generality, one can assume that

$$\det\left(B_0 \restriction P_0 X\right) \neq 0. \tag{2.14}$$

To prove this, let α be a complex number. Then the perturbation

$$\tilde{A}(z) = A(z) + \alpha z$$

is a normal perturbation of $\tilde{A}(0) = A_0$. One obtains

$$\tilde{B}(z) = \frac{1}{z}\left(A(z) + \alpha z - \lambda_0\right) \tilde{P}(z)$$

and

$$\tilde{B}(0) = \tilde{P}_0 \tilde{A}_1 \tilde{P}_0 = P_0 (A_1 + \alpha)\, P_0 = B_0 + \alpha P_0.$$

Hence,

$$\tilde{B}_0 \upharpoonright P_0X = (B_0 + \alpha) \upharpoonright P_0X.$$

Choosing α such that $-\alpha$ is not an eigenvalue of $B_0 \upharpoonright P_0X$ (for example, let $|\alpha|$ be sufficiently large), one obtains $\det \{(B_0 + \alpha) \upharpoonright P_0X\} \neq 0$.

That is, if $\det (B_0 \upharpoonright P_0X) = 0$, then the perturbation $A(z)$ is replaced by $\tilde{A}(z) = A(z) + \alpha z$, where $\det (\tilde{B}_0 \upharpoonright P_0X) \neq 0$.

Let

$$\chi(\lambda, B_0 \upharpoonright P_0X) = \det [(\lambda - B_0) \upharpoonright P_0X] = \prod_{\varrho=1}^{r} (\lambda - \lambda_\varrho^{(1)})^{\mu_\varrho}. \tag{2.15}$$

Then,

$$\lambda_\varrho^{(1)} \neq 0, \qquad \varrho = 1, 2, \ldots, r. \tag{2.16}$$

Let

$$Q_\varrho^{(1)} = \frac{1}{2\pi i} \int\limits_{(\lambda_\varrho^1)} (\lambda - B_0)^{-1} \, d\lambda, \qquad \varrho = 1, \ldots, r, \tag{2.17}$$

where, as usual, one integrates over a sufficiently small positively oriented sphere centered at $\lambda_\varrho^{(1)}$ and belonging to the resolvent set of B_0. The orthoprojections $Q_\varrho^{(1)}$ are the Riesz eigenprojections of B_0 with respect to $\lambda_\varrho^{(1)}$. One has

$$\sum_{\varrho=1}^{r} Q_\varrho^{(1)} = P_0. \tag{2.18}$$

To prove this formula, note that

$$B_0(1 - P_0) = (1 - P_0) B_0 = 0$$

and

$$\det (B_0 \upharpoonright P_0X) \neq 0.$$

Therefore,

$$1 - P_0 = \frac{1}{2\pi i} \int\limits_{(0)} (\lambda - B_0)^{-1} \, d\lambda,$$

where one integrates over a sufficiently small positively oriented sphere centered at $\lambda = 0$. Therefore, one obtains (2.18).

The spectral representation of the normal operator B_0 takes the form

$$B_0 = \sum_{\varrho=1}^{r} \lambda_\varrho^{(1)} Q_\varrho^{(1)}.$$

One obtains

Lemma 2. *Let* $\dim P_0 = n$, *and let* $e_1, e_2, \ldots, e_n$ *denote any orthonormal basis of* P_0X. *Then,*

$$\chi(\lambda, B_0 \upharpoonright P_0X) = \det \big(\lambda \delta_{\varrho\sigma} - (A_1 e_\varrho, e_\sigma)\big), \qquad \varrho, \sigma = 1, 2, \ldots, n. \tag{2.19}$$

Proof. One has

$$\chi(\lambda, B_0 \upharpoonright P_0X) = \chi(\lambda, P_0A_1P_0 \upharpoonright P_0X) = \det\,[(\lambda - P_0A_1P_0) \upharpoonright P_0X].$$

Furthermore, the relation

$$(P_0A_1P_0e_\varrho, e_\sigma) = (A_1e_\varrho, e_\sigma), \qquad \varrho, \sigma = 1, 2, \ldots, n,$$

holds, which implies the assertion. ■

Thus, the characteristic polynomial $\chi(\lambda, B_0 \upharpoonright P_0X)$ is the polynomial on the left hand side of the secular equation (2.8) of E. Schrödinger, which appeared in the elementary reduction process in Subsect. 7.2.1.

Consider the normal perturbation $B(z)$ of B_0. There are r different non-vanishing unperturbed eigenvalues $\lambda_1^{(1)}, \ldots, \lambda_r^{(1)}$. From $\lambda_\varrho^{(1)}$, the perturbation $B(z)$ generates the $\lambda_\varrho^{(1)}$-group of perturbed eigenvalues with the corresponding group projection $Q_\varrho^{(1)}(z)$, which is given by

$$Q_\varrho^1(z) = \frac{1}{2\pi i} \int\limits_{(\lambda_\varrho^1)} \big(\lambda - B(z)\big)^{-1}\, d\lambda, \qquad |z| < \delta_1 \leqq \delta. \tag{2.20}$$

These group projections have the following properties:

Theorem 1. *Let the group projections $Q_\varrho^1(z)$ of the perturbation $B(z)$ be defined by* (2.20). *Then the relations*

$$A(z)\, Q_\varrho^1(z) = Q_\varrho^1(z)\, A(z), \qquad |z| < \delta_1, \tag{2.21}$$

$$Q_\varrho^1(z)\, P(z) = P(z)\, Q_\varrho^1(z) = Q_\varrho^1(z), \qquad |z| < \delta_1, \tag{2.22}$$

$$\sum_{\varrho=1}^{r} Q_\varrho^1(z) = P(z), \tag{2.23}$$

$$\dim Q_\varrho^1(z) = \mu_\varrho, \qquad \varrho = 1, 2, \ldots, r, \tag{2.24}$$

hold.

Proof. (2.24) follows from $Q_\varrho^1(0) = Q_\varrho^1$ and from $\dim Q_\varrho^1 = \mu_\varrho$, which is implied by (2.15). Furthermore, from (2.11)

$$\big(\lambda - B(z)\big)^{-1} A(z) = A(z) \big(\lambda - B(z)\big)^{-1}$$

follows. This implies (2.21) because of (2.20). From (2.12), the relation

$$\big(\lambda - B(z)\big)^{-1} P(z) = P(z) \big(\lambda - B(z)\big)^{-1}$$

follows; hence, because of (2.20),

$$Q_\varrho^1(z)\, P(z) = P(z)\, Q_\varrho^1(z). \tag{2.25}$$

Furthermore,

$$\lambda - B(z) = \lambda - P(z)\,B(z) = \big(\lambda - B(z)\big)\,P(z) + \lambda\big(1 - P(z)\big)$$

and

$$\big(\lambda - B(z)\big)^{-1} = \big(\lambda - B(z)\big)^{-1} P(z) + \frac{1}{\lambda}\,\big(1 - P(z)\big); \tag{2.26}$$

hence,

$$\int\limits_{(\lambda_\varrho^{(1)})} \big(\lambda - B(z)\big)^{-1}\,\mathrm{d}\lambda = P(z) \int\limits_{(\lambda_\varrho^{(1)})} \big(\lambda - B(z)\big)^{-1}\,\mathrm{d}\lambda + \big(1 - P(z)\big) \int\limits_{(\lambda_\varrho^{(1)})} \frac{\mathrm{d}\lambda}{\lambda}.$$

Because $\int\limits_{(\lambda_\varrho^{(1)})} \lambda^{-1}\,\mathrm{d}\lambda = 0$,

$$Q_\varrho^1(z) = P(z)\,Q_\varrho^1(z). \tag{2.27}$$

(2.25) and (2.27) express that $Q_\varrho^1(z)$ is a subprojection of $P(z)$. Thus, (2.22) has been proved. Finally, from (2.26), the relation

$$\int\limits_{(0)} \big(\lambda - B(z)\big)^{-1}\,\mathrm{d}\lambda = \int\limits_{(0)} \big(\lambda - B(z)\big)^{-1} P(z)\,\mathrm{d}\lambda + \big(1 - P(z)\big) \int\limits_{(0)} \frac{\mathrm{d}\lambda}{\lambda}$$

follows. Note that $\big(\lambda - B(z)\big)^{-1} P(z)$, considered as a function of λ, is holomorphic at the point $\lambda = 0$. Therefore,

$$1 - P(z) = \frac{1}{2\pi \mathrm{i}} \int\limits_{(0)} \big(\lambda - B(z)\big)^{-1}\,\mathrm{d}\lambda. \tag{2.28}$$

From (2.28) and (2.20), one obtains (2.23). ■

The projections $Q_\varrho^1(z)$ suggest a refined partition of the λ_0-group of perturbed eigenvalues. This is due to the relations (2.21) and (2.23) in particular. First one has

$$\chi\big(\lambda, A(z) \restriction P(z)\,X\big) = \prod_{\varrho=1}^{r} \chi\big(\lambda, A(z) \restriction Q_\varrho^1(z)\,X\big). \tag{2.29}$$

The roots of the characteristic polynomial $\chi\big(\lambda, A(z) \restriction Q_\varrho^1(z)\,X\big)$, i.e., the eigenvalues of $A(z)$ restricted to $Q_\varrho^1(z)\,X$, are collected in a *Q_ϱ^1-group* of perturbed eigenvalues; *the Q_ϱ^1-group of perturbed eigenvalues of $A(z)$ consists exactly of the part* spec $\{A(z) \restriction Q_\varrho^1(z)\,X\}$ *of the λ_0-group of $A(z)$.* Because of (2.24), there are exactly μ_ϱ eigenvalues within the Q_ϱ^1-group, if their multiplicity is taken into account.

Theorem 2. *Let $\lambda(z)$ be a perturbed eigenvalue of the λ_0-group of $A(z)$. Then $\lambda(z)$ belongs to the Q_ϱ^1-group of $A(z)$ if and only if*

$$\lambda(z) = \lambda_0 + z\lambda_\varrho^{(1)} + O(z^2), \qquad z \to 0. \tag{2.30}$$

Proof. Let $\lambda(z)$ be an eigenvalue of the λ_0-group and let $x(z)$ be a corresponding eigenvector; i.e.,

$$A(z)\,x(z) = \lambda(z)\,x(z).$$

Then,

$$B(z)\,x(z) = \frac{\lambda(z) - \lambda_0}{z}\,x(z).$$

Assume that $\lambda(z)$ belongs to the Q_ϱ^1-group of $A(z)$; that is, let

$$Q_\varrho^1(z)\,x(z) = x(z). \tag{2.31}$$

This means that the eigenvalue $z^{-1}(\lambda(z) - \lambda_0)$ necessarily belongs to the $\lambda_\varrho^{(1)}$-group of $B(z)$; thus, (2.30) follows. Conversely, let (2.30) be valid. Then $z^{-1}(\lambda(z) - \lambda_0)$ belongs to the $\lambda_\varrho^{(1)}$-group of $B(z)$. Then there exists a corresponding eigenvector $x(z)$ such that (2.31) holds; i.e., $\lambda(z)$ belongs to the Q_ϱ^1-group of $A(z)$. ■

According to Theorem 2, the Q_ϱ^1-group is often synonymously called the $(\lambda_0 + z\lambda_\varrho^{(1)})$-group of perturbed eigenvalues of $A(z)$. $Q_\varrho^1(z)$ is then the group projection of the $(\lambda_0 + z\lambda_\varrho^{(1)})$-group. The decomposition

$$P_0 = \sum_{p=1}^{r} Q_\varrho^1$$

simultaneously yields initial information about the zero-order approximations x_0 of perturbed eigenvectors and the zero-order approximations E_0 of the perturbed eigenprojections.

Theorem 3. *Let E_0 be the zero-order approximation of a perturbed eigenprojection $E(z)$ belonging to the eigenvalue $\lambda(z)$ of the $(\lambda_0 + z\lambda_\varrho^{(1)})$-group; i.e., let $E(0) = E_0$. Then,*

$$E_0 = E_0 Q_\varrho^1 = Q_\varrho^1 E_0; \tag{2.32}$$

that is, E_0 is a subprojection of Q_ϱ^1.

In particular, if $x_0 \neq 0$ denotes the zero-order approximation of a perturbed eigenvector $x(z)$ belonging to the eigenvalue $\lambda(z)$, i.e., if $x(0) = x_0$, then,

$$Q_\varrho^1 x_0 = x_0. \tag{2.33}$$

Conversely, if (2.33) is assumed to be satisfied, then $\lambda(z)$ necessarily belongs to the $(\lambda_0 + z\lambda_\varrho^{(1)})$-group.

Proof. $Q_\varrho^1(z)$ is the group projection of the $(\lambda_0 + z\lambda_\varrho^{(1)})$-group; that is, the relation

$$E(z)\,Q_\varrho^1(z) = Q_\varrho^1(z)\,E(z) = E(z)$$

holds, which implies (2.32), in particular. Furthermore, in this case, $x_0 = E_0 x_0 = Q_\varrho^1 x_0$. Conversely, if (2.33) is assumed to be true and if $\lambda(z)$ belongs to the $(\lambda_0 + z\lambda_\sigma^{(1)})$-group with $\sigma \neq \varrho$, then $x_0 = Q_\sigma^1 x_0$; hence, $Q_\varrho^1 x_0 = Q_\varrho^1 Q_\sigma^1 x_0 = 0$ which contradicts $x_0 \neq 0$. ■

Therefore, the zero-order approximations of perturbed eigenvectors necessarily belong to $Q^1_\varrho X$, $\varrho = 1, \ldots, r$. This space can be generated by determining μ_ϱ linearly independent solutions of the equation

$$(\lambda^1_\varrho - P_0 A_1)\, x = 0, \qquad P_0 x = x.$$

The calculations of the elementary reduction process performed earlier show that there is a parallel between the determination of the first-order approximations $\lambda^{(1)}_\varrho$ of perturbed eigenvalues and the restriction of the zero-order approximations of perturbed eigenvectors and eigenprojections, i.e., the determination of the projections Q^1_ϱ.

If the objects $\lambda^{(1)}_\varrho$ and Q^1_ϱ are assumed to be given, then the group projections $Q^1_\varrho(z)$ and the mean values of the $(\lambda_0 + z\lambda^{(1)}_\varrho)$-groups can be calculated by the procedure pointed out earlier (Sect. 3.7 and 3.8; see Subsect. 7.2.4).

7.2.3. Iteration of the process

The procedure applied to $A(z)$ and λ_0 in Subsect. 7.2.2 can be applied to $B(z)$ and one of its eigenvalues $\lambda^{(1)}_\varrho$. Note that $B(z)$ is a normal perturbation of B_0. Let

$$C_\varrho(z) = \frac{1}{z}\left(B(z) - \lambda_\varrho\right) Q_\varrho(z); \qquad \lambda_\varrho = \lambda^{(1)}_\varrho, \qquad Q_\varrho = Q^1_\varrho, \tag{2.34}$$

or, upon substitution of (2.10) into (2.34),

$$C_\varrho(z) = \frac{1}{z^2}\left(A(z) - \lambda_0 - z\lambda_\varrho\right) Q_\varrho(z). \tag{2.35}$$

Then $C_\varrho(z)$ is holomorphic at the point $z = 0$ and it is a normal perturbation of $C_\varrho(0)$, where

$$C_\varrho(0) = Q_\varrho B_1 Q_\varrho. \tag{2.36}$$

Assume

$$\det\left[(Q_\varrho B_1 Q_\varrho) \restriction Q_\varrho X\right] \neq 0,$$

then one obtains a spectral representation for $C_\varrho(0)$,

$$C_\varrho(0) = \sum_{\sigma=1}^{r_\varrho} \lambda_{\varrho\sigma} Q_{\varrho\sigma},$$

where the eigenvalues $\lambda_{\varrho\sigma}$ are different from zero and where

$$\sum_{\sigma=1}^{r_\varrho} Q_{\varrho\sigma} = Q_\varrho.$$

Note that the orthoprojections $Q_{\varrho\sigma}$ reduce A_0. The perturbation $C_\varrho(z)$ generates $\lambda_{\varrho\sigma}$-groups of perturbed eigenvalues and group projections $Q_{\varrho\sigma}(z)$ which also reduce $A(z)$ and yield a further partition of the $(\lambda_0 + z\lambda_\varrho)$-groups.

In this way, $Q_{\varrho\sigma}$-groups of eigenvalues $\lambda(z)$ of $A(z)$ arise which satisfy the characteristic asymptotic relation

$$\lambda(z) = \lambda_0 + z\lambda_\varrho + z^2\lambda_{\varrho\sigma} + O(z^3), \qquad z \to 0;$$

thus, the eigenvalues $\lambda_{\varrho\sigma}$ of $C_\varrho(0)$ form the second-order approximations of the perturbed eigenvalues.

Correspondingly, the orthoprojections $Q_{\varrho\sigma}$ yield a further restriction of the zero-order approximations for perturbed eigenvectors and eigenprojections.

Note that this procedure can be iterated arbitrarily because the appearing perturbations $B(z)$, $C_\varrho(z)$, ... are always normal perturbations. In this way, one obtains a sequence of coefficients

$$\lambda_0, \lambda_{\varrho_1}, \lambda_{\varrho_1\varrho_2}, \ldots, \lambda_{\varrho_1\varrho_2\cdots\varrho_m}, \ldots$$

and projections

$$P_0, Q_{\varrho_1}, Q_{\varrho_1\varrho_2}, \ldots, Q_{\varrho_1\varrho_2\cdots\varrho_m}, \ldots,$$

which satisfy

$$P_0 \geqq Q_{\varrho_1} \geqq Q_{\varrho_1\varrho_2} \geqq \cdots;$$

that is, the eigenprojection $Q_{\varrho_1\varrho_2\cdots\varrho_m}$ of the m-th-order approximation are subprojections of the eigenprojections $Q_{\varrho_1\varrho_2\cdots\varrho_{m-1}}$ of the $(m-1)$th-order approximation. The perturbed eigenvalues $\lambda(z)$ can be represented asymptotically in the form

$$\lambda(z) = \lambda_0 + z\lambda_{\varrho_1} + z^2\lambda_{\varrho_1\varrho_2} + \cdots + z^m\lambda_{\varrho_1\varrho_2\cdots\varrho_m} + O(z^{m+1}), \qquad z \to 0. \tag{2.37}$$

They can be collected in $(\varrho_1, \varrho_2, \ldots, \varrho_m)$-groups whose group projections have the zero-order approximations $Q_{\varrho_1\varrho_2\cdots\varrho_m}$.

If there is a splitting of $\lambda_{\varrho_1\varrho_2\cdots\varrho_m}$ by the transition from the mth- to the $(m+1)$th-order approximation, then this splitting necessarily causes a decrease of the dimensions of the corresponding eigenprojections. From this fact it follows that, after finitely many steps, no further splitting occurs. Ultimately, this is the case if the dimension of the eigenprojection equals 1.

If there is no splitting after the mth step of the process then the iteration can be finished. In this case, the orthoprojections $Q_{\varrho_1\varrho_2\cdots\varrho_m}$ are the zero-order approximations of the eigenprojections of those perturbed eigenvalues $\lambda_{\varrho_1\varrho_2\cdots\varrho_m}(z)$ of $A(z)$ which have the asymptotic behavior (2.37). The group projections $Q_{\varrho_1\varrho_2\cdots\varrho_m}(z)$ represent the perturbed eigenprojections and the mean values of the $\lambda_{\varrho_1\varrho_2\cdots\varrho_m}$-groups represent the perturbed eigenvalues. Therefore, in this situation it is sufficient to continue the procedure by the process described in Sect. 3.7 and 3.8 (see Subsect. 7.2.4).

Besides the question how to perform the calculations described, the following question remains open: How can one determine that number m of steps after which no further splitting occurs. In other words, when do the group projections and mean values to be constructed coincide with the eigenprojections and eigenvalues? Of course, the practical value of such a question is rather small, because in general one calculates first-order, at best second-order approximations, without considering higher-order approximations. The question will be much more interesting if it is expanded to include the search for a quantitative criterion which is completely independent of the reduction process. If such a criterion can be presented, then the question for splitting or non-splitting at a fixed step of the reduction process can also be answered.

The problem just indicated will be treated in Subsect. 7.2.5 but it has not been solved completely.

7.2.4. Calculation of group projections and mean values of $(\lambda_0 + z\lambda_\rho)$-groups for $A(z)$

As mentioned before, the problem is calculating the usual group projections of the λ_ϱ-groups of $B(z)$ and, essentially, calculating the mean values of these λ_ϱ-groups. Knowledge of the eigenvalues λ_ϱ and of the corresponding eigenprojections Q_ϱ with respect to B_0 is assumed.

The group projection $Q_\varrho(z)$ of the $(\lambda_0 + z\lambda_\varrho)$-group of $A(z)$ coincides with the group projection of the λ_ϱ-group of $B(z)$. As such, it can be calculated by the method of Sect. 3.7. Let

$$Q_\varrho(z) = Q_\varrho + zQ_{\varrho 1} + z^2 Q_{\varrho 2} + \cdots. \tag{2.38}$$

An essential tool for the calculation is the reduced resolvent of B_0 at the point λ_ϱ. One has

$$R(\mu, B_0) = (\mu - B_0)^{-1} = \sum_{\varrho=1}^{r} \frac{Q_\varrho}{\mu - \lambda_\varrho} + \frac{1 - P_0}{\mu}. \tag{2.39}$$

Hence, one obtains, for the reduced resolvent S_ϱ,

$$S_\varrho = \sum_{\substack{\sigma=1 \\ \sigma \neq \varrho}}^{r} \frac{Q_\sigma}{\lambda_\varrho - \lambda_\sigma} + \frac{1 - P_0}{\lambda}; \tag{2.40}$$

that is, $S_\varrho = \hat{S}_\varrho + \tilde{S}_\varrho$, where $\hat{S}_\varrho$ and $\tilde{S}_\varrho = \lambda_\varrho^{-1}(1 - P_0)$ denote the parts of the reduced resolvent in P_0X and in $(1 - P_0)\,X$, respectively. One has

$$\hat{S}_\varrho P_0 = P_0 \hat{S}_\varrho = \hat{S}_\varrho$$

and

$$\hat{S}_\varrho Q_\varrho = Q_\varrho \hat{S}_\varrho = 0.$$

Since all eigenvalues λ_ϱ are semisimple (note that B_0 is normal), one can use the simplified formulas from Subsect. 3.7.2 for the determination of the coefficients $Q_{\varrho 1}, Q_{\varrho 2}, \ldots$ In the resolvent formula (3.7.7), one takes

$$S_0^{(0)} = Q_\varrho, \qquad S_0^{(-m)} = 0, \qquad S_0^{(m)} = (-1)^{m-1} S_\varrho^m, \qquad m = 1, 2, 3, \ldots \tag{2.41}$$

For example, for the first-order approximation of the group projection, one obtains the relation

$$Q_{\varrho 1} = Q_\varrho B_1 S_\varrho + S_\varrho B_1 Q_\varrho, \tag{2.42}$$

where B_1 is defined by

$$B(z) = B_0 + zB_1 + z^2B_2 + \cdots. \tag{2.43}$$

As mentioned in Subsect. 7.2.2, the coefficients $B_0, B_1, B_2, \ldots$ can be obtained from (3.7.13). For example

$$\begin{aligned} B_1 &= \sum_{k=1}^{2} \sum_{\substack{\mu_1+\cdots+\mu_k=2\\ \sigma_1+\cdots+\sigma_{k+1}=k-1\\ \mu_j\ge 1,\sigma_j\ge 0}} S_0^{(\sigma_1)}A_{\mu_1}S_0^{(\sigma_2)}A_{\mu_2}\cdots A_{\mu_k}S_0^{(\sigma_{k+1})} \\ &= S_0^{(0)}A_2S_0^{(0)} + \sum_{\sigma_1+\sigma_2+\sigma_3=1} S_0^{(\sigma_1)}A_1S_0^{(\sigma_2)}A_1S_0^{(\sigma_3)}. \end{aligned} \tag{2.44}$$

Note that in the formulas for the coefficients $B_0, B_1, B_2, \ldots$, for example in (2.44), the operators $S_0^{(m)}$ are not defined as in (2.41), but they are determined by the eigenprojection of λ_0 with respect to A_0 and the corresponding reduced resolvent. That is, in these formulas one substitutes

$$S_0^{(0)} = P_0, \qquad S_0^{(-m)} = 0, \qquad S_0^{(m)} = (-1)^{m-1}\,S_0^m, \qquad m = 1, 2, 3, \ldots \tag{2.45}$$

Then,

$$\begin{aligned} Q_\varrho B_1 S_\varrho &= Q_\varrho A_2 \hat{S}_\varrho + Q_\varrho A_1 S_0 A_1 \hat{S}_\varrho + Q_\varrho A_1 P_0 A_1 S_0 S_\varrho \\ &= Q_\varrho A_2 \hat{S}_\varrho + Q_\varrho A_1 S_0 A_1 \hat{S}_\varrho + \frac{1}{\lambda_\varrho} Q_\varrho A_1 P_0 A_1 S_0, \end{aligned}$$

because

$$S_0 S_\varrho = \sum_{\substack{\sigma=1\\ \sigma\neq\varrho}}^{r} \frac{S_0 Q_\sigma}{\lambda_\varrho - \lambda_\sigma} + \frac{S_0(1-P_0)}{\lambda_\varrho} = \frac{S_0}{\lambda_\varrho}.$$

Furthermore,

$$Q_\varrho A_1 P_0 = Q_\varrho P_0 A_1 P_0 = P_0 A_1 P_0 Q_\varrho = Q_\varrho A_1 Q_\varrho = \lambda_\varrho Q_\varrho;$$

therefore,

$$\frac{1}{\lambda_\varrho} Q_\varrho A_1 P_0 A_1 S_0 = Q_\varrho A_1 S_0;$$

that is,

$$Q_\varrho B_1 S_\varrho = Q_\varrho A_2 \hat{S}_\varrho + Q_\varrho A_1 S_0 + Q_\varrho A_1 S_0 A_1 \hat{S}_\varrho.$$

Hence, for $Q_{\varrho 1}$ the formula

$$\begin{aligned} Q_{\varrho 1} = {} & Q_\varrho A_2 \hat{S}_\varrho + Q_\varrho A_1 S_0 + Q_\varrho A_1 S_0 A_1 \hat{S}_\varrho + \hat{S}_\varrho A_2 Q_\varrho \\ & + S_0 A_1 Q_\varrho + \hat{S}_\varrho A_1 S_0 A_1 Q_\varrho \end{aligned}$$

is obtained. In this final result for the first-order approximation $Q_{\varrho 1}$, the eigenvalue λ_ϱ does not appear explicitly, but the eigenprojection Q_ϱ and the reduced resolvent $\hat{S}_\varrho$ with respect to P_0X do.

The calculation of higher-order approximations is done similar. Of course, the expressions are very complicated.

In order to calculate the mean value of the $(\lambda_0 + z\lambda_\varrho)$-group, one first calculates the mean value of the λ_ϱ-group with respect to $B(z)$, according to Subsect. 3.7.3 and 3.7.4. This mean value is denoted by $\lambda_\varrho(z)$ and one writes

$$\lambda_\varrho(z) = \lambda_\varrho + z\lambda_{\varrho 1} + z^2\lambda_{\varrho 2} + \cdots.$$

For the calculation of the coefficients $\lambda_{\varrho 1}, \lambda_{\varrho 2}, \ldots$, one uses the formula (3.7.22), where n is to be replaced by $\mu_\varrho = \dim Q_\varrho$, P_0 by Q_ϱ, S_0 by S_ϱ, and A_k by B_k. Moreover, note that λ_ϱ is semisimple. Then, for example, one obtains

$$\lambda_{\varrho 1} = \frac{1}{\mu_\varrho} \operatorname{tr} (B_1 Q_\varrho)$$

as the first-order approximation. Using the expression (2.44) for B_1, one obtains

$$B_1 Q_\varrho = P_0 A_2 Q_\varrho + S_0 A_1 P_0 A_1 Q_\varrho + P_0 A_1 S_0 A_1 Q_\varrho$$

and

$$\operatorname{tr} (B_1 Q_\varrho) = \operatorname{tr} (A_2 Q_\varrho) + \operatorname{tr} (A_1 S_0 A_1 Q_\varrho);$$

hence,

$$\lambda_{\varrho 1} = \frac{1}{\mu_\varrho} \operatorname{tr} (A_2 Q_\varrho + A_1 S_0 A_1 Q_\varrho).$$

Further coefficients can be calculated similarly; the expressions becomes more and more complicated.

For the mean value of the $(\lambda_0 + z\lambda_\varrho)$-group, one obtains the expression

$$\lambda_0 + z\lambda_\varrho(z) = \lambda_0 + z\lambda_\varrho + z^2\lambda_{\varrho 1} + \cdots. \tag{2.46}$$

If the $(\lambda_0 + z\lambda_\varrho)$-group of $A(z)$ consists of one single eigenvalue, i.e., if λ_ϱ fails to split already by the perturbation $B(z)$ of B_0, then (2.46) provides the perturbation series of this eigenvalue. In particular, this is the case if λ_ϱ has multiplicity 1; i.e., if $\mu_\varrho = 1$.

7.2.5. Splitting problems

As mentioned before, the question is still open how one can decide at a certain step m of the reduction process whether the mean values and group projections of the step m already coincide with the eigenvalues and eigenprojections or not. In other words, the question is deciding whether one has to iterate the procedure or whether one can stop it.

The basic problem can also be formulated as follows: Let $A(z)$ be a normal perturbation of A_0 and let λ_0 be an eigenvalue of A_0. We have to decide whether λ_0 splits by the perturbation $A(z)$ or not.

More precisely, one can ask what is the exact number of eigenvalues $\lambda(z)$ that result from the splitting of λ_0.

In the following paragraphs we will discuss this problem without solving it completely.

The results of Sect. 2.11, in particular Lemma 5, suggest the use of algebras for the investigation of the splitting behavior of an unperturbed eigenvalue λ_0 of A_0 under a normal perturbation $A(z)$; in particular, algebras whose operators commute with every operator of this perturbation. For simplicity, one starts with the largest algebra of this type.

Definition 1. Let $A(z)$ be a normal perturbation of $A(0) = A_0$ in $|z| < R$. The set of all operators C with the property

$$CA(z) = A(z)\, C, \qquad |z| < R, \tag{2.47}$$

is called *the commutant of the perturbation* and is denoted by $\mathscr{C}$.

Obviously, $\mathscr{C}$ is an algebra with 1. Let

$$A(z) = A_0 + zA_1 + z^2 A_2 + \cdots.$$

Then $C \in \mathscr{C}$ if and only if

$$CA_\varrho = A_\varrho C, \qquad \varrho = 0, 1, 2, 3, \ldots$$

The commutant $\mathscr{C}$ of the perturbation $A(z)$ is to be distinguished from the commutants of the single operators $A(\zeta)$, where $|\zeta| < R$. The commutant of $A(\zeta)$ is denoted by com $A(\zeta)$. Obviously,

$$\mathscr{C} \subseteq \operatorname{com} A(\zeta), \qquad |\zeta| < R, \tag{2.48}$$

but in general one has $\mathscr{C} \subset \operatorname{com} A(\zeta)$.

In general, $\mathscr{C}$ is too small to provide exact information on the splitting behavior of the unperturbed eigenvalues under the perturbation $A(z)$. To illustrate this observation consider a simple example: Let

$$A(z) = \begin{pmatrix} 0 & 0 \\ 0 & 0 \end{pmatrix} + z \begin{pmatrix} 0 & 1 \\ 1 & 0 \end{pmatrix} + z^2 \begin{pmatrix} 0 & \mathrm{i} \\ -\mathrm{i} & 0 \end{pmatrix}.$$

The perturbation $A(z)$ is normal and even selfadjoint. For an operator $\begin{pmatrix} \alpha & \beta \\ \gamma & \delta \end{pmatrix}$ of the commutant, one first obtains the condition

$$\begin{pmatrix} \beta & \alpha \\ \delta & \gamma \end{pmatrix} = \begin{pmatrix} \alpha & \beta \\ \gamma & \delta \end{pmatrix} \begin{pmatrix} 0 & 1 \\ 1 & 0 \end{pmatrix} = \begin{pmatrix} 0 & 1 \\ 1 & 0 \end{pmatrix} \begin{pmatrix} \alpha & \beta \\ \gamma & \delta \end{pmatrix} = \begin{pmatrix} \gamma & \delta \\ \alpha & \beta \end{pmatrix},$$

which implies $\alpha = \delta$, $\beta = \gamma$. A further condition is

$$\begin{pmatrix} -\beta & \alpha \\ -\alpha & \beta \end{pmatrix} = \begin{pmatrix} \alpha & \beta \\ \beta & \alpha \end{pmatrix} \begin{pmatrix} 0 & 1 \\ -1 & 0 \end{pmatrix} = \begin{pmatrix} 0 & 1 \\ -1 & 0 \end{pmatrix} \begin{pmatrix} \alpha & \beta \\ \beta & \alpha \end{pmatrix} = \begin{pmatrix} \beta & \alpha \\ -\alpha & -\beta \end{pmatrix},$$

which implies $\beta = 0$. Therefore, the commutant $\mathscr{C}$ consists of the scalar operators.

However, the commutant $\mathscr{C}$ of $A(z)$ can be used to formulate a sufficient criterion for the non-splitting of an unperturbed eigenvalue λ_0 of A_0. For this purpose, let the disc $|z| < \delta$ be chosen sufficiently small such that $0 < |z| < \delta$ consists of simple points only (see Subsect. 3.2.2). Only the point $z = 0$ can be multiple.

Lemma 3. *Let $A(z)$ be a normal perturbation of $A_0 = \sum_{\varrho=1}^{r} \lambda_\varrho P_\varrho$ in $|z| < \delta$. Let $A(z) = \sum_{\varrho=1}^{r} \sum_{\sigma=1}^{s_\varrho} \lambda_{\varrho\sigma}(z) P_{\varrho\sigma}(z)$; i.e. let $\lambda_{\varrho\sigma}(z)$ denote the perturbed eigenvalues, $\lambda_{\varrho\sigma}(0) = \lambda_\varrho$, and let $P_{\varrho\sigma}(z)$ denote the perturbed eigenprojections. Furthermore, let $\mathscr{C}$ denote the commutant of $A(z)$. Then,*

$$P_{\varrho\sigma}(z)\, C = C P_{\varrho\sigma}(z), \quad |z| < \delta, \quad C \in \mathscr{C}, \quad \varrho = 1, \ldots, r; \quad \sigma = 1, \ldots, s_\varrho. \tag{2.49}$$

Proof. Let $0 < |z| < \delta$ and let z be fixed. Then the eigenprojections $P_{\varrho\sigma}(z)$ are polynomials in $A(z)$. Hence, from (2.47), the assertion (2.49) follows for all $z \neq 0$. Taking the limit $z \to 0$, one obtains the assertion for $z = 0$, too. ■

Thus, one obtains the following criterion. (See Definition 2.11.2. See also T. Kato [5, p. 83].)

Theorem 4. *Assume that the eigenprojection P_0 of λ_0 belonging to A_0 is irreducible with respect to $\mathscr{C}$. Then there is no splitting of λ_0 under the perturbation; i.e., $(0, \lambda_0)$ is normal.*

Proof. Assume that there is a splitting of λ_0 with the corresponding eigenprojections $Q_\varrho(z)$, $\varrho = 1, \ldots, k$, $k \geq 2$, where

$$\sum_{\varrho=1}^{k} Q_\varrho(0) = P_0.$$

From (2.49),

$$CQ_\varrho(0) = Q_\varrho(0)\, C, \qquad C \in \mathscr{C}, \qquad \varrho = 1, 2, \ldots, k.$$

Therefore, the projections $Q_\varrho(0)$ are proper reducing subprojections of P_0 with respect to $\mathscr{C}$. This is a contradiction. ■

This criterion can be used within the framework of the reduction process. Of course, the question arises: is its converse also true? That is, if the eigenvalue λ_0 of A_0 does not split under the perturbation $A(z)$, does it then follow that the corresponding eigenprojection P_0 is irreducible with respect to $\mathscr{C}$? In general this will not be the case because $\mathscr{C}$ is too small. To illustrate the situation, an example will be given. Let

$$P(z) = \begin{pmatrix} 2/3 & \frac{1}{3}\,e^{iz} & \frac{1}{3}\,e^{iz} \\ \frac{1}{3}\,e^{-iz} & 2/3 & -1/3 \\ \frac{1}{3}\,e^{-iz} & -1/3 & 2/3 \end{pmatrix}.$$

$P(z)$ is a projection-valued function, holomorphic at $z = 0$, selfadjoint; i.e., $P(\varepsilon)^* = P(\varepsilon)$ for real $z = \varepsilon$. One has

$$P_0 = \begin{pmatrix} 2/3 & 1/3 & 1/3 \\ 1/3 & 2/3 & -1/3 \\ 1/3 & -1/3 & 2/3 \end{pmatrix}.$$

Note that $\dim P_0 = 2$. The eigenvalues of P_0 are 0 and 1. These eigenvalues remain invariant under the perturbation $P(z)$. Consider the eigenvalue $\lambda_0 = 1$ of multiplicity 2. It does not split under the perturbation. Let $P(z) = P_0 + zP_1 + z^2P_2 + \cdots$. Then,

$$P_1 = \begin{pmatrix} 0 & i/3 & i/3 \\ -i/3 & 0 & 0 \\ -i/3 & 0 & 0 \end{pmatrix}$$

and

$$P_2 = \frac{1}{2}\begin{pmatrix} 0 & -1/3 & -1/3 \\ -1/3 & 0 & 0 \\ -1/3 & 0 & 0 \end{pmatrix}.$$

The commutant $\mathscr{C}$ of $P(z)$ consists of exactly those operators which commute with

$$\tilde{P}_0 = \begin{pmatrix} 2 & 1 & 1 \\ 1 & 2 & -1 \\ 1 & -1 & 2 \end{pmatrix}, \quad \tilde{P}_1 = \begin{pmatrix} 0 & 1 & 1 \\ -1 & 0 & 0 \\ -1 & 0 & 0 \end{pmatrix}, \quad \tilde{P}_2 = \begin{pmatrix} 0 & 1 & 1 \\ 1 & 0 & 0 \\ 1 & 0 & 0 \end{pmatrix}.$$

An easy calculation yields: $\mathscr{C}$ consists exactly of all operators of the form

$$A(\alpha, \beta) = \begin{pmatrix} \alpha + \beta & 0 & 0 \\ 0 & \alpha & \beta \\ 0 & \beta & \alpha \end{pmatrix},$$

where α and β are arbitrary complex numbers. That is, there are two fixed projections Q_1 and Q_2, independent of α and β, where $\dim Q_1 = 2$ and $\dim Q_1 = 1$, such that

$$A(\alpha, \beta) = (\alpha + \beta)\, Q_1 + (\alpha - \beta)\, Q_2.$$

Hence, $\mathscr{C}$ has three one-dimensional irreducible projections, which are not all uniquely determined. The projection P_0, reducing $\mathscr{C}$, is not irreducible. P_0X is generated by $x_1 = (0, 1, -1)$, $x_2 = (2, 1, 1)$ and one has $A(\alpha, \beta)\, x_1 = (\alpha - \beta)\, x_1$ and $\mathring{A}(\alpha, \beta)\, x_2 = (\alpha + \beta)\, x_2$.

This example shows once more, that the commutant $\mathscr{C}$ of the perturbation is too small to yield a necessary and sufficient condition for the splitting behavior.

According to the considerations of Sect. 2.11, it is obvious that such conditions can be formulated if one takes into consideration the commutants com $A(\zeta)$ of the single operators $A(\zeta)$, $|\zeta| < \delta$.

The commutants com $A(\zeta)$ are mutually strongly connected. To derive this connection, recall Subsect. 3.4.2, in particular Theorem 3.4.1. According to this theorem, given the complete disjoint system $P_{\varrho\sigma}(z)$ of projections, holomorphic in $|z| < \delta$ there exists a transformation operator $V(z)$, which together with its inverse is holomorphic in $|z| < \delta$ and which has the property

$$P_{\varrho\sigma}(z) = V(z)\, P_{\varrho\sigma}(0)\, V(z)^{-1}, \quad |z| < \delta, \quad \varrho = 1, \ldots, r; \quad \sigma = 1, \ldots, s_\varrho. \tag{2.50}$$

Moreover, $V(0) = 1$.

Lemma 4. *The normal perturbation*

$$B(z) = V(z)^{-1}\, A(z)\, V(z), \qquad |z| < \delta, \tag{2.51}$$

of $B(0) = A_0$ *is holomorphic in* $|z| < \delta$. *The commutant* $\mathscr{D}$ *of this perturbation coincides with the commutant* com $B(\zeta)$ *for* $\zeta \neq 0$; *i.e.,*

$$\mathscr{D} = \operatorname{com} B(\zeta), \qquad 0 < |\zeta| < \delta.$$

Proof. Note that

$$B(z) = \sum_{\varrho=1}^{r} \sum_{\sigma=1}^{s_\varrho} \lambda_{\varrho\sigma}(z)\, P_{\varrho\sigma}(0).$$

Therefore, the normal perturbation $B(z)$ has constant eigenprojections $P_{\varrho\sigma}(0)$, but the splitting behavior of the eigenvalues with respect to $B(z)$ is the same as with respect to $A(z)$. One has $C \in \mathscr{D}$ if and only if $CP_{\varrho\sigma}(0) = P_{\varrho\sigma}(0)\, C$. However, this is true if and only if $CB(\zeta) = B(\zeta)\, C$ for fixed ζ, $0 < |\zeta| < \delta$, i.e., if and only if $C \in \operatorname{com} B(\zeta)$. ∎

Furthermore, $\mathscr{D} \subseteqq \operatorname{com} A(0)$ and $\mathscr{D} \subset \operatorname{com} A(0)$ if at least one of the unperturbed eigenvalues of A_0 splits under the perturbation $A(z)$.

Lemma 5. *Let* ζ *be arbitrary but fixed, where* $0 < |\zeta| < \delta$. *Then*

$$\operatorname{com} A(\zeta) = V(\zeta)^{-1}\, \mathscr{D} V(\zeta). \tag{2.52}$$

That is, if $C \in \operatorname{com} A(\zeta)$, *then* $D = V(\zeta)\, CV(\zeta)^{-1} \in \mathscr{D}$ *and, conversely, if* $D \in \mathscr{D}$, *then* $C = V(\zeta)^{-1}\, DV(\zeta) \in \operatorname{com} A(\zeta)$.

Proof. In the first case one has

$$\begin{aligned} B(\zeta)\, D &= B(\zeta)\, V(\zeta)\, CV(\zeta)^{-1} = V(\zeta)\, A(\zeta)\, CV(\zeta)^{-1} \\ &= V(\zeta)\, CA(\zeta)\, V(\zeta)^{-1} = V(\zeta)\, CV(\zeta)^{-1}\, B(\zeta) = DB(\zeta); \end{aligned}$$

that is, $D \in \operatorname{com} B(\zeta)$; hence, $D \in \mathscr{D}$. In the second case, one has

$$\begin{aligned} CA(\zeta) &= V(\zeta)^{-1}\, DV(\zeta)\, A(\zeta) = V(\zeta)^{-1}\, DB(\zeta)\, V(\zeta) \\ &= V(\zeta)^{-1}\, B(\zeta)\, DV(\zeta) = A(\zeta)\, V(\zeta)^{-1}\, DV(\zeta) = A(\zeta)\, C; \end{aligned}$$

that is, $C \in \operatorname{com} A(\zeta)$. ∎

Lemma 5 shows that the commutants com $A(\zeta)$ of the single operators $A(\zeta)$ have one and the same structure for $\zeta \neq 0$, namely the structure of $\mathcal{D}$.

According to Lemma 2.11.5, the eigenprojection P_0 of an unperturbed eigenvalue λ_0 of A_0 is an irreducible projection with respect to com $A(0)$. It is merely a reducing projection with respect to the subalgebra $\mathcal{D}$. P_0 can be decomposed into a sum of disjoint irreducible projections with respect to $\mathcal{D}$. According to Lemma 2.11.5, these irreducible parts are eigenprojections of $B(\zeta)$ for $\zeta \neq 0$. Finally, one obtains the following result:

Theorem 5. *The eigenvalue λ_0 of A_0 splits under the normal perturbation $A(z)$ of A_0 into r different perturbed eigenvalues if and only if the eigenprojection P_0 of λ_0 can be decomposed into a sum of r disjoint projections $Q_1, \ldots, Q_r$, each of them irreducible with respect to $\mathcal{D}$, $P_0 = \sum_{\varrho=1}^{r} Q_\varrho$. Moreover, denoting the perturbed eigenprojections by $P_\varrho(z)$, $\varrho = 1, \ldots, r$, one obtains $Q_\varrho = P_\varrho(0)$, $\varrho = 1, \ldots, r$.*

This criterion can be applied practically if information on the structure of the commutants com $A(\zeta)$ for $\zeta \neq 0$, hence of the commutant $\mathcal{D}$, is available. Such information can be provided by the representation theory. However, this theory lies outside the scope of this book. It should be mentioned that most of the group-theoretic methods for the determination of the splitting behavior of unperturbed eigenvalues and of the remaining rest multiplicities are based on Theorem 5. (For this topic, see the classical presentations of E. Wigner [1], H. Weyl [1], B. L. van der Waerden [1]; see also G. J. Ljubarski [1].)

7.3. Reduction process for semisimple eigenvalues

The first step of the reduction process, described in Sect. 7.2 for normal perturbations, depends merely on the assumption that the unperturbed eigenvalue is semisimple. This property follows from the normality of the perturbation, in particular from the normality of the unperturbed operator.

Let $A(z)$ be an arbitrary holomorphic perturbation of $A(0) = A_0$ in $|z| < R$ and let λ_0 be an unperturbed semisimple eigenvalue. Then the reduction process can be used for the investigation of the λ_0-group of $A(z)$. In this case, of course, the propositions implied by Rellich's theorems for normal perturbations are no longer available. In general, the perturbed eigenvalues fail to be holomorphic. However, the reduction process yields estimates for the appearing periods of cycles. In particular, one obtains sufficient conditions for the existence of holomorphic perturbed eigenvalues belonging to the λ_0-group.

For holomorphic operator-valued functions in Banach spaces, the reduction process for semisimple unperturbed eigenvalues was developed and presented by F. Wolf [1], who considered a finite number of reduction steps.

A presentation for the finite-dimensional case, treated in this section, is given by T. Kato [5].

Further investigations in this direction are due to P. Lancaster [1, 2], who also considers higher reduction steps, but under different (in some sense weaker) assumptions compared with those of F. Wolf.

Finally, some results of T. S. Motzkin, of O. Taussky [1, 2] and of O. Taussky [1] on this topic will be presented.

These results deal with a linear perturbation $A_0 + zA_1$ and are based on special assumptions, for example, the global diagonality of $A_0 + zA_1$.

7.3.1. Reduction process

The following considerations parallel those of Subsect. 7.2.2. Let $A(z) = A_0 + zA_1 + z^2A_2 + \cdots$ be a holomorphic perturbation of $A(0) = A_0$ in $|z| < R$. Let the disc $|z| < \delta$ be chosen such that $0 < |z| < \delta$ consists of simple points only. At most $z = 0$ is a multiple point. Let λ_0 be an unperturbed semisimple eigenvalue with the eigenprojection P_0. Let $\dim P_0 = n$. One has $A_0P_0 = \lambda_0P_0$. The group projection of the λ_0-group, holomorphic in $|z| < \delta$, is denoted by $P(z)$.

Lemma 1. *The operator-valued function*

$$B(z) = \frac{1}{z}\big(A(z) - \lambda_0\big)\,P(z), \qquad |z| < \delta, \tag{3.1}$$

is holomorphic in $|z| < \delta$, in particular at $z = 0$. The relations

$$B(z)\,A(z) = A(z)\,B(z), \qquad |z| < \delta, \tag{3.2}$$

$$B(z)\,P(z) = P(z)\,B(z) = B(z), \qquad |z| < \delta, \tag{3.3}$$

and

$$B_0 = B(0) = P_0A_1P_0 \tag{3.4}$$

hold.

Proof. The proof is analogous to the proof of Lemma 2.1. ■

Without loss of generality, one can assume

$$\det\,(B_0 \restriction P_0X) \neq 0; \tag{3.5}$$

that is, the eigenvalues $\lambda_1, \lambda_2, \ldots, \lambda_r$ of B_0 in P_0X are different from zero and one obtains

$$\chi(\lambda, B_0 \restriction P_0X) = \prod_{\varrho=1}^{r} (\lambda - \lambda_\varrho)^{\mu_\varrho}, \qquad \lambda_\varrho \neq 0, \qquad \varrho = 1, \ldots, r. \tag{3.6}$$

As in Subsect. 7.2.2, one can consider the perturbation $\tilde{A}(z) = A(z) + \alpha z$. This implies $\tilde{B}_0 \restriction P_0X = (B_0 + \alpha) \restriction P_0X$. Now α is chosen such that (3.5) holds with respect to the perturbation $\tilde{A}(z)$. Furthermore, one obtains

Lemma 2. *Let $e_1, e_2, \ldots, e_n$ be a basis in P_0X and let $f_1, \ldots, f_n$ denote the corresponding biorthogonal basis in $P_0^*X^*$. Then*

$$\chi(\lambda, B_0 \upharpoonright P_0X) = \det\,[\lambda\delta_{\varrho\sigma} - (A_1e_\varrho, f_\sigma)]. \tag{3.7}$$

Proof. One has

$$\chi(\lambda, B_0 \upharpoonright P_0X) = \det\,[(\lambda - P_0A_1P_0) \upharpoonright P_0X].$$

Moreover,

$$(\lambda - P_0A_1P_0)\, e_\varrho = \lambda e_\varrho - P_0A_1e_\varrho$$

and

$$\big((\lambda - P_0A_1P_0)\, e_\varrho, f_\sigma\big) = \lambda\delta_{\varrho\sigma} - (A_1e_\varrho, f_\sigma);$$

hence, the assertion follows. ■

According to Lemma 1, $B(z)$ is a holomorphic perturbation of B_0 in $|z| < \delta$. The unperturbed eigenvalues are $\lambda_1, \lambda_2, \ldots, \lambda_r$, which are different from zero. If $\dim\,(1 - P_0) > 0$, one has to add the eigenvalue 0.

The eigenprojections of $\lambda_1, \ldots, \lambda_r$ are denoted by $Q_1, \ldots, Q_r$, respectively. Then,

$$Q_\varrho = \frac{1}{2\pi i} \int\limits_{(\lambda_\varrho)} (\lambda - B_0)^{-1}\, d\lambda, \qquad \varrho = 1, \ldots, r.$$

As in Subsect. 7.2.2, on obtains

$$\sum_{\varrho=1}^{r} Q_\varrho = P_0, \tag{3.8}$$

because on the one hand,

$$B_0(1 - P_0) = (1 - P_0)\, B_0 = 0$$

and on the other hand,

$$\det\,(B_0 \upharpoonright P_0X) \neq 0,$$

which implies

$$1 - P_0 = \frac{1}{2\pi i} \int\limits_{(0)} (\lambda - B_0)^{-1}\, d\lambda.$$

From this equation, (3.8) follows.

The perturbation $B(z)$ generates the λ_ϱ-group from λ_ϱ. By $Q_\varrho(z)$ one denotes the corresponding group projection which is defined and holomorphic in a disc $|z| < \delta_1 \leqq \delta$. One has

$$Q_\varrho(z) = \frac{1}{2\pi i} \int\limits_{(\lambda_\varrho)} (\lambda - B(z))^{-1}\, d\lambda, \qquad |z| < \delta_1, \qquad \varrho = 1, \ldots, r. \tag{3.9}$$

The properties of these group projections are summarized in

Theorem 1. *For all z from* $|z| < \delta_1$, *the relations*

$$A(z)\, Q_\varrho(z) = Q_\varrho(z)\, A(z), \qquad \varrho = 1, \ldots, r, \tag{3.10}$$

$$Q_\varrho(z)\, P(z) = P(z)\, Q_\varrho(z) = Q_\varrho(z), \qquad \varrho = 1, \ldots, r, \tag{3.11}$$

$$\sum_{\varrho=1}^{r} Q_\varrho(z) = P(z) \tag{3.12}$$

and

$$\dim Q_\varrho(z) = \mu_\varrho, \qquad \varrho = 1, \ldots, r \tag{3.13}$$

hold.

Proof. The proof is analogous to the proof of Theorem 2.1. ■

The projections $Q_1(z), \ldots, Q_r(z)$ suggest a refined partition of the λ_0-group of per turbed eigenvalues.

Let $V(z)$ be a transformation operator of the complete disjoint system $Q_1(z), \ldots, Q_r(z)$, $1 - P(z)$ of projection-valued functions in $|z| < \delta_1$ (see Subsect. 3.4.2); that is, the relations

$$Q_\varrho(z) = V(z)\, Q_\varrho(0)\, V(z)^{-1}, \qquad V(0) = 1, \tag{3.14}$$

hold. Then,

$$\tilde{A}(z) = V(z)^{-1}\, A(z)\, V(z), \qquad |z| < \delta_1, \tag{3.15}$$

is a perturbation equipped with constant reducing projections $Q_1(0), \ldots, Q_r(0)$ (cf., for example, Subsect. 3.4.4). One obtains

$$\chi\big(\lambda; \tilde{A}(z) \restriction Q_\varrho(0)\, X\big) = \chi\big(\lambda; A(z) \restriction Q_\varrho(z)\, X\big), \qquad |z| < \delta_1,$$

which implies that $\chi\big(\lambda; A(z) \restriction Q_\varrho(z)\, X\big)$ is a polynomial of degree μ_ϱ whose coefficients are holomorphic in $|z| < \delta_1$. Hence, for the characteristic polynomial of $A(z)$ with respect to the group space $P(z)\, X$ of the λ_0-group, one obtains a decomposition on $|z| < \delta_1$ of the form

$$\chi\big(\lambda; A(z) \restriction P(z)\, X\big) = \prod_{\varrho=1}^{r} \chi\big(\lambda; A(z) \restriction Q_\varrho(z)\, X\big) \tag{3.16}$$

into a product of r polynomials whose coefficients are holomorphic in $|z| < \delta_1$.

The roots of the characteristic polynomial $\chi\big(\lambda; A(z) \restriction Q_\varrho(z)\, X\big)$, i.e., the eigenvalues of $A(z)$ restricted to $Q_\varrho(z)\, X$, are collected in a *Q_ϱ-group* of perturbed eigenvalues; that is, the Q_ϱ-group is exactly the part spec $\{A(z) \restriction Q_\varrho(z)\, X\}$ of the whole λ_0-group spec $\{A(z) \restriction P(z)\, X\}$ of $A(z)$. The sum of all algebraic multiplicities of the eigenvalues of the Q_ϱ-group equals $\dim Q_\varrho = \mu_\varrho$. Furthermore, the eigenvalues of the Q_ϱ-group can be characterized by their asymptotic behavior, as $z \to 0$.

Theorem 2. *Let $\lambda(z)$ be a perturbed eigenvalue of the λ_0-group of $A(z)$. Then $\lambda(z)$ belongs to the Q_ϱ-group of $A(z)$ if and only if the asymptotic relation*

$$\lambda(z) = \lambda_0 + z\lambda_\varrho + o(z), \qquad z \to 0, \tag{3.17}$$

holds, where

$$o(z) = O\left(z^{1+\frac{1}{p}}\right) \tag{3.18}$$

and where p denotes the period of that cycle of the Q_ϱ-group to which $\lambda(z)$ belongs.

Proof. Let $\lambda(z)$ be a perturbed eigenvalue of the λ_0-group of $A(z)$ and $x(z)$ a corresponding eigenvector, which is continuous at $z = 0$, where $x(0) = x_0 \neq 0$. Therefore, $A(z)\, x(z) = \lambda(z)\, x(z)$. Then,

$$B(z)\, x(z) = \frac{\lambda(z) - \lambda_0}{z}\, x(z). \tag{3.19}$$

Next, $\lambda(z)$ belongs to the Q_ϱ-group if and only if $Q_\varrho(z)\, x(z) = x(z)$. On the other hand, this is a necessary and sufficient condition for $z^{-1}(\lambda(z) - \lambda_0)$ to belong to the λ_ϱ-group of $B(z)$. Therefore, the first part of Theorem 2 has been proved. Finally, as a perturbed eigenvalue of the λ_0-group, $\lambda(z)$ belongs to a certain cycle of this λ_0-group. Denoting the period of this cycle by p, one obtains an expansion for $\lambda(z)$ of the form

$$\lambda(z) = \lambda_0 + \alpha_1 z^{1/p} + \alpha_2 z^{2/p} + \cdots + \alpha_p z + \alpha_{p+1} z^{1+\frac{1}{p}} + \cdots.$$

Then, the property that $\lambda(z)$ belongs to the Q_ϱ-group can be expressed by

$$\alpha_1 = \alpha_2 = \cdots = \alpha_{p-1} = 0$$

and $\alpha_p = \lambda_\varrho$. Then,

$$\lambda(z) = \lambda_0 + \lambda_\varrho z + \alpha_{p+1} z^{1+\frac{1}{p}} + \cdots,$$

which implies (3.18). ■

According to Theorem 2, the Q_ϱ-group is often called the $(\lambda_0 + z\lambda_\varrho)$-group of perturbed eigenvalues of $A(z)$. $Q_\varrho(z)$ is the group projection of the $(\lambda_0 + z\lambda_\varrho)$-group.

According to earlier results, the perturbed eigenvalues $\lambda(z)$ of the λ_0-group are algebroid functions that may be branched at $z = 0$, but are continuous at $z = 0$. If λ_0 is semisimple, even the derivation $\lambda'(z)$ is continuous at $z = 0$. That is, one obtains

Corollary 1. *Each perturbed eigenvalue $\lambda(z)$ of the λ_0-group has a derivation at $z = 0$ and the function $\lambda'(z)$ is continuous at $z = 0$ (if λ_0 is semisimple). That is, $\lambda(z)$ is continuous and continuously differentiable at $z = 0$. $\lambda'(0) = \lambda_\varrho$ if and only if $\lambda(z)$ belongs to the Q_ϱ-group.*

Corollary 2. *If p is a period of a cycle of the $(\lambda_0 + z\lambda_\varrho)$-group, then $1 \leqq p \leqq \mu_\varrho$.*

Proof. Let $\lambda(z)$ be a perturbed eigenvalue of the $(\lambda_0 + z\lambda_\varrho)$-group with the cycle period p. Then the irreducible polynomial belonging to $\lambda(z)$ is of degree p, but it is also a divisor of $\chi(\lambda, A(z) \restriction Q_\varrho(z)\, X)$. This fact implies the assertion. ■

In general, for the cycle periods p of the λ_0-group, one obtains the estimate

$$1 \leqq p \leqq \max(\mu_1, \mu_2, \ldots, \mu_r). \tag{3.20}$$

Let $P_{\varrho 1}(z), P_{\varrho 2}(z), \ldots, P_{\varrho l_\varrho}(z)$ denote the perturbed eigenprojections of the $(\lambda_0 + z\lambda_\varrho)$-group. In general, they have branching points at $z = 0$. One obtains

Corollary 3. *If* $\dim P_{\varrho\sigma}(z) = m_\varrho$, *the relations*

$$\sum_{\sigma=1}^{l_\varrho} P_{\varrho\sigma}(z) = Q_\varrho(z), \qquad \varrho = 1, \ldots, r, \tag{3.21}$$

and

$$\mu_\varrho = \sum_{\sigma=1}^{l_\varrho} m_{\varrho\sigma} \tag{3.22}$$

hold.

Proof. If $P_{\varrho\sigma}(z)$ is the eigenprojection of $\lambda_{\varrho\sigma}(z)$ from the $(\lambda_0 + z\lambda_\varrho)$-group with respect to $A(z)$, then $\tilde{P}_{\varrho\sigma}(z) = V(z)^{-1} P_{\varrho\sigma}(z)\, V(z)$ is the eigenprojection of $\lambda_{\varrho\sigma}(z)$ with respect to $\tilde{A}(z)$ (see (3.15)). Then, $\sum_\sigma \tilde{P}_{\varrho\sigma}(z) = Q_\varrho(0)$, hence also (3.21). (3.22) follows from (3.21). ■

Corollary 4. *Let $x(z)$ be a perturbed eigenvector belonging to an eigenvalue $\lambda(z)$ of $A(z)$ from the $(\lambda_0 + z\lambda_\varrho)$-group. Furthermore, let $x(z)$ be continuous at $z = 0$ and $x(0) = x_0 \neq 0$ (such an $x(z)$ can be selected according to* Subsect. 6.1.5). *Then,*

$$B_0 x_0 = \lambda_\varrho x_0, \qquad Q_\varrho x_0 = x_0. \tag{3.23}$$

Proof. One has $A(z)\, x(z) = \lambda(z)\, x(z)$ and $B(z)\, x(z) = z^{-1}\big(\lambda(z) - \lambda_0\big)\, x(z)$. Taking the limit $z \to 0$, one obtains $B_0 x_0 = \lambda_\varrho x_0$, hence also $Q_\varrho x_0 = x_0$. ■

7.3.2. Holomorphy criterion for perturbed eigenvalues

From the considerations of Subsect. 7.3.1, one obtains special sufficient conditions for perturbed eigenvalues to be holomorphic (see, for example, G. Porath [1]).

Theorem 3. *Let $A(z)$ be a holomorphic perturbation of $A(0) = A_0$ and let λ_0 be a semisimple unperturbed eigenvalue. Let $\chi(\lambda; B_0 \restriction P_0X)$ have a simple root μ_1; i.e., B_0 restricted to P_0X has a simple eigenvalue. Then there is a holomorphic eigenvalue*

$$\lambda(z) = \lambda_0 + \mu_1 z + \mu_2 z^2 + \cdots$$

belonging to the λ_0-group of perturbed eigenvalues, which is simple for $z \neq 0$, and there is a corresponding holomorphic perturbed eigenvector $x(z)$ with $x(0) = x_0 \neq 0$. The zero-order approximation x_0 satisfies

$$B_0 x_0 = \mu_1 x_0, \tag{3.24}$$

where x_0 is uniquely determined (up to a scalar multiple) by this relation.

Proof. Let $Q_1(z)$ be the group projection of the μ_1-group of $B(z)$. Then, according to Theorem 1, the relation $\dim Q_1(z) = 1$ is satisfied. That is, the μ_1-group of $B(z)$

consists of a single holomorphic eigenvalue $\mu_1(z)$ with the eigenprojection $Q_1(z)$. Hence, the $(\lambda_0 + z\mu_1)$-group consists of a single perturbed eigenvalue $\lambda(z) = \lambda_0 + z\mu_1(z)$, which is holomorphic, with the same perturbed eigenprojection $Q_1(z)$. Thus one also obtains the existence of a holomorphic eigenvector $rx(z)$ (see Subsect. 6.1.7). For $x(z)$, one obtains $B(z)\,x(z) = \mu_1(z)\,x(z)$; if $z = 0$, the relation (3.24) follows. Because of $\dim Q_1(z) = 1$, the vector x_0 is uniquely determined by (3.24) up to a scalar factor. ■

It should be noted that this criterion is not valid unless λ_0 is semisimple. For example, let

$$A_0 = \begin{pmatrix} 0 & 1 \\ 0 & 0 \end{pmatrix}, \qquad A_1 = \begin{pmatrix} 0 & 1 \\ 1 & 0 \end{pmatrix}$$

and let $A(z) = A_0 + zA_1$. Then, $\lambda_0 = 0$, $P_0 = 1$, $\det(\lambda - A_1) = \lambda^2 - 1$; i.e., A_1 has two simple eigenvalues, $+1$ and -1. However, one obtains $\det\big(\lambda - A(z)\big) = \lambda^2 - z(z + 1)$; that is, $\lambda(z) = \pm(z^2 + z)^{1/2}$ are perturbed eigenvalues, neither are holomorphic at $z = 0$. (Hence, Theorem 2 in G. Porath [2] must have the additional assumption: "the unperturbed eigenvalue is semisimple".)

According to Theorem 3 it is easy to formulate a sufficient condition which ensures that all perturbed eigenvalues of the λ_0-group are holomorphic.

Theorem 4. *Let $A(z)$ be a holomorphic perturbation of A_0 and let λ_0 be a semisimple unperturbed eigenvalue. Assume that the characteristic polynomial $\chi(\lambda; B_0 \restriction P_0X)$ has only simple roots, denoted by $\mu_1, \mu_2, \ldots, \mu_n$; i.e., B_0 restricted to P_0X has only simple eigenvalues $\mu_1, \ldots, \mu_n$. Then the λ_0-group consists of n different holomorphic perturbed eigenvalues*

$$\lambda_\varrho(z) = \lambda_0 + \mu_\varrho z + \mu_{\varrho 2} z^2 + \cdots, \qquad \varrho = 1, \ldots, n,$$

which are simple for $z \neq 0$. There are corresponding holomorphic eigenvectors $x_\varrho(z)$, which are uniquely determined up to holomorphic scalar factors and such that $x_\varrho(0) = x_\varrho \neq 0$. The zero-order approximations $x_1, x_2, \ldots, x_n$ satisfy the relations

$$B_0 x_\varrho = \mu_\varrho x_\varrho, \qquad \varrho = 1, 2, \ldots, n,$$

by which they are uniquely determined up to scalar multiples.

Proof. The proof follows from Theorem 3. The group projections $Q_\varrho(z)$ of the μ_ϱ-groups of $B(z)$ coincide with the eigenprojections of the perturbed (holomorphic) eigenvalues $\mu_\varrho(z)$ of $B(z)$, because μ_ϱ is simple, which implies $\dim Q_\varrho(z) = 1$. Therefore, they also coincide with the eigenprojections of the perturbed (holomorphic) eigenvalues $\lambda_0 + z\mu_\varrho(z)$ of the λ_0-group of $A(z)$. ■

7.3.3. Holomorphic diagonalization of perturbations

The results of Subsect. 7.3.2 provide a further sufficient condition (as a supplement to Theorem 1.3) for the possibility of holomorphic diagonalization of a perturbation $A(z)$.

Theorem 5. *Let $A(z)$ be a holomorphic perturbation of $A(0) = A_0$ and let A_0 be a diagonal operator, $A_0 = \sum_{\varrho=1}^{r} \lambda_\varrho P_\varrho$. Let $B_\varrho := P_\varrho A_1 P_\varrho$, $\varrho = 1, \ldots, r$. Assume that all roots of $\chi(\lambda; B_\varrho \restriction P_\varrho X)$, $\varrho = 1, \ldots, r$, are simple; i.e.,*

$$\chi(\lambda, B_\varrho \restriction P_\varrho X) = \prod_{\sigma=1}^{m_\varrho} (\lambda - \mu_{\varrho\sigma}), \qquad \varrho = 1, \ldots, r;$$

$$m_\varrho = \dim P_\varrho; \qquad \mu_{\varrho\sigma} \neq \mu_{\varrho\sigma'}, \qquad \sigma \neq \sigma';$$

that is, B_ϱ restricted to $P_\varrho X$ has only simple eigenvalues for each $\varrho = 1, \ldots, r$. Then all eigenvalues $\lambda_{\varrho\sigma}(z)$ of $A(z)$ are holomorphic and simple for $z \neq 0$. The relations

$$\lambda_{\varrho\sigma}(z) = \lambda_\varrho + z\mu_{\varrho\sigma} + O(z^2), \qquad z \to 0, \qquad \varrho = 1, \ldots, r; \qquad \sigma = 1, \ldots, m_\varrho,$$

hold. The perturbed eigenprojections $P_{\varrho\sigma}(z)$ are holomorphic at $z = 0$, and $P_{\varrho\sigma}(0) = P_{\varrho\sigma}$ is the eigenprojection of $\mu_{\varrho\sigma}$ with respect to B_ϱ on $P_\varrho X$. One has $\dim P_{\varrho\sigma} = 1$. The perturbation $A(z)$ has the spectral representation

$$A(z) = \sum_{\varrho=1}^{r} \sum_{\sigma=1}^{m_\varrho} \lambda_{\varrho\sigma}(z)\, P_{\varrho\sigma}(z), \qquad |z| > 0;$$

that is, $A(z)$ is holomorphic diagonalizable.

The mere diagonality of $A(z)$ in a neighborhood $|z| < \delta$ of $z = 0$ is neither necessary nor sufficient for the perturbed eigenvalues to be holomorphic at $z = 0$, not even for perturbations of the form $A(z) = A_0 + zA_1$ (cf. the example in Subsect. 7.1.4).

7.3.4. Implications and applications

The final remark of Subsect. 7.3.3 suggests the presentation of some results of T. S. Motzkin, O. Taussky [1, 2], and O. Taussky [1].

Let A_0 be a diagonal operator with the spectral representation $A_0 = \sum_{\varrho=1}^{r} \lambda_\varrho P_\varrho$ and let $A(z) = A_0 + zA_1$ be a linear perturbation of A_0. Let

$$\det\,[(\lambda - P_\varrho A_1 P_\varrho) \restriction P_\varrho X] = \prod_{\sigma=1}^{s_\varrho} (\lambda - \mu_{\varrho\sigma})^{m_{\varrho\sigma}}, \qquad \varrho = 1, \ldots, r. \tag{3.25}$$

Therefore, the numbers $\mu_{\varrho\sigma}$ are the eigenvalues of $P_\varrho A_1 P_\varrho$ restricted to $P_\varrho X$. One has $\dim P_\varrho = \sum_{\sigma=1}^{s_\varrho} m_{\varrho\sigma}$. According to Subsect. 7.3.1, in particular Theorem 2, for an eigenvalue $\lambda_{\varrho\sigma}(z)$ of $A(z)$ from the $(\lambda_\varrho + z\mu_{\varrho\sigma})$-group, one obtains

$$\lambda_{\varrho\sigma}(z) = \lambda_\varrho + z\mu_{\varrho\sigma} + o(z), \quad z \to 0; \quad \varrho = 1, \ldots, r; \quad \sigma = 1, \ldots, s_\varrho. \tag{3.26}$$

From (3.25), one obtains

Lemma 3. *The numbers $\mu_{\varrho\sigma}$, $\varrho = 1, \ldots, r$; $\sigma = 1, \ldots, s_\varrho$, their multiplicities taken into account, coincide with the eigenvalues of A_1 if and only if*

$$\det(\lambda - A_1) = \prod_{\varrho=1}^{r} \det[(\lambda - P_\varrho A_1 P_\varrho) \restriction P_\varrho X]. \tag{3.27}$$

In O. Taussky [1], the relation (3.27) is called "property 1" with respect to both operators A_0 and A_1.

One asks for conditions sufficient to ensure the validity of (3.27). For this purpose, the following condition is introduced:

Condition 1 (property L). *For all complex numbers α and β, the operator $\alpha A_0 + \beta A_1$ has the eigenvalues $\alpha\lambda_i + \beta\mu_i$, $i = 1, 2, \ldots, N$, where $\lambda_1, \lambda_2, \ldots, \lambda_N$ and $\mu_1, \mu_2, \ldots, \mu_N$ denote fixed N-tuples of complex numbers which are independent of α and β.*

Obviously, the numbers $\lambda_1, \lambda_2, \ldots, \lambda_N$ are the eigenvalues of A_0 ($\alpha = 1$, $\beta = 0$) and the numbers $\mu_1, \mu_2, \ldots, \mu_N$ are the eigenvalues of A_1 ($\alpha = 0$, $\beta = 1$) in a suitable ordering (where the algebraic multiplicity is taken into account).

Condition 1 may be also formulated in an inhomogeneous form:

For all complex numbers z the operator $A_0 + zA_1$ is assumed to have the eigenvalues $\lambda_i + z\mu_i$, $i = 1, 2, \ldots, N$.

Again, the numbers $\lambda_1, \ldots, \lambda_N$ are the eigenvalues of A_0. The numbers $\mu_i + z^{-1}\lambda_i$ are the eigenvalues of $A_1 + z^{-1}A_0$. Taking the limit $z \to \infty$, one obtains: The numbers $\mu_1, \ldots, \mu_N$ are the eigenvalues of A_1. Therefore, both formulations are equivalent. One obtains

Theorem 6. *Assume that Condition 1 is satisfied for the operators A_0, A_1. Moreover, let A_0 be a diagonal operator. Then (3.27) holds.*

Proof. Let $A_0 = \sum_{\varrho=1}^{r} \lambda_\varrho P_\varrho$ be the spectral representation of A_0. In order to achieve notations consistent with the beginning of Subsect. 7.3.4, the eigenvalues of $A_0 + zA_1$ are denoted by

$$\lambda_\varrho + z\mu_{\varrho\sigma}, \qquad \varrho = 1, 2, \ldots, r; \qquad \sigma = 1, 2, \ldots, s_\varrho.$$

Here, the set $\mu_{\varrho 1}, \mu_{\varrho 2}, \ldots, \mu_{\varrho s_\varrho}$ of eigenvalues of A_1 is assigned to the eigenvalue λ_ϱ of A_0. The relation (3.26) takes the form

$$\lambda_{\varrho\sigma}(z) = \lambda_\varrho + z\mu_{\varrho\sigma} \tag{3.28}$$

and the numbers $\mu_{\varrho\sigma}$ run through the spectrum of A_1 (note that here the numbers $\mu_{\varrho\sigma}$ need not be mutually different). Application of Lemma 3 yields the assertion. ∎

The next question is: how can Condition 1 be satisfied? Obviously, one has

Lemma 4. *If A_0 is a diagonal operator commuting with A_1, then Condition 1 is satisfied.*

Proof. Let $A_0 = \sum_{\varrho=1}^{r} \lambda_\varrho P_\varrho$ be the spectral representation of A_0. From $A_1A_0 = A_0A_1$, the relation $P_\varrho A_1 = A_1 P_\varrho$, $\varrho = 1, \ldots, r$ follows; hence, $P_\varrho(A_0 + zA_1) = (A_0 + zA_1) P_\varrho$, $\varrho = 1, \ldots, r$. One obtains

$$\chi(\lambda, A_0 + zA_1) = \prod_{\varrho=1}^{r} \chi\big(\lambda, (A_0 + zA_1) \restriction P_\varrho X\big).$$

That is, it is sufficient to consider $A_0 + zA_1$ on one of the spaces $P_\varrho X$. Without loss of generality, one can take $A_0 = \lambda_0 1$; A_1 is arbitrary. Then,

$$\det [\lambda - (\lambda_0 + zA_1)] = \det [(\lambda - \lambda_0) - zA_1] = z^N \det \left(\frac{\lambda - \lambda_0}{z} - A_1\right)$$
$$= z^N \prod_{\varrho=1}^{N} \left(\frac{\lambda - \lambda_0}{z} - \mu_\varrho\right) = \prod_{\varrho=1}^{N} \big(\lambda - (\lambda_0 + z\mu_\varrho)\big),$$

where the numbers $\mu_1, \mu_2, \ldots, \mu_N$ run through all eigenvalues of A_1 (the algebraic multiplicities of the eigenvalues are taken into account). ∎

Another sufficient condition for the validity of Condition 1 is the global diagonality of $A_0 + zA_1$. One has (see T. S. Motzkin and O. Taussky [2]; see T. Kato [5, p. 85])

Theorem 7. *Let $A_0 + zA_1$ be a diagonal operator for all complex numbers z; in particular, let $A_0 = \sum_{\varrho=1}^{r} \lambda_\varrho P_\varrho$ be the spectral representation of A_0. Then all perturbed eigenvalues are linear; i.e., they take the form*

$$\lambda_{\varrho\sigma}(z) = \lambda_\varrho + z\mu_{\varrho\sigma}, \qquad \varrho = 1, \ldots, r; \qquad \sigma = 1, 2, \ldots, s_\varrho. \tag{3.29}$$

(Here the numbers $\mu_{\varrho\sigma}$ are again mutually different for $\sigma = 1, \ldots, s_\varrho$, if ϱ is fixed.) The corresponding perturbed eigenprojections $P_{\varrho\sigma}(z)$ are entire functions of z.

Proof. Let $\mathfrak{P}$ denote the analytic function belonging to a perturbed eigenvalue $\lambda(z)$ (see Subsect. 3.2.3). $\mathfrak{P}$ is algebraic; there are at most finitely many branching points. If z is not the trace point of a branching point then $\lambda'(z)$ is well-defined. According to the assumption, at a trace point ζ of branching points, $A_0 + \zeta A_1$ is a diagonal operator. Hence, Theorem 2 and, in particular, Corollary 1 can be applied to $A_0 + \zeta A_1 + (z - \zeta) A_1$. This shows that each branch $\lambda(z)$ has a well-defined derivative at the point $z = \zeta$. Moreover, the function $\lambda'(z)$ is continuous in the neighborhood of $z = \zeta$.

In the neighborhood of $z = \infty$ one obtains the following behavior for the perturbed eigenvalues. One has

$$A_0 + zA_1 = z\left(A_1 + \frac{1}{z} A_0\right), \qquad z \neq 0.$$

The asymptotic behavior of perturbed eigenvalues $\mu(z)$ of $A_1 + z^{-1}A_0$ is given by the relation

$$\mu(z) = \mu + z^{-1/p}\mu_1 + o(z^{-1/p}), \qquad z \to \infty,$$

where μ is an eigenvalue of A_1 and where p denotes the period of the cycle corresponding to $\mu(z)$. For the eigenvalues $\lambda(z)$ of $A_0 + zA_1$, one obtains

$$\lambda(z) = \mu z + \mu_1 z^{1-\frac{1}{p}} + o\left(z^{1-\frac{1}{p}}\right), \qquad z \to \infty,$$

and

$$\lambda'(z) = \mu + \left(1 - \frac{1}{p}\right)\mu_1 z^{-1/p} + o(z^{-1/p}), \qquad z \to \infty.$$

Hence, the derivatives $\lambda'(z)$ are bounded in the neighborhood of $z = \infty$.

Applying Liouville's theorem to the function $\lambda'(z)$, which is holomorphic on $\mathfrak{P}$ including the branching points, and bounded at infinity, one obtains: $\lambda'(z)$ is a constant. Hence, (3.29) is valid. Note that $\mu_{\varrho\sigma} \neq \mu_{\varrho\sigma'}$, if $\sigma \neq \sigma'$.

Hence, the $(\lambda_\varrho + z\mu_{\varrho\sigma})$-group of $A_0 + zA_1$ consists of one single eigenvalue. Therefore, the group projection of this group is at the same time the eigenprojection $P_{\varrho\sigma}(z)$ belonging to the eigenvalue $\lambda_{\varrho\sigma}(z)$. Hence, $P_{\varrho\sigma}(z)$ is holomorphic at $z = 0$ (cf. also Corollary 3) and, correspondingly, holomorphic at every finite point z. ■

Moreover, one obtains

Theorem 8. *Let the assumptions of Theorem 7 be satisfied. If, additionally, the operator A_1 is diagonal, then A_0 and A_1 commute.*

Proof. Since A_1 is diagonal, Theorem 7 can be applied to $A_1 + z^{-1}A_0$. Therefore, the perturbed eigenprojections are entire functions with respect to z^{-1}. Since the eigenprojections of $A_1 + z^{-1}A_0$ and $A_0 + zA_1$ coincide for $z \neq 0$, one obtains: The perturbed eigenprojections are entire functions with respect to z as well as with respect to z^{-1}; in particular, they are holomorphic on the closed z-plane. Therefore, according to Liouville's theorem, the eigenprojections are constants; i.e., one has $P_{\varrho\sigma}(z) \equiv P_{\varrho\sigma}$. However, the eigenprojections $P_{\varrho\sigma}$ are subprojections of the eigenprojections of A_0 as well as of A_1. Since A_0 and A_1 are diagonal, this property implies that they commute. ■

Theorem 7 and 8 can be summarized in the following homogeneous form:

Theorem 9. *Let A and B be two operators such that $\alpha A + \beta B$ is diagonal for all, except possibly one, pairs (α, β), where α and β are complex numbers. Then the eigenvalues of $\alpha A + \beta B$ have the form $\alpha\lambda_i + \beta\mu_i$, where the numbers λ_i, μ_i are independent of α, β (the λ_i are the eigenvalues of A and the μ_i are the eigenvalues of B in a suitable ordering). If there is no exceptional pair, then A and B commute.*

7.3.5. Iteration of the reduction process

Next, we establish conditions under which the reduction process described in Subsect. 7.3.1 can be iterated. Since the main assumption for the application of the procedure is given by the semisimplicity of the unperturbed eigenvalue, each step of the process in turn requires this assumption. Recall that, in the case of a normal perturbation, this difficulty does not exist, because the reduction process, with respect to the auxiliary perturbations used by the process, does not exceed the class of normal perturbations. Hence, the condition of semisimplicity of the unperturbed eigenvalues is always satisfied (see Subsect. 7.2.3).

Let

$$A(z) = A_0 + zA_1 + z^2A_2 + \cdots$$

be the given perturbation. Let λ_0 be a semisimple unperturbed eigenvalue with the corresponding eigenprojection P_0 and let $P(z)$ denote the group projection of the λ_0-group. Let $\mu \neq 0$ denote one of the eigenvalues of B_0 restricted to P_0X; i.e., let $\chi(\mu, B_0 \upharpoonright P_0X) = 0$, where $B(z) = z^{-1}(A(z) - \lambda_0) P(z)$ denotes the auxiliary perturbation. Let $Q_\mu(z)$ denote the group projection of the μ-group of $B(z)$; i.e., $Q_\mu(z)$ is the group projection of the $(\lambda_0 + z\mu)$-group of $A(z)$.

If μ is a semisimple eigenvalue of B_0, then the reduction process can be applied to μ and to the perturbation $B(z)$ of B_0. This is the second step of the reduction procedure. In other words, in the following one assumes

$$\nu(\mu; B_0 \upharpoonright P_0X) = 1. \tag{3.30}$$

Let

$$B(z) = B_0 + zB_1 + z^2B_2 + \cdots. \tag{3.31}$$

The new auxiliary perturbation takes the form

$$C(z) = \frac{1}{z}\,(B(z) - \mu)\, Q_\mu(z),$$

and one obtains

$$C_0 = C(0) = Q_\mu B_1 Q_\mu.$$

The coefficients of the power series (3.31) can be calculated according to Subsect. 3.7.3. One obtains

$$B_1 = P_0A_2P_0 + P_0A_1P_0A_1S_0 + P_0A_1S_0A_1P_0 + S_0A_1P_0A_1P_0,$$

where S_0 denotes the reduced resolvent of λ_0 with respect to A_0. Because $P_0S_0 = S_0P_0 = 0$, in particular, $Q_\mu S_0 = S_0 Q_\mu = 0$, one obtains

$$C_0 = Q_\mu(A_2 + A_1S_0A_1)\, Q_\mu.$$

Let $\mu_2 \neq 0$ be an eigenvalue of C_0 restricted to $Q_\mu X$; i.e., let $\chi(\mu_2, C_0 \upharpoonright Q_\mu X) = 0$. The group projection of the μ_2-group of $C(z)$ is denoted by $R_2(z)$. Then,

$$R_2(z)\, Q_\mu(z) = Q_\mu(z)\, R_2(z) = R_2(z),$$

$$R_2(z)\, A(z) = A(z)\, R_2(z),$$

and from the $(\lambda_0 + z\mu)$-group of $A(z)$ an R_2-group can be selected. The eigenvalues of the R_2-group can be characterized as follows:

An eigenvalue $\lambda(z)$ of the $(\lambda_0 + z\mu)$-group belongs to the R_2-group if and only if

$$\lambda(z) = \lambda_0 + z\mu_1 + z^2\mu_2 + o(z^2), \qquad z \to 0, \qquad \mu_1 = \mu. \tag{3.32}$$

In this case, if $x(z)$ denotes a corresponding eigenvector, continuous at $z = 0$, where $x(0) = x_0 \neq 0$, then $R_2(z)\, x(z) = x(z)$ and, in particular, $R_2(0)\, x_0 = x_0$; i.e., one has $C_0 x_0 = \mu_2 x_0$.

Thus, for all eigenvalues $\lambda(z)$ of the $(\lambda_0 + z\mu)$-group of $A(z)$, asymptotic relations of the form (3.32) can be established, where μ_2 runs through the different eigenvalues of C_0 in $Q_\mu X$; in this way, one obtains the different R_2-groups.

Further considerations leading to the *m*th reduction step are omitted because no new perspective arises. In this context, the paper of P. Lancaster [2] is relevant.

Finally, it should be noted that the assumption "μ is a semisimple eigenvalue of B_0 in P_0X" is satisfied if

$$A_1 P_0 = \mu P_0; \tag{3.33}$$

i.e., if P_0X is an invariant subspace of A_1 and if A_1 restricted to P_0X reduces to the scalar operator μ. Then, $\chi(\lambda, B_0 \upharpoonright P_0X) = (\lambda - \mu)^{\dim P_0}$; that is, there is exactly one eigenvalue of B_0 restricted to P_0X and this eigenvalue is semisimple.

In this case, the first reduction step does not provide any splitting of the λ_0-group but, for all eigenvalues $\lambda(z)$ of the λ_0-group, the relation

$$\lambda(z) = \lambda_0 + z\mu + o(z), \qquad z \to 0,$$

holds and the second reduction step can be performed (possibly providing a splitting). Note that the method of F. Wolf [1] is based on the assumption (3.33) and on the corresponding assumptions for the higher reduction steps.

The calculation of the perturbation series for group projections and mean values of the $(\lambda_0 + z\lambda_\varrho)$-groups of $A(z)$ is omitted because it parallels that in Subsect. 7.2.4.

7.4. A reduction process for non-semisimple eigenvalues

Let $A(z)$ be a holomorphic perturbation of $A(0) = A_0$ and let λ_0 denote an unperturbed eigenvalue. If $v(\lambda_0, A_0) > 1$, then the reduction process described in § 3 is no longer applicable, because the operator-valued function

$$B(z) = \frac{1}{z}\left(A(z) - \lambda_0\right) P(z)$$

is not holomorphic at the point $z = 0$. In this general case, the problem to determine the Jordan structure of the perturbation $A(z)$ for $z \neq 0$ becomes very complicated.

The first results in this direction were obtained by M. I. Višik and L. A. Ljusternik [2] in the case of linear perturbations $A_0 + zA_1$. These authors start with a definite assumption on the cycle periods of the perturbed eigenvalues, namely that these periods coincide with the dimensions of the irreducible projections of A_0 with respect to λ_0.

Then the authors obtain sufficient conditions for the unique determination of the coefficients of the perturbation series formally established for eigenvalues and eigenvectors. In fact, they discuss the recursion formulas for the coefficients obtained from the eigenvalue equation $A(z)\, x(z) = \lambda(z)\, x(z)$ by comparison of corresponding terms. This discussion provides the sufficient conditions desired. These conditions guarantee a total splitting of λ_0 with respect to $A(z)$ at the first-order approximation. That is, if the eigenprojection P_0 of λ_0 with respect to A_0 has a Jordan decomposition consisting of r_1 irreducible projections of dimension m_1, r_2 projections of dimension $m_2, \ldots, r_p$ projections of dimension m_p, then the λ_0-group of $A(z)$ consists of r_1 cycles of period m_1, r_2 cycles of period $m_2, \ldots, r_p$ cycles of period m_p.

In 1966, V. B. Lidskii [1] presented results (with complete proofs) that extended these results. The results of M. I. Višik and L. A. Ljusternik are completely contained in Lidskii's results. Lidskii used matrix-theoretic methods and started from a transformation of the characteristic polynomial of $A(z)$.

Using another method, M. M. Vainberg and V. A. Trenogin [1] obtained the same results as M. I. Višik and L. A. Ljusternik.

In 1966, T. Sabirov also published results (without proofs) on this topic [1].

In this section, a method is presented which is based on a direct generalization of the reduction process in Sect. 7.3.

If $\nu(\lambda_0, A_0) = m_1$, then, in the first reduction step of this method, the central role is played by the operator-valued function

$$B_1(z) = \frac{1}{z}\left[\left(A(z) - \lambda_0\right) P(z)\right]^{m_1},$$

where $P(z)$ denotes the group projection of λ_0 with respect to $A(z)$. Functions of the form

$$B_\varrho(z) = \frac{1}{z}\left[\left(A(z) - \lambda_0\right) R(z)\right]^{m_\varrho}$$

have to be considered in further reduction steps if these steps are at all feasible. Here, $A(z)\, R(z) = R(z)\, A(z)$, $R(z)^2 = R(z)$ and $\nu(\lambda_0, A_0 \restriction R_0X) = m_\varrho$.

The results of V. B. Lidskii [1] are almost completely contained in these results. Moreover, they will be strengthened and refined considerably. Under the assumptions used by V. B. Lidskii, one obtains not only propositions on the perturbation series of eigenvalues and eigenvectors, but also propositions on decompositions of the group projection $P(z)$ into a sum of disjoint holomorphic reducing perturbation projections whose zero-order approximations have characteristic properties.

The results of M. I. Višik and L. A. Ljusternik are completely contained in the results announced.

In order to give some insight into the problems and results of this method, an essential theorem is first established.

Let $A(z) = A_0 + zA_1 + z^2A_2 + \cdots$ be a perturbation of A_0, let λ_0 be an eigenvalue of A_0 and let P_0 be the corresponding eigenprojection. P_0 can be decomposed into a sum of disjoint irreducible Jordan projections:

$$P_0 = \sum_{i=1}^{p} \sum_{j=1}^{r_i} P_i^j, \qquad \dim P_i^j = m_i, \qquad m_1 > m_2 > \cdots > m_p \geqq 1. \tag{4.1}$$

Let

$$P_i = \sum_{j=1}^{r_i} P_i^j, \qquad i = 1, \ldots, p. \tag{4.2}$$

The projection P_i collects those irreducible projections of A_0 with respect to λ_0 which have one and the same dimension m_i.

Construct a Jordan basis of P_0X adapted to the decomposition (4.1). Then the projections $E_1, E_2, \ldots, E_p$ are also well-defined, projecting onto the geometric eigenspaces of $P_1X, P_2X, \ldots, P_pX$, respectively. Furthermore, the subspaces $(E_1 + E_2 + \cdots + E_i) X$, $i = 1, 2, \ldots, p$, are independent of the chosen Jordan basis in P_0X. On the subspace $(E_1 + E_2 + \cdots + E_i) X$ one defines the operator

$$K_i = (E_1 D_0^{m_1-1} + E_2 D_0^{m_2-1} + \cdots + E_i D_0^{m_i-1}) A_1 \upharpoonright (E_1 + \cdots + E_i) X,$$

where $D_0 = (A_0 - \lambda_0) P_0$. Of course, K_i depends on the Jordan basis chosen in P_0X, but the *"pseudo-characteristic" polynomials*

$$\det (K_i - \lambda E_i), \qquad i = 1, 2, \ldots, p,$$

and, in particular, the numbers $\det K_i$, $i = 1, \ldots, p$, are independent of the Jordan basis. One has the following

Theorem. *Assume that*

$$\det K_1 \neq 0, \det K_2 \neq 0, \ldots, \det K_i \neq 0$$

is valid. Then there exists a decomposition of the group projection

$$P(z) = \sum_{\varrho=1}^{i} \sum_{\sigma=1}^{k_\varrho} Q_\varrho^\sigma(z) + R(z) \tag{4.3}$$

into a sum of holomorphic and, with respect to $A(z)$, reducing projections with

$$\dim Q_\varrho^\sigma = \gamma_\varrho^\sigma m_\varrho, \qquad \sigma = 1, 2, \ldots, k_\varrho.$$

If $Q_\varrho(z) = \sum_{\sigma=1}^{k_\varrho} Q_\varrho^\sigma(z)$, one has $P(z) = \sum_{\varrho=1}^{i} Q_\varrho(z) + R(z)$. In particular, at $z = 0$,

$$P_0 = \sum_{\varrho=1}^{i} Q_\varrho + R_0; \qquad Q_\varrho A_0 = A_0 Q_\varrho, \qquad R_0 A_0 = A_0 R_0;$$

where $Q_\varrho(0) = Q_\varrho$, $R(0) = R_0$. Q_ϱ is a reducing projection of A_0 which collects exactly the irreducible parts of fixed dimension m_ϱ. Moreover,

$$\nu(\lambda_0; A_0 \upharpoonright R_0X) = m_{i+1} < m_i.$$

The calculation of the projections $Q_\varrho^\sigma(z)$ proceeds by the following formulas: If

$$B_\varrho(z) = \frac{1}{z} \left[(A(z) - \lambda_0) \left(P(z) - \sum_{\sigma=1}^{\varrho-1} Q_\sigma(z) \right) \right]^{m_\varrho}$$

and

$$\det (K_\varrho - \lambda E_\varrho) = (-1)^{r_\varrho} \det K_{\varrho-1} \prod_{\sigma=1}^{k_\varrho} (\lambda - \mu_\varrho^\sigma)^{\gamma_\varrho^\sigma},$$

one obtains

$$Q_\varrho^\sigma(z) = \frac{1}{2\pi i} \int_{(\mu_\varrho^\sigma)} (\lambda - B_\varrho(z))^{-1} \,d\lambda.$$

Furthermore, it should be noted that the eigenvalues belonging to the Q_ϱ^σ-groups, corresponding to the decomposition (4.3), can be characterized asymptotically. Moreover, the zero-order approximations x_0 of corresponding eigenvectors must satisfy the equation

$$(K_\varrho - \mu_\varrho^\sigma E_\varrho)\, x_0 = 0.$$

The theorem just quoted will be derived in the course of this section from more general propositions. These propositions present sufficient conditions, such that the existence of a holomorphic reducing projection $Q(z)$, where $Q(0) = Q_0$ collects exactly those irreducible parts of A_0 whose dimensions are not less than m_i, implies the existence of a holomorphic reducing projection $R(z)$, where $R(0) = R_0$ satisfies exactly the same condition with respect to m_{i+1}. Then, the holomorphic reducing projection $R(z) - Q(z) = S(z)$ has the property that $S(0) = S_0$ collects exactly those irreducible parts of A_0 whose dimensions equal m_{i+1}.

7.4.1. Preparatory propositions

The following considerations are continuations of those in Sect. 2.9. Let A_0 be an operator, let λ_0 be an eigenvalue of A_0 and let P_0 be the corresponding eigenprojection. According to Sect. 2.9, P_0 can be decomposed, with respect to A_0, into a sum of disjoint reducing irreducible projections. By $m_1, m_2, \ldots, m_p$ one denotes the different dimensions of these projections. Let

$$m_1 > m_2 > \cdots > m_p, \qquad p \geqq 1.$$

By $r_1, r_2, \ldots, r_p$ one denotes the corresponding numbers of irreducible projections with fixed dimension. Therefore, one has r_i irreducible projections of dimension m_i. Then the decomposition takes the form

$$P_0 = \sum_{i=1}^{p} \sum_{j=1}^{r_i} P_i^j, \qquad \dim P_i^j = m_i.$$

If $\dim P_0 = n$, one has $n = \sum_{i=1}^{p} r_i m_i$. According to Sect. 2.9, a Jordan decomposition is not uniquely determined, but the numbers m_i and r_i are. Let

$$P_i = \sum_{j=1}^{r_i} P_i^j. \tag{4.4}$$

Then,

$$P_0 = \sum_{i=1}^{p} P_i. \tag{4.5}$$

The disjoint reducing projections P_i collect the irreducible parts of fixed dimension m_i "under a single covering". They are important for the holomorphic perturbations of A_0 and they have been used essentially, but not explicitly, by M. I. Višik and L. A. Ljusternik [2]. As usual, $\nu(\lambda_0, A_0)$ denotes the index of λ_0 with respect to A_0. One has $\nu(\lambda_0, A_0) = m_1$. For the corresponding eigennilpotent $D_0 = (A_0 - \lambda_0) P_0$, one obtains

$$D_0^{m_1} = 0, \qquad D_0^{m_1-1} \neq 0. \tag{4.6}$$

In the irreducible subspace $P_i^j X$ one chooses a Jordan basis $e_i^{j1}, e_i^{j2}, \ldots, e_i^{jm_i}$ such that

$$D_0 e_i^{j1} = 0, \qquad D_0 e_i^{j2} = e_i^{j1}, \ldots, D_0 e_i^{jm_i} = e_i^{jm_i-1}. \tag{4.7}$$

Hence, e_i^{j1} is an eigenvector (Jordan vector of first degree), e_i^{j2} is a Jordan vector of second degree, and, finally, $e_i^{jm_i}$ is a Jordan vector of m_ith degree. $e_i^{jm_i}$ is also called a generating or cyclic vector of $P_i^j X$. The choice of a Jordan basis $e_i^{j\varrho}$ corresponds to a decomposition of P_i^j into a sum of disjoint one-dimensional projections $P_i^{j\varrho}$:

$$P_i^j = \sum_{\varrho=1}^{m_i} P_i^{j\varrho}, \qquad \dim P_i^{j\varrho} = 1, \qquad P_i^{j\varrho} e_i^{j\varrho} = e_i^{j\varrho}. \tag{4.8}$$

Of course, the projections $P_i^{j\varrho}$ are no longer reducing with respect to A_0. The Jordan bases $e_i^{j\varrho}$, $i = 1, \ldots, p$; $j = 1, \ldots, r_i$; $\varrho = 1, \ldots, m_i$, together form a Jordan basis of A_0 in $P_0 X$.

Each vector e_i^{j1} is an eigenvector of A_0 with eigenvalue λ_0. The projection

$$E = \sum_{i=1}^{p} \sum_{j=1}^{r_i} P_i^{j1}$$

projects onto the geometric eigenspace of λ_0. As is well-known, this subspace EX is completely independent of the particular Jordan basis (but the projection E is not). Let

$$E_i = \sum_{j=1}^{r_i} P_i^{j1}, \qquad i = 1, \ldots, p.$$

Then,

$$E = \sum_{i=1}^{p} E_i.$$

That is, a Jordan basis given in $P_0 X$ implies a direct decomposition of the geometric eigenspace EX according to

$$EX = E_1 X \oplus E_2 X \oplus \cdots \oplus E_p X.$$

The subspace $E_i X$ collects exactly those eigenvectors which come from the irreducible parts of fixed dimension m_i. In general, these subspaces $E_i X$ depend on the particular Jordan basis in $P_0 X$. The following lemma is true.

Lemma 1. *The subspaces $(E_1 + E_2 + \cdots + E_i)\,X$ are independent of the particular Jordan basis in P_0X for every $i = 1, 2, \ldots, p$. That is, these subspaces are invariants of A_0. If $E'_1, E'_2, \ldots, E'_p$ denote the correspondingly defined projections with respect to a second Jordan basis, then*

$$(E_1 + E_2 + \cdots + E_i)\,X = (E'_1 + E'_2 + \cdots + E'_i)\,X, \qquad i = 1, 2, \ldots, p,$$

holds.

Proof. Let $x \in (E_1 + E_2 + \cdots + E_i)\,X$; i.e.,

$$x = \sum_{\varrho=1}^{i} \sum_{\sigma=1}^{r_\varrho} \alpha_\varrho^\sigma e_\varrho^{\sigma 1} = \sum_{\varrho=1}^{i} D_0^{m_\varrho - 1} \sum_{\sigma=1}^{r_\varrho} \alpha_\varrho^\sigma e_\varrho^{\sigma m_\varrho} = D_0^{m_i - 1} \left(\sum_{\varrho=1}^{i} D_0^{m_\varrho - m_i} \sum_{\sigma=1}^{r_\varrho} \alpha_\varrho^\sigma e_\varrho^{\sigma m_\varrho} \right).$$

If

$$y = \sum_{\varrho=1}^{i} D_0^{m_\varrho - m_i} \sum_{\sigma=1}^{r_\varrho} \alpha_\varrho^\sigma e_\varrho^{\sigma m_\varrho},$$

one obtains

$$x = D_0^{m_i - 1} y, \qquad y \in P_0X.$$

Choose a second Jordan basis. The projections defined by this second Jordan basis are indicated by a prime. That is, one has the projections $P'_1, \ldots, P'_p$, $P_1'^1, \ldots, P_p'^{r_p}$, $E'_1, \ldots, E'_p$, E'. Because $E'X = EX$, one has $x \in E'X$; i.e.,

$$x = \sum_{\varrho=1}^{i} E'_\varrho x + \sum_{\varrho=i+1}^{p} E'_\varrho x.$$

For the term $\sum_{\varrho=1}^{i} E'_\varrho x$ one obtains, as above, a representation of the form

$$\sum_{\varrho=1}^{i} E'_\varrho x = D_0^{m_i - 1} y', \qquad y' \in P_0X.$$

Hence,

$$D_0^{m_i - 1}(y - y') = \sum_{\varrho=i+1}^{p} E'_\varrho x.$$

For short, let $y - y' = u$. Then $u \in P_0X$ and

$$u = P'_1 u + P'_2 u + \cdots + P'_p u. \tag{4.9}$$

According to the definition of the projections $P'_1, P'_2, \ldots, P'_p$, one has

$$\nu(\lambda_0; A_0 \restriction P'_\varrho X) = m_\varrho, \qquad \varrho = 1, \ldots, p;$$

i.e.,

$$D_0^{m_\varrho} P'_\varrho = 0, \qquad \varrho = 1, \ldots, p.$$

In particular,

$$D_0^{m_i - 1} P'_\varrho = 0, \qquad \varrho = i + 1, i + 2, \ldots, p, \tag{4.10}$$

because $m_p < m_{p-1} < \cdots < m_{i+1} \leqq m_i - 1$. Applying $D_0^{m_i-1}$ to (4.9), one obtains

$$D_0^{m_i-1}u = P_1'(D_0^{m_i-1}u) + P_2'(D_0^{m_i-1}u) + \cdots + P_i'(D_0^{m_i-1}u). \tag{4.11}$$

Moreover,

$$D_0^{m_i-1}u = E_{i+1}'x + E_{i+2}'x + \cdots + E_p'x.$$

That is, on the one hand,

$$D_0^{m_i-1}u \in (P_1' + P_2' + \cdots + P_i')\,X$$

and, on the other hand,

$$D_0^{m_i-1}u \in (P_{i+1}' + P_{i+2}' + \cdots + P_p')\,X.$$

This implies $D_0^{m_i-1}u = 0$; i.e., one obtains

$$\sum_{\varrho=i+1}^{p} E_\varrho' x = 0;$$

hence,

$$x \in (E_1' + E_2' + \cdots + E_i')\,X.$$

Interchanging the roles of $E_1, \ldots, E_p$ and $E_1', \ldots, E_p'$, one obtains the assertion. ■

In Sect. 2.10, the operators C commuting with an operator A_0 have been characterized. In particular, if C commutes with A_0, then the eigenprojections P_0 of A_0 also commute with C. That is, all properties of operators commuting with A_0 can be studied in terms of the nilpotent operator $A_0 - \lambda_0$ in P_0X.

In the present connection, the following lemma is of interest.

Lemma 2. *Let A_0 be an operator, let λ_0 be an eigenvalue of A_0 and let P_0 be the corresponding eigenprojection. Assume that the projections $E_1, E_2, \ldots, E_p$ are defined as above with respect to a given Jordan basis of A_0 in P_0X. Assume $A_0C = CA_0$. Then the subspaces $(E_1 + E_2 + \cdots + E_i)\,X$ are invariant subspaces of C for every $i = 1, 2, \ldots, p$.*

Proof. It is sufficient to prove this lemma for a nilpotent operator $D_0 = A_0 - \lambda_0$ in X. As above, the unit operator 1 is first decomposed into a sum of disjoint, with respect to D_0 reducing, projections, $1 = \sum_{i=1}^{p} P_i$, where again the projections P_i collect the irreducible parts of fixed dimension m_i. Again $m_1 > m_2 > \cdots > m_p \geqq 1$ is assumed. First the assertion is true for $i = p$: EX is an invariant subspace with respect to C, for, if $x \in EX$, i.e., $D_0x = 0$, then $D_0Cx = CD_0x = C0 = 0$; hence, $Cx \in EX$. Let $x \in (E_1 + \cdots + E_i)\,X$. This means

$$x = E_1x + E_2x + \cdots + E_ix.$$

One has

$$E_\varrho x = D_0^{m_\varrho-1}u_\varrho, \qquad u_\varrho \in P_\varrho X.$$

Therefore,

$$x = \sum_{\varrho=1}^{i} E_\varrho x = \sum_{\varrho=1}^{i} D_0^{m_\varrho-1} u_\varrho = D_0^{m_i-1} \sum_{\varrho=1}^{i} D_0^{m_\varrho-m_i} u_\varrho. \tag{4.12}$$

Furthermore,

$$Cx = E_1 Cx + E_2 Cx + \cdots + E_p Cx.$$

Since the relation

$$E_\varrho Cx = D_0^{m_\varrho-1} v_\varrho, \qquad v_\varrho \in P_\varrho X,$$

holds,

$$Cx = \sum_{\varrho=1}^{p} D_0^{m_\varrho-1} v_\varrho. \tag{4.13}$$

From (4.12), by application of C, the equation

$$Cx = CD_0^{m_i-1} \sum_{\varrho=1}^{i} D_0^{m_\varrho-m_i} u_\varrho = D_0^{m_i-1} \sum_{\varrho=1}^{i} D_0^{m_\varrho-m_i} Cu_\varrho$$

follows. Together with (4.13), one obtains

$$D_0^{m_i-1}\left[\sum_{\varrho=1}^{i} D_0^{m_\varrho-m_i} Cu_\varrho - \sum_{\varrho=1}^{i} D_0^{m_\varrho-m_i} v_\varrho\right] = \sum_{\varrho=i+1}^{p} D_0^{m_\varrho-1} v_\varrho.$$

Replacing, for short the vector within the brackets by y, one obtains the relation

$$D_0^{m_i-1} y = \sum_{\varrho=i+1}^{p} E_\varrho Cx.$$

Therefore, on the one hand one has

$$D_0^{m_i-1} y \in (P_{i+1} + P_{i+2} + \cdots + P_p)\, X$$

and on the other hand (cf. (4.11) in the proof of Lemma 1)

$$D_0^{m_i-1} y = P_1(D_0^{m_i-1} y) + P_2(D_0^{m_i-1} y) + \cdots + P_i(D_0^{m_i-1} y);$$

that is,

$$D_0^{m_i-1} y \in (P_1 + P_2 + \cdots + P_i)\, X.$$

This implies $D_0^{m_i-1} y = 0$; i.e.,

$$\sum_{\varrho=i+1}^{p} E_\varrho Cx = 0;$$

therefore, $Cx \in (E_1 + E_2 + \cdots + E_i)\, X$. ∎

Lemma 3. *Let D be nilpotent in X. By $P_1, P_2, \ldots, P_p$ one denotes the reducing projections of D, adapted to a fixed Jordan basis, which collect the irreducible parts of fixed dimension $m_1, m_2, \ldots, m_p$, respectively. Furthermore, let C commute with D. Then*

$$\det C = \prod_{i=1}^{p} \det (P_i C P_i \upharpoonright P_i X). \tag{4.14}$$

Lemma 4. *Let D be nilpotent in X. The irreducible projections of D are assumed to have a fixed dimension m (i.e., let $p = 1$ and $m_1 = m$). Furthermore, let C commute with D. Then*

$$\det C = [\det (C \upharpoonright EX)]^m \tag{4.15}$$

holds.

Both Lemmas 3 and 4 are proved by an auxiliary theorem from matrix theory.

Theorem 1. *Let the $n \times n$-matrix Δ be an (upper) Jordan normal form. First let Δ be a matrix, formed by diagonal blocks, of the form*

$$\Delta = \{\Lambda_1, \Lambda_2, \ldots, \Lambda_p\}.$$

Furthermore, let the matrix Λ_α, $\alpha = 1, 2, \ldots, p$, be an $r_\alpha m_\alpha \times r_\alpha m_\alpha$-matrix of the form

$$\Lambda_\alpha = (\delta_{ij} \Gamma_{m_\alpha}), \qquad i, j = 1, \ldots, r_\alpha; \qquad \Gamma_{m_\alpha} = \Gamma_{m_\alpha}(0),$$

where Γ_{m_α} denotes a Jordan block of dimension m_α. That is, on the main diagonal of the matrix Λ_α one has r_α Jordan blocks of dimension m_α. Let

$$m_1 > m_2 > \cdots > m_p \geqq 1.$$

Furthermore, let an $n \times n$-matrix A be given whose elements are written, in accordance with the foregoing notation, in the form

$$a\begin{pmatrix} \alpha & i & \mu \\ \beta & j & \nu \end{pmatrix}, \qquad 1 \leqq \alpha, \beta \leqq p; \qquad \begin{matrix} 1 \leqq i \leqq r_\alpha, & 1 \leqq \mu \leqq m_\alpha, \\ 1 \leqq j \leqq r_\beta, & 1 \leqq \nu \leqq m_\beta. \end{matrix}$$

Assume that A commutes with Δ. Then

$$\det A = \prod_{\alpha=1}^{p} \prod_{\mu=1}^{m_\alpha} \det \left[a\begin{pmatrix} \alpha & i & \mu \\ \alpha & j & \mu \end{pmatrix} \right]. \tag{4.16}$$

Proof. First, let

$$A_{\alpha\beta} = \left(a\begin{pmatrix} \alpha & i & \mu \\ \beta & j & \nu \end{pmatrix} \right), \qquad \alpha, \beta \text{ fixed},$$

and

$$A_{\alpha\beta}^{ij} = \left(a\begin{pmatrix} \alpha & i & \mu \\ \beta & j & \nu \end{pmatrix} \right), \qquad \alpha, \beta, i, j \text{ fixed}.$$

Then A is a matrix formed by blocks $A_{\alpha\beta}$ which in turn are formed by subblocks $A_{\alpha\beta}^{ij}$.

According to Sect. 2.10, because Δ and $\boldsymbol{A}$ commute, the blocks $\boldsymbol{A}^{ij}_{\alpha\beta}$ are upper-triangular matrices whose upper-diagonal elements are equal. For the proof of the present theorem one uses only the property that the blocks $\boldsymbol{A}^{ij}_{\alpha\beta}$ are upper-triangular. (One additionally needs the second property to prove Lemma 4; this property leads to an additional simplification of formula (4.16).)

The property that the block $\boldsymbol{A}^{ij}_{\alpha\beta}$ is upper-triangular means that the elements have the following properties:

1. One has $a\begin{pmatrix}\alpha & i & \mu\\ \alpha & j & \nu\end{pmatrix} = 0$ if $\mu > \nu$, where i and j are arbitrary.

2. If $\alpha < \beta$, then $a\begin{pmatrix}\alpha & i & \mu\\ \beta & j & \nu\end{pmatrix} = 0$, provided that $\mu > \nu$, where i and j are arbitrary.

3. If $\alpha > \beta$, then $a\begin{pmatrix}\alpha & i & \mu\\ \beta & j & \nu\end{pmatrix} = 0$, provided that $\mu + (m_\beta - m_\alpha) > \nu$, where i and j are arbitrary.

Consider permutations π of the triples (α, i, μ). Let

$$\pi(\alpha, i, \mu) = \big(\pi_1(\alpha, i, \mu), \pi_2(\alpha, i, \mu), \pi_3(\alpha, i, \mu)\big).$$

Because of the properties 1, 2 and 3, to form the determinant of $\boldsymbol{A}$, one only needs to take into consideration the permutations π with the following property:

For each triple one has either

A. $\quad \pi_1(\alpha, i, \mu) \geqq \alpha \quad$ *and* $\quad \pi_3(\alpha, i, \mu) \geqq \mu,$

or

B. $\quad \pi_1(\alpha, i, \mu) < \alpha \quad$ *and* $\quad \pi_3(\alpha, i, \mu) \geqq \mu + (m_{\pi_1(\alpha,i,\mu)} - m_\alpha).$

In the following, one considers only permutations π with this property. For these permutations, one obtains

$$\pi_1(\alpha, i, \mu) = \alpha, \qquad \pi_3(\alpha, i, \mu) = \mu \tag{4.17}$$

for arbitrary triples (α, i, μ). To prove (4.17), assume that there exists a triple (α, i, μ) such that (4.17) is not true. That is, one has $\pi_1(\alpha, i, \mu) \neq \alpha$ or $\pi_3(\alpha, i, \mu) \neq \mu$. Either proposition leads to a contradiction.

First, let $\pi_3(\alpha, i, \mu) = \mu_1 \neq \mu$. Then, because of A and B, one necessarily has $\mu_1 > \mu$. Consider the permutation cycle generated by (α, i, μ) with respect to π. Let

$$\pi(\alpha, i, \mu) = (\alpha_1, i_1, \mu_1), \ldots, \pi^k(\alpha, i, \mu) = (\alpha_k, i_k, \mu_k).$$

Then, again because of A and B

$$\mu < \mu_1 \leqq \mu_2 \leqq \mu_3 \leqq \cdots \leqq \mu_k.$$

Therefore, the relation

$$\mu < \mu_k, \qquad k = 1, 2, 3, \ldots,$$

is always valid, that is, no k exists such that $\pi_3^k(\alpha, i, \mu) = \mu$. However, this is a contradiction. Therefore,

$$\pi_3(\alpha, i, \mu) = \mu, \qquad (\alpha, i, \mu) \text{ arbitrary}. \tag{4.18}$$

Secondly, let $\pi_1(\alpha, i, \mu) \neq \alpha$. Then $\pi_1(\alpha, i, \mu) = \alpha_1 > \alpha$, for otherwise, if $\alpha_1 < \alpha$, according to B, $\pi_3(\alpha, i, \mu) \geqq \mu + (m_{\alpha_1} - m_\alpha) \geqq \mu + 1$, which contradicts (4.18). If $\pi^k(\alpha, i, \mu) = (\alpha_k, i_k, \mu_k)$, then

$$\alpha < \alpha_1 \leqq \alpha_2 \leqq \cdots \leqq \alpha_k$$

for $k = 1, 2, 3, \ldots$, since the case B is always excluded, because of (4.18). This again contradicts the existence of a cycle with respect to the triple (α, i, μ). Therefore, (4.17) has been proved.

A permutation π equipped with the property (4.17) determines uniquely $\sum_{\alpha=1}^{p} m_\alpha$ permutations $\pi_{\alpha,\mu}$, where $\pi_{\alpha,\mu}$ is a permutation of the numbers $1, 2, \ldots, r_\alpha$:

$$\pi(\alpha, i, \mu) = \big(\alpha, \pi_{\alpha,\mu}(i), \mu\big). \tag{4.19}$$

Conversely, from $\sum_{\alpha=1}^{p} m_\alpha$ permutations $\pi_{\alpha,\mu}$, where $\pi_{\alpha,\mu}$ is a permutation of the numbers $1, 2, \ldots, r_\alpha$, one obtains a unique permutation π via (4.19) with the property (4.17), where

$$\operatorname{sgn} \pi = \prod_{\alpha=1}^{p} \prod_{\mu=1}^{m_\alpha} \operatorname{sgn} \pi_{\alpha,\mu}.$$

Hence,

$$\det \left[a \begin{pmatrix} \alpha & i & \mu \\ \beta & j & \nu \end{pmatrix} \right]$$

$$= \sum_\pi \operatorname{sgn} \pi \prod_{(\alpha,i,\mu)} a \begin{pmatrix} \alpha & i\ \mu \\ \pi(\alpha, i, \mu) \end{pmatrix} = \sum_\pi \operatorname{sgn} \pi \prod_{(\alpha,i,\mu)} a \begin{pmatrix} \alpha & i & \mu \\ \alpha & \pi_{\alpha,\mu}(i) & \mu \end{pmatrix}$$

$$= \sum_\pi \Bigg(\prod_{(\alpha,\mu)} \operatorname{sgn} \pi_{\alpha,\mu} \Bigg) \prod_{(\alpha,\mu)} \prod_i a \begin{pmatrix} \alpha & i & \mu \\ \alpha & \pi_{\alpha,\mu}(i) & \mu \end{pmatrix}$$

$$= \sum_\pi \prod_{\alpha,\mu} \operatorname{sgn} \pi_{\alpha,\mu} \prod_i a \begin{pmatrix} \alpha & i & \mu \\ \alpha & \pi_{\alpha,\mu}(i) & \mu \end{pmatrix} = \prod_{\alpha,\mu} \sum_{\pi_{\alpha,\mu}} \operatorname{sgn} \pi_{\alpha,\mu} \prod_i a \begin{pmatrix} \alpha & i & \mu \\ \alpha & \pi_{\alpha,\mu}(i) & \mu \end{pmatrix}$$

$$= \prod_{\alpha=1}^{p} \prod_{\mu=1}^{m_\alpha} \det \left[a \begin{pmatrix} \alpha & i & \mu \\ \alpha & j & \mu \end{pmatrix} \right].$$

This proves the theorem. ■

Because Δ and $\boldsymbol{A}$ commute one has, in particular,

$$\Lambda_\alpha \boldsymbol{A}_{\alpha\alpha} = \boldsymbol{A}_{\alpha\alpha} \Lambda_\alpha, \qquad \alpha = 1, \ldots, p.$$

That is, the block $A_{\alpha\alpha}$ commutes with the block Λ_α. Therefore, Theorem 1 can be applied to this case, which yields the equation

$$\det A_{\alpha\alpha} = \prod_{\mu=1}^{m_\alpha} \det \left[a \begin{pmatrix} \alpha & i & \mu \\ \alpha & j & \mu \end{pmatrix} \right]. \tag{4.20}$$

Therefore, the relation (4.16) can be written in the form

$$\det A = \prod_{\alpha=1}^{p} \det A_{\alpha\alpha}. \tag{4.21}$$

Proof of Lemma 3. First, one introduces a fixed Jordan basis with respect to D in X. Secondly, the projections, adapted to the Jordan basis, reducing with respect to D, which collect the irreducible parts of fixed dimensions $m_1, m_2, \ldots, m_p$, are denoted by $P_1, P_2, \ldots, P_p$, respectively. Then one can make the following statement: To the operator $P_\alpha C P_\beta$ on $P_\beta X$ corresponds the block $A_{\alpha\beta}$ of the matrix A, with respect to the particular Jordan basis, where A denotes the matrix corresponding to C in X with respect to the particular Jordan basis. In particular, to the operator $P_\alpha C P_\alpha$ on $P_\alpha X$ corresponds the block $A_{\alpha\alpha}$. One has

$$\det (P_\alpha C P_\alpha \upharpoonright P_\alpha X) = \det A_{\alpha\alpha}.$$

Because of (4.21), one obtains the assertion. ■

Proof of Lemma 4. Since $p = 1$, one can omit the index α. Let A denote the matrix of C on X with respect to a fixed Jordan basis and let A^{ij}, $j = 1, \ldots, r$, denote the blocks of A formed in accordance with the proof of Theorem 1, which are $m \times m$-matrices for all i, j,

$$A^{ij} = \left(a \begin{pmatrix} i & \mu \\ j & \nu \end{pmatrix} \right), \qquad i, j \text{ fixed}; \qquad \mu, \nu = 1, 2, \ldots, m.$$

Then, Theorem 1 yields the relation

$$\det A = \prod_{\mu=1}^{m} \det \left[a \begin{pmatrix} i & \mu \\ j & \mu \end{pmatrix} \right]. \tag{4.22}$$

Next, the numbers $a \begin{pmatrix} i & \mu \\ j & \mu \end{pmatrix}$, $\mu = 1, 2, \ldots, m$, form the elements of the main diagonal of the block A^{ij}. According to Sect. 2.10, these numbers coincide:

$$a \begin{pmatrix} i & \mu \\ j & \mu \end{pmatrix} = a \begin{pmatrix} i & 1 \\ j & 1 \end{pmatrix} = a_{ij}, \qquad \mu = 1, 2, \ldots, m.$$

Hence, from (4.22), one obtains the relation

$$\det A = [\det (a_{ij})]^m.$$

Remains to prove that the matrix (a_{ij}) coincides with the matrix of C on EX. It has been proved in Lemma 2 that EX is invariant with respect to C. According to the definition of the matrix A, one has

$$Ce^{i\varrho} = \sum_{j,\sigma} a\begin{pmatrix} j & \sigma \\ i & \varrho \end{pmatrix} e^{j\sigma},$$

where $e^{i\varrho}$, $i = 1, \ldots, r$; $\varrho = 1, \ldots, m$, denotes the Jordan basis which has been chosen in X. In particular,

$$Ce^{i1} = \sum_{j,\sigma} a\begin{pmatrix} j & \sigma \\ i & 1 \end{pmatrix} e^{j\sigma}.$$

Since A^{ij} is an upper-triangular matrix, one obtains

$$a\begin{pmatrix} j & \sigma \\ i & 1 \end{pmatrix} = 0, \qquad \sigma > 1;$$

hence,

$$Ce^{i1} = \sum_{j=1}^{r} a\begin{pmatrix} j & 1 \\ i & 1 \end{pmatrix} e^{j1} = \sum_{j=1}^{r} a_{ji}e^{j1},$$

which proves the assertion. ∎

Note that Theorem 1 can also be formulated in a more operator-theoretic form:

Theorem 2. *Let D be nilpotent in X and let $P_1, P_2, \ldots, P_p$ denote reducing projections of D, adapted to a fixed Jordan basis and collecting the irreducible parts of D of fixed dimension $m_1, m_2, \ldots, m_p$, respectively. Furthermore, let $E_1, E_2, \ldots, E_p$ denote the corresponding projections onto those geometric eigenspaces belonging to the dimensions $m_1, \ldots, m_p$. Finally, let C be an operator which commutes with D. Then,*

$$\det C = \prod_{i=1}^{p} [\det (P_iCP_i \restriction E_iX)]^{m_i}. \tag{4.23}$$

Note that, in general, E_iX is not an invariant subspace of C. E_iX is an invariant subspace only with respect to the operator P_iCP_i (cf. Lemma 2). Theorem 2 contains the propositions of Lemma 3 and Lemma 4 as special cases.

7.4.2. Application

This subsection contains a first application of the determinant theorems proved in Subsect. 7.4.1.

Theorem 3. *Let $A(z) = A_0 + zA_1$ be a linear perturbation of A_0. Assume that $A_0A_1 = A_1A_0$ is satisfied. Furthermore, let λ_0 be an unperturbed eigenvalue. Then the eigen-*

values of the λ_0-group of $A(z)$ have the following form:

$$\lambda_{i\varrho}(z) = \lambda_0 + z\mu_{i\varrho}, \qquad i = 1, 2, \ldots, p; \qquad \varrho = 1, 2, \ldots, r_i, \tag{4.24}$$

where the numbers $\mu_{i\varrho}$ denote the eigenvalues of $P_iA_1P_i$ restricted to E_iX (their algebraic multiplicities taken into account). Note that P_i and E_i are defined as before, with respect to A_0. Let P_0 denote the eigenprojection of λ_0 with respect to A_0. Then P_0 is simultaneously the group projection of the λ_0-group of $A(z)$.

Proof. Since $P_0A_1 = A_1P_0$, $P_0A(z) = A(z)\,P_0$. Hence, the perturbation $A(z)$ can be treated in P_0X and $(1 - P_0)\,X$ separately. With respect to P_0X, the number λ_0 is the only unperturbed eigenvalue, λ_0 is a regular point of A_0 with respect to $(1 - P_0)\,X$. Hence, $P_0P(z) = P(z)\,P_0$, where $P(z)$ denotes the group projection of the λ_0-group. Because $(1 - P_0)\,P(z) = P(z)\,(1 - P_0) = 0$, the equation $P_0P(z) = P(z)$ follows and, since $\dim P_0 = \dim P(z) = n$, $P_0 \equiv P(z)$. The operator A_0 restricted to P_0X satisfies $A_0 = \lambda_0 + D_0$. The operator $\lambda_0 + D_0 + zA_1$ commutes with D_0, hence, $(\lambda - \lambda_0) - (D_0 + zA_1)$ also commutes with D_0. Therefore, according to Lemma 3 and Theorem 2, one obtains

$$\det\,[\lambda - \lambda_0 - (D_0 + zA_1)] = \prod_{i=1}^{p} \det\left[(\lambda - \lambda_0 - P_i(D_0 + zA_1)\,P_i) \upharpoonright P_iX\right];$$

hence,

$$\chi(\lambda, A(z)) = \prod_{i=1}^{p} [\det\,\{(\lambda - \lambda_0 - P_i(D_0 + zA_1)\,P_i) \upharpoonright E_iX\}]^{m_i},$$

where P_0X is replaced by X and where $A_0 = \lambda_0 + D_0$ is assumed to hold. Then $P_iD_0P_i$ restricted to E_iX is the zero operator; hence,

$$\chi(\lambda, A(z)) = \prod_{i=1}^{p} [\det\,\{(\lambda - \lambda_0 - zP_iA_1P_i) \upharpoonright E_iX\}]^{m_i}$$

$$= z^n \prod_{i=1}^{p} \left[\det\left\{\left(\frac{\lambda - \lambda_0}{z} - P_iA_1P_i\right) \upharpoonright E_iX\right\}\right]^{m_i}.$$

If $z^{-1}(\lambda - \lambda_0) = \mu$, then,

$$\chi(\lambda, A(z)) = z^n \prod_{i=1}^{p} [\det\,\{(\mu - P_iA_1P_i) \upharpoonright E_iX\}]^{m_i}.$$

One has

$$\det\,[(\mu - P_iA_1P_i) \upharpoonright E_iX] = \chi(\mu, P_iA_1P_i \upharpoonright E_iX).$$

Denoting the eigenvalues of $P_iA_1P_i$ with respect to E_iX by $\mu_{i1}, \mu_{i2}, \ldots, \mu_{ir_i}$ (their algebraic multiplicity taken into account), one finally obtains

$$\chi(\lambda, A(z)) = \prod_{i=1}^{p} \left[\prod_{\varrho=1}^{r_i} (\lambda - \lambda_0 - z\mu_{i\varrho})\right]^{m_i}.$$

This proves the theorem. ∎

7.4.3. Preparatory propositions (continuation)

In order to introduce the reduction process, the following lemma is also needed.

Lemma 5. *Let K be nilpotent in X; i.e., let $K^l = 0$ for some l. Furthermore, let A be an arbitrary operator and let S be defined by*

$$S = \sum_{\varrho+\sigma=m-1} K^\varrho A K^\sigma, \qquad m > l. \tag{4.25}$$

Then S is also nilpotent; one has $S^l = 0$.

Proof. In (4.25), only those terms are different from zero whose exponents ϱ and σ satisfy $0 \leqq \varrho \leqq l-1$ and $0 \leqq \sigma \leqq l-1$. This means $0 \leqq m-1-\varrho \leqq l-1$ or $\varrho \geqq m-l$. That is, one obtains the following restriction for the index ϱ:

$$m - l \leqq \varrho \leqq l - 1.$$

If $m - l > l - 1$, then S is zero. Consider a sum

$$S = \sum_{\alpha,\beta} K^\alpha A K^\beta, \qquad \alpha \geqq 1, \qquad \beta \geqq 1.$$

Next, consider

$$K^\alpha A K^{\beta+\varrho} A K^{m-1-\varrho}.$$

If $\beta + \varrho \geqq l$, then the term vanishes. Therefore, let $\beta + \varrho < l$ or $-\beta - \varrho > -l$; i.e. $m - 1 - \varrho > m - 1 - l + \beta$; i.e., $m - 1 - \varrho \geqq m - l + \beta \geqq \beta + 1 \geqq 2$. In particular,

$$S^2 = \sum_{\alpha,\gamma} K^\alpha A_{\alpha\gamma} K^\gamma, \qquad \alpha \geqq 1, \qquad \gamma \geqq 2.$$

Furthermore, consider product terms of the form

$$K^\alpha A_{\alpha\gamma} K^{\gamma+\varrho} A K^{m-1-\varrho}.$$

If $\gamma + \varrho \geqq l$, then the term vanishes. Therefore, let $\gamma + \varrho < l$, i.e. $-\gamma - \varrho > -l$. Then, $m - 1 - \varrho > m - 1 - l + \gamma$, i.e., one has $m - 1 - \varrho \geqq m - l + \gamma \geqq \gamma + 1 \geqq 3$, that is,

$$S^3 = \sum_{\alpha,\gamma} K^\alpha A_{\alpha\gamma} K^\gamma, \qquad \alpha \geqq 1, \qquad \gamma \geqq 3.$$

After finitely many steps, one obtains $S^l = 0$. ■

7.4.4. Reduction operator and its properties

After the preparations collected in Subsect. 7.4.1 and 7.4.3, one can generalize the reduction process for non-semisimple unperturbed eigenvalues. A central role in this reduction process is played by the "reduction operator" whose properties are derived in this subsection.

Note that the reduction process will be dealt with on an arbitrary "level", i.e., for an arbitrary dimension of the irreducible parts of the unperturbed eigenvalue.

Let $A(z) = A_0 + zA_1 + z^2A_2 + \cdots$ be a holomorphic perturbation of A_0 and let λ_0 be an unperturbed eigenvalue with eigenprojection P_0. Let $P(z)$ be the group projection of the λ_0-group.

First, the existence of a holomorphic reducing projection $R(z)$ of $A(z)$ which is a subprojection of $P(z)$, is assumed. That is, the relations

$$R(z)\,A(z) = A(z)\,R(z); \qquad R(z)\,P(z) = P(z)\,R(z) = R(z), \qquad |z| < \delta, \tag{4.26}$$

are assumed to be valid; in particular, at $z = 0$, one obtains

$$R_0A_0 = A_0R_0, \qquad R_0P_0 = P_0R_0 = R_0. \tag{4.27}$$

Such projections always exist; for example, one can take $R(z) = P(z)$.

The index of λ_0 in R_0X is denoted by m; i.e., one has

$$\nu(\lambda_0, A_0 \upharpoonright R_0X) = m \leqq \nu(\lambda_0, A_0 \upharpoonright P_0X). \tag{4.28}$$

For the sake of completeness, a further assumption with respect to R_0 is noted, which will play a role in Subsect. 7.4.7, but which will not be used in Subsect. 7.4.4 to 7.4.6:

$Q_0 = P_0 - R_0$ reduces A_0. Denoting by d the dimension of an irreducible part of A_0 restricted to Q_0X, one requires $d > m$.

The reduction operator $B(z)$ is defined by

$$B(z) = \frac{1}{z}\left[\left(A(z) - \lambda_0\right) R(z)\right]^m, \qquad 0 < |z| < \delta. \tag{4.29}$$

Then one obtains

Lemma 6. *The operator $B(z)$ is holomorphic in $|z| < \delta$, in particular at the point $z = 0$. One has*

$$B_0 = B(0) = \sum_{\varrho+\sigma=m-1} D_0^\varrho R_0A_1R_0D_0^\sigma, \qquad D_0 = A_0 - \lambda_0, \tag{4.30}$$

and

$$A(z)\,B(z) = B(z)\,A(z), \qquad |z| < \delta. \tag{4.31}$$

Furthermore,

$$B(z)\,R(z) = R(z)\,B(z) = B(z), \qquad |z| < \delta. \tag{4.32}$$

Proof. $B(z)$ is holomorphic in $0 < |z| < \delta$. Because of (4.28), one has $[(A_0 - \lambda_0)\,R_0]^m = 0$; that is, the operator-valued function $\left[\left(A(z) - \lambda_0\right) R(z)\right]^m$, which is holomorphic in $|z| < \delta$, vanishes at $z = 0$. This implies that $B(z)$ is holomorphic in the whole region $|z| < \delta$. One obtains

$$B_0 = \frac{\mathrm{d}}{\mathrm{d}z}\left[R(z)\left(A(z) - \lambda_0\right) R(z)\right]_{z=0}^m = \sum_{\varrho+\sigma=m-1} D_0^\varrho R_0A_1R_0D_0^\sigma.$$

(4.31) and (4.32) follow immediately from (4.29). ■

Fixing a Jordan basis in R_0X with respect to A_0 also fixes the projections which collect the irreducible parts of fixed dimension and the projections onto the corresponding geometric eigenspaces.

By E_m one denotes the projection onto the geometric eigenspace of those eigenvectors which come from the irreducible parts of maximal dimension m. Let $\dim E_m = r$; that is, r denotes the number of irreducible parts, with respect to A_0 restricted to R_0X, of dimension m.

According to Lemma 1, the subspace E_mX is independent of the special Jordan basis of A_0 in R_0X. One has

Lemma 7. *The subspace E_mX is invariant with respect to the operator B_0; i.e.,*

$$E_mB_0E_m = B_0E_m, \tag{4.33}$$

and one obtains

$$B_0 \upharpoonright E_mX = D_0^{m-1}R_0A_1R_0 \upharpoonright E_mX = E_mD_0^{m-1}A_1E_m \upharpoonright E_mX. \tag{4.34}$$

Proof. According to Lemma 6, one has $A_0B_0 = B_0A_0$. Then (4.33) follows, because of Lemma 2. The relation (4.34) follows from (4.33). ■

The calculation of the characteristic polynomial of B_0 in R_0X is important for a reduction. One has

Lemma 8. *The characteristic polynomial $\chi(\lambda, B_0 \upharpoonright R_0X)$ is given by*

$$\chi(\lambda, B_0 \upharpoonright R_0X) = \lambda^l[\chi(\lambda, B_0 \upharpoonright E_mX)]^m, \qquad l + rm = \dim R_0. \tag{4.35}$$

Proof. Choose a Jordan basis of A_0 in R_0X. One denotes by P_m the projection which collects the irreducible parts of maximal dimension m and by P_d the projection which collects the irreducible parts of dimension d. Then,

$$R_0 = P_m + \sum_{d<m} P_d.$$

Since $A_0B_0 = B_0A_0$, in particular for the operators restricted to R_0X, Lemma 3 is applicable and one obtains

$$\chi(\lambda, B_0 \upharpoonright R_0X) = \chi(\lambda, P_mB_0P_m \upharpoonright P_mX) \prod_{d<m} \chi(\lambda, P_dB_0P_d \upharpoonright P_dX). \tag{4.36}$$

One calculates $\chi(\lambda, P_mB_0P_m \upharpoonright P_mX)$. For this purpose one applies Lemma 4 to the space P_mX, to the nilpotent operator $A_0 - \lambda_0$ in P_mX and to the operator $\lambda - P_mB_0P_m$ in P_mX which commutes with $A_0 - \lambda_0$. One obtains

$$\det[(\lambda - P_mB_0P_m) \upharpoonright P_mX] = (\det[(\lambda - B_0) \upharpoonright E_mX])^m;$$

that is,

$$\chi(\lambda, P_mB_0P_m \upharpoonright P_mX) = [\chi(\lambda, B_0 \upharpoonright E_mX)]^m. \tag{4.37}$$

One determines $\chi(\lambda, P_dB_0P_d \upharpoonright P_dX)$ for $d < m$. One has

$$P_dB_0P_d = \sum_{\varrho+\sigma=m-1} (P_dD_0)^\varrho P_dA_1P_d(P_dD_0)^\sigma .$$

However, $(P_dD_0)^d = 0$. Therefore, Lemma 5 is applicable. This lemma says that $(P_dB_0P_d)^d = 0$; i.e., $P_dB_0P_d$ is nilpotent. Hence,

$$\chi(\lambda, P_dB_0P_d \upharpoonright P_dX) = \lambda^{\dim P_d}, \qquad d < m. \tag{4.38}$$

Substituting the results (4.37) and (4.38) into (4.36), one obtains (4.35). ■

The following fact should be emphasized: Even though the projection P_m depends on the choice of the Jordan basis in R_0X, from (4.37) one knows that $\chi(\lambda, P_mB_0P_m \upharpoonright P_mX)$ is independent of the choice of the Jordan basis. For every choice of the projection P_m one always obtains one and the same characteristic polynomial $[\chi(\lambda, B_0 \upharpoonright E_mX)]^m$.

7.4.5. Reduction process

In this subsection a condition is formulated sufficient for the construction of a characteristic holomorphic reducing projection of $A(z)$ if one starts from the assumptions of Subsect. 7.4.4.

Let $R(z)$ be a holomorphic reducing projection of $A(z)$ in $|z| < \delta$ with the properties of Subsect. 7.4.4. Let $B(z)$ be defined by (4.29). The condition mentioned takes the following form:

Condition 1. *Assume that*

$$\det (B_0 \upharpoonright E_mX) \neq 0 \tag{4.39}$$

is valid.

In other words: The characteristic polynomial $\chi(\lambda, B_0 \upharpoonright E_mX)$ is assumed to have only nonvanishing roots $\mu_1, \mu_2, \ldots, \mu_k$; i.e., the relation

$$\chi(\lambda, B_0 \upharpoonright E_mX) = \prod_{\varrho=1}^{k} (\lambda - \mu_\varrho)^{\gamma_\varrho}, \qquad \mu_\varrho \neq 0, \qquad \varrho = 1, 2, \ldots, k, \tag{4.40}$$

is assumed to hold, where $1 \leqq k \leqq r = \dim E_mX$ and $r = \sum_{\varrho=1}^{k} \gamma_\varrho$. The numbers $\mu_1, \mu_2, \ldots, \mu_k$ are the eigenvalues of B_0 in E_mX. According to (4.35), these numbers coincide with the nonvanishing eigenvalues of B_0 restricted to R_0X.

The eigenprojections of $\mu_1, \ldots, \mu_k$ with respect to B_0 are denoted by $Q_1, Q_2, \ldots, Q_k$, respectively. Then,

$$Q_\varrho = \frac{1}{2\pi i} \int_{(\mu_\varrho)} (\lambda - B_0)^{-1} \, d\lambda, \qquad \varrho = 1, 2, \ldots, k. \tag{4.41}$$

The projections $Q_1, \ldots, Q_k$ form a disjoint system. Let

$$Q_0 = \sum_{\varrho=1}^{k} Q_\varrho . \tag{4.42}$$

Then the projection Q_0 is a subprojection of R_0; i.e., the relation

$$Q_0 R_0 = R_0 Q_0 = Q_0 \tag{4.43}$$

holds. From Lemma 6, (4.32), one obtains $(1 - R_0) B_0 = B_0(1 - R_0) = 0$. Together with (4.41), this implies $(1 - R_0) Q_0 = Q_0(1 - R_0) = 0$, and (4.43) follows.

The holomorphic perturbation $B(z)$ of B_0 generates the μ_ϱ-groups from the non-vanishing eigenvalues μ_ϱ. By $Q_\varrho(z)$ one denotes the corresponding group projections; i.e., one writes

$$Q_\varrho(z) = \frac{1}{2\pi i} \int_{(\mu_\varrho)} (\lambda - B(z))^{-1} \, d\lambda, \qquad |z| < \delta_1, \qquad \varrho = 1, \dots, k. \tag{4.44}$$

The properties of these group projections are collected in the following theorem.

Theorem 4. *Let the assumptions of Subsect.* 7.4.4 *and Condition* 1 *be satisfied. The group projections of the μ_ϱ-groups with respect to $B(z)$ are given by* (4.44). *Then, for the disjoint system of group projections $Q_1(z), \dots, Q_k(z)$, which are holomorphic in $|z| < \delta_1$, the following properties are satisfied*

$$A(z) Q_\varrho(z) = Q_\varrho(z) A(z), \qquad \varrho = 1, 2, \dots, k, \tag{4.45}$$

$$Q_\varrho(z) R(z) = R(z) Q_\varrho(z) = Q_\varrho(z), \qquad \varrho = 1, 2, \dots, k, \tag{4.46}$$

$$\dim Q_\varrho(z) = \gamma_\varrho m, \qquad \varrho = 1, \dots, k. \tag{4.47}$$

If

$$Q(z) = \sum_{\varrho=1}^{k} Q_\varrho(z); \qquad Q_0 = Q(0) = \sum_{\varrho=1}^{k} Q_\varrho, \tag{4.48}$$

then,

$$Q(z) R(z) = R(z) Q(z) = Q(z), \tag{4.49}$$

$$\dim Q(z) = rm \tag{4.50}$$

and

$$\nu(\lambda_0, A_0 \restriction (R_0 - Q_0) X) < m; \tag{4.51}$$

that is, after a suitable Jordan basis in R_0X has been chosen, the reducing projection Q_0 with respect to A_0 collects the irreducible parts of A_0 restricted to R_0X with maximal dimension m.

Proof. From (4.44) and (4.31) (see Lemma 6), (4.45) follows. (4.47) is obtained from (4.35) (see Lemma 8) and (4.40). Next one proves (4.46). According to (4.32) (see Lemma 6), one has

$$(1 - R(z)) B(z) = B(z) (1 - R(z)) = 0.$$

Because of (4.44), this implies

$$(1 - R(z)) Q_\varrho(z) = Q_\varrho(z) (1 - R(z)) = 0;$$

that is, (4.46) holds.

(4.49) is obtained from (4.46), and (4.50) from (4.47), (4.48) and $r = \sum_{\varrho=1}^{k} \gamma_\varrho$. The proof of (4.51) follows: $A_0Q_0 = Q_0A_0$ holds, as well as $B_0Q_0 = Q_0B_0$. In particular,

$$\chi(\lambda, B_0 \upharpoonright Q_0X) = \left(\prod_{\varrho=1}^{k} (\lambda - \mu_\varrho)^{\gamma_\varrho}\right)^m = \left(\chi(\lambda, B_0 \upharpoonright E_mX)\right)^m. \tag{4.52}$$

Let $R_0 - Q_0 = S_0$. Then $\dim S_0 = \dim R_0 - rm = l$ and the relation

$$\chi(\lambda, B_0 \upharpoonright S_0X) = \lambda^l$$

follows. Since Q_0 reduces A_0, R_0X can be decomposed into a direct sum of irreducible parts with respect to A_0 which is compatible with the given reducing projection Q_0. One denotes by P_m and P_d, $d < m$, those projections of this decomposition which collect the irreducible parts of fixed dimension m and $d < m$, respectively. Then,

$$R_0 = P_m + \sum_{d<m} P_d, \tag{4.53}$$

and this decomposition is compatible with Q_0, that is,

$$Q_0P_m = P_mQ_0, \qquad Q_0P_d = P_dQ_0, \qquad d < m.$$

Hence,

$$P_m = P_mQ_0 + P_mS_0, \qquad P_d = P_dQ_0 + P_dS_0.$$

These equations express that one has to take into account the possibility of finding P_m-parts and P_d-parts in Q_0X as well as in S_0X.

Restrict both operators, A_0 and B_0 to the (reducing) subspace Q_0X. Applying Lemma 8 to the operator $\lambda - B_0$ which commutes with $A_0 - \lambda_0$, one obtains

$$\chi(\lambda, B_0 \upharpoonright Q_0X)$$
$$= \chi\big(\lambda, (P_mQ_0)\, B_0(P_mQ_0) \upharpoonright (P_mQ_0)\, X\big) \prod_{d<m} \chi\big(\lambda, (P_dQ_0)\, B_0(P_dQ_0) \upharpoonright (P_dQ_0)\, X\big). \tag{4.54}$$

Because of Lemma 5,

$$\chi\big(\lambda, (P_dQ_0)\, B_0(P_dQ_0) \upharpoonright (P_dQ_0)\, X\big) = \lambda^{\dim(P_dQ_0)}, \tag{4.55}$$

since

$$P_dQ_0B_0P_dQ_0 = \sum_{\varrho+\sigma=m-1} (P_dQ_0D_0)^\varrho\, P_dQ_0A_1P_dQ_0(P_dQ_0D_0)^\sigma$$

is nilpotent in $(P_dQ_0)\, X$. From (4.52), (4.54) and (4.55), one obtains

$$\dim P_dQ_0 = 0, \qquad d < m;$$

i.e.,

$$P_dQ_0 = 0, \qquad d < m.$$

This means: there are no P_d-parts of dimension $d < m$ in the subspace Q_0X. Hence,

$$P_d = P_dS_0 = S_0P_d, \qquad d < m;$$

i.e.,

$$P_d \leqq S_0, \qquad d < m.$$

Moreover, because of (4.53),

$$Q_0 = P_mQ_0 = Q_0P_m;$$

i.e.,

$$Q_0 \leqq P_m.$$

Note that

$$\dim Q_0 = rm = \dim P_m.$$

Hence,

$$Q_0 = P_m.$$

This means: the projection Q_0 is a projection which collects the m-dimensional irreducible parts, with respect to a suitable Jordan basis. That is, in S_0X there are only parts with dimensions $d < m$. Hence, (4.51) follows. ■

Because $Q_0 = \sum_{\varrho=1}^{k} Q_\varrho$ and $A_0Q_\varrho = Q_\varrho A_0$, the reducing projection Q_ϱ collects exactly γ_ϱ irreducible parts of dimension m.

Therefore, Condition 1 has the following consequence: If it is satisfied, then there is a distinct Jordan decomposition of the unperturbed operator A_0 restricted to R_0X such that certain groups of irreducible parts with maximal dimension behave holomorphically under the perturbation $A(z)$, that is, these groups appear as zero-order approximations of holomorphic reducing projections with respect to $A(z)$.

Theorem 4 expresses a definite reduction procedure. Concerning the iteration of this procedure, property (4.51) is of special importance. From this property, one obtains the basic assumption that the reduction procedure can be repeated, by means of the reducing projection $R(z) - Q(z)$, on a lower-dimensional level (of the irreducible parts). Of course, an additional condition, similar to Condition 1, must be satisfied for the repetition the process.

It is essential to put these conditions in a form which is independent of the reduction steps performed before. The formulation of these conditions should not be related to distinct Jordan decompositions of A_0 in certain reducing subspaces generated by previous reduction steps.

For this purpose, certain operators will be introduced which are essentially due to V. B. Lidskii [1]. These operators can be related, as desired, to Condition 1 of the reduction process.

7.4.6. Lidskii operators

Let A_0 be an operator on X, λ_0 an eigenvalue of A_0, P_0 the corresponding eigenprojection. By the choice of a Jordan basis in P_0X, the projections $E_1, E_2, \ldots, E_p$ are fixed. Let A_1 be an arbitrary operator on X. Let

$$\begin{aligned}
K_1 &= E_1 D_0^{m_1-1} A_1 E_1,\\
K_2 &= (E_1 D_0^{m_1-1} + E_2 D_0^{m_2-1})\, A_1(E_1 + E_2),\\
&\vdots\\
K_i &= (E_1 D_0^{m_1-1} + E_2 D_0^{m_2-1} + \cdots + E_i D_0^{m_i-1})\, A_1(E_1 + \cdots + E_i),
\end{aligned}$$

where $D_0 = A_0 - \lambda_0$ and $1 \leqq i \leqq p$. K_i is a well-defined linear operator on X.

Obviously, K_i *has the invariant subspace* $(E_1 + E_2 + \cdots + E_i)\, X$.

Since in the following K_i is to be considered first of all in the subspace $(E_1 + \cdots + E_i)X$, one uses the short notation K_i for $K_i \restriction (E_1 + \cdots + E_i)\, X$.

The operator K_i has the following operator matrix form with respect to $(E_1 + E_2 + \cdots + E_i)\, X$:

$$K_i \sim \begin{pmatrix} E_1 D_0^{m_1-1} A_1 E_1, & E_1 D_0^{m_1-1} A_1 E_2, & \ldots, & E_1 D_0^{m_1-1} A_1 E_i\\ E_2 D_0^{m_2-1} A_1 E_1, & E_2 D_0^{m_2-1} A_1 E_2, & \ldots, & E_2 D_0^{m_2-1} A_1 E_i\\ \vdots & & & \\ E_i D_0^{m_i-1} A_1 E_1, & E_i D_0^{m_i-1} A_1 E_2, & \ldots, & E_i D_0^{m_i-1} A_1 E_i \end{pmatrix}. \tag{4.56}$$

This means, if $x = \sum_{\varrho=1}^{i} x_\varrho \in (E_1 + \cdots + E_i)\, X$, then $K_i x = \sum_{\sigma=1}^{i} \left(\sum_{\varrho=1}^{i} E_\sigma D_0^{m_\sigma-1} A_1 E_\varrho x \right)$. Determine the matrix of K_i with respect to a given basis in $(E_1 + \cdots + E_i)\, X$. Let the eigenvectors e_ϱ^σ, $\varrho = 1, \ldots, i$; $\sigma = 1, \ldots, r_\varrho$, be the elements of this basis. This basis is a part of a Jordan basis $e_\varrho^{\sigma\varkappa}$ of A_0 in P_0X, where $e_\varrho^{\sigma 1} = e_\varrho^\sigma$. The corresponding biorthogonal basis in $P_0^*X^*$ is denoted by $f_\varrho^{\sigma\varkappa}$. Then,

$$(f_{\varrho'}^{\sigma'\varkappa'}, e_\varrho^{\sigma\varkappa}) = \delta_{\varrho\varrho'}\delta_{\sigma\sigma'}\delta_{\varkappa\varkappa'}.$$

The part of the biorthogonal basis belonging to the vectors e_ϱ^σ, $\varrho = 1, \ldots, i$; $\sigma = 1, \ldots, r_\varrho$, consists of the vectors $f_\varrho^{\sigma 1} = f_\varrho^\sigma$, $\varrho = 1, \ldots, i$; $\sigma = 1, \ldots, r_\varrho$. As is well-known, these vectors are generating (cyclic) vectors of A_0^* in $P_0^*X^*$; namely, f_ϱ^σ is generating with respect to some irreducible part of dimension m_ϱ. For the eigenvectors g_ϱ^σ of A_0^* in $P_0^*X^*$ belonging to these generating vectors, one obtains

$$g_\varrho^\sigma = (D_0^*)^{m_\varrho-1} f_\varrho^\sigma, \qquad \varrho = 1, \ldots, i; \qquad \sigma = 1, \ldots, r_\varrho.$$

Then,

$$K_i e_\varrho^\sigma = \sum_{\varrho', \sigma'} (K_i e_\varrho^\sigma, f_{\varrho'}^{\sigma'})\, e_{\varrho'}^{\sigma'},$$

and one has to calculate the matrix elements $(K_i e_\varrho^\sigma, f_{\varrho'}^{\sigma'})$. One obtains

$$\begin{aligned}
(K_i e_\varrho^\sigma, f_{\varrho'}^{\sigma'}) &= \sum_{\mu=1}^{i} (E_\mu D_0^{m_\mu-1} A_1 e_\varrho^\sigma, f_{\varrho'}^{\sigma'}) = (D_0^{m_{\varrho'}-1} A_1 e_\varrho^\sigma, f_{\varrho'}^{\sigma'})\\
&= \left(A_1 e_\varrho^\sigma, (D_0^*)^{m_{\varrho'}-1} f_{\varrho'}^{\sigma'}\right) = (A_1 e_\varrho^\sigma, g_{\varrho'}^{\sigma'}).
\end{aligned}$$

Therefore, the following lemma holds.

Lemma 9. *The matrix of K_i in $(E_1 + \cdots + E_i)\,X$, with respect to the basis e_ϱ^σ, $\varrho = 1, \ldots, i$; $\sigma = 1, \ldots, r_\varrho$, takes the form*

$$(A_1 e_{\varrho'}^{\sigma'}, g_\varrho^\sigma), \qquad \varrho, \varrho' = 1, \ldots, i; \qquad \sigma = 1, \ldots, r_\varrho; \qquad \sigma' = 1, \ldots, r_{\varrho'},$$

where the vectors g_ϱ^σ are the eigenvectors of A_0^* with respect to $\bar{\lambda}_0$ in $P_0^* X^*$ taken from a biorthogonal basis belonging to e_ϱ^σ.

V. B. Lidskii [1] used the operators K_i in the matrix form given by Lemma 9.

By means of the matrix of K_i one can verify important invariance properties of K_i. Note that the operators K_i depend on the choice of a Jordan basis of A_0 in $P_0 X$, i.e., on the choice of the operators $E_1, \ldots, E_i$. The following lemma is true.

Lemma 10. *The operator $K_i - \lambda E_i$ restricted to $(E_1 + \cdots + E_i)\,X$, which depends on an arbitrary parameter λ, has an invariant determinant, i.e.,*

$$\det(K_i - \lambda E_i), \qquad i = 1, 2, \ldots, p, \tag{4.57}$$

is independent of the choice of the Jordan basis in $P_0 X$.

Proof. Let e_ϱ^σ denote a starting basis in $(E_1 + \cdots + E_i)\,X$ and let $\tilde{e}_\varrho^\sigma$ be a second transformed basis, where

$$\tilde{e}_\varrho^\sigma = \sum_{\varrho', \sigma'} \alpha_{\varrho\varrho'}^{\sigma\sigma'} e_{\varrho'}^{\sigma'}, \qquad A = (\alpha_{\varrho\varrho'}^{\sigma\sigma'}). \tag{4.58}$$

Furthermore, let f_ϱ^σ denote the biorthogonal basis assigned to e_ϱ^σ and $\tilde{f}_\varrho^\sigma$ the biorthogonal basis assigned to $\tilde{e}_\varrho^\sigma$. If

$$\tilde{f}_\varrho^\sigma = \sum_{\varrho', \sigma'} \beta_{\varrho\varrho'}^{\sigma\sigma'} f_{\varrho'}^{\sigma'}, \qquad B = (\beta_{\varrho\varrho'}^{\sigma\sigma'}), \tag{4.59}$$

then,

$$B = (A^*)^{-1},$$

where A^* denotes the adjoint matrix, $A^* = \bar{A}^\mathsf{T}$. A can be written as a block matrix $A = (A_{\varrho\varrho'})$, $A_{\varrho\varrho'} = (\alpha_{\varrho\varrho'}^{\sigma\sigma'})$, where $A_{\varrho\varrho'}$ is an $r_\varrho \times r_{\varrho'}$-matrix. Because of Lemma 1,

$$A_{\varrho\varrho'} = 0, \qquad \varrho < \varrho';$$

that is, A takes the form

$$A = \begin{pmatrix} A_{11} & 0 & 0 & \cdots & 0 \\ A_{21} & A_{22} & 0 & \cdots & 0 \\ \vdots & & & & \\ A_{i1} & A_{i2} & A_{i3} & \cdots & A_{ii} \end{pmatrix},$$

because $(E_1 + \cdots + E_\varrho)\,X$ is invariant under a change of the Jordan basis; i.e., the transformed basis vector $\tilde{e}_\varrho^\sigma$ must belong to $(E_1 + \cdots + E_\varrho)\,X$, $\varrho = 1, 2, \ldots, n$.

Hence,

$$A^* = \begin{pmatrix} A_{11}^* & A_{21}^* \cdots A_{i1}^* \\ 0 & A_{22}^* \cdots A_{i2}^* \\ \vdots & \\ 0 & 0 \quad \cdots A_{ii}^* \end{pmatrix}$$

and

$$B = (A^*)^{-1} = \begin{pmatrix} (A_{11}^*)^{-1} & * & \cdots * \\ 0 & (A_{22}^*)^{-1} & \cdots * \\ \vdots & & \\ 0 & 0 & \cdots (A_{ii}^*)^{-1} \end{pmatrix},$$

where the blocks denoted by $*$ are not relevant. Writing $B = (B_{\varrho\varrho'})$, where $B_{\varrho\varrho'} = (B_{\varrho\varrho'}^{\sigma\sigma'})$, one obtains

$$B_{\varrho\varrho'} = 0, \qquad \varrho > \varrho'.$$

Therefore, the formula (4.59) takes the form

$$\tilde{f}_\varrho^\sigma = \sum_{\varrho' \geqq \varrho} \sum_{\sigma'} \beta_{\varrho\varrho'}^{\sigma\sigma'} f_{\varrho'}^{\sigma'}. \tag{4.60}$$

One has

$$g_\varrho^\sigma = (D_0^*)^{m_\varrho - 1} f_\varrho^\sigma$$

and, correspondingly,

$$\tilde{g}_\nu^\sigma = (D_0^*)^{m_\varrho - 1} \tilde{f}_\varrho^\sigma.$$

Therefore, for $\varrho < \varrho'$,

$$(D_0^*)^{m_\varrho - 1} f_{\varrho'}^{\sigma'} = (D_0^*)^{m_\varrho - m_{\varrho'}} (D_0^*)^{m_{\varrho'} - 1} f_{\varrho'}^{\sigma'} = 0,$$

because $m_\varrho - m_{\varrho'} > 0$. Therefore, applying $(D_0^*)^{m_\varrho - 1}$ to (4.60), one obtains

$$\tilde{g}_\varrho^\sigma = \sum_{\sigma'=1}^{r_\varrho} \beta_{\varrho\varrho}^{\sigma\sigma'} g_\varrho^{\sigma'}, \qquad (\beta_{\varrho\varrho}^{\sigma\sigma'}) = B_{\varrho\varrho} = (A_{\varrho\varrho}^*)^{-1}. \tag{4.61}$$

Hence, the transition matrix Δ from g_ϱ^σ to $\tilde{g}_\varrho^\sigma$ takes the form

$$\Delta = \begin{pmatrix} (A_{11}^*)^{-1} & 0 & \cdots 0 \\ 0 & (A_{22}^*)^{-1} & \cdots 0 \\ \vdots & & \\ 0 & 0 & \cdots (A_{ii}^*)^{-1} \end{pmatrix}.$$

Consider the matrix

$$\Phi = \big((A_1 e^{\sigma'}_{\varrho'}, g^{\sigma}_{\varrho})\big), \qquad \varrho, \varrho' = 1, \ldots, i; \qquad \sigma = 1, \ldots, r_\varrho; \qquad \sigma' = 1, \ldots, r_{\varrho'}.$$

This matrix can be written as a block matrix

$$\Phi = (\Phi_{\varrho\varrho'}), \qquad \Phi_{\varrho\varrho'} = \big((A_1 e^{\sigma'}_{\varrho'}, g^{\sigma}_{\varrho})\big).$$

$\Phi_{\varrho\varrho'}$ is an $r_\varrho \times r_{\varrho'}$-matrix. Let $\varphi^{\sigma\sigma'}_{\varrho\varrho'} = (A_1 e^{\sigma'}_{\varrho'}, g^{\sigma}_{\varrho})$ and $\tilde{\varphi}^{\sigma\sigma'}_{\varrho\varrho'} = (A_1 \tilde{e}^{\sigma'}_{\varrho'}, \tilde{g}^{\sigma}_{\varrho})$; furthermore $\tilde{\Phi} = (\tilde{\varphi}^{\sigma\sigma'}_{\varrho\varrho'})$. Substituting (4.58) and (4.61) into the expression for $\varphi^{\sigma\sigma'}_{\varrho\varrho'}$, one obtains the transformation formula

$$\tilde{\varphi}^{\sigma\sigma'}_{\varrho\varrho'} = \sum_{\sigma'''\varrho''\sigma''} \bar{\beta}^{\sigma\sigma'''}_{\varrho\varrho} \varphi^{\sigma'''\sigma''}_{\varrho\varrho''} \alpha^{\sigma'\sigma''}_{\varrho'\varrho''}$$

or

$$\tilde{\Phi} = \bar{\varLambda} \Phi A^{\mathsf{T}};$$

i.e.,

$$\tilde{\Phi} = \begin{pmatrix} (A^{\mathsf{T}}_{11})^{-1} & 0 & \cdots 0 \\ 0 & (A^{\mathsf{T}}_{22})^{-1} & \cdots 0 \\ \vdots & & \\ 0 & 0 & \cdots (A^{\mathsf{T}}_{ii})^{-1} \end{pmatrix} \Phi \begin{pmatrix} A^{\mathsf{T}}_{11} & A^{\mathsf{T}}_{21} & \cdots A^{\mathsf{T}}_{i1} \\ 0 & A^{\mathsf{T}}_{22} & \cdots A^{\mathsf{T}}_{i2} \\ \vdots & & \\ 0 & 0 & \cdots A^{\mathsf{T}}_{ii} \end{pmatrix}.$$

Furthermore,

$$\begin{pmatrix} (A^{\mathsf{T}}_{11})^{-1} & 0 & \cdots 0 \\ 0 & (A^{\mathsf{T}}_{22})^{-1} & \cdots 0 \\ \vdots & & \\ 0 & 0 & \cdots (A^{\mathsf{T}}_{ii})^{-1} \end{pmatrix} \begin{pmatrix} 0 & 0 \cdots 0 & 0 \\ 0 & 0 \cdots 0 & 0 \\ \vdots & & \\ 0 & 0 \cdots 0 & 0 \\ 0 & 0 \cdots 0 & E \end{pmatrix} \begin{pmatrix} A^{\mathsf{T}}_{11} & A^{\mathsf{T}}_{21} & \cdots A^{\mathsf{T}}_{i1} \\ 0 & A^{\mathsf{T}}_{22} & \cdots A^{\mathsf{T}}_{i2} \\ \vdots & & \\ 0 & 0 & \cdots A^{\mathsf{T}}_{ii} \end{pmatrix}$$

$$= \begin{pmatrix} 0 & 0 \cdots 0 & 0 \\ 0 & 0 \cdots 0 & 0 \\ \vdots & & \\ 0 & 0 \cdots 0 & 0 \\ 0 & 0 \cdots 0 & E \end{pmatrix},$$

where E denotes the $r_i \times r_i$-unit matrix. Thus, if

$$\begin{pmatrix} 0 & 0 \cdots 0 & 0 \\ 0 & 0 \cdots 0 & 0 \\ \vdots & & \\ 0 & 0 \cdots 0 & 0 \\ 0 & 0 \cdots 0 & E \end{pmatrix} = \{0, 0, \ldots, E\} = E_i,$$

one obtains

$$(\tilde{\Phi} - \lambda E_i) = \bar{\varLambda}(\Phi - \lambda E_i)\, A^{\mathsf{T}},$$

and this implies

$$\det(\tilde{\Phi} - \lambda E_i) = \det \bar{\Delta} \det(\Phi - \lambda E_i) \det A^{\mathsf{T}}$$
$$= \prod_{\varrho=1}^{i} \det (A_{\varrho\varrho}^{\mathsf{T}})^{-1} \prod_{\varrho=1}^{i} \det A_{\varrho\varrho}^{\mathsf{T}} \det(\Phi - \lambda E_i) = \det(\Phi - \lambda E_i).$$

But the matrix $(\Phi - \lambda E_i)$ is the matrix of the operator $K_i - \lambda E_i$, formed with respect to the starting basis e_ϱ^σ. Correspondingly, $(\tilde{\Phi} - \lambda E_i)$ is the matrix of the operator $\tilde{K}_i - \lambda \tilde{E}_i$ formed analogously with respect to the transformed basis $\tilde{e}_\varrho^\sigma$. ■

The polynomial $\det(K_i - \lambda E_i)$, $i = 1, 2, \ldots, p$, is called the *pseudo-characteristic polynomial* of the operator K_i. In particular, the numbers $\det K_i$, $i = 1, 2, \ldots, p$, are independent of the choice of the Jordan basis in P_0X.

7.4.7. Condition 1 and the Lidskii operators

In this subsection the treatment of the reduction process will be continued. The assumptions of Subsect. 7.4.4 and 7.4.5 are assumed to be satisfied and Condition 1 will be investigated.

For short, let $Q(z) = P(z) - R(z)$. According to (4.28), one has $\nu(\lambda_0, A_0 \upharpoonright R_0X) = m$. For the dimensions d of the irreducible parts of A_0 in Q_0X, one obtains $d > m$. The number m is one of the dimensions of the irreducible parts of A_0 in P_0X. Let $m = m_i$, where $m_1 > m_2 > \cdots > m_p \geqq 1$ denotes the sequence of all such dimensions.

The reduction operator $B(z)$ satisfies Lemma 6. In particular, from (4.32),

$$R_0B_0 = B_0R_0;$$

hence,

$$R_0B_0Q_0 = Q_0B_0R_0 = 0;$$

i.e., one has

$$\sum_{\varrho+\sigma=m-1} D_0^\varrho Q_0 A_1 R_0 D_0^\sigma = 0.$$

Therefore,

$$D_0^{m-1} Q_0 A_1 R_0 E = 0,$$

and, more generally,

$$D_0^{d-1} Q_0 A_1 R_0 E = 0, \qquad d \geqq m.$$

Choosing a Jordan basis, adapted to the decomposition $P_0 = Q_0 + R_0$, one obtains

$$E_\varrho D_0^{m_\varrho - 1} Q_0 A_1 R_0 E_\sigma = 0, \qquad \varrho < i, \qquad \sigma \geqq i,$$

where the operators $E_1, E_2, \ldots, E_p$ are defined as before, with respect to the Jordan basis, or finally

$$E_\varrho D_0^{m_\varrho - 1} A_1 E_\sigma = 0, \qquad \varrho < i, \qquad \sigma \geqq i. \tag{4.62}$$

Consider the operator K_i on $(E_1 + E_2 + \cdots + E_i)\, X$. According to (4.56), it takes the operator matrix form

$$\begin{pmatrix} E_1 D_0^{m_1-1} A_1 E_1 & \cdots E_1 D_0^{m_1-1} A_1 E_{i-1} & E_1 D_0^{m_1-1} A_1 E_i \\ E_2 D_0^{m_2-1} A_1 E_1 & \cdots E_2 D_0^{m_2-1} A_1 E_{i-1} & E_2 D_0^{m_2-1} A_1 E_i \\ \vdots & & \\ E_{i-1} D_0^{m_{i-1}-1} A_1 E_1 & \cdots E_{i-1} D_0^{m_{i-1}-1} A_1 E_{i-1} & E_{i-1} D_0^{m_{i-1}-1} A_1 E_i \\ E_i D_0^{m_i-1} A_1 E_i & \cdots E_i D_0^{m_i-1} A_1 E_{i-1} & E_i D_0^{m_i-1} A_1 E_i \end{pmatrix}.$$

Choosing a Jordan basis, adapted to the decomposition $P_0 = Q_0 + R_0$, by means of (4.62), one obtains the matrix

$$\begin{pmatrix} E_1 D_0^{m_1-1} A_1 E_1 & \cdots E_1 D_0^{m_1-1} A_1 E_{i-1} & 0 \\ E_2 D_0^{m_2-1} A_1 E_1 & \cdots E_2 D_0^{m_2-1} A_1 E_{i-1} & 0 \\ \vdots & & \\ E_{i-1} D_0^{m_{i-1}-1} A_1 E_1 & \cdots E_{i-1} D_0^{m_{i-1}-1} A_1 E_{i-1} & 0 \\ E_i D_0^{m_i-1} A_1 E_1 & \cdots E_i D_0^{m_i-1} A_1 E_{i-1} & E_i D_0^{m_i-1} A_1 E_i \end{pmatrix}$$

representing K_i. From this representation, the relation

$$\det K_i = \det K_{i-1} \det (E_i D_0^{m_i-1} A_1 E_i \upharpoonright E_i X), \qquad i = 1, \ldots, p, \tag{4.63}$$

follows, where one takes $\det K_0 = 1$. Taking into account (4.34), one obtains

$$\det K_i = \det K_{i-1} \det (B_0 \upharpoonright E_m X), \qquad E_m = E_i. \tag{4.64}$$

The determinants $\det K_i$ and $\det K_{i-1}$ are completely independent of the Jordan basis chosen in $P_0 X$. That is, formula (4.64) enables us to formulate a condition which implies Condition 1, without explicit knowledge of the operators R_0 and B_0. This condition takes the form:

Condition 2. *Assume that the relations*

$$\det K_{i-1} \neq 0, \qquad \det K_i \neq 0 \tag{4.65}$$

are satisfied.

If Condition 2 is true, then one obtains

$$\det (B_0 \upharpoonright E_m X) = \frac{\det K_i}{\det K_{i-1}} \neq 0, \tag{4.66}$$

and Theorem 4 can be applied.

Together with (4.64), one obtains additionally

$$\det (K_i - \lambda E_i) = \det K_{i-1} \det [(B_0 - \lambda) \upharpoonright E_m X]. \tag{4.67}$$

Therefore, if Condition 2 is satisfied, one obtains

$$\chi(\lambda, B_0 \upharpoonright E_m X) = \frac{(-1)^r}{\det K_{i-1}} \det (K_i - \lambda E_i); \tag{4.68}$$

i.e., the characteristic polynomial of B_0 restricted to $E_i X$ is linked with the pseudo-characteristic polynomial $\det (K_i - \lambda E_i)$.

7.4.8. Holomorphic decompositions of group projections

This subsection contains essential results of the reduction theory. First, according to the considerations of Subsect. 7.4.7, one obtains a variant of Theorem 4. This variant allows the reduction by the reduction operator $B(z)$ if Condition 2 is satisfied.

Theorem 5. *Let $A(z) = A_0 + zA_1 + z^2A_2 + \cdots$ be a holomorphic perturbation of A_0 in $|z| < R$. Furthermore, let λ_0 be an unperturbed eigenvalue with eigenprojection P_0. The group projection of the λ_0-group is denoted by $P(z)$. Assume that $R(z)$ is a reducing holomorphic projection of $A(z)$ and a subprojection of $P(z)$. Let $P(z) = Q(z) + R(z)$. Let $\nu(\lambda_0, A_0 \upharpoonright R_0X) = m_i$ and let the dimensions d of irreducible parts of A_0 restricted to Q_0X satisfy the inequality $d > m_i$. Furthermore, let*

$$B(z) = \frac{1}{z} [(A(z) - \lambda_0) R(z)]^{m_i}, \qquad |z| < \delta.$$

Assume that the relations

$$\det K_i \neq 0, \qquad \det K_{i-1} \neq 0 \tag{4.69}$$

are satisfied. Then all roots of the pseudo-characteristic polynomial $\det (K_i - \lambda E_i)$ *are different from zero. Let*

$$\det (K_i - \lambda E_i) = (-1)^{r_i} \det K_{i-1} \prod_{\varrho=1}^{k} (\lambda - \mu_\varrho)^{\gamma_\varrho}, \qquad r_i = \sum_{\varrho=1}^{k} \gamma_\varrho, \tag{4.70}$$

where r_i denotes the degree of $\det (K_i - \lambda E_i)$, *and*

$$S_\varrho(z) = \frac{1}{2\pi i} \int_{(\mu_\varrho)} (\lambda - B(z))^{-1} \, d\lambda, \qquad |z| < \delta_1, \qquad \varrho = 1, \ldots, k. \tag{4.71}$$

The operators $S_1(z), \ldots, S_k(z)$ form a system of holomorphic disjoint projections, which have the properties

$$A(z) S_\varrho(z) = S_\varrho(z) A(z), \tag{4.72}$$

$$S_\varrho(z) R(z) = R(z) S_\varrho(z) = S_\varrho(z), \qquad \varrho = 1, \ldots, k, \tag{4.73}$$

$$\dim S_\varrho(z) = \gamma_\varrho m_i. \tag{4.74}$$

Furthermore, if

$$S(z) = \sum_{\varrho=1}^{k} S_\varrho(z), \qquad S_0 = \sum_{\varrho=1}^{k} S_\varrho(0), \tag{4.75}$$

one obtains

$$S(z)\, R(z) = R(z)\, S(z) = S(z), \tag{4.76}$$

$$\dim S(z) = r_i m_i \tag{4.77}$$

and

$$\nu\big(\lambda_0, A_0 \upharpoonright (R_0 - S_0)\, X\big) < m_i; \tag{4.78}$$

i.e., the reducing projection S_0 of A_0 collects exactly the irreducible parts of fixed dimension m_i.

Proof. The proof follows from Subsect. 7.4.7 and Theorem 4. ∎

If $i = 1$, then one again takes $\det K_0 = 1$. This case is the starting case, where $R(z) = P(z)$ and where the projection S_0 collects the irreducible parts of maximal dimension m_1.

Of course, the significance of Theorem 5 is more or less methodical because the hypothetical reducing operator $R(z)$ appears in the assumptions of this theorem. By successive application of Theorem 5, under suitable assumptions, one obtains explicit results. For example, one can formulate

Theorem 6. *Let $A(z) = A_0 + zA_1 + z^2A_2 + \cdots$ be a holomorphic perturbation of A_0 and λ_0 an unperturbed eigenvalue of A_0 with the group projection $P(z)$. Assume that*

$$\det K_1 \neq 0, \qquad \det K_2 \neq 0, \ldots, \det K_i \neq 0, \qquad 1 \leqq i \leqq p, \tag{4.79}$$

is satisfied, where i is a fixed number. Then, in a neighborhood $|z| < \delta$, there is a decomposition of the group projection

$$P(z) = \sum_{\varrho=1}^{i} \sum_{\sigma=1}^{k_\varrho} Q_\varrho^\sigma(z) + R(z), \qquad |z| < \delta, \tag{4.80}$$

into a sum of holomorphic disjoint projections, reducing with respect to $A(z)$, where

$$\dim Q_\varrho^\sigma(z) = \gamma_\varrho^\sigma m_\varrho, \qquad \varrho = 1, 2, \ldots, i; \qquad \sigma = 1, 2, \ldots, k_\varrho.$$

If $Q_\varrho(z) = \sum_{\sigma=1}^{k_\varrho} Q_\varrho^\sigma(z)$, one has $P(z) = \sum_{\varrho=1}^{i} Q_\varrho(z) + R(z)$. In particular, at $z = 0$ one obtains

$$P_0 = \sum_{\varrho=1}^{i} Q_\varrho + R_0, \qquad Q_\varrho A_0 = A_0 Q_\varrho, \qquad R_0 A_0 = A_0 R_0,$$

where $Q_\varrho(0) = Q_\varrho$, $R(0) = R_0$. The projection Q_ϱ, reducing with respect to A_0, collects exactly the irreducible parts of fixed dimension m_ϱ. Moreover, one has

$$\nu(\lambda_0, A_0 \upharpoonright R_0 X) = m_{i+1} < m_i.$$

The numbers k_ϱ and $\gamma_\varrho^{(\sigma)}$ are the number and multiplicity, respectively, of the (because of (4.79) non-vanishing) roots of the pseudo-characteristic polynomial $\det(K_\varrho - \lambda E_\varrho)$; *i.e., one has*

$$\det(K_\varrho - \lambda E_\varrho) = (-1)^{r_\varrho} \det K_{\varrho-1} \prod_{\sigma=1}^{k_\varrho} (\lambda - \mu_\varrho^\sigma)^{\gamma_\varrho^\sigma}.$$

If

$$B_\varrho(z) = \frac{1}{z}\left[\left(A(z) - \lambda_0\right)\left(P(z) - \sum_{\sigma=1}^{\varrho-1} Q_\sigma(z)\right)\right]^{m_\varrho},$$

then the projections $Q_\varrho^\sigma(z)$ are determined by the formula

$$Q_\varrho^\sigma(z) = \frac{1}{2\pi i} \int_{(\mu_\varrho^\sigma)} (\lambda - B_\varrho(z))^{-1} \, d\lambda. \tag{4.81}$$

Proof. The proof consists of i applications of Theorem 5. At the first step, one has to take $i = 1$ (i denotes the index in Theorem 5). At the second step, this index is equal to 2; furthermore, $R(z) = P(z) - Q_1(z)$. Assumption (4.79) ensures that Theorem 5 can be applied at every step. The additional assumptions for application of Theorem 5 are provided by the results obtained by application of Theorem 5 at the previous step. ■

The main result of Theorem 6 is yielding in sufficient conditions (4.79), which are independent of a special choice of a Jordan basis in P_0X, such that a holomorphic decomposition (4.80) of the group projection $P(z)$ exists. Moreover, the zero-order approximations $Q_\varrho^\sigma(0) = Q_\varrho^\sigma$ of the holomorphic decomposition operators can be characterized in a simple way in terms of the Jordan decomposition of P_0. In some sense, these zero-order approximations Q_ϱ^σ restrict the Jordan bases of A_0 in P_0X which are compatible with the perturbation $A(z)$. Namely, the Jordan bases which are compatible with the perturbation $A(z)$ are exactly those which are adapted to the decomposition $P_0 = \sum_{\varrho=1}^{i} \sum_{\sigma=1}^{k_\varrho} Q_\varrho^\sigma + R_0$.

7.4.9. Subgroups of perturbed eigenvalues and their asymptotics

In the following analysis, (4.79) of Theorem 6 is assumed to be satisfied. Then there is a decomposition of $P(z)$ of the form (4.80). This decomposition suggests a partition of the λ_0-group of perturbed eigenvalues into several subgroups.

As in Subsect. 7.3.1 one proves that the characteristic polynomials $\chi(\lambda, A(z) \restriction Q_\varrho^\sigma(z) X)$ and $\chi(\lambda, A(z) \restriction R(z) X)$ are polynomials of degree $\gamma_\varrho^{(\sigma)} m_\varrho$ and $\dim R_0$, respectively, whose coefficients are holomorphic in $|z| < \delta$ (use Subsect. 3.5.2 and 3.5.4 for this proof). Then, the relation

$$\chi(\lambda, A(z) \restriction P(z) X) = \prod_{\varrho=1}^{i} \prod_{\sigma=1}^{k_\varrho} \chi(\lambda, A(z) \restriction Q_\varrho^\sigma(z) X) \, \chi(\lambda, A(z) \restriction R(z) X) \tag{4.82}$$

represents a decomposition of χ into a product of polynomials whose coefficients are holomorphic in $|z| < \delta$.

The part spec $\{A(z) \upharpoonright Q_\varrho^\sigma(z) X\}$ of the λ_0-group of $A(z)$, i.e., the set of all roots of $\chi(\lambda, A(z) \upharpoonright Q_\varrho^\sigma(z) X)$, is called *the Q_ϱ^σ-group* of perturbed eigenvalues. Correspondingly, the part spec $\{A(z) \upharpoonright R(z) X\}$ is called the *R-group* of perturbed eigenvalues. The eigenvalues of the Q_ϱ^σ-groups and of the R-group can be characterized by certain asymptotic properties.

Theorem 7. *Assume that Condition* (4.79) *of Theorem* 6 *is satisfied with respect to the perturbation $A(z)$. Let $\lambda(z)$ denote a perturbed eigenvalue of the λ_0-group. Then: $\lambda(z)$ belongs to the Q_ϱ^σ-group if and only if*

$$\lambda(z) = \lambda_0 + (\mu_\varrho^\sigma z)^{1/m_\varrho} + o(z^{1/m_\varrho}), \qquad z \to 0. \tag{4.83}$$

$\lambda(z)$ belongs to the R-group if and only if

$$\lambda(z) = \lambda_0 + o(z^{1/m_i}), \qquad z \to 0. \tag{4.84}$$

Proof. Let $x(z)$ denote a perturbed eigenvector belonging to the eigenvalue $\lambda(z)$, where $x(0) = x_0 \neq 0$. Assume that $\lambda(z)$ belongs to the Q_ϱ^σ-group. Then $Q_\varrho^\sigma(z)\, x(z) = x(z)$ and, in particular, $Q_\varrho(z)\, x(z) = x(z)$. Together with $\big(A(z) - \lambda_0\big)\, x(z) = \big(\lambda(z) - \lambda_0\big)\, x(z)$, one obtains

$$\big(A(z) - \lambda_0\big) \left(P(z) - \sum_{\sigma=1}^{\varrho-1} Q_\sigma(z)\right) x(z) = \big(\lambda(z) - \lambda_0\big)\, x(z)$$

and

$$\left[\big(A(z) - \lambda_0\big) \left(P(z) - \sum_{\sigma=1}^{\varrho-1} Q_\sigma(z)\right)\right]^{m_\varrho} x(z) = \big(\lambda(z) - \lambda_0\big)^{m_\varrho}\, x(z);$$

that is,

$$B_\varrho(z)\, x(z) = \frac{\big(\lambda(z) - \lambda_0\big)^{m_\varrho}}{z}\, x(z), \qquad z \neq 0. \tag{4.85}$$

Because of $Q_\varrho^\sigma(z)\, x(z) = x(z)$, the number $z^{-1}\big(\lambda(z) - \lambda_0\big)^{m_\varrho}$ is an eigenvalue of the μ_ϱ^σ-group of $B_\varrho(z)$ and $x(z)$ is a corresponding eigenvector. Therefore $z^{-1}\big(\lambda(z) - \lambda_0\big)^{m_\varrho} \to \mu_\varrho^\sigma$ as $z \to 0$. Hence (4.83) holds for some branch of $(\mu_\varrho^\sigma z)^{1/m_\varrho}$. Assume that $\lambda(z)$ belongs to the R-group. Then $R(z)\, x(z) = x(z)$. One has

$$B_i(z) = \frac{1}{z} \left[\big(A(z) - \lambda_0\big) \big(Q_i(z) + R(z)\big)\right]^{m_i}.$$

Therefore,

$$B_i(z)\, x(z) = \frac{1}{z} \big(\lambda(z) - \lambda_0\big)^{m_i}\, x(z).$$

One has

$$\chi\big(\lambda, B_i(0) \upharpoonright (Q_i + R_0)\, X\big) = \lambda^{\dim R_0} \big[\chi\big(\lambda, B_i(0) \upharpoonright E_i X\big)\big]^{m_i},$$

in particular $\chi(\lambda, B_i(0) \upharpoonright R_0X) = \lambda^{\dim R_0}$. Then, $z^{-1}(\lambda(z) - \lambda_0)^{m_i} \to 0$ as $z \to 0$. Hence (4.84) holds. ■

Theorem 7 suggests the question whether to every branch of $(\mu_\varrho^\sigma z)^{1/m_\varrho}$ corresponds an eigenvalue $\lambda(z)$ of the Q_ϱ^σ-group with the asymptotic behavior (4.83). One obtains

Lemma 11. *Let g denote the degree of a prime factor of $\chi(\lambda, A(z) \upharpoonright Q_\varrho^\sigma(z)\, X)$ with respect to the decomposition into prime factors in $\mathscr{H}_{|z|<\varepsilon}[\lambda]$, where ε is sufficiently small (cf. Subsect. 3.3.6). Then m_ϱ is a divisor of g; i.e., there is a natural number k such that $g = km_\varrho$.*

Proof. Let $\lambda(z)$ denote an eigenvalue belonging to the cycle defined by the prime factor of degree g. Then there is a Puiseux series expansion for $\lambda(z)$ in $|z| < \varepsilon$ of the type

$$\lambda(z) = \lambda_0 + \alpha_1 z^{1/g} + \alpha_2 z^{2/g} + \cdots, \qquad |z| < \varepsilon.$$

On the other hand, as an eigenvalue of the Q_ϱ^σ-group, $\lambda(z)$ has an asymptotic expansion of the form

$$\lambda(z) = \lambda_0 + (\mu_\varrho^\sigma)^{1/m_\varrho}\, z^{1/m_\varrho} + o(z^{1/m_\varrho}), \qquad z \to 0.$$

Hence, the existence of a number k with the properties

$$\frac{k}{g} = \frac{1}{m_\varrho}, \qquad \alpha_1 = \alpha_2 = \cdots = \alpha_{k-1} = 0, \qquad \alpha_k = (\mu_\varrho^\sigma)^{1/m_\varrho},$$

follows. One obtains $\lambda(z) = \lambda_0 + (\mu_\varrho^\sigma)^{1/m_\varrho}\, z^{1/m\varrho} + \alpha_{k+1} z^{1/m\varrho + 1/g} + \cdots$ and Lemma 11 has been proved. ■

Lemma 11 answers the question asked before: If $z^{1/g}$ runs through all g branches of $z^{1/g}$, then z^{1/m_ϱ} runs through all m_ϱ branches of z^{1/m_ϱ}. That is, all m_ϱ branches of $(\mu_\varrho^\sigma z)^{1/m_\varrho}$ already appear in the different perturbed eigenvalues of only a single cycle within the Q_ϱ^σ-group.

The contents of Lemma 11 can also be expressed as follows: *The periods of cycles of the Q_ϱ^σ-group are multiples of m_ϱ.*

In connection with Theorem 6, from Lemma 11, one obtains the following result:

Corollary 1. *Assume that Condition (4.79) of Theorem 6 is satisfied with respect to the perturbation $A(z)$. If μ_ϱ^σ is a simple root of $\det(K_\varrho - \lambda E_\varrho)$, i.e., if $\gamma_\varrho^\sigma = 1$, then the Q_ϱ^σ-group of perturbed eigenvalues consists of only a single cycle with period m_ϱ, that is, the eigenvalues $\lambda_\alpha(z)$ of the Q_ϱ^σ-group can be expanded into Puiseux series of the form*

$$\lambda_\alpha(z) = \lambda_0 + (\mu_\varrho^\sigma z)^{1/m_\varrho} + \sum_{\nu=2}^{\infty} \gamma_\nu z^{\nu/m_\varrho}, \qquad |z| < \delta, \tag{4.86}$$

where α runs from 1 to m_ϱ and where the m_ϱ different eigenvalues are assigned to the m_ϱ different values of z^{1/m_ϱ}.

Proof. According to Theorem 6, one has $\dim Q_\varrho^\sigma(z) = m_\varrho$. On the other hand, according to Lemma 11, m_ϱ is a divisor of g, where g is the degree of any prime factor of $\chi(\lambda, A(z) \upharpoonright Q_\varrho^\sigma(z)\, X)$. Because $1 \leqq g \leqq m_\varrho$, $g = m_\varrho$; i.e., $\chi(\lambda, A(z) \upharpoonright Q_\varrho^\sigma(z)\, X)$ itself is irreducible. Hence, one obtains Corollary 1 (for example, because of Theorem 3.3.6; see also Subsect. A 5.5). ■

The case described by Corollary 1 is a case of "total splitting" within a fixed Q_ϱ^σ-group. This total splitting is due to the simplicity of the root μ_ϱ^σ. The zero-order approximation $Q_\varrho^\sigma = Q_\varrho^\sigma(0)$ of the holomorphic projection $Q_\varrho^\sigma(z)$ is an irreducible projection with respect to A_0 in this case.

In general, under the assumption of Theorem 6, the projection Q_ϱ^σ is the sum of γ_ϱ^σ disjoint irreducible projections with respect to A_0 of dimension m_ϱ.

7.4.10. Zero-order approximations of perturbed eigenvectors of Q_ρ^σ-groups

In this subsection, some properties of zero-order approximations of perturbed eigenvectors are derived under the assumptions (4.79) of Theorem 6. Note that, at first one only knows that $x_0 \in EX$ if $x(z)$ is a perturbed eigenvector with $x(0) = x_0 \neq 0$.

In the following lemma, ker $\{K_i - \mu_0 E_i\}$ denotes the kernel of $K_i - \mu_0 E_i$ restricted to $(E_1 + \cdots + E_i)\, X$.

Lemma 12. *Let the index i be fixed,* $1 \leqq i \leqq p$. *Furthermore, let* det $(K_i - \mu_0 E_i) = 0$. *Then* ker $\{K_i - \mu_0 E_i\}$ *is independent of the special choice of the Jordan basis of* A_0 *in* P_0X, *i.e., independent of the special choice of the operators* K_i *and* E_i.

Proof. Assume that a basis transformation

$$\tilde{e}_\varrho^\sigma = \sum_{\varrho',\sigma'} \alpha_{\varrho\varrho'}^{\sigma\sigma'} e_{\varrho'}^{\sigma'}, \qquad A = (\alpha_{\varrho\varrho'}^{\sigma\sigma'})$$

is given in $(E_1 + E_2 + \cdots + E_i)\, X$. The matrix of the operator K_i with respect to the first basis e_ϱ^σ is denoted by Φ; the matrix of the corresponding operator $\tilde{K}_i$ with respect to the second basis $\tilde{e}_\varrho^\sigma$ is denoted by $\tilde{\Phi}$; moreover, let

$$E_i = \begin{pmatrix} 0 & 0 \\ 0 & E \end{pmatrix},$$

where E denotes the unit matrix of order r_i and where E_i is an $(r_1 + r_2 + \cdots + r_i) \times (r_1 + r_2 + \cdots + r_i)$-matrix (cf. the proof of Lemma 10). Then,

$$(\tilde{\Phi} - \mu_0 E_i) = \bar{A}(\Phi - \mu_0 E_i)\, A^{\mathsf{T}}.$$

If $x = \sum_{\varrho,\sigma} \beta_\varrho^\sigma e_\varrho^\sigma = \sum_{\varrho,\sigma} \tilde{\beta}_\varrho^\sigma \tilde{e}_\varrho^\sigma$, then

$$(\tilde{\Phi} - \mu_0 E_i)\, \tilde{\beta} = \bar{A}(\Phi - \mu_0 E_i)\, A^{\mathsf{T}} \tilde{\beta} = \bar{A}(\Phi - \mu_0 E_i)\, \beta,$$

where β and $\tilde{\beta}$ represent columns with elements β_ϱ^σ and $\tilde{\beta}_\varrho^\sigma$, respectively. Hence, the relation

$$(\tilde{\Phi} - \mu_0 E_i)\, \tilde{\beta} = 0$$

is valid if and only if

$$(\Phi - \mu_0 E_i)\,\beta = 0.$$

In other words, the relation

$$(\tilde{K}_i - \mu_0 \tilde{E}_i)\,x = 0, \qquad x \in (E_1 + \cdots + E_i)\,X$$

is valid if and only if

$$(K_i - \mu_0 E_i)\,x = 0, \qquad x \in (E_1 + \cdots + E_i)\,X.$$

This means

$$\ker\,(\tilde{K}_i - \mu_0 \tilde{E}_i) = \ker\,(K_i - \mu_0 E_i).$$

Lemma 12 has been proved. ■

Lemma 12 can be used, under the assumptions of Theorem 6, to formulate characteristic relations for the zero-order approximations of perturbed eigenvectors without explicit use of the class of Jordan bases compatible with the perturbation, i.e., without explicit use of the decomposition (4.80).

Let $\lambda(z)$ be a perturbed eigenvalue of the Q_ϱ^σ-group and let $x(z)$ denote a corresponding perturbed eigenvector with $x(0) = x_0 \neq 0$. Then,

$$A(z)\,x(z) = \lambda(z)\,x(z), \qquad Q_\varrho^\sigma(z)\,x(z) = x(z).$$

Furthermore,

$$B_\varrho(z)\,x(z) = \frac{(\lambda(z) - \lambda_0)^{m_\varrho}}{z}\,x(z).$$

If $B_\varrho(0) = B_\varrho$, then

$$B_\varrho x_0 = \mu_\varrho^\sigma x_0, \qquad Q_\varrho^\sigma x_0 = x_0.$$

In particular, $Q_\varrho x_0 = x_0$ and, because $A_0 x_0 = \lambda_0 x_0$, one has

$$E_\varrho x_0 = x_0, \qquad x_0 \in E_\varrho X.$$

Note that $E_\varrho X$ denotes the geometric eigenspace of A_0 restricted to $Q_\varrho X$. $Q_\varrho A_0 = A_0 Q_\varrho$, and the projection Q_ϱ is well-defined by the properties of the perturbation $A(z)$. Q_ϱ is "compatible" with the perturbation $A(z)$; i.e., it is the zero-order approximation of a holomorphic reducing projection and it collects all irreducible parts of A_0 with fixed dimension m_ϱ. Furthermore, one has

$$B_\varrho = \sum_{\alpha+\beta=m_\varrho-1} D_0^\alpha (Q_\varrho + \tilde{R}_0)\,A_1 (Q_\varrho + \tilde{R}_0)\,D_0^\beta,$$

where $\tilde{R}_0$ denotes the rest projection $\tilde{R}_0 = \sum_{\sigma=\varrho+1}^{i} Q_\sigma + R_0$. Therefore,

$$B_\varrho x_0 = D_0^{m_\varrho - 1} (Q_\varrho + \tilde{R}_0)\,A_1 x_0.$$

According to Lemma 7, $E_\varrho X$ is invariant with respect to B_ϱ; i.e., $E_\varrho B_\varrho E_\varrho = B_\varrho E_\varrho$. Therefore,

$$B_\varrho x_0 = E_\varrho D_0^{m_\varrho - 1} A_1 E_\varrho x_0 = \mu_\varrho^\sigma x_0, \qquad E_\varrho x_0 = x_0.$$

The last equation may be written in the form

$$(K_\varrho - \mu_\varrho^\sigma E_\varrho)\, x_0 = 0. \tag{4.87}$$

Using the operator matrix form (4.56) of K_ϱ, one obtains

$$K_\varrho x_0 = \begin{pmatrix} K_{\varrho-1} & 0 \\ * & E_\varrho D_0^{m_\varrho-1} A_1 E_\varrho \end{pmatrix} \begin{pmatrix} 0 \\ x_0 \end{pmatrix} = \begin{pmatrix} 0 \\ E_\varrho D_0^{m_\varrho-1} A_1 E_\varrho x_0 \end{pmatrix} = \mu_\varrho^\sigma \begin{pmatrix} 0 \\ x_0 \end{pmatrix}$$
$$= \mu_\varrho^\sigma x_0 = \mu_\varrho^\sigma E_\varrho x_0.$$

According to Lemma 12, equation (4.87) has an invariant meaning independent of the special choice of the Jordan basis. In other word, equation (4.87), which is satisfied by the vector x_0, is independent of the special choice of the operators K_ϱ and E_ϱ. It is an equation in the invariant subspace $(E_1 + E_2 + \cdots + E_\varrho)\, X$. One obtains

Theorem 8. *Assume that Condition* (4.79) *of Theorem* 6 *is satisfied for the perturbation* $A(z)$. *If* $x(z)$ *denotes a perturbed eigenvector of the* Q_ϱ^σ*-group with* $x(0) = x_0 \neq 0$, *then the zero-order approximation* x_0 *satisfies the equation*

$$(K_\varrho - \mu_\varrho^\sigma E_\varrho)\, x_0 = 0, \qquad x_0 \in (E_1 + E_2 + \cdots + E_\varrho)\, X, \tag{4.88}$$

where the operators K_ϱ *and* E_ϱ *are defined with respect to any Jordan basis in* P_0X.

Equation (4.88) has at least one and at most γ_ϱ^σ linearly independent solutions. Thus one obtains the following corollary.

Corollary 2. *Assume that the assumptions of Theorem* 8 *are satisfied. Furthermore, assume that the number* μ_ϱ^σ *is a simple root of* $\det\,(K_\varrho - \lambda E_\varrho)$. *Then there is one and only one linearly independent zero-order approximation* x_0 *with respect to the perturbed eigenvectors* $x(z)$ *of the* Q_ϱ^σ*-group, and* x_0 *is uniquely determined* (*up to a scalar factor*) *by the equation*

$$(K_\varrho - \mu_\varrho^\sigma E_\varrho)\, x_0 = 0.$$

7.4.11. Special cases

The framework given by Theorem 6 could be weakened. This will not be done here, because new insights and points of view would not appear. Some special cases will be considered.

The considerations of the foregoing subsections suggest that the situation becomes especially simple if the roots in question are simple roots, that is, if a total splitting appears. In this case one obtains more precise insights into the structure of the perturbed eigenvalues as well as the zero-order approximations of perturbed reducing projections and eigenvectors.

Corollary 3. *Let*

$$\det K_1 \neq 0, \qquad \det K_2 \neq 0, \ldots, \det K_p \neq 0. \tag{4.89}$$

Moreover, assume that all roots of all pseudo-characteristic polynomials $\det(K_\varrho - \lambda E_\varrho)$ *are simple; i.e., let*

$$\det(K_\varrho - \lambda E_\varrho) = (-1)^{r_\varrho} \det K_{\varrho-1} \prod_{\sigma=1}^{r_\varrho} (\lambda - \mu_\varrho^\sigma), \qquad \varrho = 1, \ldots, p.$$

Then there is a decomposition of the group projection $P(z)$ *into a sum of holomorphic disjoint projections* $Q_\varrho^\sigma(z)$, *reducing* $A(z)$,

$$P(z) = \sum_{\varrho=1}^{p} \sum_{\sigma=1}^{r_\varrho} Q_\varrho^\sigma(z), \qquad |z| < \delta, \tag{4.90}$$

where $\dim Q_\varrho^\sigma(z) = m_\varrho$. *Moreover, the projections* $Q_\varrho^\sigma = Q_\varrho^\sigma(0)$ *are irreducible with respect to* A_0 *and the* Q_ϱ^σ*-group of perturbed eigenvalues consists of only a single cycle of period* m_ϱ; *i.e., there are Puiseux series expansions of the form*

$$\lambda_{\varrho\alpha}^\sigma(z) = \lambda_0 + (\mu_\varrho^\sigma z)^{1/m_\varrho} + \sum_{\nu=2}^{\infty} \gamma_\nu z^{\nu/m_\varrho}, \qquad |z| < \delta, \qquad \alpha = 1, 2, \ldots, m_\varrho,$$

and the zero-order approximation x_ϱ^σ *of the perturbed eigenvectors of the* Q_ϱ^σ*-group is uniquely determined (up to a scalar factor) by the equation*

$$(K_\varrho - \mu_\varrho^\sigma E_\varrho)\, x_\varrho^\sigma = 0.$$

Proof. The proof follows from Theorem 6 and from Corollaries 1 and 2. ∎

Corollary 3 describes a case of *total splitting at the first-order approximation*. In this case, $A(z)$ is a diagonal operator for $0 < |z| < \delta$.

In Theorem 6 and in the considerations after this theorem, A_0 restricted to P_0X was assumed equipped with the most general Jordan structure. The following remarks select some special cases.

Condition 3. *The irreducible parts of* A_0 *restricted to* P_0X *are assumed to have a fixed dimension* m.

This condition is equivalent with $p = 1$. If the number of irreducible parts of dimension m is denoted by r, then $n = rm$. In this case, there is only one geometric eigenspace EX, where $\dim EX = r$. Correspondingly, there is only one Lidskii operator, namely

$$K = ED_0^{m-1} A_1 E,$$

and there is only one operator-valued function

$$B(z) = \frac{1}{z} \left[\left(A(z) - \lambda_0\right) P(z)\right]^m.$$

The assumptions of Theorem 6 take the simple form

$$\det K \neq 0.$$

There is only one pseudo-characteristic polynomial and this is a "proper" characteristic polynomial

$$\det (K - \lambda E) = (-1)^r \prod_{\sigma=1}^{k} (\lambda - \mu_\sigma)^{\gamma_\sigma}, \qquad \mu_\sigma \neq 0, \qquad \sigma = 1, \ldots, k. \tag{4.91}$$

According to Theorem 6, to the decomposition (4.91) corresponds a decomposition of the group projection

$$P(z) = Q_1(z) + Q_2(z) + \cdots + Q_k(z),$$

where

$$Q_\varrho(z) = \frac{1}{2\pi i} \int_{(\mu_\varrho)} (\lambda - B(z))^{-1} \, d\lambda$$

and $\dim Q_\varrho(z) = \gamma_\varrho m$. The perturbed eigenvalues $\lambda(z)$ of the Q_ϱ-group are characterized by the asymptotic relation

$$\lambda(z) = \lambda_0 + (\mu_\varrho z)^{1/m} + o(z^{1/m}),$$

and the zero-order approximations x_0 of perturbed eigenvectors of the Q_ϱ-group satisfy the equation

$$(K - \mu_\varrho E)\, x_0 = 0, \qquad x_0 \in EX.$$

A further specialization is given by

Condition 4. *The unperturbed eigenvalue λ_0 of A_0 is assumed prime; i.e., it has the geometric multiplicity* 1; *i.e., P_0 itself is irreducible.*

Then $r = 1$; i.e., the geometric eigenspace EX is one-dimensional. In this case, K is a one-dimensional (scalar) operator; i.e.,

$$Kx_0 = D_0^{n-1} P_0 A_1 x_0 = \mu x_0, \qquad x_0 \in EX, \qquad \dim P_0 = n,$$

where μ is a number. The assumption in Theorem 6 means

$$\mu \neq 0. \tag{4.92}$$

In other words,

$$(A_1 e, f) \neq 0,$$

where $e \in EX$ and $f \in E^*X^*$ (i.e., f is an eigenvector of A_0^* with respect to $\bar{\lambda}_0$).

According to Theorem 6, assumption (4.92) leads to total splitting: The λ_0-group consists of a single cycle of period n,

$$\lambda_\sigma(z) = \lambda_0 + (\mu z)^{1/n} + \sum_{\alpha=2}^{\infty} \gamma_\alpha z^{\alpha/n}, \qquad |z| < \delta;$$

the eigenvalues $\lambda_\sigma(z)$ are simple for $0 < |z| < \delta$.

7.4.12. Example

The framework for the investigations of Subsect. 7.4.8 to 7.4.11 is essentially fixed by Theorem 6. If λ_0 is prime, then the contents of the assumptions of Theorem 6 are very simple: There are only the alternatives $\mu \neq 0$ and $\mu = 0$, and Theorem 6 treats the case $\mu \neq 0$. In this subsection, an example is given for the other case, $\mu = 0$.

Let $X = \mathbb{C}^n$ and $A_0 = \Gamma_n(\lambda_0)$, $n \geqq 2$. Furthermore, let $A_1 = (a_{\nu\mu})$ be any $n \times n$-matrix. Let $A(z) := A_0 + zA_1$. A Jordan basis $e_1, e_2, \ldots, e_n$ of A_0 is given by $e_1 = (1, 0, \ldots, 0)$, $e_2 = (0, 1, 0, \ldots, 0)$, $\ldots$, $e_n = (0, 0, \ldots, 0, 1)$, where e_1 is an eigenvector and where e_n is a cyclic (generating) vector. With the scalar product $(x, y) = \sum x_i\bar{y}_i$, the corresponding biorthogonal basis is $f_1 = e_1, f_2 = e_2, \ldots, f_n = e_n$, where f_1 is cyclic (generating) with respect to A_0^* and f_n is an eigenvector. One assumes

$$\mu = (A_1e_1, f_n) = a_{n1} = 0.$$

Then,

$$\det\big(\lambda - A(z)\big) = (\lambda - \lambda_0)^n + \cdots + (-1)\,(a_{n-1,1} + a_{n,2})\, z(\lambda - \lambda_0).$$

If, additionally,

$$c = a_{n-1,1} + a_{n,2} \neq 0,$$

then, for the equation

$$\mu^n + \cdots + (-1)\, cz\mu = 0,$$

one obtains the Puiseux-Newton-diagram shown in Fig. 5 (see Sect. A 7).

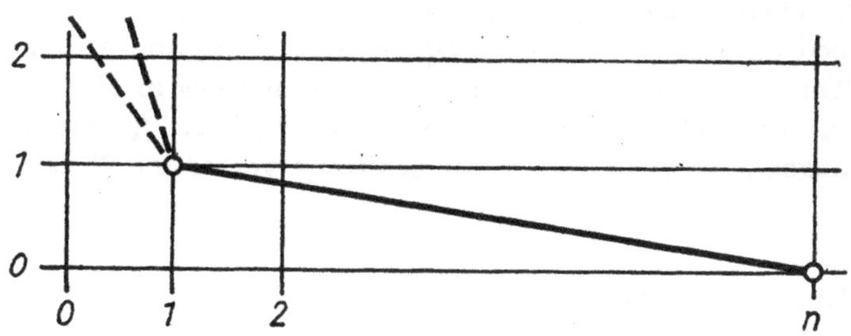

Fig. 5

Therefore, possible solutions $\mu = \mu(z)$ are

$$\mu_1(z) = \varepsilon_1 z^{\frac{1}{n-1}} + \cdots; \qquad \mu_2(z) = \delta_1 z + \delta_2 z^2 + \cdots.$$

One obtains $\varepsilon_1 = c^{1/(n-1)}$. Therefore, the λ_0-group of $A(z)$ consists of two cycles; one of them has the period $n - 1$, the other one has the period 1,

$$\lambda_\sigma(z) = \lambda_0 + (cz)^{\frac{1}{n-1}} + o\left(z^{\frac{1}{n-1}}\right), \qquad z \to 0, \qquad \sigma = 1, \ldots, n-1,$$

$$\lambda_0(z) = \lambda_0 + \lambda_1 z + \lambda_2 z^2 + O(z^3), \qquad z \to 0.$$

In this connection, cf. M. M. Vainberg and V. A. Trenogin [1, p. 448, Theorem 32.4].

7.4.13. Calculation of the perturbation series by recursion

As in the reduction theory for semisimple eigenvalues, the question arises how to calculate the perturbation series, given qualitatively for perturbed eigenvalues and eigenvectors, for example in the case of total splitting in the first-order approximation within a fixed Q_ϱ^σ-group. The coefficients of these perturbation series can be calculated recursively.

To avoid complications in the formulas, the corresponding calculation will be given here only for the case of a prime unperturbed eigenvalue λ_0. This recursive method of calculating the coefficients of perturbation series for non-semisimple unperturbed eigenvalues goes back to M. I. Višik and L. A. Ljusternik [2]. These authors have to derive from the formal series, all those properties which are given here a priori because of Theorem 6 and its implications. For the following, see also M. M. Vainberg and V. A. Trenogin [1, p. 450 sqq.].

For short, one considers only linear perturbations $A(z) = A_0 + zA_1$. First Condition 4 holds and, in particular, (4.92). If $\dim P_0 = n$, the perturbed eigenvalues take the form

$$\lambda_\sigma(z) = \lambda_0 + \mu^{1/n} z^{1/n} + \sum_{\varrho=2}^{\infty} \lambda_\varrho z^{\varrho/n}, \qquad \sigma = 1, 2, \ldots, n, \tag{4.93}$$

and the perturbed eigenvectors can be expanded into a series of the corresponding form

$$x_\sigma(z) = x_0 + \sum_{\varrho=1}^{\infty} x_\varrho z^{\varrho/n}, \qquad \sigma = 1, \ldots, n. \tag{4.94}$$

One has to calculate the coefficients $\lambda_2, \lambda_3, \ldots$ and $x_1, x_2, x_3, \ldots$. $x_0 \neq 0$ is an arbitrary, but fixed, unperturbed eigenvector. Let $e_1 = D_0^{n-1} P_0 A_1 x_0 = \mu x_0$ and $e_\varrho = D_0^{n-\varrho} P_0 A_1 x_0$ for $\varrho = 2, 3, \ldots, n$. The vectors $e_1, e_2, \ldots, e_n$ form a Jordan basis in $P_0 X$, where e_1 is an eigenvector and e_n is a cyclic (generating) vector. By $f_1, f_2, \ldots, f_n$, one denotes the corresponding biorthogonal basis in $P_0^* X^*$. An arbitrary vector $x \in X$ can be represented in the form

$$x = \sum_{\varrho=1}^{n} (x, f_\varrho)\, e_\varrho + (1 - P_0)\, x.$$

Note that, from the beginning, the coefficients $x_1, x_2, \ldots$ are not uniquely defined, because one can multiply $x_\sigma(z)$ by an arbitrary scalar power series in $z^{1/n}$ without disturbing the character of $x_\sigma(z)$ as an analytic eigenvector. Since $(x_0, f_1) = \mu^{-1}(e_1, f_1)$

$= \mu^{-1}$, one imposes (as is done by the normalization rule of M. I. VIŠIK and L. A. LJUSTERNIK [2]) the condition

$$\big(x(z), f_1\big) = \mu^{-1}$$

for a perturbed eigenvector $x(z)$. For the coefficients $x_1, x_2, x_3, \ldots$ in (4.93), the relations

$$(x_\varrho, f_1) = 0, \qquad \varrho = 1, 2, 3, \ldots,$$

follow and one obtains

$$x_\varrho = (x_\varrho, f_2)\, e_2 + (x_\varrho, f_3)\, e_3 + \cdots + (x_\varrho, f_n)\, e_n + (1 - P_0)\, x_\varrho;$$

that is, the vectors x_ϱ do not possess any e_1-part. This normalization implies that the coefficients $x_1, x_2, x_3, \ldots$ are uniquely determined.

Using the coefficients $\lambda_1, \lambda_2, \lambda_3, \ldots$, one forms the expressions

$$\alpha_\varrho^{(k)} = \sum_{\substack{i_1+i_2+\cdots+i_\varrho=k \\ 1\leq i_\sigma\leq k}} \lambda_{i_1}\lambda_{i_2} \cdots \lambda_{i_\varrho}.$$

These expressions appear already in the paper of M. I. VIŠIK and L. A. LJUSTERNIK [2] (cf. also M. M. VAINBERG and V. A. TRENOGIN [1, p. 450sqq.]). The symbol $\alpha_\varrho^{(k)}$ is a function of $\lambda_1, \lambda_2, \ldots, \lambda_{k-\varrho+1}$,

$$\alpha_\varrho^{(k)} = \alpha_\varrho^{(k)}(\lambda_1, \lambda_2, \ldots, \lambda_{k-\varrho+1}).$$

The relation

$$\left(\sum_{\alpha=1}^{\infty} \lambda_\alpha z^\alpha\right)^\varrho = \sum_{k=\varrho}^{\infty} \alpha_\varrho^{(k)} z^k$$

holds in the sense of an identity of formal power series. Besides the symbols $\alpha_\varrho^{(k)}$, one needs the expressions

$$\gamma_n^{(n+k)} = \sum_{\substack{i_1+i_2+\cdots+i_n=n+k \\ 1\leq i_\varrho\leq k}} \lambda_{i_1}\lambda_{i_2} \cdots \lambda_{i_n}, \qquad n = 1, 2, 3, \ldots, \qquad k = 1, 2, 3, \ldots$$

For example, one has $\gamma_n^{(n+1)} = 0$, $\gamma_n^{(n+2)} = \binom{n}{2} \lambda_1^{n-2}\lambda_2^2$. The symbol $\gamma_n^{(n+k)}$ is a function of $\lambda_1, \lambda_2, \ldots, \lambda_k$,

$$\gamma_n^{(n+k)} = \gamma_n^{(n+k)}(\lambda_1, \lambda_2, \ldots, \lambda_k).$$

One has

$$\alpha_n^{(n+k)} = n\lambda_1^{n-1}\lambda_{k+1} + \gamma_n^{(n+k)}, \qquad n = 1, 2, 3, \ldots, \qquad k = 1, 2, 3, \ldots \tag{4.95}$$

From the eigenvalue equation $A(z)\, x_\sigma(z) = \lambda_\sigma(z)\, x_\sigma(z)$, by comparing the coefficients, one obtains the following relations

$$(A_0 - \lambda_0)\, x_k = \lambda_1 x_{k-1} + \lambda_2 x_{k-2} + \cdots + \lambda_{k-1} x_1 + \lambda_k x_0, \qquad 1 \leqq k \leqq n - 1,$$

$$(A_0 - \lambda_0)\, x_k = \lambda_1 x_{k-1} + \lambda_2 x_{k-2} + \cdots + \lambda_{k-1} x_1 + \lambda_k x_0 - A_1 x_{k-n}, \qquad n \leqq k.$$

According to these relations,

$$(A_0 - \lambda_0)\, x_1 = \lambda_1 x_0 = \lambda_1 \mu^{-1} D_0^{n-1} P_0 A_1 x_0 = \lambda_1^{-n} \lambda_1 e_1;$$

hence,

$$x_1 = \lambda_1^{-n} \lambda_1 e_2.$$

Furthermore,

$$(A_0 - \lambda_0)\, x_2 = \lambda_1 x_1 + \lambda_2 x_0 = \lambda_1^{-n}(\lambda_1^2 e_2 + \lambda_2 e_1);$$

hence,

$$x_2 = \lambda_1^{-n}(\lambda_1^2 e_3 + \lambda_2 e_2).$$

If $3 < n$, then one obtains

$$x_3 = \lambda_1^{-n}(\lambda_1^3 e_4 + 2\lambda_1\lambda_2 e_3 + \lambda_3 e_2).$$

By induction,

$$(A_0 - \lambda_0)\, x_k = \lambda_1^{-n}(\alpha_k^{(k)} e_k + \alpha_{k-1}^{(k)} e_{k-1} + \cdots + \alpha_1^{(k)} e_1), \qquad 0 < k < n,$$

and

$$x_k = \lambda_1^{-n}(\alpha_k^{(k)} e_{k+1} + \alpha_{k-1}^{(k)} e_k + \cdots + \alpha_1^{(k)} e_2), \qquad 0 < k < n.$$

Furthermore,

$$\begin{aligned}(A_0 - \lambda_0)\, x_n &= \lambda_1^{-n}(\alpha_n^{(n)} e_n + \alpha_{n-1}^{(n)} e_{n-1} + \cdots + \alpha_1^{(n)} e_1) - A_1 x_0\\ &= e_n - A_1 x_0 + \lambda_1^{-n}(\alpha_{n-1}^{(n)} e_{n-1} + \cdots + \alpha_1^{(n)} e_1)\\ &= P_0 A_1 x_0 - A_1 x_0 + \lambda_1^{-n}(\cdots) = -Q_0 A_1 x_0 + \lambda_1^{-n}(\cdots),\end{aligned}$$

where $Q_0 = 1 - P_0$. Denoting by R_0 the reduced resolvent $(\lambda_0 - A_0)^{-1}$, which is well-defined on $Q_0 X$, one obtains

$$x_n = R_0 Q_0 A_1 x_0 + \lambda_1^{-n}(\alpha_{n-1}^{(n)} e_n + \cdots + \alpha_1^{(n)} e_2).$$

Furthermore,

$$\begin{aligned}(A_0 - \lambda_0)\, x_{n+1} = {}& \lambda_1 Q_0 x_n - Q_0 A_1 x_1 + e_n\big(\lambda_1^{-n}\alpha_n^{(n+1)} - (A_1 x_1, f_n)\big)\\ &+ e_{n-1}\big(\lambda_1^{-n}\alpha_{n-1}^{(n+1)} - (A_1 x_1, f_{n-1})\big) + \cdots + e_2\big(\lambda_1^{-n}\alpha_2^{(n+1)} - (A_1 x_1, f_2)\big)\\ &+ e_1\big(\lambda_1^{-n}\alpha_1^{(n+1)} - (A_1 x_1, f_1)\big).\end{aligned}$$

The condition that this equation has a solution is simultaneously an equation for determining the coefficient λ_2. One obtains

$$\lambda_1^{-n}\alpha_n^{(n+1)} - (A_1 x_1, f_n) = 0;$$

hence,

$$\lambda_2 = \frac{1}{n}\, \lambda_1 (A_1 x_1, f_n).$$

Then,

$$\begin{aligned}x_{n+1} = {}& R_0(Q_0 A_1 x_1 - \lambda_1 Q_0 x_n) + e_n\big(\lambda_1^{-n}\alpha_{n-1}^{(n+1)} - (A_1 x_1, f_{n-1})\big)\\ &+ e_{n-1}\big(\lambda_1^{-n}\alpha_{n-2}^{(n+1)} - (A_1 x_1, f_{n-2})\big) + \cdots + e_2\big(\lambda_1^{-n}\alpha_1^{(n+1)} - (A_1 x_1, f_1)\big),\end{aligned}$$

and, inductively, for $k \geqq 1$,

$$\begin{aligned}(A_0 - \lambda_0)\, x_{n+k} &= \lambda_1 Q_0 x_{n+k-1} + \lambda_2 Q_0 x_{n+k-2} + \cdots + \lambda_k Q_0 x_n - Q_0 A_1 x_k \\ &+ e_n \left(\lambda_1^{-n} \alpha_n^{(n+k)} - \left[(A_1 x_k, f_n) + \sum_{\beta=1}^{n-1} \sum_{\sigma=\beta}^{k-1} \alpha_\beta^{(\sigma)} (A_1 x_{k-\sigma}, f_{n-\beta})\right]\right) \\ &+ \sum_{\varrho=1}^{n-1} e_\varrho \left(\lambda_1^{-n} \alpha_\varrho^{(n+k)} - \left[(A_1 x_k, f_\varrho) + \sum_{\beta=1}^{\varrho-1} \sum_{\sigma=\beta}^{k-1} \alpha_\beta^{(\sigma)} (A_1 x_{k-\sigma}, f_{\varrho-\beta})\right]\right),\end{aligned}$$

where the sum $\sum_{\beta=1}^{\varrho-1}$ vanishes for $\varrho = 1$. For $\beta > k - 1$, the sum $\sum_{\sigma=\beta}^{k-1}$ vanishes; it vanishes in particular for $k = 1$. The sum $\sum_{\beta=1}^{n-1}$ vanishes for $n = 1$. The equation which determines λ_{k+1} uniquely takes the form

$$\lambda_1^{-n} \alpha_n^{(n+k)} = (A_1 x_k, f_n) + \sum_{\beta=1}^{n-1} \sum_{\sigma=\beta}^{k-1} \alpha_\beta^{(\sigma)} (A_1 x_{k-\sigma}, f_{n-\beta}),$$

where $n \geqq 1$, $k \geqq 1$. Because of (4.95), one obtains

$$\lambda_{k+1} = \frac{\lambda_1}{n} \left[-\lambda_1^{-n} \gamma_n^{(n+k)} + (A_1 x_k, f_n) + \sum_{\beta=1}^{n-1} \sum_{\sigma=\beta}^{k-1} \alpha_\beta^{(\sigma)} (A_1 x_{k-\sigma}, f_{n-\beta})\right]. \tag{4.96}$$

Furthermore, for $k \geqq 1$ one obtains

$$\begin{aligned}x_{n+k} &= R_0 (Q_0 A_1 x_k - [\lambda_1 Q_0 x_{n+k-1} + \lambda_2 Q_0 x_{n+k-2} + \cdots + \lambda_k Q_0 x_n]) \\ &+ \sum_{\varrho=2}^{n} e_\varrho \left(\lambda_1^{-n} \alpha_{\varrho-1}^{(n+k)} - \left[(A_1 x_k, f_{\varrho-1}) + \sum_{\beta=1}^{\varrho-2} \sum_{\sigma=\beta}^{k-1} \alpha_\beta^{(\sigma)} (A_1 x_{k-\sigma}, f_{\varrho-\beta-1})\right]\right).\end{aligned} \tag{4.97}$$

By these formulas, the coefficients $\lambda_1, \lambda_2, \lambda_3, \ldots$ and $x_1, x_2, x_3, \ldots$ of the Puiseux series (4.93), (4.94) are determined recursively and uniquely.

7.5. Bifurcation equations

Some results of the reduction theory proved in the foregoing sections can also be obtained by another method, namely the method of bifurcation equations.

Originally, this method was developed for non-linear problems (E. Schmidt [1]), but it can also be applied to problems of linear perturbation theory. For example, these applications are presented in the monograph of M. M. Vainberg and V. A. Trenogin [1, pp. 440—461]. For this reason (and also for brevity), only a short sketch of this method will be given in this book. Furthermore, this method provides only results about the splitting behavior and the function-theoretic dependence of perturbed eigenvalues and eigenvectors. Results on holomorphic decomposition of group projections, as proved in § 4, remain completely outside the scope of this method. The unperturbed Jordan vectors appear only as auxiliary objects and they are taken into account in order to explain, for example, why there may exist many more splitting eigenvalues than unperturbed linearly independent eigenvectors.

In some cases (for example, if the unperturbed eigenvalue is simple), this method can be used to determine perturbed eigenvalues and eigenvectors iteratively (P. ROSENBLOOM [1]).

7.5.1. Derivation of the bifurcation equations

The starting point is an eigenvalue equation

$$A_0 x_0 = \lambda_0 x_0, \qquad x_0 \neq 0.$$

Let

$$\dim \ker (A_0 - \lambda_0) = \dim \ker D_0 = r > 0;$$

r denotes the geometric multiplicity of the unperturbed eigenvalue λ_0. Let $A(z) = A_0 + H(z)$ be a holomorphic perturbation of A_0. A perturbed eigenvalue $\lambda(z) = \lambda_0 + \mu(z)$ and a corresponding perturbed eigenvector $x(z)$ together satisfy the perturbed eigenvalue equation $A(z)\, x(z) = \lambda(z)\, x(z)$, or

$$(A_0 - \lambda_0)\, x(z) = -H(z)\, x(z) + \mu(z)\, x(z). \tag{5.1}$$

Let $x_1, x_2, \ldots, x_r$ be a basis of $\ker (A_0 - \lambda_0)$ and $f_1, f_2, \ldots, f_r$ any corresponding biorthogonal basis. Moreover, let $g_1, g_2, \ldots, g_r$ be a basis of $\{\mathrm{ima}\,(A_0 - \lambda_0)\}^\perp = \ker (A_0 - \lambda_0)^*$ and $y_1, y_2, \ldots, y_r$ any corresponding biorthogonal basis. In accordance with E. Schmidt, one introduces the operator

$$\tilde{B}_0 = (A_0 - \lambda_0) + \sum_{i=1}^{r} (\cdot, f_i)\, y_i. \tag{5.2}$$

It satisfies $\tilde{B}_0 x_i = y_i$, $i = 1, 2, \ldots, r$. Moreover, $\tilde{B}_0$ is invertible. Let $\tilde{B}_0^{-1} = C$. The perturbed eigenvalue equation (5.1) is equivalent with

$$\tilde{B}_0 x(z) = -H(z)\, x(z) + \mu(z)\, x(z) + \sum_{i=1}^{r} \xi_i y_i, \qquad \xi_i = \big(x(z), f_i\big),$$

and with

$$x(z) = -CH(z)\, x(z) + \mu(z)\, Cx(z) + \sum_{i=1}^{r} \xi_i x_i.$$

If $|z|$ and also $|\mu(z)|$ are sufficiently small, then

$$x(z) = \sum_{i=1}^{r} \xi_i \big(1 + CH(z) - \mu(z)\, C\big)^{-1} x_i.$$

Substituting this expression into the equations

$$\xi_i = \big(x(z), f_i\big), \qquad i = 1, \ldots, r,$$

one obtains a linear homogeneous system of equations for the ξ_i,

$$\sum_{j=1}^{r} a_{ij}(z, \mu)\, \xi_i = 0, \qquad i = 1, 2, \ldots, r, \tag{5.3}$$

where the coefficients $a_{ij}(z, \mu)$ are holomorphic functions of both complex variables z and μ in a neighborhood of the point $(z, \mu) = (0, 0)$. For the existence of a nontrivial solution of (5.3), the equation

$$\det [a_{ij}(z, \mu)] = 0 \tag{5.4}$$

is necessary and sufficient. This equation is called *bifurcation equation*. It provides the perturbed eigenvalues $\mu = \mu(z)$ as algebroid functions in a neighborhood of $z = 0$.

Discussing the bifurcation equation, one can obtain propositions on the behavior of perturbed eigenvalues and eigenvectors. A collection of such propositions, all of which were also obtained in Sect. 7.4, can be found in the monograph of M. M. Vainberg and V. A. Trenogin [1, pp. 445—455].

7.5.2. Simple unperturbed eigenvalue

In the case of a simple unperturbed eigenvalue one obviously has $\dim P_0 = 1$. Let $x_0 \neq 0$ be an unperturbed eigenvector. According to Subsect. 7.1.2 one chooses a normalization for a perturbed eigenvector, for example the linear one. Therefore, if f_0 is an eigenvector of A_0^* with respect to $\bar{\lambda}_0$ and with $(x_0, f_0) = 1$, the normalization rule reads

$$\big(x(z), f_0\big) = 1,$$

or, equivalently,

$$\big(x(z) - x_0, f_0\big) = 0.$$

Since $(y, f_0) = 0$ if and only if $P_0 y = 0$, this relation is equivalent with

$$P_0\big(x(z) - x_0\big) = 0;$$

that is, one obtains

$$x(z) - x_0 = y(z) \in (1 - P_0)\, X.$$

The eigenvalue equation $\big(A(z) - \lambda(z)\big)\, x(z) = 0$ can be written in the form

$$(A_0 - \lambda_0)\, y(z) = \big(\mu(z) - H(z)\big)\, x(z). \tag{5.5}$$

Multiplying (5.5) from the left by the reduced resolvent S_0 of A_0 with respect to λ_0 and using

$$(\lambda_0 - A_0)\, S_0 = S_0(\lambda_0 - A_0) = 1 - P_0,$$

one obtains

$$y(z) = S_0\big(H(z) - \mu(z)\big)\, x(z)$$

or

$$x(z) - x_0 = S_0\big(H(z) - \mu(z)\big)\, x(z).$$

Hence,

$$x(z) = \left[1 - S_0\big(H(z) - \mu(z)\big)\right]^{-1} x_0$$

and, moreover,

$$y(z) = \left(\left[1 - S_0\big(H(z) - \mu(z)\big)\right]^{-1} - 1\right) x_0 .$$

Therefore,

$$y(z) = \left[1 - S_0\big(H(z) - \mu(z)\big)\right]^{-1} S_0\big(H(z) - \mu(z)\big)\, x_0$$

and because $S_0 x_0 = 0$, one finally obtains

$$y(z) = \left[1 - S_0\big(H(z) - \mu(z)\big)\right]^{-1} S_0 H(z)\, x_0 . \tag{5.6}$$

That is, if $\mu(z)$ is already known, then (5.6) is a convenient formula to calculate the perturbed eigenvector.

However, (5.5) also implies

$$\big(\mu(z)\, x(z) - H(z)\, x(z), f_0\big) = 0 .$$

Hence,

$$\mu(z) = \big(H(z)\, x(z), f_0\big) . \tag{5.7}$$

Substituting (5.7) into (5.5), one obtains

$$(A_0 - \lambda_0)\, y(z) = \left[\big(H(z)\, x(z), f_0\big) - H(z)\right] x(z)$$

and

$$y(z) = S_0\left[H(z) - \big(H(z)\, x(z), f_0\big)\right] x(z) .$$

From the last equation, one can extract a nonlinear iteration scheme for the perturbed eigenvector $x = x(z)$ at a point z:

$$x_{n+1} = x_0 + S_0[H - (Hx_n, f_0)]\, x_n , \qquad n = 0, 1, 2, \ldots, \tag{5.8}$$

the convergence of which can be verified for sufficiently small $|z|$ by Banach's fixed point theorem, for example (cf. P. Rosenbloom [1]).

Such an iteration scheme can also be used for the calculation of perturbed eigenvalues and eigenvectors for semisimple unperturbed eigenvalues if it is combined with the reduction process presented in the foregoing sections.

8. Numerical analysis for perturbation series – convergence radii and error estimates

The concept of group projection for a λ_0-group of a perturbed eigenvalue was defined in Subsect. 3.4.3. The perturbation series for group projections and mean values of λ_0-groups were calculated in Sect. 3.8. Recall that the problems of estimating convergence radii and finding error estimates were unresolved.

In this chapter, such estimates for the perturbation series will be presented. Essentially, there are two methods for this purpose.

The starting point for the first method is the second Neumann series

$$\big(\lambda - A(z)\big)^{-1} = (\lambda - A_0)^{-1} \sum_{n=1}^{\infty} \big[\big(A(z) - A_0\big)(\lambda - A_0)^{-1}\big]^n, \qquad (0.1)$$

together with conditions for its convergence. By the Cauchy-Riesz integral, these conditions are transmitted to the group projection

$$P(z) = \frac{1}{2\pi i} \int\limits_{(\lambda_0)} \big(\lambda - A(z)\big)^{-1} \, d\lambda$$

and to its perturbation series, hence also to the perturbation series for the mean value $\lambda(z)$ (see B. v. Sz.-Nagy [1, 2]; T. Kato [1, 3, 4, 5]. For the case of a selfadjoint unperturbed operator, see also F. W. Schäfke [1, 2] and, especially, [3]).

If the unperturbed eigenvalue λ_0 is simple, then this approach provides estimates for the perturbed eigenvalue $\lambda(z)$. If λ_0 is not simple, then this method has to be combined with the reduction process to obtain estimates for the perturbed eigenvalues.

For example, if λ_0 is semisimple and if $\chi(\lambda, P_0 A_1 P_0 \restriction P_0 X)$ has a simple root λ_1, i.e., if there is a simple "first-order approximation" λ_1, then this method can be applied to the perturbation $z^{-1}\big(A(z) - \lambda_0\big) P(z)$ with the unperturbed and *simple* eigenvalue λ_1. Then the result is an estimate for *the* perturbed eigenvalue which is characterized by the expansion $\lambda(z) = \lambda_0 + z\lambda_1 + \cdots$ and an estimate for the corresponding eigenprojection. (This method was essentially used by B. v. Sz.-Nagy [1, p. 362 sqq.] to obtain estimates for perturbed eigenvalues when the unperturbed eigenvalue is a multiple one. See als F. Rellich [4].)

The second method is called the *method of majorizing series*. This method can be used in several ways. As in the first method, one may obtain conditions for the convergence of the second Neumann series (0.1) by the construction of assigned majorizing series. By function-theoretic considerations, one can derive further majorizing series from a majorizing series for the second Neumann series, for example for $P(z)$, $\lambda(z)$ and other spectral objects. From these majorizing series, one can obtain convergence conditions and error estimates.

It is also possible to derive majorizing series for a given perturbation series directly. For example, one can start with assumptions which ensure a definite splitting behavior for λ_0, according to the reduction process (for example, let λ_0 be semisimple and assume the existence of a simple first-order approximation). Then, the qualitative form of the corresponding perturbation series for the perturbed eigenvalue is known. Thus, it is possible to establish recursion formulas for the coefficients of that perturbation series. From these recursion formulas one can sometimes construct majorizing series representing simple functions, which, for example, satisfy algebraic equations. In this way, one obtains estimates for convergence radii and error estimates for the perturbation series in question. (This method was used first in perturbation theory by F. Rellich [4]. Later, it was extended and refined by J. Schröder [1, 2, 3]. See also M. K. Gavurin [1, 2], B. V. Loginov [1], B. V. Loginov and V. E. Pospeev [1], M. I. Višik and L. A. Ljusternik [2], H. Baumgärtel [5].) This variant of the method of majorizing series is called the *recursion method of majorizing series*.

Let $A(z) = A_0 + zA_1 + z^2A_2 + \cdots$ be a selfadjoint perturbation; let λ_0 be simple and assume that the relation

$$\|A_n\| \leqq ac^{n-1}, \qquad n = 1, 2, 3, \ldots, \qquad a > 0, \qquad c > 0,$$

holds. Furthermore, let d denote the distance from λ_0 to the rest spectrum of A_0. By the latter method, F. Rellich [4, p. 361] obtains the following estimate for the convergence radius r of the perturbed eigenvalue $\lambda(z)$:

$$r \geqq (c + 16ad^{-1})^{-1}. \tag{0.2}$$

In particular, if $A(z) = A_0 + zA_1$ is a linear perturbation, i.e., if $c = 0$, then one obtains the estimate $r \geqq (1/16)\, da^{-1}$.

The estimate (0.2) of F. Rellich has been improved by the first method described above. B. v. Sz.-Nagy obtained the estimate $r \geqq (c + 4ad^{-1})^{-1}$; finally, B. v. Sz.-Nagy [2], T. Kato [4] and F. W. Schäfke [3] obtained the estimate $r \geqq (c + 2ad^{-1})^{-1}$.

The estimate (0.2) can also be improved by application of the method of majorizing series and its recursion variant.

For example, let $A(z) = A_0 + zA_1$ be any linear perturbation (i.e., not necessarily normal or selfadjoint), let λ_0 be simple, $A_0x_0 = \lambda_0x_0$ and $A_0^*x_0 = \bar{\lambda}_0x_0$ (where X is assumed to be unitary such that $X = X^*$). Then, by the recursion method of majorizing series, one can improve the estimate to $r \geqq (1/6)\, (\|A_1\| \cdot \|S_0\|)^{-1}$. Here S_0 denotes the reduced resolvent. (See H. Baumgärtel [5]; Loginov [1] obtains $r \geqq (1/8)\, (\|A_1\| \cdot \|S_0\|)^{-1}$, by means of the same method.) Using the method of majorizing series, T. Kato [5, p. 93] improves the estimate to $r \geqq (1/4)\, (\|A_1\| \cdot \|S_0\|)^{-1}$.

If A_0 is normal, then $\|S_0\| = d^{-1}$ so that the estimates just mentioned can be compared directly with (0.2) and its improvements.

The recursion method of majorizing series seems to be more flexible than the other methods, at least it is still applicable in cases where the splitting behavior is more or less complicated (for example, see Sect. 8.3, where the perturbation series are investigated in the case of a prime unperturbed eigenvalue).

Besides the methods for estimating convergence radii and obtaining error estimates described here, there are further methods which will be omitted in this book to limit its size (see, for example, T. KATO [5, p. 93sqq. and p. 97sqq.], J. SCHRÖDER [1, 2, 3]).

8.1. Estimates for group projections and mean values (first method)

8.1.1. Radii of convergence

Let $A(z)$ be a holomorphic perturbation of A_0, given in $|z| < R$, and let λ_0 be an eigenvalue of A_0. If $A(z) = A_0 + H(z)$, then $H(z) = zA_1 + z^2A_2 + \cdots$. The second Neumann series

$$\left(\lambda - A(z)\right)^{-1} = (\lambda - A_0)^{-1} \sum_{n=0}^{\infty} [H(z)\,(\lambda - A_0)^{-1}]^n, \qquad \lambda \in \operatorname{res} A_0, \tag{1.1}$$

is convergent if

$$\|H(z)\,(\lambda - A_0)^{-1}\| < 1. \tag{1.2}$$

This condition is satisfied if

$$\sum_{n=1}^{\infty} |z|^n\, \|A_n(\lambda - A_0)^{-1}\| < 1. \tag{1.3}$$

Let $\lambda \in \operatorname{res} A_0$ be fixed. Furthermore, let the positive number $r(\lambda) > 0$ be chosen such that

$$\sum_{n=1}^{\infty} r(\lambda)^n\, \|A_n(\lambda - A_0)^{-1}\| = 1. \tag{1.4}$$

One obtains: if $|z| < r(\lambda)$, then (1.3) is valid.

Let Γ be any piecewise smooth and bounded path belonging to the resolvent set $\operatorname{res} A_0$. Then,

$$r_0 = \min_{\lambda \in \Gamma} r(\lambda) > 0. \tag{1.5}$$

To prove this inequality, note that the condition

$$\sum_{n=1}^{\infty} |z|^n\, \|A_n\| < \|(\lambda - A_0)^{-1}\|^{-1} \tag{1.6}$$

implies (1.3) in any case. One has

$$\min_{\lambda \in \Gamma} \|(\lambda - A_0)^{-1}\|^{-1} = d > 0. \tag{1.7}$$

Choosing $\varepsilon > 0$ such that the relation

$$\sum_{n=1}^{\infty} \varepsilon^n \|A_n\| = d$$

is valid, one obtains the relation (1.6) for all z with $|z| < \varepsilon$; hence, (1.3) is true for all $\lambda \in \Gamma$. Thus (1.5) has been proved.

Let $0 < \delta < r_0$. One obtains: *The second Neumann series is uniformly absolutely convergent with respect to* (z, λ) *from* $K_\delta \times \Gamma$, *where* $K_\delta = \{z : |z| < \delta\}$. One has

$$\sum_{n=1}^{\infty} |z|^n \|A_n(\lambda - A_0)^{-1}\| = \sum_{n=1}^{\infty} \left(\frac{|z|}{r_0}\right)^n r_0^n \|A_n(\lambda - A_0)^{-1}\|.$$

Now,

$$\sum_{n=1}^{\infty} r_0^n \|A_n(\lambda - A_0)^{-1}\| \leqq 1, \qquad \lambda \in \Gamma;$$

hence,

$$\sum_{n=1}^{\infty} \left(\frac{|z|}{r_0}\right) r_0^n \|A_n(\lambda - A_0)^{-1}\| = \frac{|z|}{r_0} \sum_{n=1}^{\infty} \left(\frac{|z|}{r_0}\right)^{n-1} r_0^n \|A_n(\lambda - A_0)^{-1}\|$$

$$\leqq \frac{\delta}{r_0} \sum_{n=1}^{\infty} r_0^n \|A_n(\lambda - A_0)^{-1}\| \leqq \frac{\delta}{r_0} < 1.$$

This implies the uniform absolute convergence of (1.1) in $K_\delta \times \Gamma$. Therefore one can calculate the integral

$$\int_\Gamma \left(\lambda - A(z)\right)^{-1} d\lambda$$

by integrating the terms of the second Neumann series and by reordering the series obtained as a power series in z. One obtains

Theorem 1. *Let Γ be a simply closed piecewise smooth path belonging to* res A_0, *separating the eigenvalue λ_0 which lies inside Γ from the remainder of the spectrum of A_0. Then the power series for the group projection*

$$P(z) = \frac{1}{2\pi i} \int_\Gamma \left(\lambda - A(z)\right)^{-1} d\lambda$$

converges at least for $|z| < r_0 = \min r(\lambda)$.

Since $\lambda(z) = n^{-1} \operatorname{tr} \left(A(z) P(z)\right)$, r_0 is a lower bound for the radius of convergence of the perturbation series of the mean value $\lambda(z)$ and for the radius of convergence of the power series of the operator-valued function $\left(A(z) - \lambda_0\right) P(z)$. (Note that this function plays a role in the reduction theory if λ_0 is semisimple. By applying the same consideration to this function one obtains lower bounds for the perturbation series of subprojections and corresponding mean values. See F. Rellich [4], B. v. Sz.-Nagy [1].) Therefore, if λ_0 is simple, then r_0 is a lower bound for the radius of convergence of the perturbation series of the perturbed eigenvalue $\lambda(z)$.

8.1.2. Special lower bounds for the radii of convergence

Let $1/c$ be the radius of convergence of $H(z)$, $0 \leqq c < \infty$ and let

$$\left\|\frac{H(z)}{z}\right\| \leqq \|A_1\| + |z| \cdot \|A_2\| + \cdots \leqq a, \qquad a > 0.$$

Then, as is well-known,

$$\|A_n\| \leqq ac^{n-1}, \qquad n = 1, 2, 3, \ldots$$

One has

$$\sum_{n=1}^{\infty} |z|^n \cdot \|A_n\| \cdot \|(\lambda - A_0)^{-1}\| \leqq \|(\lambda - A_0)^{-1}\| \cdot \sum_{n=1}^{\infty} |z|^n ac^{n-1}$$

$$= a\,|z| \cdot \|(\lambda - A_0)^{-1}\| \sum_{n=1}^{\infty} (c\,|z|)^{n-1} = \frac{a\,|z| \cdot \|(\lambda - A_0)^{-1}\|}{1 - c\,|z|}.$$

Furthermore, $a\,|z| \cdot \|(\lambda - A_0)^{-1}\|\,(1 - c\,|z|)^{-1} < 1$ is valid if and only if

$$|z| < (c + a\,\|(\lambda - A_0)^{-1}\|)^{-1}.$$

According to Theorem 1, one obtains

$$r_0 \geqq \min_{\lambda \in \Gamma} \left(c + a\|(\lambda - A_0)^{-1}\|\right)^{-1} = \frac{1}{c + a \max\limits_{\lambda \in \Gamma} \|(\lambda - A_0)^{-1}\|}. \tag{1.8}$$

The last number on the right hand side is a special lower bound for the radius of convergence of $P(z)$, $\lambda(z)$ and also of further perturbation series which can be derived from the former ones (this lower bound can be found in T. Kato [4, 5] and in B. v. Sz.-Nagy [2]). This bound is the best possible one in the sense that there are examples where the radius of convergence of $P(z)$ and $\lambda(z)$ is exactly equal to this bound.

For example, let $X = \mathbb{C}^2$ be equipped with the unitary norm, and let

$$A(z) = \begin{pmatrix} 1 & z \\ z & -1 \end{pmatrix} = \begin{pmatrix} 1 & 0 \\ 0 & -1 \end{pmatrix} + z \begin{pmatrix} 0 & 1 \\ 1 & 0 \end{pmatrix}. \tag{1.9}$$

In this case, one has $c = 0$, $a = \|A_1\| = 1$ (unitary norm). Let $\Gamma = \{\lambda : |\lambda - 1| = 1\}$. Then

$$\max_{\lambda \in \Gamma} \|(\lambda - A_0)^{-1}\| = 1;$$

hence, $\left(c + a \max\limits_{\lambda \in \Gamma} \|(\lambda - A_0)^{-1}\|\right)^{-1} = 1$. On the other hand, the perturbed eigenvalue corresponding to $\lambda_0 = 1$ is given by $\lambda(z) = (1 + z^2)^{1/2}$ and its radius of convergence is $r = 1$.

8.1.3. Further specialization if the unperturbed operator A_0 is normal

If A_0 is normal, then $\|(\lambda - A_0)^{-1}\| = \big(\operatorname{dis}(\lambda, \operatorname{spec} A_0)\big)^{-1}$, where $\operatorname{dis}(\lambda, \operatorname{spec} A_0)$ denotes the distance from the regular point λ to the spectrum $\operatorname{spec} A_0$ (see Lemma 2.7.2). The lower bound (1.8) for the radius of convergence then reads

$$\left[c + a \max_{\lambda \in \Gamma} \frac{1}{\operatorname{dis}(\lambda, \operatorname{spec} A_0)}\right]^{-1} = \left[c + \frac{a}{\min\limits_{\lambda \in \Gamma} \operatorname{dis}(\lambda, \operatorname{spec} A_0)}\right]^{-1}.$$

Denoting by d the distance from λ_0 to the rest spectrum,

$$d = \operatorname{dis}(\lambda_0, \operatorname{spec} A_0 \setminus \{\lambda_0\})$$

and choosing, in particular,

$$\Gamma = \left\{\lambda : |\lambda - \lambda_0| = \frac{d}{2}\right\},$$

one obtains the number

$$r_0 \geqq \left(c + \frac{2a}{d}\right)^{-1}$$

as a lower bound for the radius of convergence of $P(z)$ (see F. W. Schäfke [3]). In particular, if $A(z) = A_0 + zA_1$, hence $c = 0$ and $a = \|A_1\|$, one obtains the following theorem (see T. Kato [5], F. W. Schäfke [3]).

Theorem 2. *Let X be unitary, $A(z) = A_0 + zA_1$, and let A_0 be normal. Then the perturbation series for $P(z)$ and $\lambda(z)$, with respect to an unperturbed eigenvalue λ_0 of A_0, are convergent at least for*

$$|z| < \frac{d}{2\,\|A_1\|}, \tag{1.10}$$

where d denotes the distance from λ_0 to the remainder of the spectrum.

The special bound (1.10) is the best possible one in the sense mentioned before, as is shown by example (1.9). If A_0 is normal, then

$$d = \|S_0\|^{-1}, \tag{1.11}$$

where S_0 denotes the reduced resolvent of A_0 with respect to λ_0. (See Lemma 2.7.2. In general, one only has $\|S_0\|^{-1} \leqq d$.) Hence, the bound (1.10) can also be written in the form

$$\frac{1}{2}\,(\|A\|_1 \cdot \|S_0\|)^{-1}.$$

Later, bounds of this type (but with scalar factors smaller than 1/2) will also be obtained for operators A_0 which are not normal (see Subsect. 8.3.5; T. Kato [5, p. 91], H. Baumgärtel [5], B. V. Loginov [1]).

8.1.4. Estimates of the coefficients and errors for the mean value

Additionally, assume that the simply closed piecewise smooth path Γ separating λ_0 from the remainder of the spectrum spec $A_0 \setminus \{\lambda_0\}$ is convex. If $|z| < r_0$, then the perturbed eigenvalues $\lambda_\varrho(z)$ and also their mean value $\lambda(z)$ lie inside Γ. If

$$\varrho = \max_{\lambda \in \Gamma} |\lambda - \lambda_0|,$$

one obtains

$$|\lambda(z) - \lambda_0| < \varrho, \qquad |z| < r_0.$$

If

$$\lambda(z) = \lambda_0 + z\lambda_1 + z^2\lambda_2 + \cdots,$$

then the estimates

$$|\lambda_n| \leqq \frac{\varrho}{r_0^n}, \qquad n = 1, 2, 3, \ldots, \tag{1.12}$$

follow. This implies an error estimate for $\lambda(z) - \sum_{\varrho=0}^{n} \lambda_\varrho z^\varrho$. One obtains

$$\left|\lambda(z) - \sum_{\varrho=0}^{n} \lambda_\varrho z^\varrho\right| \leqq \sum_{\varrho=n+1}^{\infty} |\lambda_\varrho| \cdot |z|^\varrho \leqq \varrho \sum_{\sigma=n+1}^{\infty} \frac{1}{r_0^\varrho} |z|^\sigma = \frac{\varrho \left(\frac{|z|}{r_0}\right)^{n+1}}{1 - \frac{|z|}{r_0}} = \frac{\varrho}{r_0^n} \frac{|z|^{n+1}}{r_0 - |z|}.$$

In particular, if A_0 is normal and $A(z) = A_0 + zA_1$, then $r_0 = d/2a$ and $\varrho = d/2$, and one obtains

$$|\lambda_1| \leqq a, \qquad |\lambda_n| \leqq a^n \left(\frac{2}{d}\right)^{n-1}, \qquad n = 2, 3, \ldots,$$

8.2. Method of majorizing series

For this method see T. Kato [5, p. 89sqq.].

8.2.1. Majorizing series for $H(z)\,(\lambda - A_0)^{-1}$

The operator $H(z)\,(\lambda - A_0)^{-1}$ appears in the second Neumann series. One has

$$H(z)\,(\lambda - A_0)^{-1} = \sum_{n=1}^{\infty} z^n A_n (\lambda - A_0)^{-1} = \sum_{n=1}^{\infty} z^n A_n \sum_{k=-\nu}^{\infty} C_k (\lambda - \lambda_0)^k$$

$$= \sum_{n=1}^{\infty} \sum_{k=-\nu}^{\infty} A_n C_k z^n (\lambda - \lambda_0)^k,$$

where ν denotes the index of λ_0, A_n and C_k are operator-valued coefficients. A scalar double power series with nonnegative coefficients α_{nk}

$$\Phi(\lambda - \lambda_0, z) = \sum_{n=1}^{\infty} \sum_{k=-\nu}^{\infty} \alpha_{nk} z^n (\lambda - \lambda_0)^k$$

is said to be a ***majorizing series*** for $H(z)\,(\lambda - A_0)^{-1}$ if the estimates

$$\|A_n C_k\| \leqq \alpha_{nk}, \qquad n = 1, 2, 3, \ldots; \qquad k = -\nu, -\nu + 1, \ldots, \tag{2.1}$$

are valid. The inequalities (2.1) imply the relation

$$\|H(z)\,(\lambda - A_0)^{-1}\| \leqq \Phi(|\lambda - \lambda_0|, |z|). \tag{2.2}$$

Obviously, the second Neumann series converges if

$$\Phi(|\lambda - \lambda_0|, |z|) < 1.$$

Choosing Γ as the sphere $|\lambda - \lambda_0| = \varrho$, one obtains a lower bound for the radius of convergence of $P(z)$ and $\lambda(z)$ by determining the smallest positive root of the equation

$$\Phi(\varrho, r) = 1.$$

Note that $\Phi(\varrho, 0) = 0$. In a given case, ϱ should be chosen such that this root r is as large as possible.

8.2.2. Majorizing series for $\lambda(z) - \lambda_0$

From a majorizing series $\Phi(\lambda - \lambda_0, z)$ for $H(z)\,(\lambda - A_0)^{-1}$, one can derive a majorizing series for $\lambda(z) - \lambda_0$. For this purpose, one starts from

$$\lambda(z) - \lambda_0 = \frac{1}{2\pi i n} \operatorname{tr} \int_\Gamma \sum_{p=1}^{\infty} \frac{1}{p} [H(z)\,(\lambda - A_0)^{-1}]^p \,\mathrm{d}\lambda$$

$$= -\frac{1}{2\pi i n} \operatorname{tr} \int_\Gamma \log\left(1 - H(z)\,(\lambda - A_0)^{-1}\right) \mathrm{d}\lambda$$

(see Subsect. 3.8.4). The Laurent expansion of $(\lambda - A_0)^{-1}$ takes the form (see (2.6.8))

$$(\lambda - A_0)^{-1} = \frac{P_0}{\lambda - \lambda_0} + \sum_{\varrho=2}^{\nu(\lambda_0)} \frac{D_0^{\varrho-1}}{(\lambda - \lambda_0)^\varrho} + \sum_{n=0}^{\infty} (-1)^n S_0^{n+1} (\lambda - \lambda_0)^n.$$

Hence, only those terms which contain at least one factor P_0 or D_0 contribute to the integral $\int_\Gamma \cdots \mathrm{d}\lambda$. One has $\operatorname{ran} P_0 = n$ and $\operatorname{ran} D_0 < n$; moreover, for any operator C, the estimate $\operatorname{tr} C \leqq \operatorname{ran} C \cdot \|C\|$ holds. To prove this, note that $\operatorname{tr} C = \sum_{\varrho=1}^{N} \mu_\varrho$, where $\mu_1, \ldots, \mu_N$ denote the eigenvalues. Let $n(0)$ be the algebraic multiplicity of the eigenvalue 0 (if 0 is not an eigenvalue, then $n(0) = 0$). One has $N - n(0) \leqq N - \dim\ker C = \operatorname{ran} C$ and $|\mu_\varrho| \leqq \|C\|$; hence, $\operatorname{tr} C \leqq \left(N - n(0)\right) \|C\| \leqq \operatorname{ran} C \cdot \|C\|$. Therefore, for the terms and their operator coefficients under the integral $\int_\Gamma \cdots \mathrm{d}\lambda$, one obtains

$$\operatorname{tr} C \leqq n\,\|C\|.$$

Using these estimates in the series for $\lambda(z) - \lambda_0$, one can drop the factor $1/n$ and replace the operator coefficients of the expansion of the integrand by their norms. Replacing these norms by the coefficients α_{nk} of the majorizing series Φ of $H(z)(\lambda - A_0)^{-1}$, one obtains the following majorizing function for $\lambda(z) - \lambda_0$:

$$\Psi(z) = \frac{1}{2\pi i} \int\limits_{|\lambda-\lambda_0|=\varrho} \sum_{p=1}^{\infty} \frac{1}{p} [\Phi(\lambda - \lambda_0, z)]^p_k \, d\lambda$$

$$= -\frac{1}{2\pi i} \int\limits_{|\lambda-\lambda_0|=\varrho} \log\big(1 - \Phi(\lambda - \lambda_0, z)\big) \, d\lambda. \tag{2.3}$$

Integration by parts yields

$$\Psi(z) = \frac{1}{2\pi i} \int\limits_{|\lambda-\lambda_0|=\varrho} (\lambda - \lambda_0) \frac{\big(1 - \Phi(\lambda - \lambda_0, z)\big)'}{1 - \Phi(\lambda - \lambda_0, z)} \, d(\lambda - \lambda_0). \tag{2.4}$$

As is well-known, this integral equals the sum of all zeros minus the sum of all poles of the function $1 - \Phi(\lambda - \lambda_0, z)$ inside $|\lambda - \lambda_0| = \varrho$. The only pole of this function is at $\lambda - \lambda_0 = 0$; i.e., it does not contribute to the sum. Thus one obtains (see T. Kato [5, p. 90])

Theorem 1. *A majorizing series for $\lambda(z) - \lambda_0$ is obtained by the Taylor series of the sum of zeros $\mu(z)$ of the function $1 - \Phi(\mu, z)$ in a neighborhood of $\mu = 0$ for sufficiently small $|z|$ (their multiplicities taken into account). This majorizing series and also the corresponding series for $P(z)$ and $\lambda(z)$ are convergent for $|z| < r$, where r is the smallest positive root of $\Phi(\varrho, r) = 1$ and where ϱ is arbitrary, provided that the sphere $|\lambda - \lambda_0| = \varrho$ separates the eigenvalue λ_0 from the remainder of the spectrum.*

8.2.3. Example

For this example, see T. Kato [5, pp. 90, 91]. Let $A(z) = A_0 + zA_1$ and let λ_0 be semi-simple; i.e., $D_0 = (A_0 - \lambda_0) P_0 = 0$. Then,

$$(\lambda - A_0)^{-1} = \frac{P_0}{\lambda - \lambda_0} + \sum_{k=0}^{\infty} (-1)^k (\lambda - \lambda_0)^k S_0^{k+1}.$$

The conditions

$$\|A_1 P_0\| \leqq \alpha_{1,-1}, \qquad \|A_1 S_0^{k+1}\| \leqq \alpha_{1,k}, \qquad k = 0, 1, 2, \ldots,$$

have to be satisfied. Because $S_0 P_0 = P_0 S_0 = 0$, one has $S_0^{k+1} = S_0(S_0 - \alpha P_0)^k$ for all numbers α. If

$$\alpha_{1,-1} = \|A_1 P_0\|, \qquad \alpha_{1,k} = \|A_1 S_0\| \cdot \|S_0 - \alpha P_0\|^k,$$

one obtains

$$\Phi(\mu, z) = z \sum_{k=-1}^{\infty} \alpha_{1,k} \mu^k = z \left(\frac{\|A_1 P_0\|}{\mu} + \|A_1 S_0\| \sum_{k=0}^{\infty} \mu^k \|S_0 - \alpha P_0\|^k \right);$$

hence,

$$\Phi(\mu, z) = z\left(\frac{\|A_1P_0\|}{\mu} + \frac{\|A_1S_0\|}{1 - \mu\,\|S_0 - \alpha P_0\|}\right).$$

For short,

$$\|A_1P_0\| = p, \qquad \|A_1S_0\| = q, \qquad \|S_0 - \alpha P_0\| = s.$$

Then,

$$\Phi(\mu, z) = z\left(\frac{p}{\mu} + \frac{q}{1 - s\mu}\right).$$

If $|z|$ is small, then there is only a single root $\mu = \Psi(z)$ of the equation $1 - \Phi(\mu, z) = 0$ which lies in the neighborhood of $\mu = 0$. According to Theorem 1, this power series $\Psi(z)$ is a majorizing series for $\lambda(z) - \lambda_0$. One obtains

$$\Psi(z) = pz + \frac{1}{2s}\left[1 - (ps + q)\,z - \Omega(z)\right],$$

where

$$\Omega(z) = \left[(1 - (ps + q)\,z)^2 - 4pqsz^2\right]^{1/2}.$$

The coefficients α_n of the power series expansion $\Psi(z) = \sum_{n=1}^{\infty} \alpha_n z^n$ are upper bounds for the numbers $|\lambda_n|$, where $\lambda(z) - \lambda_0 = \sum_{n=1}^{\infty} \lambda_n z^n$. Furthermore, one has $\left|\lambda(z) - \sum_{\varrho=0}^{n} \lambda_\varrho z^\varrho\right| \leqq \sum_{\varrho=n+1}^{\infty} \alpha_\varrho |z|^\varrho$. For example,

$$|\lambda_1| \leqq p$$

and

$$|\lambda(z) - \lambda_0 - \lambda_1 z| \leqq \frac{1}{2s}\left[1 - (ps + q)\,|z| - \Omega(|z|)\right].$$

If ϱ is given, then the equation $\Phi(\varrho, r) = 1$ yields the (unique) root

$$r = \frac{\varrho(1 - s\varrho)}{p + (q - sp)\,\varrho} = \left(\frac{p}{\varrho} + \frac{q}{1 - s\varrho}\right)^{-1};$$

ϱ can be chosen arbitrarily. A maximum value of r,

$$r = \frac{1}{[(ps)^{1/2} + q^{1/2}]} \tag{2.5}$$

is obtained if

$$\varrho = \left(\frac{p}{s}\right)^{1/2} \left((ps)^{1/2} + q^{1/2}\right)^{-1}. \tag{2.6}$$

This value of ϱ is admissible. One has

$$S_0 = \sum_{\mu \neq \lambda_0} \left(\frac{P_\mu}{\lambda_0 - \mu} + \sum_{n=1}^{\nu_\mu - 1} \frac{D_\mu^n}{(\lambda_0 - \mu)^{n+1}}\right),$$

where μ runs through all eigenvalues of A_0 different from λ_0. If x denotes an eigenvector of A_0 belonging to the eigenvalue μ, i.e., if $P_\mu x = x$, $D_\mu x = 0$ and $P_0 x = 0$, then

$$(S_0 - \alpha P_0)\, x = \frac{x}{\lambda_0 - \mu}.$$

One obtains

$$\frac{1}{|\lambda_0 - \mu|} \leqq \|S_0 - \alpha P_0\|$$

for all μ. This result implies

$$s^{-1} = \|S_0 - \alpha P_0\|^{-1} \leqq \operatorname{dis}\,(\lambda_0, \operatorname{spec} A_0 \setminus \{\lambda_0\}) = d.$$

From (2.6),

$$\varrho < s^{-1} \leqq d;$$

hence, ϱ is admissible.

(2.5) can also be obtained directly as the radius of convergence of the power series expansions of $\Omega(z)$ and $\Psi(z)$ in the neighborhood of $z = 0$.

8.2.4. Other majorizing series

From

$$P(z) = \frac{1}{2\pi i} \int\limits_\Gamma (\lambda - A_0)^{-1} [1 + H(z)(\lambda - A_0)^{-1}]^{-1} \, d\lambda$$

and from the majorizing function $\Phi(\lambda - \lambda_0, z)$ for $H(z)(\lambda - A_0)^{-1}$, one obtains a majorizing function for $P(z)$, namely

$$\frac{1}{2\pi i} \int\limits_{|\lambda - \lambda_0| = \varrho} \Phi_1(\lambda - \lambda_0) \left(1 - \Phi(\lambda - \lambda_0, z)\right)^{-1} d\lambda,$$

where $\Phi_1(\lambda - \lambda_0)$ denotes a majorizing series of $(\lambda - A_0)^{-1}$.

If λ_0 is simple and if one normalizes the perturbed eigenvector $x(z)$ by $\left(x(z), f_0\right) = 1$, then one obtains

$$x(z) - x_0 = \left[1 - S_0 H(z) + S_0\left(\lambda(z) - \lambda_0\right)\right]^{-1} S_0 H(z)\, x_0.$$

If $\Phi_2(z)$ and $\Phi_3(z)$ are majorizing series for $S_0 H(z)$ and $H(z)\, x_0$, respectively, then

$$\|S_0\| \left(1 - \Phi_2(z) - \|S_0\|\, \Psi(z)\right)^{-1} \Phi_3(z) \tag{2.7}$$

is a majorizing series for $x(z) - x_0$. If $|z|$ is sufficiently small, more precisely, if z is chosen such that (2.7) is less than $\|x_0\|$, then $x(z)$ is necessarily different from zero.

8.3. Recursive majorizing series

As mentioned before, this method can be used if the qualitative splitting behavior is known and if recursion formulas for the coefficients of the perturbation series are given. In this section, this method will be applied to the following case: Let $A(z) = A_0 + zA_1$ and assume that the unperturbed eigenvalue λ_0 is prime with algebraic multiplicity n. Then one has $D_0^{n-1}P_0A_1x_0 = \mu x_0$ (see Subsect. 7.4.11). If $\mu \neq 0$, then λ_0 splits into a single cycle of perturbed eigenvalues $\lambda_\sigma(z)$, $\sigma = 1, \ldots, n$ (see Subsect. 7.4.11). In this case, the coefficients of the perturbation series can be calculated recursively (see Subsect. 7.4.13). From these recursion formulas, majorizing series will be derived. For the following discussion, see F. Rellich [4], J. Schröder [1, 2, 3], M. I. Višik and L. A. Ljusternik [2], B. V. Loginov [1], H. Baumgärtel [5].

8.3.1. First relation for a majorizing function

To determine majorizing series for the power series

$$\lambda(z) = \lambda_0 + \mu^{1/n}z^{1/n} + \lambda_2 z^{2/n} + \cdots,$$

$$x(z) = x_0 + x_1 z^{1/n} + x_2 z^{2/n} + \cdots, \qquad \big(x(z), f_1\big) = \mu^{-1},$$

one starts with an estimate of $|\lambda_{k+1}|$ by means of the relation (7.4.96), where the relations $|\gamma_n^{(n+k)}(\lambda)| \leqq \gamma_n^{(n+k)}(|\lambda|)$ and $|\alpha_\beta^{(\sigma)}(\lambda)| \leqq \alpha_\beta^{(\sigma)}(|\lambda|)$ are taken into account. The argument $|\lambda|$ indicates that the moduli $|\lambda_1|, |\lambda_2|, |\lambda_3|, \ldots$ appear as arguments of $\gamma_n^{(n+k)}$ and $\alpha_\beta^{(\sigma)}$. One obtains

$$|\lambda_{k+1}| \leqq \frac{|\lambda_1|}{n}\left(|\lambda_1|^{-n}\gamma_n^{(n+k)}(|\lambda|) + \|A_1\|\cdot\|x_k\|\cdot\|f_n\| + \sum_{\beta=1}^{n-1}\sum_{\sigma=\beta}^{k-1}\alpha_\beta^{(\sigma)}(|\lambda|)\,\|A_1\|\cdot\|x_{k-\sigma}\|\cdot\|f_{n-\beta}\|\right)$$

$$= \frac{|\lambda_1|}{n}\left(|\lambda_1|^{-n}\gamma_n^{(n+k)}(|\lambda|) + \|A_1\|\left[\|x_k\|\cdot\|f_n\| + \sum_{\sigma=1}^{k-1}\sum_{\beta=1}^{\min(\sigma,n-1)}\alpha_\beta^{(\sigma)}(|\lambda|)\,\|x_{k-\sigma}\|\cdot\|f_{n-\beta}\|\right]\right)$$

$$\leqq \frac{|\lambda_1|}{n}\Bigg(|\lambda_1|^{-n}\,\gamma_n^{(n+k)}(|\lambda|)$$

$$+ \|A_1\|\left[\|x_k\|\cdot\|f_n\| + \max_{\beta=1,\ldots,n-1}\|f_{n-\beta}\|\sum_{\sigma=1}^{k-1}\|x_{k-\sigma}\|\sum_{\beta=1}^{\min(\sigma,n-1)}\alpha_\beta^{(\sigma)}(|\lambda|)\right]\Bigg).$$

If $\max\limits_{\beta=0,1,\ldots,n-1}\|f_{n-\beta}\| = F$ and $\sum\limits_{\beta=1}^{\min(\sigma,n-1)}\alpha_\beta^{(\sigma)}(|\lambda|) = \beta_\sigma$, $\sigma = 1, 2, 3, \ldots$, $\beta_0 = 1$, one obtains

$$|\lambda_{k+1}| \leqq \frac{|\lambda_1|}{n}\left(|\lambda_1|^{-n}\,\gamma_n^{(n+k)}(|\lambda|) + F\,\|A_1\|\sum_{\sigma=0}^{k-1}\beta_\sigma\,\|x_{k-\sigma}\|\right), \qquad k = 1, 2, 3, \ldots \tag{3.1}$$

Let $u_1, u_2, u_3, \ldots$ be any majorizing sequence for $\|x_1\|, \|x_2\|, \|x_3\|, \ldots$; i.e., let $\|x_\varrho\| \leqq u_\varrho$ for $\varrho = 1, 2, 3, \ldots$ Then one defines a sequence $\mu_1, \mu_2, \mu_3, \ldots$ of positive numbers by $\mu_1 = |\lambda_1|$, where $\lambda_1 = \mu^{1/n}$, and by the equation

$$\mu_{k+1} = \frac{|\lambda_1|}{n}\left(|\lambda_1|^{-n}\,\gamma_n^{(n+k)}(\mu) + F\,\|A_1\|\sum_{\sigma=0}^{k-1}\beta_\sigma(\mu)\,u_{k-\sigma}\right), \qquad k = 1, 2, 3, \ldots \tag{3.2}$$

According to this definition, one obtains, for example,

$$\mu_2 = \frac{|\lambda_1|}{n} F \, \|A_1\| \, u_1,$$

$$\mu_3 = \frac{|\lambda_1|}{n} \left(|\lambda_1|^{-n} \gamma_n^{(n+2)}(\mu_1, \mu_2) + F \, \|A_1\| \left(u_2 + \beta_1(\mu_1) \, u_1\right)\right).$$

Obviously, if $|\lambda_\varrho| \leqq \mu_\varrho$ for $\varrho = 1, 2, \ldots, k$, then $\gamma_n^{(n+k)}(|\lambda|) \leqq \gamma_n^{(n+k)}(\mu)$ and $\beta_\sigma(|\lambda|) \leqq \beta_\sigma(\mu)$, $\sigma = 0, 1, 2, \ldots, k-1$; hence, $|\lambda_{k+1}| \leqq \mu_{k+1}$. Therefore, the sequence $\mu_1, \mu_2, \mu_3, \ldots$ is a majorizing sequence with respect to the sequence $|\lambda_1|, |\lambda_2|, |\lambda_3|, \ldots$

Next, one derives a manageable relation between the (formal) power series

$$\mu(z) = \mu_1 + \mu_2 z + \mu_3 z^2 + \cdots \tag{3.3}$$

and

$$u(z) = u_1 z + u_2 z^2 + u_3 z^3 + \cdots. \tag{3.4}$$

One obtains

$$\mu_1 + \sum_{k=1}^{\infty} \mu_{k+1} z^k = \mu_1 + \frac{|\lambda_1|}{n} \left(|\lambda_1|^{-n} \sum_{k=1}^{\infty} \gamma_n^{(n+k)}(\mu) \, z^k + F \, \|A_1\| \sum_{k=1}^{\infty} z^k \sum_{\sigma=0}^{k-1} \beta_\sigma(\mu) \, u_{k-\sigma} \right)$$

$$= \mu_1 + \frac{|\lambda_1|^{1-n}}{n} \sum_{k=1}^{\infty} \gamma_n^{(n+k)}(\mu) \, z^k + \frac{|\lambda_1|}{n} F \, \|A_1\| \sum_{k=1}^{\infty} z^k \sum_{\sigma=0}^{k-1} \beta_\sigma(\mu) \, u_{k-\sigma}.$$

One has

$$\sum_{k=1}^{\infty} z^k \sum_{\sigma=1}^{k-1} u_{k-\sigma} \beta_\sigma = \left(\sum_{a=1}^{\infty} u_a z^a \right) \left(\sum_{b=0}^{\infty} \beta_b z^b \right).$$

Furthermore, one has $\beta_0 = 1$ and

$$\beta_1 = \alpha_1^{(1)},$$

$$\beta_2 = \alpha_1^{(2)} + \alpha_2^{(2)},$$

$$\beta_3 = \alpha_1^{(3)} + \alpha_2^{(3)} + \alpha_3^{(3)},$$

$$\vdots$$

$$\beta_{n-1} = \alpha_1^{(n-1)} + \alpha_2^{(n-1)} + \cdots + \alpha_{n-1}^{(n-1)},$$

$$\beta_n = \alpha_1^{(n)} + \alpha_2^{(n)} + \cdots + \alpha_{n-1}^{(n)},$$

$$\vdots$$

$$\beta_{n+\varrho} = \alpha_1^{(n+\varrho)} + \alpha_2^{(n+\varrho)} + \cdots + \alpha_{n-1}^{(n+\varrho)}, \qquad \varrho = 1, 2, 3, \ldots$$

Hence,

$$\sum_{b=0}^{\infty} \beta_b z^b = 1 + \sum_{b=1}^{\infty} \left(\sum_{c=1}^{\min(b,n-1)} \alpha_c^{(b)} \right) z^b = 1 + \sum_{c=1}^{n-1} \sum_{b=c}^{\infty} \alpha_c^{(b)} z^b$$

$$= 1 + \sum_{c=1}^{n-1} z^c \sum_{b=c}^{\infty} \alpha_c^{(b)} z^{b-c} = 1 + \sum_{c=1}^{n-1} z^c \sum_{\varrho=0}^{\infty} \alpha_c^{(c+\varrho)} z^\varrho,$$

which implies

$$\sum_{b=0}^{\infty} \beta_b z^b = 1 + \sum_{c=1}^{n-1} \big(z\mu(z)\big)^c = \frac{1 - \big(z\mu(z)\big)^n}{1 - z\mu(z)}.$$

Moreover,

$$\sum_{k=1}^{\infty} \gamma_n^{(n+k)}(\mu)\, z^k = \big(\mu(z)\big)^n + (n-1)\,|\lambda_1|^n - n\,|\lambda_1|^{n-1}\,\mu(z).$$

Therefore,

$$\mu(z) = \mu_1 + \frac{|\lambda_1|^{1-n}}{n}\left[\big(\mu(z)\big)^n + (n-1)\,|\lambda_1|^n - n\,|\lambda_1|^{n-1}\,\mu(z)\right]$$

$$+ \frac{|\lambda_1|}{n} F\,\|A_1\|\,\frac{1-(z\mu)^n}{1-z\mu}\sum_{a=1}^{\infty} u_a z^a,$$

and, finally,

$$\mu(z) - \frac{|\lambda_1|^{1-n}}{2n}\big(\mu(z)\big)^n - \left(1 - \frac{1}{2n}\right)|\lambda_1| = \frac{|\lambda_1|}{2n} F\,\|A_1\|\,\frac{1-(z\mu)^n}{1-z\mu}\sum_{a=1}^{\infty} u_a z^a. \tag{3.5}$$

Therefore, if a convergent majorizing series $\sum_{a=1}^{\infty} u_a z^a$ is given, then (3.5) provides an algebraic relation for a majorizing series $\mu(z)$. This relation takes the form

$$f(\mu) = czg(z, \mu).$$

If $z = 0$, then $f(\mu) = 0$, and $\mu = |\lambda_1|$ is a simple root of this equation, because $f'(\mu) = 1 - 2^{-1}|\,\lambda_1|^{1-n}\,\mu^{n-1}$; hence, $f'(|\lambda_1|) = 1/2$. That is, $\mu(z)$ is a convergent majorizing series, which can be obtained by solving the equation (3.5) explicitly for $\mu = \mu(z)$ in a neighborhood of $z = 0$, $\mu = |\lambda_1|$.

8.3.2. Second relation for a majorizing series

A further relation linking a majorizing function $u(z)$ and a majorizing function $\mu(z)$ can be obtained as follows. The starting relation is

$$(A_0 - \lambda_0)\, x_{k+1} = \lambda_1 x_k + \lambda_2 x_{k-1} + \cdots + \lambda_k x_1 + \lambda_{k+1} x_0 - A_1 x_{k+1-n}, \tag{3.6}$$

which is valid for all $k = 0, 1, 2, \ldots$, if one puts $x_{k+1-n} = 0$ for $k + 1 < n$ (for the following cf. Subsect. 7.4.13, where the notations used here have been introduced). The right hand side of (3.6) is always of the form

$$x = \alpha_1 e_1 + \alpha_2 e_2 + \cdots + \alpha_{n-1} e_{n-1} + y,$$

where $\alpha_1, \ldots, \alpha_{n-1}$ denote complex numbers and where $Q_0 y = y$; i.e., one has $x \in$ spa $\{e_1, e_2, \ldots, e_{n-1}\} \oplus Q_0 X$. Then,

$$(A_0 - \lambda_0)^{-1}\, x = \alpha_1 e_2 + \cdots + \alpha_{n-1} e_n - S_0 y,$$

where S_0 denotes the reduced resolvent $(\lambda_0 - A_0)^{-1}$ which is well-defined on Q_0X. The space spa $(e_1, e_2, \ldots, e_{n-1}) \oplus Q_0X$ is mapped one-to-one and linearly onto spa $(e_2, e_3, \ldots, e_n) \oplus Q_0X$ by $(A_0 - \lambda_0)^{-1}$. The norm of this operator will be denoted by K. Then, the estimate

$$\|x_{k+1}\| \leqq K(|\lambda_1| \cdot \|x_k\| + |\lambda_2| \cdot \|x_{k-1}\| + \cdots + |\lambda_{k+1}| \cdot \|x_0\| + \|A_1\| \cdot \|x_{k+1-n}\|)$$

$$= K \sum_{\substack{\sigma+\varrho=k+1 \\ \sigma\geqq 1, \varrho\geqq 0}} |\lambda_\sigma| \cdot \|x_\varrho\| + K \|A_1\| \cdot \|x_{k+1-n}\|, \qquad k = 0, 1, 2, \ldots, \tag{3.7}$$

follows. If a majorizing sequence μ_ϱ for $|\lambda_\varrho|$ is given, then a majorizing sequence u_ϱ for $\|x_\varrho\|$ can be defined by $u_0 = \|x_0\|$ and

$$u_{k+1} = K \left(\sum_{\substack{\sigma+\varrho=k+1 \\ \sigma\geqq 1, \varrho\geqq 0}} \mu_\sigma u_\varrho + \|A_1\| \, u_{k+1-n} \right), \qquad k = 0, 1, 2, \ldots \tag{3.8}$$

Because of (3.7), one obtains successively the majorizing property of the sequence u_ϱ; i.e., the inequalities $\|x_\varrho\| \leqq u_\varrho$ hold for $\varrho = 1, 2, 3, \ldots$ One obtains a relation between the (formal) power series $u(z)$ and $z\mu(z)$ in the following way:

$$\sum_{k=0}^{\infty} u_{k+1} z^{k+1} = K \sum_{k=0}^{\infty} \left(\sum_{\sigma+\varrho=k+1} \mu_\sigma u_\varrho \right) z^{k+1} + K \|A_1\| \sum_{k=0}^{\infty} u_{k+1-n} z^{k+1}$$

$$= K \sum_{\sigma=1}^{\infty} \mu_\sigma z^\sigma \sum_{\varrho=0}^{\infty} u_\varrho z^\varrho + K \|A_1\| \, z^n \sum_{\alpha=0}^{\infty} u_\alpha z^\alpha;$$

i.e.,

$$\sum_{\varrho=1}^{\infty} u_\varrho z^\varrho = K \sum_{\varrho=0}^{\infty} u_\varrho z^\varrho \left(\sum_{\sigma=1}^{\infty} \mu_\sigma z^\sigma + \|A_1\| \, z^n \right).$$

Therefore,

$$u(z) = K\big(u_0 + u(z)\big) \big(z\mu(z) + \|A_1\| \, z^n\big)$$

or, equivalently,

$$u(z) = u_0 \frac{Kz\big(\mu(z) + \|A_1\| \, z^{n-1}\big)}{1 - Kz\big(\mu(z) + \|A_1\| \, z^{n-1}\big)}. \tag{3.9}$$

Therefore, if a convergent majorizing series $\mu(z)$ is given, one obtains a corresponding majorizing series $u(z)$ by (3.9).

8.3.3. Combination

The conditions (3.5) and (3.9) must be combined. That is, one investigates majorizing functions $\mu(z)$ and $u(z)$ which satisfy both conditions; the additional conditions $\mu(0) = |\lambda_1|$, $u_0 = \|x_0\|$ must also be taken into account. A majorizing function $\mu(z)$ satisfying both conditions fulfils the algebraic equation

$$\mu(z) - \frac{|\lambda_1|^{1-n}}{2n} (\mu(z))^n - \left(1 - \frac{1}{2n}\right) |\lambda_1|$$

$$= \frac{|\lambda_1|}{2n} F \|A_1\| \frac{1 - (z\mu(z))^n}{1 - z\mu(z)} u_0 K z \frac{\mu(z) + \|A_1\| z^{n-1}}{1 - Kz(\mu(z) + \|A_1\| z^{n-1})}. \tag{3.10}$$

This equation can be solved by a power series $\mu(z) = |\lambda_1| + \mu_2 z + \mu_3 z^2 + \cdots$ in the neighborhood of $z = 0$, $\mu = |\lambda_1|$, because $\mu = |\lambda_1|$ is a simple root of this equation at $z = 0$.

This power series is a majorizing series. Namely, fixing $u(z)$ according to (3.9), one obtains the following recursive system for the coefficients $\mu_1, \mu_2, \mu_3, \ldots$ and $u_1, u_2, u_3, \ldots$:

$$u_k = K \left(\sum_{\substack{\sigma+\varrho=k \\ \sigma\geq 1, \varrho \geq 0}} \mu_\sigma u_\varrho + \|A_1\| u_{k-n} \right), \qquad k = 1, 2, 3, \ldots,$$

$$\mu_{k+1} = \frac{|\lambda_1|}{n} \left(|\lambda_1|^{-n} \gamma_n^{(n+k)}(\mu) + F \|A_1\| \sum_{\sigma=0}^{k-1} \beta_\sigma(\mu) u_{k-\sigma} \right), \qquad k = 1, 2, 3, \ldots$$

Additional conditions are $\mu_1 = |\lambda_1|$ and $u_0 = \|x_0\|$. To determine u_1, one needs u_0 and μ_1; to determine μ_2, one needs μ_1 and u_1; to determine u_2, one needs μ_1, μ_2, u_0, u_1; etc. By this recursive system the coefficients are uniquely defined. By the foregoing considerations, one successively obtains the majorizing property of the coefficients.

Using the relations (3.10) and (2.9), one can estimate the convergence radius of the majorizing functions $\mu(z)$ and $u(z)$ by function-theoretic tools.

Such an estimate will be sketched. In particular, the case $n = 1$ will be considered.

8.3.4. Estimates of radii of convergence of majorizing series

If

$$c = \frac{|\lambda_1|}{2n} K F u_0 \|A_1\|,$$

$$g(\mu, z) = \frac{1 - (z\mu)^n}{1 - z\mu} \frac{\mu + \|A_1\| z^{n-1}}{1 - Kz(\mu + \|A_1\| z^{n-1})}$$

and

$$f(\mu) = \mu - \frac{|\lambda_1|^{1-n}}{2n} \mu^n - \left(1 - \frac{1}{2n}\right) |\lambda_1|,$$

(3.10) takes the form

$$f(\mu) = czg(\mu, z), \tag{3.11}$$

and the function to be considered is

$$h(\mu, z) = f(\mu) - czg(\mu, z).$$

One has $h(\mu, z) - h(\mu, 0) = -czg(\mu, z)$. A slight transformation with respect to $f(\mu)$ gives

$$f(\mu) = (\mu - |\lambda_1|)\left(1 - \frac{1}{2n}\left[1 + \frac{\mu}{|\lambda_1|} + \left(\frac{\mu}{|\lambda_1|}\right)^2 + \cdots + \left(\frac{\mu}{|\lambda_1|}\right)^{n-1}\right)\right].$$

One writes $|\lambda_1| = \alpha$, $\|A_1\| = a$. The following result is well-known (for example, see L. Bieberbach [1, p. 197sqq.]): Let $\mu(z)$ be the power series solution of $h(\mu, z) = 0$ defined in the neighborhood of $z = 0$, $\mu = \alpha$. Furthermore, let $\varrho > 0$. Then $\mu(z)$ is convergent in a circle $|z| \leqq R$ such that

$$|h(\mu, z) - h(\mu, 0)| < |h(\mu, 0)|, \qquad |\mu - \alpha| = \varrho, \qquad |z| \leqq R.$$

Note that ϱ is a number which is smaller than the distance from α to the nearest zero $\tilde{\mu}$ of $h(\mu, 0)$; i.e., $0 < \varrho < |\tilde{\mu} - \alpha|$. Then,

$$A(\varrho) = \min_{|\mu-\alpha|=\varrho} \left|1 - \frac{1}{2n}\left(1 + \frac{\mu}{\alpha} + \left(\frac{\mu}{\alpha}\right)^2 + \cdots + \left(\frac{\mu}{\alpha}\right)^{n-1}\right)\right| > 0;$$

hence,

$$|f(\mu)| = |h(\mu, 0)| \geqq \varrho A(\varrho) \quad \text{for} \quad |\mu - \alpha| = \varrho.$$

Consider $g(\mu, z)$, the denominator $1 - Kaz^n - Kz\mu$ first. Let

$$r(z, \varrho) = \min_{|\mu-\alpha|=\varrho} |1 - Kaz^n - Kz\mu|.$$

One obtains

$$r(z, \varrho) = \big|\,|1 - Kaz^n - Kz\alpha| - K\varrho\,|z|\,\big|. \tag{3.12}$$

For example, $r(0, \varrho) = 1$. Consider the equation

$$|1 - Kaz^n - Kz\alpha| = K\varrho\,|z|. \tag{3.13}$$

Let $z_0 = z_0(\varrho)$ be a root of this equation with minimal modulus. One has $|z_0(\varrho)| > 0$. Choosing R such that

$$0 < R < |z_0(\varrho)|,$$

one obtains

$$\min_{\substack{|z|\leqq R\\ |\mu-\alpha|=\varrho}} |1 - Kaz^n - Kz\mu| = \min_{|z|\leqq R} r(z, \varrho) > 0.$$

If $\min_{|z|\leqq R} r(z, \varrho) = D(R, \varrho)$ the estimate

$$\max_{\substack{|z|\leqq R\\ |\mu-\alpha|=\varrho}} |g(\mu, z)| \leqq \left(1 + R(\alpha + \varrho) + \cdots + [R(\alpha + \varrho)]^{n-1}\right) \frac{(\alpha + \varrho) + aR^{n-1}}{D(R, \varrho)}$$
$$= H(R, \varrho)$$

holds. Consider the inequality

$$c\,|z|\,H(R, \varrho) < \varrho A(\varrho).$$

If $cRH(R, \varrho) < \varrho A(\varrho)$, then the power series $\mu(z)$ is convergent for $|z| \leqq R = R_1$. Otherwise, one chooses R_1 such that $0 < R_1 < (\varrho A(\varrho))/(cH(R, \varrho))$. As a lower bound for the radius of convergence of $\mu(z)$, both results together provide the number

$$R_1 = \min\left(R, \frac{\varrho A(\varrho)}{cH(R, \varrho)}\right).$$

Hence, the original power series $\lambda(z) = \lambda_0 + \lambda_1 z^{1/n} + \lambda_2 z^{2/n} + \cdots$ is convergent at least for $|z|^{1/n} < R_1$, i.e., for $|z| < R_1^n$.

For $u(z)$, the following expressions are available:

$$u(z) = u_0 Kz \frac{\mu(z) + az^{n-1}}{1 - Kz(\mu(z) + az^{n-1})} \tag{3.14}$$

and

$$u(z) = f(\mu(z)) \frac{2n}{\alpha aF} \frac{1 - z\mu(z)}{1 - (z\mu(z))^n}. \tag{3.15}$$

From (3.15), one obtains the convergence of the power series $u(z)$ for $|z| \leqq \min\left(R_1, r/(\alpha + \varrho)^{1/n}\right)$, where $0 < r < 1$ is arbitrarily chosen. Hence, a lower bound for the convergence radius is given by

$$\min\left(R_1, (\alpha + \varrho)^{-1/n}\right).$$

From the relation (3.14) an estimate can also be obtained. First (3.12) holds; in particular, one has $r(z, 0) = |1 - Kaz^n - Kz\alpha|$. Let ϱ vary, namely let $0 \leqq \varrho \leqq \varrho_0$, where ϱ_0 is chosen such that $0 < \varrho_0 < |\tilde{\mu} - \alpha|$. Again, let $z_0(\varrho)$ be a root of (3.13) of smallest modulus for all ϱ with $0 \leqq \varrho \leqq \varrho_0$. One has $\min\limits_{0 \leqq \varrho \leqq \varrho_0} |z_0(\varrho)| > 0$. Let $0 < R_0 < \min\limits_{0 \leqq \varrho \leqq \varrho_0} |z_0(\varrho)|$. Then $\tilde{D}(R_0, \varrho_0) := \min\limits_{\substack{|z| \leqq R_0 \\ 0 \leqq \varrho \leqq \varrho_0}} r(z, \varrho) > 0$. For all $|z| \leqq R_0$, the values of $\mu(z)$ belong to the circle $|\mu - \alpha| \leqq \varrho_0$. Then (3.14) yields the convergence of $u(z)$ at least in the circle

$$|z| < \min(R_0, R_1).$$

8.3.5. Special case, $n = 1$

Let λ_0 be simple; i.e., the algebraic multiplicity $n = n(\lambda_0) = 1$. Then the recursion method of majorizing series leads to bounds which can be compared with bounds obtained by other methods.

One has $F = \|f_1\|$, $u_0 = \|x_0\|$, $a = \|A_1\|$, $c = (\alpha/2)\, Ka\, \|x_0\| \cdot \|f_1\|$. Furthermore,

$$f(\mu) = \frac{1}{2}(\mu - \alpha), \qquad g(\mu, z) = \frac{\mu + \alpha}{1 - Kz(\mu + a)}.$$

The equation (3.11) for the majorizing series $\mu(z)$ now reads

$$\mu - \alpha = 2cz \frac{\mu + a}{1 - Kz(\mu + a)}.$$

This equation is explicitly solvable with respect to μ. One obtains

$$\mu + a = -\frac{1}{2}\left(d - \frac{1}{Kz}\right) \pm \left(-\frac{b}{Kz} + \frac{1}{4}\left(d - \frac{1}{Kz}\right)^2\right)^{1/2}$$

$$d = \frac{2c}{K} - b, \qquad b = \alpha + a.$$

For the power series solution in the neighborhood of $z = 0$, $\mu = \alpha$, one obtains

$$\mu(z) = \frac{1}{2Kz}\left[1 - (d + 2a)\,Kz - \left(1 - 2Kz(d + 2b) + d^2K^2z^2\right)^{1/2}\right].$$

By direct calculation it follows that its radius of convergence is given by $R = \{(2c)^{1/2} + K^{1/2}(\alpha + a)^{1/2}\}^{-2}$; i.e.,

$$R = K^{-1}[(\alpha a\, \|f_1\| \cdot \|x_0\|)^{1/2} + (\alpha + a)^{1/2}]^{-2}.$$

If $\|x_0\| = 1$ and $(x_0, f_1) = \mu^{-1}$, then $|(x_0, f_1)| = \alpha^{-1}$ and, because $|(x_0, f_1)| \leqq \|f_1\|$,

$$\|f_1\| = \frac{1}{\alpha\gamma}, \qquad \gamma = \frac{|(x_0, f_1)|}{\|f_1\|}, \qquad 0 < \gamma \leqq 1.$$

Hence,

$$R = \|S_0\|^{-1}\left[\left(\frac{a}{\gamma}\right)^{1/2} + (\alpha + a)^{1/2}\right]^{-2}, \qquad K = \|S_0\|.$$

One has $P_0A_1x_0 = \mu x_0$, which implies $(P_0A_1x_0, f_1) = \mu(x_0, f_1) = 1$; i.e.,

$$1 = |(P_0A_1x_0, f_1)| = |(A_1x_0, f_1)| \leqq \|A_1\| \cdot \|f_1\| = \frac{\|A_1\|}{\alpha\gamma};$$

hence,

$$\alpha \leqq \frac{a}{\gamma}.$$

Therefore,

$$R \geqq (\|S_0\| \cdot \|A_1\|)^{-1}\,[\gamma^{-1/2} + (1 + \gamma^{-1})^{1/2}]^{-2}. \tag{3.16}$$

This bound for the radius of convergence of the eigenvalue $\lambda(z)$ is not as good as that of T. Kato (see (2.5)). In Kato's case (see Subsect. 8.3.3), with the use of

$$p = \|A_1P_0\| \leqq \|A_1\| \cdot \|P_0\| = \tilde{p},$$

$$q = \|A_1S_0\| \leqq \|A_1\| \cdot \|S_0\| = \tilde{q},$$

$$s = \|S_0\|$$

and of $\|P_0\| = \gamma^{-1}$, one obtains the bound

$$(\|S_0\| \cdot \|A_1\|)^{-1}\,[\gamma^{-1/2} + 1]^{-2}$$

for the radius of convergence of $\lambda(z)$.

Concerning $\|P_0\| = \gamma^{-1}$, note that $x = (x, f_1)\, e_1 + \cdots$ and $P_0 x = (x, f_1)\, e_1$; hence, $\|P_0 x\| = |(x, f_1)| \cdot \|e_1\|$; i.e., $\|P_0\| = \sup_{\|x\|=1} |(x, f_1)| \cdot \|e_1\| = \|e_1\| \cdot \|f_1\|$. On the other hand, since $(e_1, f_1) = 1$, one obtains $(e_1/\|e_1\|, f_1/\|f_1\|) = \gamma = (\|e_1\| \cdot \|f_1\|)^{-1}$.

If $X = H$ is unitary and $e_1 = f_1$ (i.e., A_0 and A_0^* have a common eigenvector with respect to λ_0 resp. $\bar{\lambda}_0$), which implies $\gamma = 1$, the, from (3.16), one obtains the bound

$$(\|S_0\| \cdot \|A_1\|)^{-1} \left(1 + \sqrt{2}\right)^{-2} = \frac{1}{3 + 2\sqrt{2}} (\|S_0\| \cdot \|A_1\|)^{-1} > \frac{1}{6} \frac{1}{\|S_0\| \cdot \|A_1\|}.$$

The latter bound is not as good as the bounds of T. Kato and B. v. Sz.-Nagy, but it improves those of B. V. Loginov and F. Rellich.

Because $u(z) = \big(\mu(z) - \alpha\big)\, (\alpha a F)^{-1}$, one obtains for the majorizing series $u(z)$ of $x(z)$ the same convergence bound as for $\mu(z)$.

8.3.6. Iterative methods for calculating eigenvalues and eigenvectors

This subsection contains some remarks concerning other methods for determining error estimates and lower bounds for the radii of convergence of perturbed eigenvalues and eigenvectors.

Since 1955, some authors have developed iterative procedures (for example, Newton's procedure and certain generalizations) to calculate perturbed spectral objects by other than power series expansion methods. These authors have emphasized numerical advantages of iterative procedures (for example the same algorithm at every step); see P. Rosenbloom [1], J. Schröder [3], et al. However, at first this approach was only successful for the simplest perturbation problems, for example, for the calculation of the perturbed eigenvalue and the perturbed eigenvector if the unperturbed eigenvalue is simple. In this connection, one could mention that the iterative procedure of P. Rosenbloom [1] is based on formula (7.5.8) and on Banach's fixed point theorem. Furthermore, one could mention the iterative procedure of J. Schröder [3], which can also be applied to nonanalytic perturbations and which is also suitable for more general perturbation equations than $(A_0 + A_1)\, x = \lambda x$. Later, Schröder's procedure was generalized to unperturbed eigenvalues λ_0 of arbitrary finite algebraic multiplicity n. This generalization could be handled, because the primary aim of the calculation was the perturbed algebraic eigenspace and a corresponding matrix of the perturbed operator in this space, rather than the spectral objects (see U. Pallaske [1]). These approaches, however, lie beyond the scope of this presentation and will not be discussed here.

9. Polynomial operator-valued functions

The results of Ch. 6 on the Jordan structure of analytic perturbations $A(z)$ of A_0, given in $|z| < R$, can be applied to the theory of polynomial operator-valued functions $A_0 + \lambda A_1 + \lambda^2 A_2 + \cdots + \lambda^n A_n$. In this case one obtains structural properties if one studies analytic perturbations $A_\varrho(z)$, $\varrho = 0, 1, 2, \ldots, n$, of the unperturbed operators A_ϱ.

These possibilities for applying the original theory have been emphasized by I. C. Gochberg. The results obtained by this application to be presented in the following discussion strengthen results of V. M. Eni [2, 4]. (Further references to V. M. Eni are [1, 3, 5].) Eni obtained similar results for holomorphic operator-valued functions $A(\lambda) = A_0 + \lambda A_1 + \lambda^2 A_2 + \cdots$. The following stronger results can be generalized similarly but this generalization will be omitted here.

Moreover, the problems and results of Ch. 7 (reduction theory) can be transferred to analytic perturbations of polynomial operator-valued functions. This topic is also omitted here.

9.1. Basic concepts for polynomial operator-valued functions

Let X be a complex linear space of finite dimension $N = \dim X$. Let $A_0, A_1, \ldots, A_n$ denote operators from $\mathscr{L}(X)$. In the following discussion one considers polynomial operators $L(\lambda)$ of the form

$$L(\lambda) = A_0 + \lambda A_1 + \lambda^2 A_2 + \cdots + \lambda^n A_n. \tag{1.1}$$

First, the most important definitions and concepts for polynomial operator-valued functions (for brevity, polynomial operators) will be summarized.

9.1.1. Characteristic values and eigenvectors

A complex number λ_0 is said to be a *characteristic value* of $L(\lambda)$ and a vector $x_0 \in X$ is said to be a corresponding *eigenvector* if the equation

$$L(\lambda_0)\, x_0 = 0 \tag{1.2}$$

holds, where $x_0 \neq 0$.

If λ_0 is not a characteristic value, i.e., if the equation $L(\lambda_0)\,x = 0$ is only solvable by $x = 0$, then λ_0 is called *regular*. In this case $L(\lambda_0)$ is invertible; i.e., $L(\lambda_0)^{-1} \in \mathcal{L}(X)$
In the following discussion, one considers only polynomial operators $L(\lambda)$ with regular points. If $\det L(\lambda) \equiv 0$ for all λ, then each λ is a characteristic value.

For example, the polynomial operator $L(\lambda) = \lambda - A_0$ has regular points for every operator A_0. The set of all characteristic values coincides with the spectrum of A_0 in this case.

Let λ_0 be a characteristic value. Then the set of all corresponding eigenvectors together with the zero vector forms a linear space, the *geometric eigenspace* of λ_0. It is the kernel $\ker L(\lambda_0)$ of $L(\lambda_0)$. Its dimension $\dim \ker L(\lambda_0)$ is called the *geometric multiplicity* of λ_0. This dimension coincides with the maximal number of linearly independent eigenvectors belonging to the characteristic value λ_0.

9.1.2. Keldyš chains, multiplicity of eigenvectors and canonical systems

Let λ_0 be a characteristic value and x_0 a corresponding eigenvector. The vectors x_0, $x_1, x_2, \ldots, x_k$ from X are called a *Keldyš chain* belonging to λ_0 and x_0 of *length* $k + 1$ if

$$\sum_{p=0}^{j} \frac{1}{p!} L^{(p)}(\lambda_0)\, x_{j-p} = 0, \qquad j = 1, 2, \ldots, k. \tag{1.3}$$

The maximal length of Keldyš chains belonging to x_0 and λ_0 is called the *multiplicity of the eigenvector* x_0 and is denoted by $m(x_0)$.

Other names for "Keldyš chains" are "Jordan chains" (see H. Langer [1]), or simply "chains" (K.-H. Förster [1]). The vectors $x_1, x_2, \ldots, x_k$ are often called *the vectors assigned to* x_0. Basic propositions of this theory are due to M. V. Keldyš [1].

The maximal multiplicity of eigenvectors belonging to λ_0 is called the *index* $\nu(\lambda_0)$ of λ_0.

A vector system

$$x_{j0}, x_{j1}, \ldots, x_{j,m_j-1}, \qquad j = 1, 2, \ldots, \dim\ker L(\lambda_0), \tag{1.4}$$

is said to be a *canonical system* of Keldyš chains belonging to the characteristic value λ_0 if the following properties are satisfied:

I. x_{j0} is an eigenvector of multiplicity m_j, $j = 1, 2, \ldots, \dim\ker L(\lambda_0)$, and these eigenvectors x_{j0} generate $\ker L(\lambda_0)$.

II. The vectors $x_{j0}, x_{j1}, \ldots, x_{jm_j-1}$ form a Keldyš chain belonging to x_{j0}.

III. m_1 coincides with the index $\nu(\lambda_0)$.

IV. m_j, $j = 2, 3, \ldots, \dim\ker L(\lambda_0)$, coincides with the maximum of all indices of λ_0 with respect to all possible complementary subspaces in $\ker L(\lambda_0)$ of the subspace generated by $x_{10}, x_{20}, \ldots, x_{j-1,0}$.

The number $n(\lambda_0) = \sum_{i=1}^{\dim\ker L(\lambda_0)} m_i$ is called the *algebraic multiplicity* of the characteristic value λ_0.

9.1.3. Motivation of the concept "Keldyš chain"

Let a vector differential equation of order n,

$$A_0 x(t) + A_1 x'(t) + \cdots + A_n x^{(n)}(t) = 0, \qquad A_i \in \mathscr{L}(X), \tag{1.5}$$

be given in X. The function

$$x(t) = e^{\lambda_0 t} x_0, \qquad x_0 \in X,$$

is a solution of (1.5) if and only if

$$(A_0 + \lambda_0 A_1 + \cdots + \lambda_0^n A_n)\, x_0 = 0,$$

i.e.,

$$L(\lambda_0)\, x_0 = 0,$$

that is, if λ_0 is a characteristic value of $L(\lambda)$ and if x_0 is a corresponding eigenvector. For short,

$$A_0 x + A_1 x' + \cdots + A_n x^{(n)} = \mathscr{L}(x). \tag{1.6}$$

Then,

$$\mathscr{L}(e^{\lambda t} x) = L(\lambda)\, e^{\lambda t} x,$$

$$\mathscr{L}\left(\frac{t}{1!} e^{\lambda t} x\right) = \left(\frac{t}{1!} L(\lambda) + L'(\lambda)\right) e^{\lambda t} x,$$

$$\mathscr{L}\left(\frac{t^2}{2!} e^{\lambda t} x\right) = \left(\frac{t^2}{2!} L(\lambda) + \frac{t}{1!} L'(\lambda) + \frac{1}{2!} L''(\lambda)\right) e^{\lambda t} x;$$

finally,

$$\mathscr{L}\left(\frac{t^r}{r!} e^{\lambda t} x\right) = \left(\frac{t^r}{r!} L(\lambda) + \frac{t^{r-1}}{(r-1!)} \frac{1}{1!} L'(\lambda) + \frac{t^{r-2}}{(r-2)!} \frac{1}{2!} L''(\lambda) + \cdots + \frac{t}{1!} \frac{1}{(r-1)!} L^{(r-1)}(\lambda) + \frac{1}{r!} L^{(r)}(\lambda)\right) e^{\lambda t} x,$$

and one obtains

$$\begin{aligned}
\mathscr{L}\left(\sum_{\varrho=0}^{k} \frac{t^\varrho}{\varrho!} e^{\lambda t} x_{k-\varrho}\right) &= \sum_{\varrho=0}^{k} \mathscr{L}\left(\frac{t^\varrho}{\varrho!} e^{\lambda t} x_{k-\varrho}\right) = \sum_{\varrho=0}^{k} \sum_{\nu=0}^{\varrho} \frac{t^\nu}{\nu!} \frac{1}{(\varrho-\nu)!} L^{(\varrho-\nu)}(\lambda)\, e^{\lambda t} x_{k-\varrho} \\
&= \sum_{\nu=0}^{k} \frac{t^\nu}{\nu!} e^{\lambda t} \sum_{\varrho=\nu}^{k} \frac{1}{(\varrho-\nu)!} L^{(\varrho-\nu)}(\lambda)\, x_{k-\varrho} \\
&= \sum_{\varrho=0}^{k} \frac{t^\nu}{\nu!} e^{\lambda t} \sum_{\mu=0}^{k-\nu} \frac{1}{\mu!} L^{(\mu)}(\lambda)\, x_{k-\nu-\mu};
\end{aligned}$$

i.e.,

$$\mathscr{L}\left(\sum_{\varrho=0}^{k} \frac{t^\varrho}{\varrho!} e^{\lambda t} x_{k-\varrho}\right) = \sum_{\nu=0}^{k} \frac{t^\nu}{\nu!} e^{\lambda t} \sum_{\mu=0}^{k-\nu} \frac{1}{\mu!} L^{(\mu)}(\lambda)\, x_{k-\nu-\mu}. \tag{1.7}$$

Therefore, a function of the form

$$x(t) = \sum_{\varrho=0}^{k} \frac{t^\varrho}{\varrho!} e^{\lambda t} x_{k-\varrho}$$

is a solution of the equation $\mathscr{L}(x) = 0$ if and only if

$$\sum_{\mu=0}^{\sigma} \frac{1}{\mu!} L^{(\mu)}(\lambda)\, x_{\sigma-\mu} = 0, \qquad \sigma = 0, 1, 2, \ldots, k;$$

i.e., if λ is a characteristic value, x_0 is a corresponding eigenvector, and $x_1, x_2, \ldots, x_k$ is a corresponding Keldyš chain belonging to x_0.

9.2. Assignment of linear polynomial operators

In this section one assigns linear polynomial operators $\tilde{A} + \lambda\tilde{B}$ to arbitrary polynomial operators $L(\lambda) = \sum_{\varrho=0}^{n} \lambda^\varrho A_\varrho$ (see, for example, I. C. Gochberg and M. G. Krein [2, p. 326], V. M. Eni [2]). The relations between the Keldyš chains of $L(\lambda)$ and $\tilde{A} + \lambda\tilde{B}$ are investigated. For linear polynomial operators of the special form $1 + \lambda C$, a connection between Keldyš chains and Jordan chains of C (i.e., Jordan bases in irreducible subspaces) will be established.

9.2.1. Assignment

Introduce the space $X^n = X \times X \times \cdots \times X$ (n times) and consider on X^n the linear polynomial operator

$$\tilde{A} + \lambda\tilde{B} = \begin{pmatrix} A_0 & A_1 \cdots A_{n-1} \\ 0 & 1 \;\cdots 0 \\ \cdots & \cdots \\ 0 & 0 \;\cdots 1 \end{pmatrix} + \lambda \begin{pmatrix} 0 \cdots & 0 & A_n \\ -1 \cdots & 0 & 0 \\ \cdots & \cdots & \cdots \\ 0 \cdots & -1 & 0 \end{pmatrix}.$$

Let $f \in X^n$, $f = \{f_0, f_1, \ldots, f_{n-1}\}$. Then one obtains

$$(\tilde{A} + \lambda\tilde{B}) \begin{pmatrix} f_0 \\ f_1 \\ \vdots \\ f_{n-1} \end{pmatrix} = \begin{Bmatrix} \sum_{\varrho=0}^{n-1} A_\varrho f_\varrho + \lambda A_n f_{n-1} \\ f_1 - \lambda f_0 \\ f_2 - \lambda f_1 \\ \vdots \\ f_{n-1} - \lambda f_{n-2} \end{Bmatrix}.$$

Hence, the equation

$$(\tilde{A} + \lambda_0\tilde{B})\, f = 0 \tag{2.1}$$

is satisfied if and only if

$$\sum_{\varrho=0}^{n} \lambda_0^\varrho A_\varrho f_0 = 0,$$

i.e.,

$$L(\lambda_0)\, f_0 = 0. \tag{2.2}$$

Hence:

λ_0 is a characteristic value for $L(\lambda)$ if and only if λ_0 is a characteristic value for $\tilde{A} + \lambda\tilde{B}$. An eigenvector f of $\tilde{A} + \lambda\tilde{B}$ satisfies the relations

$$f_1 = \lambda_0 f_0,\ f_2 = \lambda_0^2 f_0,\ \ldots,\ f_{n-1} = \lambda_0^{n-1} f_0, \tag{2.3}$$

where (2.2) holds; i.e., f_0 is an eigenvector for $L(\lambda)$ belonging to λ_0.

9.2.2. Assignment of Keldyš chains

The correspondence between Keldyš chains belonging to $L(\lambda) = \sum_{\varrho=0}^{n} \lambda^\varrho A_\varrho$ and $\tilde{L}(\lambda) = \tilde{A} + \lambda\tilde{B}$ can be described as follows:

Let $f^0, f^1, f^2, \ldots, f^k$ be a Keldyš chain of length $k + 1$ for $\tilde{L}(\lambda)$ belonging to the characteristic value λ_0. Then,

$$\tilde{L}(\lambda_0)\, f^0 = 0 \tag{2.4}$$

and

$$\tilde{L}(\lambda_0)\, f^j + \tilde{L}'(\lambda_0)\, f^{j-1} = 0, \qquad j = 1, 2, \ldots, k, \tag{2.5}$$

where $\tilde{L}'(\lambda) = \tilde{B}$. (2.4) is equivalent with the equations

$$\sum_{\varrho=0}^{n} \lambda_0^\varrho A_\varrho f_\varrho^0 = 0, \qquad f_\varrho^0 = \lambda_0^\varrho f_0^0, \qquad \varrho = 1, 2, \ldots, n-1, \tag{2.6}$$

where the coordinates of f^0 are denoted by $f_0^0, f_1^0, \ldots, f_{n-1}^0$. The equations (2.5) are equivalent with the equations

$$\sum_{\varrho=0}^{n-1} A_\varrho f_\varrho^j + \lambda_0 A_n f_{n-1}^j + A_n f_{n-1}^{j-1} = 0, \tag{2.7}$$

$$\left.\begin{aligned} f_1^j - \lambda_0 f_0^j \quad - f_0^{j-1} &= 0, \\ f_2^j - \lambda_0 f_1^j \quad - f_1^{j-1} &= 0, \\ \cdots\cdots\cdots\cdots\cdots\cdots \\ f_{n-1}^j - \lambda_0 f_{n-2}^j - f_{n-2}^{j-1} &= 0, \end{aligned}\quad j = 1, 2, \ldots, k, \right\} \tag{2.8}$$

where the coordinates of f^j are denoted by $f_0^j, f_1^j, \ldots, f_{n-1}^j$. The equations (2.8) are satisfied if and only if the relation

$$f_m^j = \sum_{\varrho=0}^{\min(m,j)} \binom{m}{\varrho} \lambda_0^{m-\varrho} f_0^{j-\varrho}, \qquad m = 1, 2, \ldots, n-1; \qquad j = 1, 2, \ldots, k, \tag{2.9}$$

is valid. (Note that the upper summation index in (2.9) can be ignored if $\lambda_0^{m-\varrho}$ and $f_0^{j-\varrho}$ are taken to be zero for negative "exponents".) Using (2.9) and $f_m^0 = \lambda_0^m f_0^0$, $m = 1$,

2, ..., $n - 1$ (see (2.6)), one can write equation (2.7) in the form

$$\sum_{\varrho=0}^{n-1} A_\varrho \sum_{\sigma=0}^{(\min \varrho, j)} \binom{\varrho}{\sigma} \lambda_0^{\varrho-\sigma} f_0^{j-\sigma} + \lambda_0 A_n \sum_{\varrho=0}^{\min(n-1,j)} \binom{n-1}{\varrho} \lambda_0^{n-\varrho-1} f_0^{j-\varrho}$$
$$+ A_n \sum_{\varrho=0}^{\min(n-1,j-1)} \binom{n-1}{\varrho} \lambda_0^{n-1-\varrho} f_0^{j-1-\varrho} = 0.$$

This equation is satisfied if and only if

$$\sum_{\sigma=0}^{j} \left(\sum_{\varrho=0}^{n-1} A_\varrho \binom{\varrho}{\sigma} \lambda_0^{\varrho-\sigma} \right) f_0^{j-\sigma} + A_n \sum_{\varrho=0}^{\min(n-1,j)} \binom{n-1}{\varrho} \lambda_0^{n-\varrho} f_0^{j-\varrho}$$
$$+ A_n \sum_{\varrho=0}^{\min(n-1,j-1)} \binom{n-1}{\varrho} \lambda_0^{n-1-\varrho} f_0^{j-1-\varrho} = 0;$$

or, equivalently, if the equation

$$\sum_{\varrho=0}^{j} \left(\sum_{\sigma=0}^{n-1-\varrho} \binom{\varrho+\sigma}{\varrho} \lambda_0^\sigma A_{\varrho+\sigma} \right) f_0^{j-\varrho} + A_n \left(\sum_{\varrho=0}^{j} \binom{n-1}{\varrho} \lambda_0^{n-\varrho} f_0^{j-\varrho} \right.$$
$$\left. + \sum_{\varrho=0}^{j-1} \binom{n-1}{\varrho} \lambda_0^{n-1-\varrho} f_0^{j-1-\varrho} \right) = 0$$

is valid. The coefficient of A_n in the last equation is equal to

$$\lambda_0^n f_0^j + \sum_{\varrho=1}^{j} \left[\binom{n-1}{\varrho} + \binom{n-1}{\varrho-1} \right] \lambda_0^{n-\varrho} f_0^{j-\varrho} = \sum_{\varrho=0}^{j} \binom{n}{\varrho} \lambda_0^{n-\varrho} f_0^{j-\varrho}.$$

Finally, one obtains the equation

$$\sum_{\varrho=0}^{\min(j,n)} \left(\sum_{\sigma=0}^{n-\varrho} \binom{\varrho+\sigma}{\varrho} \lambda_0^\sigma A_{\varrho+\sigma} \right) f_0^{j-\varrho} = 0, \qquad j = 1, 2, \ldots, k. \tag{2.10}$$

One has

$$\frac{1}{\varrho!} L^{(\varrho)}(\lambda_0) = \sum_{\sigma=0}^{n-\varrho} \binom{\varrho+\sigma}{\varrho} \lambda_0^\sigma A_{\varrho+\sigma}, \qquad \varrho = 0, 1, 2, \ldots, n,$$

in particular $(n!)^{-1} L^{(n)}(\lambda_0) = A_n$. Moreover, $L^{(\varrho)}(\lambda_0) = 0$ for $\varrho = n+1, n+2, \ldots$, hence equation (2.10) can be written in the form

$$\sum_{\varrho=0}^{j} \frac{1}{\varrho!} L^{(\varrho)}(\lambda_0) f_0^{j-\varrho} = 0, \qquad j = 1, 2, \ldots, k. \tag{2.11}$$

Therefore, the vectors $f_0^0, f_0^1, \ldots, f_0^k$ form a Keldyš chain belonging to the characteristic value λ_0 and to the eigenvector f_0^0 with respect to $L(\lambda)$.

These conclusions are invertible. If a Keldyš chain $f_0^0, f_0^1, \ldots, f_0^k$ belonging to λ_0 and f_0^0 (as the eigenvector of the chain) with respect to $L(\lambda)$ is given, then first, by equations (2.9) and (2.6), the vectors $f_m^j, f_m^0, m = 1, 2, \ldots, n-1$, can be defined. Secondly, from $L(\lambda_0) f_0^0 = 0$, together with $f_m^0 = \lambda_0^m f_0^0$, $m = 1, \ldots, n-1$, the equation $\tilde{L}(\lambda_0) f^0 = 0$ follows. Equation (2.7) is satisfied, because (2.11) and (2.9) are valid. (2.9) is equivalent with (2.8); (2.7) and (2.8) together yield (2.5); that is, $f^0, f^1, \ldots, f^k$ form a Keldyš chain of $\tilde{L}(\lambda)$ belonging to the same characteristic value λ_0.

9.2.3. Determination of canonical systems

Let a characteristic value λ_0 of $L(\lambda)$ be given. By means of the considerations of Subsect. 9.2.2, the problem of finding Keldyš chains belonging to λ_0 and $L(\lambda)$ has been reduced to the problem of constructing Keldyš chains belonging to λ_0 and to the *linear* polynomial operator $\tilde{L}(\lambda)$.

Furthermore, from Subsect. 9.2.2 it is obvious that the corresponding Keldyš chains have the same length. Hence the corresponding eigenvectors have the same multiplicity. Then, the indices of λ_0 with respect to $L(\lambda)$ and $\tilde{L}(\lambda)$ coincide.

If a canonical system of Keldyš chains belonging to λ_0 and $L(\lambda)$ is given, obviously, the assigned Keldyš chains form a canonical system belonging to λ_0 and $\tilde{L}(\lambda)$. Conversely, a canonical system belonging to λ_0 and $\tilde{L}(\lambda)$ yields an assigned canonical system belonging to λ_0 and $L(\lambda)$.

Therefore, determining a canonical system belonging to $L(\lambda)$ and to a given characteristic value λ_0 it is sufficient to determine a canonical system belonging to λ_0 and $\tilde{L}(\lambda) = \tilde{A} + \lambda\tilde{B}$. In other words, it is sufficient to study the determination of canonical systems for linear polynomial operators.

9.2.4. Reduction of the problem to linear polynomial operators of the form $1 + \lambda C$

Let $A + \lambda B$ be a linear polynomial operator. Assume that λ_0 is a regular point of $A + \lambda B$; i.e., $(A + \lambda_0 B)^{-1}$ exists. Then,

$$A + \lambda B = A + \lambda_0 B + (\lambda - \lambda_0)\, B .$$

If $A + \lambda_0 B = A_0$ and $\lambda - \lambda_0 = \mu$, $A_0 + \mu B = A + \lambda B$ is a linear polynomial operator with the regular point $\mu = 0$. A canonical system of $A_0 + \mu B$ belonging to μ_1 is at the same time a canonical system of $A + \lambda B$ belonging to $\lambda_1 = \lambda_0 + \mu_1$.

Therefore, it is sufficient to consider linear polynomial operators $L(\lambda) = A + \lambda B$, where $\lambda = 0$ is regular; i.e., A^{-1} exists. Then,

$$A^{-1}L(\lambda) = 1 + \lambda A^{-1}B .$$

Obviously, λ_0 is a characteristic value of $L(\lambda)$ if and only if λ_0 is a characteristic value of $A^{-1}L(\lambda)$. Moreover, the geometric eigenspaces of $L(\lambda)$ and $A^{-1}L(\lambda)$ coincide for $\lambda = \lambda_0$.

If $f_0, f_1, \ldots, f_k$ denotes a Keldyš chain of $L(\lambda)$ for $\lambda = \lambda_0$, i.e., if

$$L(\lambda_0)\, f_0 = 0 , \qquad (A + \lambda_0 B)\, f_j + B f_{j-1} = 0 , \qquad j = 1, 2, \ldots, k ,$$

then the relation

$$A^{-1}L(\lambda_0)\, f_0 = 0 , \qquad (1 + \lambda_0 A^{-1}B)\, f_j + A^{-1}B f_{j-1} = 0 , \qquad j = 1, 2, \ldots, k ,$$

is valid; i.e., $f_0, f_1, \ldots, f_k$ is at the same time a Keldyš chain of $A^{-1}L(\lambda)$ for $\lambda = \lambda_0$. The converse is also true.

Therefore, to determine canonical systems of $L(\lambda)$ for $\lambda = \lambda_0$, it is sufficient to construct canonical systems of $A^{-1}L(\lambda)$. In the following discussion, the linear polynomial operators are assumed to have the form $L(\lambda) = 1 + \lambda C$. The problem is finding a canonical system of Keldyš chains for $L(\lambda)$ and for a given characteristic value λ_0.

9.2.5. Keldyš chains of $L(\lambda) = 1 + \lambda C$

Let λ_0 be a characteristic value of $L(\lambda) = 1 + \lambda C$. That is, a vector $x_0 \neq 0$ with $(1 + \lambda_0 C)\, x_0 = 0$ is assumed to exist. Necessarily, $\lambda_0 \neq 0$, and one has $Cx_0 = -(1/\lambda_0)\, x_0$. Hence, if λ_0 is a characteristic value of $L(\lambda)$, then $-1/\lambda_0$ is an eigenvalue of C. Conversely, if $\mu_0 \neq 0$ is an eigenvalue of C, then $-1/\mu_0$ is a characteristic value of $L(\lambda)$. In other words, there is a one-to-one correspondence between the characteristic values of $L(\lambda)$ and the nonvanishing eigenvalues of C via $\mu = -1/\lambda$.

Let $x_0, x_1, \ldots, x_k$ be a Keldyš chain of $L(\lambda)$ of length $k + 1$ belonging to λ_0. Then, the relation

$$(1 + \lambda_0 C)\, x_0 = 0, \qquad (1 + \lambda_0 C)\, x_j + Cx_{j-1} = 0, \qquad j = 1, 2, \ldots, k,$$

is assumed to hold. This means that the equations

$$\left.\begin{aligned} (1 + \lambda_0 C)\, x_0 &= 0, \\ (1 + \lambda_0 C)\, x_1 &= -Cx_0, \\ (1 + \lambda_0 C)\, x_2 &= -Cx_1, \\ &\cdots\cdots \\ (1 + \lambda_0 C)\, x_k &= -Cx_{k-1}, \end{aligned}\right\} \tag{2.12}$$

are valid. Consequently, the vector x_j is a Jordan vector of C of degree $j + 1$ belonging to the eigenvalue $-1/\lambda_0$. Namely,

$$(1 + \lambda_0 C)^2\, x_1 = (1 + \lambda_0 C)\, (-Cx_0) = -C(1 + \lambda_0 C)\, x_0 = 0,$$

but

$$(1 + \lambda_0 C)\, x_1 \neq 0.$$

If, on the contrary, $(1 + \lambda_0 C)\, x_1 = 0$, then $Cx_0 = 0$ follows; hence, $x_0 = 0$ and this contradicts $x_0 \neq 0$. Correspondingly,

$$(1 + \lambda_0 C)^{j+1}\, x_j = (1 + \lambda_0 C)^j\, (-Cx_{j-1}) = -C(1 + \lambda_0 C)^j\, x_{j-1} = 0,$$

but

$$(1 + \lambda_0 C)^j\, x_j \neq 0.$$

Assuming $(1 + \lambda_0 C)^j\, x_j = 0$, the equation

$$(1 + \lambda_0 C)^{j-1}\, (-Cx_{j-1}) = -C(1 + \lambda_0 C)^{j-1}\, x_{j-1} = \cdots = (-C)^j\, x_0 = 0$$

follows. However, $(-C)^j\, x_0 = \lambda_0^{-j} x_0$ hence $x_0 = 0$ follows and this contradicts $x_0 \neq 0$. In particular, *the vector x_k is a Jordan vector of C of degree $k + 1$ belonging to the eigenvalue $-1/\lambda_0$.*

Furthermore, one obtains: *The vectors $x_0, x_1, \ldots, x_k$ are linearly independent.*

To prove this assertion, let $\sum_{\varrho=0}^{k} \alpha_\varrho x_\varrho = 0$ be a nontrivial linear relation and let σ be the maximal index with $\alpha_\sigma \neq 0$. Then $\sum_{\varrho=0}^{\sigma} \alpha_\varrho x_\varrho = 0$. Hence, $\sum_{\varrho=0}^{\sigma} \alpha_\varrho (1 + \lambda_0 C)^\sigma\, x_\varrho = 0$, which implies $\alpha_\sigma (1 + \lambda_0 C)^\sigma\, x_\sigma = 0$; i.e., $\alpha_\sigma = 0$ in contradiction to $\alpha_\sigma \neq 0$.

One obtains: *The linear span $H = \operatorname{spa}\{x_0, x_1, \ldots, x_k\}$ of the Keldyš chain $x_0, \ldots, x_k$ has the dimension $k + 1$.*

H is an invariant subspace with respect to C, according to equation (2.12). For example, one has

$$Cx_0 = -(1/\lambda_0)\, x_0, \qquad Cx_1 = -(1/\lambda_0)\, x_1 - (1/\lambda_0)\, Cx_0 = -(1/\lambda_0)\, x_1 + (1/\lambda_0)^2\, x_0$$

etc. The vectors

$$x_k, (1 + \lambda_0 C)\, x_k, (1 + \lambda_0 C)^2\, x_k, \ldots, (1 + \lambda_0 C)^k\, x_k \tag{2.13}$$

form a basis in H. On the one hand, the relation $(1 + \lambda_0 C)^\varrho\, x_k \in H$, $\varrho = 0, 1, 2, \ldots, k$, is valid. On the other hand, the vectors (2.13) are linearly independent because x_k is a Jordan vector of degree $k + 1$. Since $(1 + \lambda_0 C)^\varrho\, x_k = \lambda_0^\varrho (C + \lambda_0^{-1})^\varrho\, x_k$, the vectors $(1 + \lambda_0 C)^\varrho\, x_k$, up to the factor λ_0^ϱ, are the vectors of the Jordan basis belonging to x_k. In the following discussion, the name *Jordan basis* is also used for the vectors $(1 + \lambda_0 C)^\varrho\, x_k$, $\varrho = 0, 1, \ldots, k$. Sometimes they are called a *Jordan chain.*

Therefore, *to every Keldyš chain $x_0, x_1, \ldots, x_k$ corresponds the Jordan chain* (2.13), *which spans the same subspace H.*

Let a Jordan chain of length $k + 1$ for C be given, belonging to the eigenvalue $-1/\lambda_0$ of C. This Jordan chain is written in the form

$$y, (1 + \lambda_0 C)\, y, \ldots, (1 + \lambda_0 C)^k\, y, \tag{2.14}$$

where

$$(1 + \lambda_0 C)^{k+1}\, y = 0, \qquad (1 + \lambda_0 C)^k\, y \neq 0.$$

The vectors (2.14) of the Jordan chain are linearly independent. The dimension of their linear span H equals $k + 1$.

In H a Keldyš chain of length $k + 1$ can be constructed by the following formulas: Let

$$x_0 = (1 + \lambda_0 C)^k\, y, \qquad x_0 \neq 0, \qquad (1 + \lambda_0 C)\, x_0 = 0, \tag{2.15}$$

and

$$x_j = \lambda_0^{-j} \sum_{\varrho=0}^{j} (-1)^{j+\varrho} \binom{j-1}{\varrho-1} (1 + \lambda_0 C)^{k-\varrho}\, y, \qquad j = 1, 2, \ldots, k. \tag{2.16}$$

One has to verify that these vectors $x_0, x_1, \ldots, x_k$ form a Keldyš chain. One has

$$(1 + \lambda_0 C)\, x_{j+1} + Cx_j = (1 + \lambda_0 C) \left(x_{j+1} + \frac{1}{\lambda_0}\, x_j\right) - \frac{1}{\lambda_0}\, x_j .$$

Now,

$$x_{j+1} + \frac{1}{\lambda_0}\, x_j = \frac{1}{\lambda_0^{j+1}} \left((1 + \lambda_0 C)^{k-j-1}\, y + \sum_{\varrho=2}^{j} (-1)^{j+\varrho+1} \binom{j-1}{\varrho-2} (1 + \lambda_0 C)^{k-\varrho}\, y\right);$$

hence,

$$(1 + \lambda_0 C)\, x_{j+1} + Cx_j = \frac{1}{\lambda_0^{j+1}} \Bigg[(1 + \lambda_0 C)^{k-j}\, y + \sum_{\varrho=1}^{j-1} (-1)^{j+\varrho} \binom{j-1}{\varrho-1} (1 + \lambda_0 C)^{k-\varrho}\, y$$
$$- \sum_{\varrho=1}^{j} (-1)^{j+\varrho} \binom{j-1}{\varrho-1} (1 + \lambda_0 C)^{k-\varrho}\, y\Bigg] = 0.$$

One obtains:

To every Jordan chain (2.14) *corresponds the Keldyš chain defined by* (2.15), (2.16) *which generates the same subspace H.*

9.2.6. Canonical systems of $L(\lambda) = 1 + \lambda C$

The considerations of Subsect. 9.2.5 provide an approach to the determination of a canonical system of $1 + \lambda C$.

Let λ_0 be a characteristic value of $L(\lambda)$. Then $-1/\lambda_0$ is an eigenvalue of C. The algebraic multiplicity of $-1/\lambda_0$ is denoted by m. Let P_0 denote the eigenprojection of $-1/\lambda_0$. Furthermore, let

$$P_0 = P_1 + P_2 + \cdots + P_l$$

be a decomposition of P_0 into a sum of disjoint reducing irreducible projections of C with respect to $-1/\lambda_0$. Let $\dim P_\varrho = m_\varrho$, $\varrho = 1, 2, \ldots, l$, where

$$m_1 \geqq m_2 \geqq \cdots \geqq m_l, \qquad \sum_{\varrho=1}^{l} m_\varrho = m.$$

One chooses a Jordan chain

$$x_\varrho, (1 + \lambda_0 C) x_\varrho, \ldots, (1 + \lambda_0 C)^{m_\varrho - 1} x_\varrho$$

as a basis in every subspace $P_\varrho X$. This is always possible. One obtains:

In every subspace $P_\varrho X$ there exists a basis formed by a Keldyš chain $y_\varrho, y_\varrho^1, \ldots, y_\varrho^{m_\varrho-1}$ of length m_ϱ, where y_ϱ is an eigenvector of $-1/\lambda_0$ with respect to C; i.e., y_ϱ is an eigenvector with respect to $L(\lambda) = 1 + \lambda C$ belonging to the characteristic value λ_0.

The set of all these Keldyš chains for $\varrho = 1, 2, \ldots, l$ forms a canonical system of $L(\lambda)$ with respect to λ_0.

Conversely, if λ_0 is a characteristic value of $L(\lambda) = 1 + \lambda C$ and if a canonical system $y_\varrho, y_\varrho^1, \ldots, y_\varrho^{n_\varrho-1}$, $\varrho = 1, 2, \ldots, l$ is given, where l denotes the geometric multiplicity of λ_0, then every Keldyš chain $y_\varrho, y_\varrho^1, \ldots, y_\varrho^{n_\varrho-1}$ forms a basis in some irreducible reducing subspace of C; $y_\varrho^{n_\varrho-1}$ is a Jordan vector of degree n_ϱ in this subspace, which generates a Jordan basis for this subspace.

Denoting the disjoint system of projections belonging to this decomposition of the algebraic eigenspace of C with respect to $-1/\lambda_0$ by $P_1', P_2', \ldots, P_j'$, one obtains $\dim P_\varrho' = n_\varrho$ and, moreover, $n_\varrho = m_\varrho$ (in a suitable ordering) for $\varrho = 1, 2, \ldots, l$. Therefore, the lengths of Keldyš chains of a canonical system are uniquely determined (and so are the number of different Keldyš chains belonging to a given canonical system, which coincides with the geometric multiplicity).

Therefore, to determine a canonical system of $L(\lambda) = 1 + \lambda C$ belonging to the characteristic value λ_0, it is sufficient to determine a disjoint system of irreducible reducing projections P_ϱ for C belonging to the eigenvalue $-1/\lambda_0$ and to choose a cyclic (generating) vector f_ϱ (i.e., a Jordan vector of maximal degree) in every subspace $P_\varrho X$, respectively. Then formulas (2.15) and (2.16) yield a canonical system.

9.3. Analytic perturbation theory for polynomial operator-valued functions

In the following discussion one considers analytic perturbations of polynomial operator-valued functions. Let $L(\lambda) = \sum_{\varrho=0}^{r} \lambda^\varrho A_\varrho$ be a polynomial operator. Furthermore, let the holomorphic functions $A_\varrho(z)$, $\varrho = 1, 2, \ldots, r$, $|z| < R$, be perturbations of $A_\varrho(0) = A_\varrho$. Then, the polynomial operator $L(\lambda, z) = \sum_{\varrho=0}^{r} \lambda^\varrho A_\varrho(z)$ is called an *analytically perturbed polynomial operator* with the unperturbed polynomial operator $L(\lambda, 0) = L(\lambda) = \sum_{\varrho=0}^{r} \lambda^\varrho A_\varrho$.

As in the case of a holomorphic perturbation $A(z)$ for a single operator A_0, one investigates the perturbed assigned objects, first of all the perturbed characteristic values and the perturbed canonical systems of Keldyš chains also for perturbed polynomial operators $L(\lambda, z)$. Using both the results cf. Ch. 6 and 7 and the considerations of Sect. 9.1 and 9.2, one can easyly obtain some insight into the behavior of the perturbed objects. After that, the Gochberg-Krein inequalities are transferred to the present case.

9.3.1. Perturbation theorem

Let X be a complex linear space of finite dimension $\dim X = N$. Let $A_0(z), A_1(z), \ldots, A_r(z)$ be holomorphic perturbations (i.e., holomorphic operator-valued functions with values from $\mathscr{L}(X)$) of $A_i(0) = A_i$, $i = 0, 1, \ldots, r$, given on $|z| < R$. Consider the polynomial operator $L(\lambda) = \sum_{\varrho=0}^{r} \lambda^\varrho A_\varrho$ and its perturbation $L(\lambda, z) = \sum_{\varrho=0}^{r} \lambda^\varrho A_\varrho(z)$.

Theorem 1. *Let λ_0 be an isolated characteristic value of $L(\lambda)$ of algebraic multiplicity $n = n(\lambda_0)$. Let Γ be a sphere, centered at λ_0, such that all numbers $\lambda \neq \lambda_0$ lying inside Γ or on Γ are regular with respect to $L(\lambda)$. Then there exists an $\varepsilon > 0$, such that the perturbed polynomial operator $L(\lambda, z)$ satisfies the following properties for $|z| < \varepsilon$:*

I. *The number s of different characteristic values of $L(\lambda, z)$ lying in the interior of Γ is constant for $z \neq 0$. There exist p numbers $g_1, g_2, \ldots, g_p$ such that the characteristic values of $L(\lambda, z)$ lying in the interior of Γ form p groups of g_α numbers $\lambda_{\alpha\sigma}(z)$, $\sigma = 1, 2, \ldots, g_\alpha$, $\alpha = 1, \ldots, p$, respectively, and such that the functions $\lambda_{\alpha\sigma}(z)$, where α is fixed, represent the g_α branches of a single analytic function over $|z| < \varepsilon$, where $z = 0$ is the only branching point which is algebroid and of order g_α. Moreover, $\sum_{\alpha=1}^{p} g_\alpha = s$, and one has Puiseux series expansions of the form*

$$\lambda_{\alpha\sigma}(z) = \lambda_0 + \sum_{\mu=1}^{\infty} a_{\alpha\mu} z^{\mu/g_\alpha}, \qquad \alpha = 1, 2, \ldots, p, \qquad |z| < \varepsilon, \tag{3.1}$$

where the g_α different values of the root z^{1/g_α} yield the characteristic values $\lambda_{\alpha 1}(z), \lambda_{\alpha 2}(z), \ldots, \lambda_{\alpha g_\alpha}(z)$.

II. *The algebraic multiplicity $n(\lambda_{\alpha\sigma}(z))$ of the characteristic number $\lambda_{\alpha\sigma}(z)$ is constant for $z \neq 0$, $n(\lambda_{\alpha\sigma}(z)) =: m_\alpha$, $\alpha = 1, \ldots, p$; $\sigma = 1, \ldots, g_\alpha$, $z \neq 0$; one has $\sum_{\alpha=1}^{p} g_\alpha m_\alpha = n$.*

III. *To every α, $1 \leqq \alpha \leqq p$, corresponds a number l_α and l_α numbers $m_\alpha^{(1)} \geqq m_\alpha^{(2)} \geqq \cdots \geqq m_\alpha^{(l_\alpha)} \geqq 1$ with $\sum_{\varrho=1}^{l_\alpha} m_\alpha^{(\varrho)} = m_\alpha$, which are uniquely determined and have the following properties:*

(a) *To every number $m_\alpha^{(\varrho)}$ and to every σ, $1 \leqq \sigma \leqq g_\alpha$, correspond $m_\alpha^{(\varrho)}$ vector-valued functions $g_{\alpha\sigma}^{\varrho\varkappa}(z)$, $\varkappa = 1, 2, \ldots, m_\alpha^{(\varrho)}$, forming, for $0 < |z| < \varepsilon$, a Keldyš chain of length $m_\alpha^{(\varrho)}$, which belongs to the characteristic value $\lambda_{\alpha\sigma}(z)$ (one may assume that $g_{\alpha\sigma}^{\varrho 1}(z)$ is the eigenvector of the Keldyš chain).*

(b) *The vector-valued functions $g_{\alpha\sigma}^{\varrho\varkappa}(z)$ can be expanded into Puiseux series of the form*

$$g_{\alpha\sigma}^{\varrho\varkappa}(z) = g_\alpha^{\varrho\varkappa} + \sum_{\mu=1}^{\infty} g_{\alpha\mu}^{\varrho\varkappa} z^{\mu/g_\alpha}, \qquad |z| < \varepsilon; \tag{3.2}$$

that is, they have an algebroid branching point at $z = 0$ of order g_α and they are holomorphic in $0 < |z| < \varepsilon$.

(c) *The l_α Keldyš chains $g_{\alpha\sigma}^{\varrho\varkappa}(z)$, $\varrho = 1, 2, \ldots, l_\alpha$, obtained in this way form a canonical system of Keldyš chains belonging to the characteristic value $\lambda_{\alpha\sigma}(z)$.*

IV. *The geometric multiplicity $h_{\alpha\sigma}(z) = \dim \ker L(\lambda_{\alpha\sigma}(z), z)$ of the characteristic value $\lambda_{\alpha\sigma}(z)$ coincides with l_α for $z \neq 0$.*

Proof. At first the proof will be given for $r = 1$, i.e., for linear polynomial operators. In this case, one has $L(\lambda, z) = A_0(z) + \lambda A_1(z)$, $L(\lambda, 0) = L(\lambda) = A_0 + \lambda A_1$. Without loss of generality, one can assume that $\lambda = 0$ is regular; i.e., A_0^{-1} exists (cf. Subsect. 9.2.4). Hence, there exists a circle $|z| < R'$, such that $A_0(z)$ is invertible in $|z| < R'$; i.e., $A_0(z)^{-1}$ exists for $|z| < R'$ and is holomorphic in that circle. Consider the linear polynomial

$$M(\lambda, z) = A_0(z)^{-1} L(\lambda, z) = 1 + \lambda A_0(z)^{-1} A_1(z), \qquad |z| < R'.$$

$M(\lambda, z)$ and $L(\lambda, z)$ have the same characteristic values and the same canonical systems. Hence, it is sufficient to verify the assertions of the theorem for the polynomial $M(\lambda, z)$. For short, let $A_0(z)^{-1} A_1(z) = C(z)$. $C(z)$ is holomorphic in $|z| < R'$ and one has $M(\lambda, z) = 1 + \lambda C(z)$.

Recall that the characteristic value λ_0 of $L(\lambda, 0)$ is also a characteristic value of $M(\lambda, 0)$; i.e., $-1/\lambda_0 =: \mu_0$ is an eigenvalue of $C(0) = C_0$. Moreover, the algebraic multiplicity $n(\lambda_0)$ of λ_0 with respect to $L(\lambda, 0)$ coincides with the algebraic multiplicity $n(\mu_0)$ of the eigenvalue μ_0 with respect to C_0, $n(\lambda_0) = n(\mu_0)$. One can apply the results of Ch. 6 (in particular, Theorem 6.1.2) to the perturbation $C(z)$ of C_0 with respect to the eigenvalue μ_0. One obtains:

(i) perturbed eigenvalues $\mu_{\alpha\sigma}(z)$ of μ_0 of the form

$$\mu_{\alpha\sigma}(z) = \mu_0 + \sum_{\varrho=1}^{\infty} c_{\alpha\varrho} z^{\varrho/g_\alpha}, \qquad |z| < \varepsilon_1, \qquad \alpha = 1, 2, \ldots, p;\ \sigma = 1, 2, \ldots, g_\alpha;$$

(ii) corresponding eigenprojections $P_{\alpha\sigma}(z)$ with expansions of the form

$$P_{\alpha\sigma}(z) = \sum_{\varrho=-N_\alpha}^{\infty} P_{\alpha\varrho} z^{\varrho/g_\alpha}, \qquad 0 < |z| < \varepsilon_1, \qquad \dim P_{\alpha\sigma}(z) = m_\alpha;$$

(iii) irreducible reducing projections $P^{\varrho}_{\alpha\sigma}(z)$, $\varrho = 1, 2, \ldots, l_\alpha$, with expansions of the form

$$P^{\varrho}_{\alpha\sigma}(z) = \sum_{\mu=-M^{\varrho}_{\alpha}} P^{\varrho}_{\alpha\mu} z^{\mu/g_\alpha}, \qquad 0 < |z| < \varepsilon_1, \qquad \dim P^{\varrho}_{\alpha\sigma}(z) = m^{(\varrho)}_{\alpha};$$

(iv) Jordan bases $e^{\varrho\varkappa}_{\alpha\sigma}(z)$, $\varkappa = 1, 2, \ldots, m^{(\varrho)}_{\alpha}$, in the irreducible subspaces $P^{\varrho}_{\alpha\sigma}(z)\, X$, also with expansions of the form

$$e^{\varrho\varkappa}_{\alpha\sigma}(z) = e^{\varrho\varkappa}_{\alpha} + \sum_{\mu=1}^{\infty} e^{\varrho\varkappa}_{\alpha\mu} z^{\mu/g_\alpha}, \qquad |z| < \varepsilon_1, \qquad \varkappa = 1, 2, \ldots, m^{(\varrho)}_{\alpha}, \tag{3.3}$$

where the vectors $e^{\varrho 1}_{\alpha\sigma}(z)$ are eigenvectors of $C(z)$ belonging to the eigenvalue $\mu_{\alpha\sigma}(z)$, respectively. Hence, its geometric multiplicity equals l_α.

The results (i) to (iv), together with the considerations of Subsect. 9.3.2 and 9.3.4 to 9.3.6, yield the results of the theorem: The perturbed characteristic values in the neighborhood of λ_0 are those of the form $\lambda_{\alpha\sigma}(z) = -1/\mu_{\alpha\sigma}(z)$. Therefore, for $\lambda_{\alpha\sigma}(z)$, one obtains expansions of the form (3.1) in a neighborhood $|z| < \varepsilon$. Hence, the number of different characteristic values in the neighborhood of λ_0 is constant for $0 < |z| < \varepsilon$ and equal to $s = \sum_{\alpha=1}^{p} g_\alpha$.

Furthermore, according to Sect. 2.5, the Jordan bases $e^{\varrho\varkappa}_{\alpha\sigma}(z)$ in the irreducible subspaces $P^{\varrho}_{\alpha\sigma}(z)\, X$ can be transformed into Keldyš chains, beginning with the eigenvectors $e^{\varrho 1}_{\alpha\sigma}(z)$, in accordance with formula (2.6). Thus, in $P^{\varrho}_{\alpha\sigma}(z)\, X$, one obtains a basis $g^{\varrho 1}_{\alpha\sigma}(z), g^{\varrho 2}_{\alpha\sigma}(z), \ldots, g^{\varrho m^{(\varrho)}_{\alpha}}_{\alpha\sigma}(z)$, which represents a Keldyš chain of length $m^{(\varrho)}_{\alpha}$ with the eigenvector $g^{\varrho 1}_{\alpha\sigma}(z)$. From (3.3) and (2.16), one obtains expansions of the form (3.2) for the vectors $g^{\varrho\varkappa}_{\alpha\sigma}(z)$.

Finally, the system consisting of all these Keldyš chains forms a canonical system belonging to the characteristic value $\lambda_{\alpha\sigma}(z)$. Hence, $n(\lambda_{\alpha\sigma}(z)) = \sum_{\varrho=1}^{l_\alpha} m^{(\varrho)}_{\alpha} = m_\alpha$ in particular; i.e., the algebraic multiplicities of the perturbed characteristic values are constant. Moreover, the number of the (perturbed) Keldyš chains is constant and equal to l_α; in other words, the geometric multiplicity of $\lambda_{\alpha\sigma}(z)$ equals l_α. The lengths of the (perturbed) Keldyš chains are constant and equal to $m^{(\varrho)}_{\alpha}$.

Again, the equation $n(\lambda_0) = \sum_{\alpha=1}^{p} g_\alpha m_\alpha$ expresses the fact that the sum of the algebraic multiplicities of all perturbed characteristic values equals the algebraic multiplicity of λ_0.

Hence, the assertions of the theorem have been proved for linear polynomial operators. Recall that these assertions follow from the correspondence between Jordan chains and Keldyš chains for linear polynomial operators of the form $1 + \lambda C$.

Finally, the propositions are transferred to the case of arbitrary polynomial operators. For this purpose, one forms

$$\tilde{A}(z) = \begin{pmatrix} A_0(z) & A_1(z) & \cdots & A_{n-1}(z) \\ 0 & 1 & \cdots & 0 \\ \cdots & \cdots & \cdots & \cdots \\ 0 & 0 & \cdots & 1 \end{pmatrix}, \qquad \tilde{B}(z) = \begin{pmatrix} 0 & 0 & \cdots & 0 & A_n(z) \\ -1 & 0 & \cdots & 0 & 0 \\ \cdots & \cdots & \cdots & \cdots & \cdots \\ 0 & 0 & \cdots & -1 & 0 \end{pmatrix},$$

and one considers the linear polynomial operator $\tilde{L}(\lambda, z) = \tilde{A}(z) + \lambda\tilde{B}(z)$. According to Subsect. 9.2.1, $L(\lambda, z)$ and $\tilde{L}(\lambda, z)$ have the same characteristic values. Therefore, the characteristic value λ_0 of $L(\lambda, 0)$ is also a characteristic value of $\tilde{L}(\lambda, 0)$, and the characteristic values of $L(\lambda, z)$ lying in the interior of Γ coincide with the characteristic values of $\tilde{L}(\lambda, z)$ with the same property. Hence, the assertions I are true, because they are already valid for linear polynomial operators. From Subsect. 9.2.3 it follows that the algebraic multiplicities of the characteristic values of $L(\lambda, z)$ and $\tilde{L}(\lambda, z)$ coincide. Hence II follows also for $L(\lambda, z)$. Furthermore, from Subsect. 9.2.2 it follows that the "first coordinates" of a canonical system of $\tilde{L}(\lambda, z)$ belonging to the characteristic value $\lambda_{\alpha\sigma}(z)$ form a canonical system of $L(\lambda, z)$ belonging to the same characteristic value. This implies the assertions III (a) to (c). Finally, assertion IV is a direct consequence of III (c). ■

9.4. Connection of perturbed and unperturbed canonical systems

In Subsect. 6.2.2, the Gochberg-Krein inequalities were derived. These inequalities connect the Jordan structure of A_0 (with respect to an eigenvalue λ_0) with the Jordan structure of the perturbation $A(z)$ of A_0 (with respect to the λ_0-group). Using the assignment of linear polynomial operators $\tilde{L}(\lambda, z)$ to $L(\lambda, z)$, and the assignment of a polynomial operator of type $1 + \lambda C$ to $\tilde{L}(\lambda, z)$, one can transfer these inequalities to the situation of analytic perturbations for polynomial operators. These inequalities express a connection between canonical systems of $L(\lambda, 0)$ belonging to λ_0 and canonical systems of $L(\lambda, z)$ belonging to the perturbed characteristic values $\lambda_{\alpha\sigma}(z)$.

9.4.1. Notations

Let $L(\lambda, z)$ be a polynomial operator and let λ_0 be an isolated characteristic value of the unperturbed polynomial operator $L(\lambda, 0)$. To the polynomial operators $L(\lambda, 0)$ and $L(\lambda, z)$ can be assigned a set of objects (cf. Theorem 1) which can be reinterpreted according to the assignment of $\tilde{L}(\lambda, z)$ and $L(\lambda, z)$, and of $1 + \lambda C(z)$ and $\tilde{L}(\lambda, z)$, respectively. This reinterpretation refers to the holomorphic perturbation $C(z)$ of $C(0) = C_0$. For convenience, these objects are listed once more:

(a) $n(\lambda_0)$, the algebraic multiplicity of λ_0 (the algebraic multiplicity of $\mu_0 = -1/\lambda_0$ with respect to C_0).

(b) $\dim \ker L(\lambda_0, 0) = h(\lambda_0)$, the geometric multiplicity of λ_0 (the geometric multiplicity of μ_0).

(c) $k_1, k_2, \ldots, k_{n(\lambda_0)}$, the lengths of Keldyš chains from a canonical system of $L(\lambda, 0)$ belonging to λ_0 (the dimensions of irreducible projections from a complete disjoint system of such projections for C_0 with respect to μ_0).

(d) $m_1, m_2, \ldots, m_p$, $m_\alpha = n(\lambda_{\alpha\sigma}(z))$, the algebraic multiplicities of the perturbed characteristic values $\lambda_{\alpha\sigma}(z)$ (the algebraic multiplicities of the perturbed eigenvalues $\mu_{\alpha\sigma}(z) = -1/\lambda_{\alpha\sigma}(z)$).

(e) $h(\lambda_{\alpha\sigma}(z)) = l_\alpha$, the geometric multiplicity of the perturbed characteristic value $\lambda_{\alpha\sigma}(z)$, i.e., the number of different Keldyš chains of a canonical system of $L(\lambda, z)$ belonging to $\lambda_{\alpha\sigma}(z)$ (the geometric multiplicity of the perturbed eigenvalue $\mu_{\alpha\sigma}(z)$, i.e., the number of different irreducible projections of a corresponding complete disjoint system for $C(z)$ with respect to $\mu_{\alpha\sigma}(z)$).

(f) $m^{(1)}, m^{(2)}, \ldots, m_\alpha^{(l_\alpha)}$, the lengths of the Keldyš chains of a canonical system of $L(\lambda, z)$ belonging to $\lambda_{\alpha\sigma}(z)$ (the dimensions of the irreducible projections of $C(z)$ with respect to $\mu_{\alpha\sigma}(z)$).

The relations

$$\sum_{\varrho=1}^{h(\lambda_0)} k_\varrho = n(\lambda_0) \tag{4.1}$$

and

$$\sum_{\varrho=1}^{l_\alpha} m_\alpha^{(\varrho)} = m_\alpha, \qquad \alpha = 1, 2, \ldots, p, \tag{4.2}$$

hold.

Denoting as before (see Subsect. 6.2.1) the number of irreducible subspaces of C_0 with respect to μ_0 of fixed dimension ϱ by $\varkappa(\mu_0, \varrho)$ and taking $d(\mu_0, q) = \dim \ker \{(C_0 - \mu_0)^q\}$, one obtains (see Subsect. 6.2.1)

$$d(\mu_0, q) = \sum_{\varrho=1}^{q} \varrho\varkappa(\mu_0, \varrho) + q \sum_{\varrho=q+1}^{n(\lambda_0)} \varkappa(\mu_0, \varrho), \qquad q = 1, 2, \ldots, n(\lambda_0).$$

In particular, $d(\mu_0, 1) = h(\lambda_0)$. Correspondingly, one obtains for the perturbed objects the equation

$$d(\mu_{\alpha\sigma}(z), q) = \sum_{\varrho=1}^{q} \varrho\varkappa(\mu_{\alpha\sigma}(z), \varrho) + q \sum_{\varrho=q+1}^{m_\alpha} \varkappa(\mu_{\alpha\sigma}(z), \varrho), \qquad q = 1, 2, \ldots, m_\alpha.$$

The number $\varkappa(\mu_0, \varrho)$ is also the number of Keldyš chains of fixed length ϱ in a canonical system of $L(\lambda, 0)$ with respect to λ_0. Correspondingly, $\varkappa(\mu_{\alpha\sigma}(z), \varrho)$ is the number of Keldyš chains of fixed length ϱ in a canonical system of $L(\lambda, z)$ with respect to $\lambda_{\alpha\sigma}(z)$.

9.4.2. Gochberg-Krein inequalities

According to the Gochberg-Krein inequalities (see Subsect. 6.2.2), one has

$$d(\mu_0, q) \geqq d(\mu_{\alpha\sigma}(z), q), \quad q = 1, 2, \ldots, n(\lambda_0); \quad \alpha = 1, \ldots, p; \quad \sigma = 1, \ldots, g_\alpha;$$
$$0 < |z| < \varepsilon. \tag{4.3}$$

Hence, one obtains the following connection between the numbers of unperturbed and perturbed Keldyš chains of fixed lengths:

Theorem 1. *The inequalities*

$$\sum_{\varrho=1}^{q} \varrho\varkappa(\mu_0, \varrho) + q \sum_{\varrho=q+1}^{n(\lambda_0)} \varkappa(\mu_0, \varrho) \geqq \sum_{\varrho=1}^{q} \varrho\varkappa(\mu_{\alpha\sigma}(z), q) + q \sum_{\varrho=q+1}^{m_\alpha} \varkappa(\mu_{\alpha\sigma}(z), q) \tag{4.4}$$

are valid for $q = 1, 2, 3, \ldots, n(\lambda_0)$, *where the second sum on the right hand side vanishes for* $m_\alpha \leqq q \leqq n(\lambda_0)$.

For $q = 1$, this inequality (4.4) has the simple form

$$h(\lambda_0) \geqq l_\alpha, \qquad \alpha = 1, 2, \ldots, p. \tag{4.5}$$

This means that a canonical system of $L(\lambda, z)$ belonging to the perturbed characteristic value $\lambda_{\alpha\sigma}(z)$ cannot contain more Keldyš chains than a canonical system of $L(\lambda, 0)$ with respect to the unperturbed characteristic value λ_0.

In particular, if $h(\lambda_0) = 1$, then $l_\alpha = 1$. One obtains

Corollary 1. *If a canonical system of* $L(\lambda, 0)$ *belonging to* λ_0 *consists of a single Keldyš chain, then each perturbed canonical system of* $L(\lambda, z)$, *belonging to any characteristic value* $\lambda_{\alpha\sigma}(z)$ *arising from* λ_0 *by perturbation, also consists of a single Keldyš chain.*

10. Analytic perturbations in Banach spaces

In this chapter, some consequences for analytic perturbations in infinite-dimensional spaces will be considered which can be derived from the results obtained in the foregoing chapters. Initially, some concepts from the theory of Banach spaces will be sketched; subsequently, this theory will be assumed known (see the standard monograph of N. Dunford and J. T. Schwartz [1]; for the purpose of this book in particular, see T. Kato [5], and F. Riesz and B. v. Sz.-Nagy [1]). Thus, this chapter lies somewhat outside the scope of the previous considerations.

Historically, the theory of analytic perturbations in Hilbert spaces, in particular the theory of selfadjoint real-analytic perturbations of selfadjoint operators, marks the beginning of perturbation theory. The first problems which were solved in this connection are those centered around the investigation of the qualitative and quantitative behavior of isolated eigenvalues of finite multiplicity and their corresponding eigenvectors (see E. Schrödinger [1], L. Rayleigh [1], F. Rellich [1—5]). Though F. Rellich already considered the finite-dimensional case in [1], his investigations mainly concerned selfadjoint real-analytic perturbations of selfadjoint operators. The study of perturbations acting on finite-dimensional spaces was merely auxiliary.

In general, the results obtained for perturbations acting on finite-dimensional spaces can be transferred to a corresponding situation with respect to a Banach space. For this reason the applications to Banach spaces are discussed in this book. The basic assumption underlying such a transfer are the explanation of the concept of a holomorphic operator-valued function and the restriction of the consideration to perturbations of isolated eigenvalues of finite algebraic multiplicity.

Because of possible applications, not only bounded perturbations of bounded operators will be discussed, but one starts with closed operators. As mentioned before, the crucial problem is defining holomorphic operator-valued functions. In this connection, we refer to T. Kato [5, p. 366sqq.].

Sect. 10.1 recalls the concepts of holomorphic vector-valued functions and of bounded-holomorphic operator-valued functions (i.e., holomorphic operator-valued functions whose values are bounded operators) on Banach spaces.

Sect. 10.2 recalls some concepts in connection with closed operators on Banach spaces and their elementary spectral properties (see, for example, T. Kato [5]). In Sect. 10.3, the concept of holomorphic operator-valued functions whose values are closed operators is introduced. A criterion for an operator-valued function to be holomorphic is given and special types of holomorphic operator-valued functions are discussed (see T. Kato [5, p. 366sqq.]).

In Sect. 10.4 a link is given between the finite-dimensional considerations and the holomorphic perturbations of finite eigenvalue systems. This link is the basis for transferring the results from the theory for finite-dimensional spaces (Ch. 5 to 9) the theory of holomorphic perturbations of closed operators acting on Banach spaces. Note that one is always concerned with propositions on perturbations of finite eigenvalue systems and of isolated eigenvalues of finite algebraic multiplicity.

The method of transferring the results from the theory for finite-dimensional spaces will be described only for the main results of Ch. 6 and 7.

10.1. Holomorphic vector- and operator-valued functions on Banach spaces

10.1.1. Banach spaces and bounded operators

In the following discussion, X denotes the basic Banach space. X is linear, normed and complete with respect to its norm $\|\cdot\|$. The corresponding *adjoint space*, i.e., the space of all bounded antilinear forms on X, is denoted by X^*. The set of all bounded operators from X into itself is denoted by $\mathscr{L}(X)$. (By $\mathscr{L}(X, Y)$ one denotes the set of all bounded operators from X into Y, where Y is a second Banach space.) $\mathscr{L}(X)$, equipped with the operator norm $\|\cdot\|$, is a Banach space. With respect to $\mathscr{L}(X)$ and to operator-valued sequences, one has the concepts "norm convergence", "strong convergence" and "weak convergence".

If $A \in \mathscr{L}(X)$, then the *adjoint operator* A^* from $\mathscr{L}(X^*)$ is defined by $(Ax, f) = (x, A^*f)$ for all $x \in X, f \in X^*$. One has $\|A\| = \|A^*\|$.

A *projection* $P \in \mathscr{L}(X)$ is defined by $P^2 = P$.

10.1.2. Holomorphic vector-valued functions

Let G be a region of the complex plane and let $u(z)$ be a function from G into X. This function is said to be *strongly holomorphic* in G if it is strongly differentiable at every point $z \in G$; i.e., if there is a vector-valued function $u'(z)$ (called the derivative) from G into X such that

$$\left\| \frac{u(z+h) - u(z)}{h} - u'(z) \right\| \to 0, \qquad h \to 0,$$

for all sequences $h \to 0$.

On the other hand, $u(z)$ is said to be *weakly holomorphic* in G if the scalar function $\big(u(z), f\big)$ is holomorphic in G for every $f \in X^*$.

Obviously, each strongly holomorphic vector-valued function is also weakly holomorphic. But the converse is also true: Each vector-valued function $u(z)$, weakly holomorphic in G, is also strongly holomorphic in G (see for example T. Kato [5, p. 139]).

Therefore, there is only a single concept of holomorphic vector-valued functions and this concept can be interpreted as the holomorphy of all coordinates $\big(u(z), f\big)$.

10.1.3. Holomorphic operator-valued functions

Let G be a region of the complex plane and let $A(z)$ be a function from G into $\mathscr{L}(X)$. According to the three concepts of convergence in $\mathscr{L}(X)$, one obtains (formally) three concepts of holomorphy. The function $A(z)$ is said to be ***holomorphic in norm*** in G if $A(z)$ is differentiable with respect to the operator norm at every point $z \in G$; i.e., if there is an operator-valued function $A'(z)$ such that

$$\left\| \frac{A(z+h) - A(z)}{h} - A'(z) \right\| \to 0$$

for all sequences $h \to 0$. Moreover, one can define strong and weak holomorphy. For example, $A(z)$ is said to be ***weakly holomorphic*** if all scalar functions $\big(A(z)\, x, f\big)$, $x \in X$, $f \in X^*$, are holomorphic.

Again, all these three concepts of holomorphy coincide; i.e., if $A(z)$ is weakly holomorphic in G, then it is also holomorphic in norm in G.

Therefore, for operator-valued functions on Banach spaces with values which are bounded operators, there is only a single concept of holomorphy which can be interpreted in such a way that all matrix elements $\big(A(z)\, x, f\big)$ are holomorphic for all $x \in X$, $f \in X^*$. (One defines holomorphic operator-valued functions with values from $\mathscr{L}(X, Y)$ similarly.)

Let $A(z)$ be holomorphic on a disc $|z| < R$. Then $A(z)$ can be expanded into a power series

$$A(z) = A_0 + zA_1 + z^2A_2 + \cdots, \qquad |z| < R, \tag{1.1}$$

which is convergent in norm, at least for $|z| < R$. Conversely, a power series (1.1) convergent in norm for $|z| < R$ represents a holomorphic operator-valued function on G.

A holomorphic operator-valued function (1.1) on $|z| < R$ whose values are bounded operators is called a ***bounded-holomorphic perturbation*** of the unperturbed bounded operator A_0. $A_1 = A'(0)$ is called the ***first-order approximation*** of the perturbation. The simplest bounded-holomorphic perturbations are the linear ones, $A(z) = A_0 + zA_1$.

If $A(z)$ is holomorphic in G and if the operator $A(z_0)^{-1}$ exists at the point $z_0 \in G$, then the operator $A(z)^{-1} \in \mathscr{L}(X)$ exists for sufficiently small $|z - z_0|$ and, moreover, $A(z)^{-1}$ is holomorphic in the neighborhood of $z = z_0$. One has

$$\big(A(z)^{-1}\big)' = -A(z)^{-1}\, A'(z)\, A(z)^{-1}.$$

An operator-valued function $A(z)$ is called ***real-analytic*** if it is holomorphic on an open interval of the real line.

10.2. Closed operators and elementary spectral properties

10.2.1. Definition and elementary properties

The ***domain*** of a linear operator A is denoted by dom A. dom A is mapped linearly into X by A.

A is said to be ***closed*** if $u_n \in \operatorname{dom} A$, $u_n \to u$ and $Au_n \to v$ imply $u \in \operatorname{dom} A$ and $Au = v$. Each bounded operator $B \in \mathscr{L}(X)$ is closed. The set of all closed operators from X into X is denoted by $\mathscr{C}(X)$. One has $\mathscr{L}(X) \subset \mathscr{C}(X)$. A criterion for A to be closed can be obtained with the use of the concept "graph" of A.

The set of all ordered pairs $\{u, Au\} \in X \times X$, where $u \in \operatorname{dom} A$ is called the *graph* gra A. gra A is a linear manifold. A is closed if and only if gra A is a subspace of $X \times X$.

If A is closed and $\operatorname{dom} A = X$, then A is bounded (closed graph theorem).

If dom A is dense in X, then the *adjoint operator* A^* is uniquely defined by

$$(Au, g) = (u, A^*g), \qquad u \in \operatorname{dom} A, \qquad g \in \operatorname{dom} A^*.$$

A^* is closed. If X is reflexive, then A^{**} exists and one has $A^{**} = A$.

If $A, B \in \mathscr{C}(X)$, $\operatorname{dom} A \subseteqq \operatorname{dom} B$ and $Ax = Bx$, $x \in \operatorname{dom} A$, then one writes $A \subseteqq B$.

If $A \in \mathscr{C}(X)$ and $B \in \mathscr{L}(X)$ then A and B are called *commuting* if from $u \in \operatorname{dom} A$ the relation $Bu \in \operatorname{dom} A$ follows and if the relation

$$BAu = ABu, \qquad u \in \operatorname{dom} A \tag{2.1}$$

holds. This relation is equivalent with

$$BA \subseteqq AB. \tag{2.2}$$

The subspace M of X is called a *reducing subspace* of $A \in \mathscr{C}(X)$ if there is a further subspace N of X with $X = M \oplus N$, such that

$$P \operatorname{dom}(A) \subseteqq \operatorname{dom}(A), \qquad A(M \cap \operatorname{dom}(A)) \subseteqq M, \qquad A(N \cap \operatorname{dom}(A)) \subseteqq N, \tag{2.3}$$

where P denotes the projection onto M along N. This is equivalent with

$$PA \subseteqq AP; \tag{2.4}$$

i.e., A and P commute.

The decomposition $X = M \oplus N$ into reducing subspaces of A implies a decomposition of A into the parts A_M and A_N of A belonging to the subspaces M and N, respectively. A_M is an operator acting on the Banach space M with $\operatorname{dom} A_M = M \cap \operatorname{dom} A$, where $A_M u = Au \in M$ for $u \in M \cap \operatorname{dom} A$. The corresponding proposition is true for A_N. If A is closed, then A_M and A_N are closed.

Correspondingly, a decomposition of A into finitely many parts is defined if a complete disjoint system of projections $P_1, P_2, \ldots, P_r$ with the property

$$P_i A \subseteqq AP_i, \qquad i = 1, 2, \ldots, r,$$

is given. Then A_{P_i} is a closed operator acting on $P_i X$ where $\operatorname{dom} A_{P_i} = P_i X \cap \operatorname{dom} A$. If dom A is dense in X, then dom A_{P_i} is dense in $P_i X$ for all $i = 1, 2, \ldots, r$.

10.2.2. Spectrum and resolvent for closed operators

Let $A \in \mathscr{C}(X)$. A complex number λ_0 is called an *eigenvalue* of A if $\dim \ker (A - \lambda_0) > 0$; i.e., if the eigenvalue equation $Ax = \lambda_0 x$ is nontrivially solvable. Recall that in the finite-dimensional case the spectrum of A is exhausted by the set of all eigenvalues. The situation in the case of an arbitrary Banach space is much more complicated.

The concept of the resolvent does not require any modification. Together with A, $\lambda - A$ is also closed. If $(\lambda - A)^{-1}$ exists and if, moreover, $(\lambda - A)^{-1} \in \mathscr{L}(X)$, then λ is called a *regular point* of A. The set of all regular points is called the *resolvent set* and is denoted by $\operatorname{res} A$. The operator-valued function $(\lambda - A)^{-1}$ whose values are from $\mathscr{L}(x)$ is called the *resolvent* of A. For the resolvent, one obtains the first Neumann series (see (2.1.3)) which implies that $\operatorname{res} A$ is open and that $(\lambda - A)^{-1}$ is piecewise holomorphic on $\operatorname{res} A$. The complement of $\operatorname{res} A$ with respect to the complex plane is called the *spectrum*; it is denoted by $\operatorname{spec} A$.

If $A \in \mathscr{L}(X)$, then $\operatorname{res} A$ contains the exterior of the circle $|\lambda| = \operatorname{spr} A = \inf_{n \geqq 1} \|A^n\|^{1/n} = \lim_{n \to \infty} \|A^n\|^{1/n}$; the resolvent is holomorphic at $\lambda = \infty$. If $A \in \mathscr{C}(X)$ and unbounded, and if $(\lambda - A)^{-1}$ is holomorphic in the neighborhood of $\lambda = \infty$, then $\lambda = \infty$ is an essential singularity for $(\lambda - A)^{-1}$.

Sometimes it is convenient to include the point $\lambda = \infty$ either in the spectrum or in the resolvent set. If $A \in \mathscr{L}(X)$, then one requires $\{\infty\} \subset \operatorname{res} A$. If A is unbounded, then the point $\lambda = \infty$ is always a point of $\operatorname{spec} A$.

10.2.3. Decomposition theorem

For the following theorem, see F. Riesz and B. v. Sz.-Nagy [1, p. 403] and T. Kato [5, p. 178]. Assume that $\operatorname{spec} A$ can be decomposed into two isolated parts σ_1 (which is bounded) and σ_2, such that there exists a simply closed piecewise smooth path Γ belonging to $\operatorname{res} A$, containing σ_1 in its interior, and σ_2 in its exterior. In this case a decomposition according to Subsect. 10.2.1 can be constructed. One has

Theorem 1. *Let* $\operatorname{spec} A = \sigma_1 \cup \sigma_2$ *and let* $\Gamma \subset \operatorname{res} A$ *be a simply closed piecewise smooth positively oriented path separating* σ_1 *(which is bounded) from* σ_2. *If*

$$P = \frac{1}{2\pi i} \int_\Gamma (\lambda - A)^{-1} \, d\lambda,$$

then the decomposition $X = M_1 \oplus M_2$, *where* $M_1 = PX$ *and* $M_2 = (1 - P) X$, *yields a decomposition of* A *into parts* A_{M_1} *and* A_{M_2}, *where* $\operatorname{spec} A_{M_1} = \sigma_1$ *and* $\operatorname{spec} A_{M_2} = \sigma_2$. *Moreover,* $A_{M_1} \in \mathscr{L}(M_1)$, *but* A_{M_2} *can be unbounded. Furthermore,*

$$PA \subseteqq AP = \frac{1}{2\pi i} \int_\Gamma \lambda(\lambda - A)^{-1} \, d\lambda \in \mathscr{L}(X).$$

Obviously, this theorem can be generalized if a decomposition of $\operatorname{spec} A$ into the union of finitely many parts $\sigma_0, \sigma_1, \sigma_2, \ldots, \sigma_n$ is given.

10.2.4. Isolated points of the spectrum

If λ_0 is an isolated point of the spectrum $\operatorname{spec} A$, then $\operatorname{spec} A = \{\lambda_0\} \cup (\operatorname{spec} A \setminus \{\lambda_0\})$ is a decomposition of $\operatorname{spec} A$ into two isolated parts $\sigma_1 = \{\lambda_0\}$, $\sigma_2 = \operatorname{spec} A \setminus \{\lambda_0\}$. Hence, Theorem 1 can be applied to this situation where, for example, a positively oriented sphere centered at λ_0 with sufficiently small radius is chosen as Γ.

λ_0 is the only point of the spectrum of the bounded operator $A_{M_1} \in \mathcal{L}(M_1)$; i.e., $A_{M_1} - \lambda_0$ is quasinilpotent, and one obtains the following expansion for the resolvent $(\lambda - A_{M_1})^{-1}$ (the first Neumann series in the neighborhood of $\lambda = \infty$):

$$(\lambda - A_{M_1})^{-1} = \sum_{n=0}^{\infty} (\lambda - \lambda_0)^{-n-1} (A_{M_1} - \lambda_0)^n, \qquad 0 < |\lambda - \lambda_0| < \infty.$$

This is an expansion in $\mathcal{L}(M_1)$. With respect to $\mathcal{L}(X)$, one obtains the corresponding expansion

$$(\lambda - A)^{-1} P = \frac{P}{\lambda - \lambda_0} + \sum_{n=1}^{\infty} \frac{D^n}{(\lambda - \lambda_0)^{n+1}},$$

where

$$D = (A - \lambda_0) P = \frac{1}{2\pi i} \int_{\Gamma} (\lambda - \lambda_0)(\lambda - A)^{-1} \, d\lambda \in \mathcal{L}(X).$$

On the other hand, $(\lambda - A_{M_2})^{-1}$ is holomorphic at $\lambda = \lambda_0$; hence, it can be expanded into a Taylor series (first Neumann series)

$$(\lambda - A_{M_2})^{-1} = \sum_{n=0}^{\infty} (\lambda - \lambda_0)^n (-1)^n [(\lambda_0 - A_{M_2})^{-1}]^{n+1}$$

or

$$(\lambda - A)^{-1}(1 - P) = \sum_{n=0}^{\infty} (\lambda - \lambda_0)^n (-1)^n S_0^{n+1},$$

where

$$S_0 = (\lambda_0 - A_{M_2})^{-1}(1 - P) = \lim_{\lambda \to \lambda_0} (\lambda - A)^{-1}(1 - P).$$

$(\lambda - A)^{-1}(1 - P)$ is called the *reduced resolvent* of A with respect to λ_0. From these two expansions, one obtains

$$(\lambda - A)^{-1} = \frac{P}{\lambda - \lambda_0} + \sum_{n=1}^{\infty} \frac{D^n}{(\lambda - \lambda_0)^{n+1}} + \sum_{n=0}^{\infty} (\lambda - \lambda_0)^n (-1)^n S_0^{n+1}; \tag{2.5}$$

i.e., the Laurent series expansion of the resolvent at $\lambda = \lambda_0$. S_0 has analogous properties as in the finite-dimensional case (cf. Subsect. 2.6.2). One has

$$S_0 = \frac{1}{2\pi i} \int_{\Gamma} (\lambda - A)^{-1} \frac{d\lambda}{\lambda - \lambda_0} \in \mathcal{L}(X)$$

and

$$S_0 A \subseteq A S_0 \in \mathcal{L}(X), \qquad (\lambda - A) S_0 = 1 - P, \qquad S_0 P = P S_0 = 0.$$

The Laurent series (2.5) of the resolvent differs from the corresponding series in the finite-dimensional case: the principal part may be an infinite series. However, if

$\dim M_1 < \infty$, i.e., if $\dim P < \infty$, then the principal part is a finite sum, because $D_{M_1} = A_{M_1} - \lambda_0$ is nilpotent. Correspondingly, $D = (A - \lambda_0) P$ is nilpotent.

In this case, λ_0 is an *eigenvalue* of A, because λ_0 belongs to the spectrum of the finite-dimensional operator A_{M_1}; i.e., it is an eigenvalue of A_{M_1}, hence also an eigenvalue of A.

An eigenvalue λ_0 of A which is an isolated point of spec A is called an *isolated eigenvalue* of A.

If $\dim P < \infty$, where λ_0 is an isolated eigenvalue of A, then $\dim P = \dim M_1$ is called the *algebraic multiplicity* of λ_0. P is the *eigenprojection* belonging to λ_0 and D is the corresponding *eigennilpotent*.

10.2.5. Finite systems of eigenvalues

Assume that a finite set $\lambda_1, \lambda_2, \ldots, \lambda_r$ of isolated points of spec A is given. Then one obtains, according to (2.5), the expansion

$$(\lambda - A)^{-1} = \sum_{\varrho=1}^{r} \left[\frac{P_\varrho}{\lambda - \lambda_\varrho} + \sum_{n=1}^{\infty} \frac{D_\varrho^n}{(\lambda - \lambda_\varrho)^{n+1}} \right] + (\lambda - A)^{-1} P_0, \tag{2.6}$$

where $P_\varrho = (2\pi i)^{-1} \int_{\Gamma_\varrho} (\lambda - A)^{-1} \, d\lambda$, $D_\varrho = (A - \lambda_\varrho) P_\varrho$, $P_0 = 1 - \sum_{\varrho=1}^{r} P_\varrho$. The function $(\lambda - A)^{-1} P_0 = P_0(\lambda - A)^{-1}$ is holomorphic at $\lambda = \lambda_\varrho$, $\varrho = 1, \ldots, r$. If the condition $\dim P_\varrho < \infty$, $\varrho = 1, \ldots, r$, is again satisfied, then the points λ_ϱ from spec A are eigenvalues, and P_ϱ, D_ϱ are the corresponding eigenprojections and eigennilpotents, respectively. One has

$$PA \subseteqq AP; \qquad AP = \sum_{\varrho=1}^{r} (\lambda_\varrho P_\varrho + D_\varrho), \qquad P = \sum_{\varrho=1}^{r} P_\varrho; \tag{2.7}$$

that is, the part of A in PX has a spectral representation (cf. Sect. 2.5).

A finite system of isolated points of spec A with $\dim P < \infty$ (P the corresponding Riesz projection) is called a *finite eigenvalue system*. The following considerations deal with the perturbation theory of such systems (see Sect. 10.4).

10.2.6. Adjoint operator

res A^* and spec A^* are obtained from res A and spec A by reflection with respect to the real line. One has

$$(\lambda - A^*)^{-1} = [(\bar{\lambda} - A)^{-1}]^*, \qquad \lambda \in \text{res}\, A. \tag{2.8}$$

Therefore, the spectral properties of A outlined in Subsect. 10.2.2 to 10.2.5 have their counterparts in corresponding properties of A^*. Their explicit notation is omitted (cf. T. Kato [5, p. 184]).

10.3. Holomorphic operator-valued functions whose values are from $\mathscr{C}(X)$

To increase the potential for applications, one considers not only holomorphic operator-valued functions whose values are bounded operators (see Subsect. 10.1.3) but also functions whose values are merely closed operators.

For short a holomorphic operator-valued function $B(z)$ whose values are from $\mathscr{L}(X)$ is called a *bounded-holomorphic* operator-valued function (see Subsect. 10.1.3).

10.3.1. Definition of holomorphic operator-valued functions

In this subsection, only a special case is considered (see T. Kato [5, p. 366], F. Rellich [3]). Let $A(z)$ be an operator-valued function whose values are from $\mathscr{C}(X)$ defined on a region G of the complex plane. $A(z)$ is said to be *holomorphic* on G if there are two bounded-holomorphic operator-valued functions $U(z)$ and $V(z)$, whose values are from $\mathscr{L}(Z, X)$, where Z is a Banach space, such that $U(z)$ maps Z one-to-one onto $\operatorname{dom} A(z)$ and

$$A(z)\, U(z) = V(z). \tag{3.1}$$

This definition contains the former one: If $A(z) \in \mathscr{L}(X)$ for all $z \in G$, then $A(z)$ is holomorphic if and only if $A(z)$ is bounded-holomorphic.

As usual, $A(z)$ is called holomorphic at z_0 if $A(z)$ is holomorphic in a neighborhood of z_0.

Let the operator-valued function $A(z) \in \mathscr{C}(X)$ be defined in a neighborhood of $z = 0$. Let $\lambda_0 \in \operatorname{res} A(0)$. Then, $A(z)$ is holomorphic at $z = 0$ if and only if, first, $\lambda_0 \in \operatorname{res} A(z)$ for sufficiently small $|z|$ and, secondly, $\big(\lambda_0 - A(z)\big)^{-1}$ is bounded-holomorphic with respect to z. In this case, $(\lambda - A(z))^{-1}$ is bounded-holomorphic with respect to both variables (λ, z) jointly, where $\lambda \in \operatorname{res} A(0)$ and $|z|$ is sufficiently small (depending on λ) (see F. Rellich [3], T. Kato [5, p. 367]).

10.3.2. Holomorphic operator-valued functions of type A

There are several types of holomorphy (see T. Kato [5, p. 375sqq.]). An important type, also for applications, is given by type A. For the other types we refer to T. Kato [5, p. 393sqq.].

An operator-valued function $A(z) \in \mathscr{C}(X)$, which is defined in a region G, is called *holomorphic of type A* if $\operatorname{dom} A(z) \equiv D$ is independent of z and if $A(z)\, u$ is a holomorphic vector-valued function on G for all $u \in D$. In this case, there exists a Taylor expansion for $A(z)\, u$ at every point $z_0 \in G$:

$$A(z)\, u = A_0 u + (z - z_0)\, A_1 u + (z - z_0)^2\, A_2 u + \cdots, \qquad u \in D. \tag{3.2}$$

The series (3.2) is strongly convergent in some disc $|z - z_0| < R$. A_0 is closed and the operators $A_1, A_2, \ldots$ are linear with domain D.

A holomorphic operator-valued function of type A is holomorphic in the sense of Definition 1; let $Z = D$ with the norm $\|u\| + \|A_0 u\|$ and let U be the operator that maps $u \in Z$ into $u \in X$.

Next one formulates a criterion for holomorphy of type A of an operator-valued function (see F. RELLICH [3], T. KATO [5, p. 377], B. v. SZ.-NAGY [2]). Let $A_0 \in \mathscr{C}(X)$ and $\operatorname{dom} A_0 = D$. Furthermore, let $A_1, A_2, \ldots$ be linear operators acting on X whose domains contain D. Assume that constants $a, b, c > 0$ exist such that

$$\|A_n u\| \leqq c^{n-1} (a \|u\| + b \|A_0 u\|), \qquad u \in D, \qquad n = 1, 2, 3, \ldots \tag{3.3}$$

Then, the series

$$A(z)\, u = A_0 u + z A_1 u + z^2 A_2 u + \cdots \tag{3.4}$$

defines an operator $A(z)$ with domain D for $|z| < 1/c$. Moreover, if $|z| < 1/(b + c)$, then $A(z)$ is closable and the corresponding closures of $A(z)$ form a holomorphic operator-valued function of type A.

In particular, this criterion can be applied if $A_2 = A_3 = \cdots = 0$. Then one can take $c = 0$ and the only remaining condition is

$$\|A_1 u\| \leqq a \|u\| + b \|A_0 u\|; \qquad u \in D.$$

10.4. Holomorphic and closed perturbations of closed operators

Let $A(z)$ be a holomorphic operator-valued function whose values are from $\mathscr{C}(X)$ defined on a disc $|z| < R$. Such a function is called a *holomorphic and closed perturbation* of the unperturbed closed operator $A(0)$. For the application of the finite-dimensional theory, it is necessary to explain the behavior of isolated parts of the spectrum under such perturbations.

10.4.1. Behavior of isolated parts of the spectrum

Let $\operatorname{spec} A(0) = \sigma_1 \cup \sigma_2$ be a decomposition of $\operatorname{spec} A(0)$ into two isolated parts and let Γ be a simply closed path separating σ_1 and σ_2; let σ_1 be bounded and $\Gamma \subset \operatorname{res} A(0)$.

If $|z|$ is sufficiently small, say $|z| < \delta$, then $\Gamma \subset \operatorname{res} A(z)$. (Note that $A(z)$ converges in a generalized sense to $A(0)$ as $z \to 0$; see T. KATO [5, p. 206]; isolated parts of the spectrum, such as σ_1, σ_2, are upper semicontinuous with respect to this convergence, see T. KATO [5, p. 212].) In other words, if $|z| < \delta$, then $\operatorname{spec} A(z)$ is also decomposed by Γ into two isolated parts $\sigma_1(z)$ and $\sigma_2(z)$. Introduce the Riesz projection

$$P(z) = \frac{1}{2\pi i} \int_\Gamma \bigl(\lambda - A(z)\bigr)^{-1} \mathrm{d}\lambda, \qquad |z| < \delta. \tag{4.1}$$

$P(z)$ is bounded-holomorphic in $|z| < \delta$. If $P(z)\, X = M_1(z)$ and $\bigl(1 - P(z)\bigr)\, X = M_2(z)$, one obtains the decomposition

$$X = M_1(z) \oplus M_2(z), \qquad |z| < \delta,$$

of the Banach space X. Since $A(z)$ and $P(z)$ commute and $P(z)\, A(z) \subseteq A(z)\, P(z)$, one obtains a decomposition of the operator $A(z)$ into two parts $A(z)_{M_1(z)}$ and $A(z)_{M_2(z)}$

acting on $M_1(z)$ and $M_2(z)$, respectively. $A(z)_{M_1(z)}$ is bounded, dom $A(z)_{M_2(z)}$ = dom $A(z)$ $\cap M_2(z)$, spec $A(z)_{M_1(z)} = \sigma_1(z)$, and spec $A(z)_{M_2(z)} = \sigma_2(z)$.

Thus, the eigenvalue problem for $A(z)$ in X is reduced to the eigenvalue problems for $A(z)_{M_1(z)}$ in $M_1(z)$ and $A(z)_{M_2(z)}$ in $M_2(z)$, but, in general, the subspaces $M_1(z)$ and $M_2(z)$ vary with z, a very inconvient feature. By passing to another perturbation similar to $A(z)$, this inconvenience can be dropped (see Subsect. 3.5.4). The subspaces M_1 and M_2 will be fixed. This procedure is often applied in perturbation theory.

Since $P(z)$ is bounded-holomorphic in $|z| < \delta$, there is a transformation operator $U(z)$ which is holomorphic in $|z| < \delta$ together with its inverse $U(z)^{-1}$ and which satisfies

$$P(z) = U(z)\, P(0)\, U(z)^{-1}, \qquad U(0) = 1. \tag{4.2}$$

For the construction of $U(z)$, see Sect. 3.5, where it is performed for a finite-dimensional X. The results can be transferred to the case of a Banach space X. Since $P(z)$ and $A(z)$ commute, and $P(z)\, A(z) \subseteq A(z)\, P(z)$, it follows that the operator

$$\tilde{A}(z) = U(z)^{-1}\, A(z)\, U(z) \tag{4.3}$$

commutes with $P(0)$, i.e.,

$$P(0)\, \tilde{A}(z) \subseteq \tilde{A}(z)\, P(0).$$

That is, the subspaces $M_1(0) = P(0)\, X$ and $M_2(0) = \big(1 - P(0)\big)\, X$ decompose $\tilde{A}(z)$ for all $|z| < \delta$. In other words, $\tilde{A}(z)$ can be considered separately on the Banach spaces $M_1(0)$ (where it is bounded) and $M_2(0)$. One has

$$\operatorname{spec}\big(\tilde{A}(z)_{M_1(0)}\big) = \sigma_1(z), \qquad \operatorname{spec}\big(\tilde{A}(z)_{M_2(0)}\big) = \sigma_2(z).$$

Moreover, the eigenvalue problems for $A(z)$ in $M_1(z)$ and $\tilde{A}(z)$ in $M_1(0)$ are equivalent.

10.4.2. Case of a finite eigenvalue system

Let $\sigma_1 = \{\lambda_1, \lambda_2, \ldots, \lambda_r\}$ and let dim $P(0) < \infty$; i.e., σ_1 is assumed to be a finite eigenvalue system of $A(0)$. Then dim $M_1(0)$ = dim $P(0)\, X < \infty$ and, because dim $P(z)$ = dim $P(0)$ for $|z| < \delta$, also dim $M_1(z) < \infty$ (see, for example, B. v. Sz.-Nagy [2]).

As in Subsect. 10.4.1

$$\tilde{A}(z)\, P(0) = U(z)^{-1}\, A(z)\, P(z)\, U(z), \qquad |z| < \delta,$$

is a bounded-holomorphic operator-valued function, because the restriction of $\tilde{A}(z)$ to $M_1(0)$ is a (bounded-) holomorphic operator-valued function on the finite-dimensional space $M_1(0)$. Already from this fact, the structure of $\sigma_1(z)$ follows. The properties of the eigenprojections and eigennilpotents of $A(z)$ restricted to $M_1(z)$ follow from the corresponding properties of the eigenprojections and eigennilpotents of $\tilde{A}(z)$ restricted to $M_1(0)$, together with the transformation formulas between both objects. For example if

$$\tilde{A}(z)\, P(0) = \sum_{\varrho} \big(\lambda_\varrho(z)\, \tilde{P}_\varrho(z) + \tilde{D}_\varrho(z)\big), \quad 0 < |z| < \delta,$$

then one obtains

$$A(z)\,P(z) = \sum_{\varrho} \big(\lambda_{\varrho}(z)\,P_{\varrho}(z) + D_{\varrho}(z)\big), \qquad 0 < |z| < \delta,$$

where

$$P_{\varrho}(z) = U(z)\,\tilde{P}_{\varrho}(z)\,U(z)^{-1}, \qquad D_{\varrho}(z) = U(z)\,\tilde{D}_{\varrho}(z)\,U(z)^{-1}, \qquad 0 < |z| < \delta. \tag{4.4}$$

In particular, this relation is true if σ_1 consists of a single isolated eigenvalue of finite algebraic multiplicity.

Finally, one obtains a transformation of the perturbation $A(z)$ via $U(z)$, which yields a new perturbation $\tilde{A}(z)$ and, simultaneously, a transformation of the corresponding spectral objects (see (4.4)). By this transformation the eigenvalue problem for $A(z)$ restricted to $M_1(z)$ reduces to the eigenvalue problem for $\tilde{A}(z)$ restricted to $M_1(0)$.

This situation is suitable for transferring all results of the analytic perturbation theory on a finite-dimensional space to the case of holomorphic and closed perturbations of isolated eigenvalues of finite multiplicity with respect to a closed unperturbed operator.

The transfer of the results will be demonstrated in the following discussion by some remarkable examples.

10.4.3. Transfer of some results of the finite-dimensional theory

First, the basic perturbation theorem, due to T. Kato [4], B. v. Sz.-Nagy [2] and F. Wolf [1], will be formulated.

Theorem 1. *Let $A(z) \in \mathscr{C}(X)$ be a holomorphic and closed perturbation of the unperturbed operator $A(0) = A_0$ in $|z| < R$. Let λ_0 be an isolated eigenvalue of A_0 of finite algebraic multiplicity. Then there is a (positively oriented) sphere Γ centered at λ_0 and a disc $|z| < \delta$, such that the spectrum of $A(z)$ in the interior of Γ consists of finitely many eigenvalue cycles of finite period. Denoting these eigenvalues (inside Γ) by $\lambda_{\alpha\sigma}(z)$, one obtains expansions of the form*

$$\lambda_{\alpha\sigma}(z) = \lambda_0 + \sum_{\mu=1}^{\infty} \lambda_{\alpha\mu} z^{\mu/g_\alpha}, \qquad \sigma = 1, 2, \ldots, g_\alpha;\ \alpha = 1, \ldots, r;\ |z| < \delta, \tag{4.5}$$

where r denotes the number of different eigenvalue cycles and $g_1, g_2, \ldots, g_r$ denote their periods. For the corresponding eigenprojections

$$P_{\alpha\sigma}(z) = \frac{1}{2\pi i} \int_{\Gamma_{\alpha\sigma}} \big(\lambda - A(z)\big)^{-1}\,d\lambda \tag{4.6}$$

(where $\Gamma_{\alpha\sigma}$ denotes a sufficiently small positively oriented sphere centered at $\lambda_{\alpha\sigma}(z)$), one obtains corresponding expansions of the form

$$P_{\alpha\sigma}(z) = \sum_{\mu=-N_\alpha}^{\infty} P_{\alpha\mu} z^{\mu/g_\alpha}, \qquad |z| < \delta, \tag{4.7}$$

and similarly for the eigennilpotents $D_{\alpha\sigma}(z) = \big(A(z) - \lambda_{\alpha\sigma}(z)\big) P_{\alpha\sigma}(z)$.

$$P(z) = \sum_{\alpha=1}^{r} \sum_{\sigma=1}^{g_\alpha} P_{\alpha\sigma}(z) = \frac{1}{2\pi i} \int_{\Gamma} \big(\lambda - A(z)\big)^{-1} \,d\lambda \tag{4.8}$$

is the group projection of the λ_0*-group of* $A(z)$*. One has* $\dim P(z) = n$*, where* n *denotes the algebraic multiplicity of* λ_0.

Proof. Subsect. 3.4.13 contains the corresponding finite-dimensional result. This result will first be applied to the holomorphic perturbation $\tilde{A}(z) \restriction P(0)X$ of $\tilde{A}(0) \restriction P(0)X$ acting on the finite-dimensional space $P(0)\,X$, where λ_0 is the unperturbed eigenvalue (see Subsect. 10.4.1 and 10.4.2). This already implies the expansions (4.5) for the perturbed eigenvalues. The corresponding eigenprojections are denoted by $Q_{\alpha\sigma}(z)$. One has $Q_{\alpha\sigma}(z) \in \mathscr{L}\big(P(0)\,X\big)$ and for the projection-valued function $Q_{\alpha\sigma}(z)$ one obtains an expansion of the form (4.7), according to Subsect. 3.4.13. Then the eigenprojections of $\tilde{A}(z)$ in X are given by $\tilde{P}_{\alpha\sigma}(z) = Q_{\alpha\sigma}(z)\,P(0)$ and, furthermore, the eigenprojections of $A(z)$ in X by

$$P_{\alpha\sigma}(z) = U(z)\,\tilde{P}_{\alpha\sigma}(z)\,U(z)^{-1}.$$

This result implies the expansions (4.7) for the projections $P_{\alpha\sigma}(z)$. Since $\sum_{\alpha,\sigma} \tilde{P}_{\alpha\sigma}(z) = P(0)$, one obtains $\sum_{\alpha,\sigma} P_{\alpha\sigma}(z) = P(z)$, hence (4.8). ■

This basic theorem now can be completed and supplemented by further propositions, in accordance with the foregoing investigations, in particular those of Ch. 6 and 7.

Theorem 2. *Let* $A(z) \in \mathscr{C}(X)$ *be a holomorphic and closed perturbation of the unperturbed operator* $A(0) = A_0$*, given in* $|z| < R$*. Let* λ_0 *be an isolated eigenvalue of* A_0 *of finite algebraic multiplicity. The perturbed eigenvalues of the* λ_0*-group are denoted by* $\lambda_{\alpha\sigma}(z)$ *and the corresponding eigenprojections (according to Theorem 1) by* $P_{\alpha\sigma}(z)$*. Then there exists a dotted disc* $0 < |z| < \delta$*, such that the eigenprojections* $P_{\alpha\sigma}(z)$ *can be decomposed into a sum of disjoint reducing irreducible projections* $P^{\varrho}_{\alpha\sigma}(z)$ *with respect to* $A(z)\,P(z)$ *of the form*

$$P_{\alpha\sigma}(z) = \sum_{\varrho=1}^{l_\alpha} P^{\varrho}_{\alpha\sigma}(z), \qquad 0 < |z| < \delta, \tag{4.9}$$

where the irreducible projections $P^{\varrho}_{\alpha\sigma}(z)$ *have expansions of the form*

$$P^{\varrho}_{\alpha\sigma}(z) = \sum_{\mu=-N^{\varrho}_{\alpha}}^{\infty} P^{\varrho}_{\alpha\mu} z^{\mu/g_\alpha}, \qquad 0 < |z| < \delta. \tag{4.10}$$

One can choose Jordan bases $f^{\varrho\varkappa}_{\alpha\sigma}(z)$*,* $\varkappa = 1, 2, \ldots, m^{\varrho}_{\alpha} = \dim P^{\varrho}_{\alpha\sigma}(z)$*, in the irreducible subspaces* $P^{\varrho}_{\alpha\sigma}(z)\,X$ *(where, for example,* $f^{\varrho 1}_{\alpha\sigma}(z)$ *is an eigenvector and* $f^{\varrho m^{\varrho}_{\alpha}}_{\alpha\sigma}(z)$ *is a generating vector), such that the Jordan vectors* $f^{\varrho\varkappa}_{\alpha\sigma}(z)$ *are continuous at* $z = 0$ *and such that there are Puiseux series expansions of the form*

$$f^{\varrho\varkappa}_{\alpha\sigma}(z) = f^{\varrho\varkappa}_{\alpha} + \sum_{\mu=1}^{\infty} f^{\varrho\varkappa}_{\alpha\mu} z^{\mu/g_\alpha}, \qquad |z| < \delta. \tag{4.11}$$

Proof. The assertions of the theorem are reduced, according to the proof of Theorem 1, to Theorem 6.1.2, where the corresponding result for operators acting on finite-dimensional spaces has been proved. ■

The essential content of Theorem 2 is that *the bounded-holomorphic operator* $A(z)\,P(z)$, *considered on* $0 < |z| < \delta$, *has an invariant Jordan structure.*

The results of the reduction theory obtained in Ch. 7 can also be transferred to the case of holomorphic and closed perturbations $A(z) \in \mathscr{C}(X)$. As an example, the counterpart of Theorem 7.4.6 will be formulated.

Theorem 3. *Let* $A(z) \in \mathscr{C}(X)$, *be a, holomorphic and closed perturbation of the unperturbed operator* $A(0) = A_0$ *in* $|z| < R$. *Let* λ_0 *be an isolated eigenvalue of* A_0 *of finite algebraic multiplicity n. Furthermore, let* $P(z)$ *denote the group projection of the* λ_0*-group. Assume that the bounded-holomorphic operator* $A(z)\,P(z)$ *has the power series expansion*

$$A(z)\,P(z) = A_0P_0 + zA_1 + z^2A_2 + \cdots, \qquad A_i \in \mathscr{L}(X), \qquad i = 1, 2, \ldots \tag{4.12}$$

Let (in correspondence with the notation in Ch. 7)

$$K_i = (E_1D_0^{m_1-1} + E_2D_0^{m_2-2} + \cdots + E_iD_0^{m_i-1})\,A_1 \restriction (E_1 + \cdots + E_i)\,X,$$
$$i = 1, 2\,\ldots, p.$$

The operators $K_1, K_2, \ldots, K_p$ *act on finite-dimensional spaces. Suppose that the relation*

$$\det K_1 \neq 0, \det K_2 \neq 0, \ldots, \det K_l \neq 0, \tag{4.13}$$

is valid, where $1 \leq l \leq p$. *Then the group projection* $P(z)$ *of the* λ_0*-group of* $A(z)$ *can be decomposed holomorphically in a disc* $|z| < \delta$ *into a sum of disjoint reducing projections* $Q_{\varrho\sigma}(z)$, $R(z)$ *of the form*

$$P(z) = \sum_{\varrho=1}^{l} \sum_{\sigma=1}^{k_\varrho} Q_{\varrho\sigma}(z) + R(z), \qquad |z| < \delta, \tag{4.14}$$

where the projections $Q_{\varrho\sigma}(z)$ *are defined as in Theorem* 7.4.6 *and where the further properties of the* $Q_{\varrho\sigma}$*-groups of perturbed eigenvalues, which were derived in Sect.* 7.4, *are also valid.*

Proof. The assertions of the theorem are reduced, according to the proof of Theorem 1, to Theorem 7.4.6, where the corresponding result for operators acting on finite-dimensional spaces is formulated. From (4.12), one obtains for the transformed perturbation $\tilde{A}(z)\,P(0) = U(z)^{-1}\,A(z)\,P(z)\,U(z)$ the power series expansion

$$\tilde{A}(z)\,P(0) = A_0P_0 + zP_0A_1P_0 + O(z^2).$$

One applies Theorem 7.4.6 to this perturbation. ■

The special results which are related to selfadjoint and normal perturbations on unitary spaces (see Sect. 3.6 and 3.7) can be transferred to results for corresponding perturbations on Hilbert spaces. The explicit formulation will be omitted.

The results of Chap. 9 on polynomial operator-valued functions can be generalized to polynomial operator-valued functions $\sum_{\varrho=0}^{m} \lambda^\varrho A_\varrho$, where the operators A_ϱ act on a Banach space X, if one assumes that λ_0 is a characteristic value of finite algebraic multiplicity. The results then obtained generalize results of V. Eni [2, 4]. This author formulates his results also for power series $\sum_{\varrho=0}^{\infty} \lambda^\varrho A_\varrho$ (see [4]). The results of Ch. 9 can also be transferred to this case.

Supplement. Analytic perturbations for linear operators depending on several complex variables

S 1. Introduction

This supplement to the English edition is essentially a translation of the author's article [7].

As before, let $\mathscr{C}(X)$ denote the set of all closed operators acting on a Banach space X. Let $A(z)$ be a holomorphic operator-valued function, defined on a circle $|z| < r$, $r > 0$, whose values are from $\mathscr{C}(X)$. Consider the perturbation $A(z)$ of the unperturbed operator $A_0 = A(0)$. The problem is investigating the properties of the spectrum spec A_0 and of the spectral objects under this perturbation. For an isolated eigenvalue λ_0 of A_0 of finite algebraic multiplicity $m(\lambda_0)$, this problem has been discussed in detail in Ch. 10.

The corresponding problem for holomorphic perturbations $A(z_1, z_2, \ldots, z_n)$ of the unperturbed operator $A(0, 0, \ldots, 0)$, depending on several complex variables has been considered by a few authors (see F. Rellich [1], F. Riesz and B. v. Sz.-Nagy [1, p. 360], T. Kato [1, p. 116]), and the results obtained did not provide a satisfactory insight into the structure of these perturbations.

In particular, if the unperturbed eigenvalue λ_0 is simple, i.e., if $m(\lambda_0) = 1$, then one can generalize the basic results for the case of only one variable, $n = 1$, to the general case of several variables, $n > 1$. One obtains the following result (for example, see T. Kato [1, p. 117]):

In the neighborhood of λ_0 there exists exactly one perturbed eigenvalue $\lambda(z)$ which is holomorphic in a sufficiently small polydisc centered at $z = 0$, i.e.,

$$\lambda(z) = \lambda_0 + \sum_{|\alpha|=1}^{\infty} \lambda_\alpha z^\alpha, \qquad \alpha = (\alpha_1, \ldots, \alpha_n), \qquad z^\alpha = z_1^{\alpha_1} z_2^{\alpha_2} \cdots z_n^{\alpha_n},$$

$$|\alpha| = \alpha_1 + \cdots + \alpha_n,$$

where $|z_i| < \delta$, $i = 1, 2, \ldots, n$, and $\delta > 0$ sufficiently small. The corresponding eigenprojection $P(z)$ is also holomorphic in that region and there exists a holomorphic nonvanishing eigenvector, for example, $P(z)\, x_0$, where $A_0 x_0 = \lambda_0 x_0$, $x_0 \neq 0$.

However, the basic theorem of F. Rellich (for example, see T. Kato [1, pp. 71, 120]) does not hold in the case of several holomorphic perturbation parameters. This can be seen by the following example (see T. Kato [1, p. 116], F. Rellich [1] or F. Riesz and B. v. Sz.-Nagy [1, p. 361]). Let

$$T(z_1, z_2) = \begin{pmatrix} z_1 & z_2 \\ z_2 & -z_1 \end{pmatrix}.$$

Its eigenvalues, which are given by

$$\lambda_\pm(z_1, z_2) = \pm(z_1^2 + z_2^2)^{1/2},$$

are not holomorphic at $(z_1, z_2) = (0, 0)$.

In the case of finite algebraic multiplicity $m(\lambda_0) < \infty$, a series of results can be obtained by generalization of corresponding results for the case $n = 1$ to the general case $n > 1$. For example, the following results are true (note that the corresponding results for the case $n = 1$ have been proved in Ch. 3, 5, 6 and 10):

I. The perturbed eigenvalues $\lambda(z)$, where $\lambda(0) = \lambda_0$, are algebroid, i.e., a perturbed eigenvalue $\lambda(z)$ satisfies an equation of the form

$$\lambda^N + a_1(z)\,\lambda^{N-1} + \cdots + a_N(z) = 0, \qquad N = m(\lambda_0), \tag{1}$$

where the functions $a_1(z), \ldots, a_N(z)$ are holomorphic with respect to $z = (z_1, z_2, \ldots, z_n)$ in the neighborhood of $z = 0$.

II. The corresponding eigenprojections $P_{\alpha\varrho}(z)$ can be represented in the form

$$P_{\alpha\varrho}(z) = C_{\alpha 0}(z) + C_{\alpha 1}(z)\,\lambda_{\alpha\varrho}(z) + \cdots + C_{\alpha,g_\alpha-1}(z)\,\lambda_{\alpha\varrho}(z)^{g_\alpha-1}, \tag{2}$$

where the eigenvalue $\lambda_{\alpha\varrho}(z)$ is a zero of an irreducible factor π_α of the left hand side of (1) and where the $C_{\alpha\varkappa}(z)$ are operator-valued functions, which are meromorphic in the neighborhood of $z = 0$ and holomorphic at all those points where the $\lambda_{\alpha\varrho}(z)$ are holomorphic.

III. Except for some analytic manifold of dimension less than n, the nilpotent operator

$$\Bigl(A(z) - \sum_{\alpha,\varrho} \lambda_{\alpha\varrho}(z)\,P_{\alpha\varrho}(z)\Bigr) \upharpoonright P(z)\,X$$

has a constant Jordan normal form in the neighborhood of $z = 0$ (i.e., the Jordan normal form is independent of z); in other words: The operator $A(z)$ restricted to $P(z)\,X$ ($P(z)$ denotes the Riesz projection of $A(z)$ with respect to λ_0) has an "invariant Jordan structure".

In the following discussion, propositions I—III will be proved for isolated eigenvalues of finite algebraic multiplicity. Other results on analytic perturbations (for several variables) will also be proved.

Some questions concerning holomorphic operator-valued functions for several complex variables are dealt with by S. G. Krein and V. P. Trofimov [1].

S 2. Holomorphic operator-valued functions of several complex variables

Let $z = (z_1, z_2, \ldots, z_n)$ and let $K \subset \mathbb{C}^n$ be the polydisc $K = \{z: |z_i| < r, r > 0, i = 1, 2, \ldots, n\}$.

Definition 1. The operator-valued function $A(z)$ from K into $\mathscr{C}(X)$ is said to be *holomorphic of type A* in K if dom $A(z) = D$ is independent of z and if, for all $f \in D$, the vector-valued function $A(z)\,f$ is holomorphic in K.

Note that only holomorphic operator-valued functions of type A will be considered (cf. T. KATO [1, p. 375]). For a general definition and for other types of holomorphic functions, see T. KATO [1, pp. 366, 393].

The holomorphy in K of the function $A(z)\,f$ is equivalent with the existence of a (strongly) convergent power series expansion

$$A(z)\,f = A_0 f + \sum_{|\alpha|=1}^{\infty} z^\alpha A_\alpha f, \qquad |z_i| < r, \qquad i = 1, 2, \ldots, n, \tag{3}$$

where the A_α are linear operators with dom $A_\alpha \supseteqq D$ and where $A_0 = A(0)$.

Let $A_0 \in \mathscr{C}(X)$ and let A_α, $|\alpha| \geqq 1$, be linear operators with dom $A_\alpha \supseteqq$ dom A_0. If the operators A_α satisfy inequalities of the form

$$\|A_\alpha f\| \leqq c^{|\alpha|} (a\,\|f\| + b\,\|A_0 f\|); \qquad f \in \operatorname{dom} A_0; \qquad c > 0; \qquad a, b \geqq 0, \tag{4}$$

then the series $\sum_{|\alpha|=1}^{\infty} z^\alpha A_\alpha f$ converges strongly for $|z_i| < 1/c$, $i = 1, 2, \ldots, n$. If we put $B(z)\,f = \sum_{|\alpha|=1}^{\infty} z^\alpha A_\alpha f$, one obtains

$$\|B(z)\,f\| \leqq (a\,\|f\| + b\,\|A_0 f\|) \prod_{\varrho=1}^{n} \frac{c\,|z_\varrho|}{1 - c\,|z_\varrho|}.$$

Moreover, if $|z_i| < 1/\left(\sqrt[n]{b} + c\right)$, $i = 1, 2, \ldots, n$, then, according to a theorem of T. KATO [1, p. 190], the operator $A(z) = A_0 + B(z)$ is closed.

The condition (4) is also necessary for the function $A(z)$ to be holomorphic (of type A), see T. KATO [1, p. 378]. Therefore, in the following discussion this condition is always assumed satisfied.

If the function $A(z) \in \mathscr{C}(X)$ is holomorphic (of type A) in a neighborhood of the point $z = 0$ and if $\Gamma \subset \operatorname{res} A_0$ is assumed compact, then $\Gamma \subset \operatorname{res} A(z)$ for all z in a sufficiently small neighborhood of the point $z = 0$ (see T. KATO [1, pp. 368, 212]). The condition (4) can be used to estimate this neighborhood (see T. KATO [1, p. 379] for the corresponding estimate for $n = 1$).

Theorem 1. *Let $A(z) \in \mathscr{C}(X)$ be holomorphic of type A in the neighborhood of the point $z = 0$. The condition (4) is assumed satisfied. Furthermore, let $\Gamma \subset \operatorname{res} A_0$ be compact and let*

$$\gamma = \sup_{\lambda \in \Gamma} \|(\lambda - A_0)^{-1}\|, \qquad \delta = \sup_{\lambda \in \Gamma} \|A_0(\lambda - A_0)^{-1}\|. \tag{5}$$

Then $\Gamma \subset \operatorname{res} A(z)$ for all z from the polydisc $K = \{z \colon |z_i| < 1/\left(c + \sqrt[n]{a\gamma + b\delta}\right), i = 1, \ldots, n\}$. Moreover, the resolvent $\left(\lambda - A(z)\right)^{-1}$ is bounded-holomorphic with respect to all variables $(\lambda, z_1, \ldots, z_n) \in \Gamma \times K$ jointly.

Proof. Let $\lambda \in \operatorname{res} A_0$. Then the equation

$$\left(\lambda - A(z)\right)^{-1} = (\lambda - A_0)^{-1} \sum_{\varrho=0}^{\infty} \left[\sum_{|\alpha|=1}^{\infty} z^\alpha A_\alpha (\lambda - A_0)^{-1}\right]^\varrho \tag{6}$$

is valid, at least formally. Since

$$\|A_\alpha(\lambda - A_0)^{-1}\| \leqq c^{|\alpha|}\big(a\,\|(\lambda - A_0)^{-1}\| + b\,\|A_0(\lambda - A_0)^{-1}\|, \tag{7}$$

a majorizing series for the right-hand side of the equation (6) is given by

$$\gamma \sum_{\varrho=0}^{\infty} \left[\sum_{|\alpha|=1}^{\infty} |z_1|^{\alpha_1} \cdots |z_n|^{\alpha_n}\, c^{|\alpha|}(a\gamma + b\delta)\right]^{\varrho} = \gamma \sum_{\varrho=0}^{\infty} \left[(a\gamma + b\delta) \prod_{\sigma=1}^{n} \frac{c\,|z_\sigma|}{1 - c\,|z_\sigma|}\right]^{\varrho},$$

which is convergent for $|z_i| < 1/\big(c + \sqrt[n]{a\gamma + b\delta}\big)$, $i = 1, 2, \ldots, n$. Consequently, one obtains $\Gamma \subset \operatorname{res} A(z)$; finally, the holomorphy of the resolvent $\big(\lambda - A(z)\big)^{-1}$ follows. ■

For bounded-holomorphic functions, see T. Kato [1, p. 365].

Next, let $\lambda_0 \in \operatorname{spec} A_0$ be an isolated point of $\operatorname{spec} A_0$ and let $\Gamma \subset \operatorname{res} A_0$ be a positively oriented sphere centered at λ_0 with sufficiently small radius, separating λ_0 from the remaining part $\operatorname{spec} A_0 \setminus \{\lambda_0\}$ of $\operatorname{spec} A_0$. The numbers γ and δ are chosen according to (5). Then one obtains

Theorem 2. *The group projection*

$$P(z) = (2\pi i)^{-1} \int_\Gamma \big(\lambda - A(z)\big)^{-1}\,d\lambda \tag{8}$$

of the λ_0-group of perturbed eigenvalues is bounded-holomorphic for all z with

$$|z_i| < r = 1/\big(c + \sqrt[n]{a\gamma + b\delta}\big), \qquad i = 1, 2, \ldots, n. \tag{9}$$

One has

$$P(z)\,A(z) \subseteqq A(z)\,P(z), \qquad |z_i| < r, \qquad i = 1, 2, \ldots, n, \tag{10}$$

where the operator-valued function $A(z)\,P(z)$ is bounded-holomorphic, and

$$\operatorname{spec}\{A(z) \restriction P(z)\,X\} = \operatorname{spec}\{A(z)\} \cap G, \tag{11}$$

where G denotes the disc with the boundary Γ.

Proof. The assertion that $P(z)$ is holomorphic in the polydisc (9) follows from Theorem 1. Since $P(z)$ commutes with the resolvent $\big(\lambda - A(z)\big)^{-1}$, (10) is valid. (11) follows from the definition (8) of the projection $P(z)$ (cf. T. Kato [1, pp. 173, 178]). ■

Remark. If the algebraic multiplicity $m(\lambda_0)$ is finite, $m(\lambda_0) = N < \infty$, then, from $\dim P(0) = N$ together with the holomorphy of the operator $P(z)$, the equation

$$\dim P(z) = N, \qquad z \in K, \tag{12}$$

follows.

Theorem 3. *The operator-valued function*

$$U(z) = P(z)\,P(0) + \big(1 - P(z)\big)\big(1 - P(0)\big), \qquad z \in K, \tag{13}$$

is holomorphic in the polydisc K, where

$$U(0) = 1. \tag{14}$$

Moreover, there exists a sufficiently small polydisc $K' = \{z\colon |z_i| < r' \leqq r, r' > 0, i = 1, 2, \dots, n\}$, such that $U(z)$ is boundedly invertible for all $z \in K'$; i.e., $U(z)^{-1} \in \mathcal{L}(X)$ exists. The operator-valued function $U(z)^{-1}$ is bounded-holomorphic in K'. Moreover, the relation

$$P(z) = U(z)\, P(0)\, U(z)^{-1}, \qquad z \in K', \tag{15}$$

is valid.

Proof. Obviously, $U(z)$ is holomorphic in K and $U(0) = 1$. Consequently, there exists a sufficiently small polydisc K' such that the bounded inverse operator $U(z)^{-1}$ exists for all $z \in K'$, where the function $U(z)^{-1}$ is continuous in K' with respect to the operator norm. Since

$$\frac{\partial}{\partial z_i} U(z)^{-1} = -U(z)^{-1} \left\{\frac{\partial}{\partial z_i} U(z)\right\} U(z)^{-1}, \qquad z \in K', \tag{16}$$

the operator-valued function $U(z)^{-1}$ is holomorphic with respect to every variable z_i, $i = 1, \dots, n$, where $|z_i| < r'$. Consequently, the function $U(z)^{-1}$ is holomorphic in the polydisc K'. Equation (15) follows from (13). ■

Theorem 4. *The operator-valued function*

$$C(z) = U(z)^{-1}\, A(z)\, U(z), \qquad z \in K', \tag{17}$$

commutes with the constant projection $P(0) = P_0$; i.e.,

$$P_0 C(z) \subseteqq C(z)\, P_0, \qquad z \in K'. \tag{18}$$

$C(z)\, P_0$ is a bounded-holomorphic operator-valued function, where

$$\operatorname{spec}\, \{C(z) \restriction P_0 X\} = \operatorname{spec}\, \{A(z) \restriction P(z)\, X\} = \operatorname{spec} A(z) \cap G. \tag{19}$$

Proof. (18) follows from (15) and (10). From Theorem 2, i.e., since $A(z)\, P(z)$ is bounded-holomorphic, it follows that $C(z)\, P_0 = U(z)^{-1}\, A(z)\, U(z)\, P_0 = U(z)^{-1}\, A(z) \times P(z)\, U(z)$ is also bounded-holomorphic. Since the operators $C(z) \restriction P_0X$ and $A(z) \restriction P(z)\, X$ are similar, one obtains (19). ■

Remark. If the algebraic multiplicity N of the eigenvalue λ_0 is finite, i.e., if $N < \infty$, then the identity $\dim P_0X = N$ follows and the values of the holomorphic operator-valued function $C(z) \restriction P_0X \in \mathcal{L}(P_0X)$ are operators acting on the finite-dimensional space P_0X. One has:

$\lambda \in \operatorname{spec}\, \{C(z) \restriction P_0X\}$ *if and only if* $\chi(\lambda; C(z) \restriction P_0X) = 0$, *where* $\chi(\lambda; C(z) \restriction P_0X)$ *denotes the characteristic polynomial of the operator* $C(z) \restriction P_0X$.

Recall that

$$\chi(\lambda; C(z) \restriction P_0X) = \lambda^N + a_1(z)\, \lambda^{N-1} + \dots + a_N(z), \tag{20}$$

where the coefficients $a_1(z), \ldots, a_N(z)$ are holomorphic (scalar) functions in the polydisc K'. Therefore, one obtains

Theorem 5. *Let $A_0 \in \mathscr{L}(X)$ and let λ_0 be an isolated eigenvalue of A_0 with finite algebraic multiplicity N. Let $A(z)$ be a holomorphic perturbation (of type A) of the operator A_0, given on the polydisc $K = \{z: |z_i| < r, i = 1, \ldots, n\}$. Then, to every sphere $\Gamma = \{\lambda: |\lambda - \lambda_0| = \varrho, \varrho > 0\}$ centered at λ_0 and separating λ_0 from the remaining spectrum of the operator A_0 correspond a polydisc $K' = \{z: |z_i| < r' \leqq r, i = 1, \ldots, n, r' > 0\}$ and scalar functions $a_1(z), \ldots, a_N(z)$, which are holomorphic in K', such that, for every $z \in K'$, the relation $\lambda \in \operatorname{spec} A(z) \cap \{\lambda: |\lambda - \lambda_0| < \varrho\}$ holds if and only if*

$$\lambda^N + a_1(z)\, \lambda^{N-1} + \cdots + a_N(z) = 0. \tag{21}$$

Corollary 1. *Assume that the assumptions of Theorem 5 are satisfied. Furthermore, if $N = 1$, i.e., if λ_0 is a simple eigenvalue of the operator A_0, then there exists exactly one perturbed eigenvalue $\lambda(z)$ in the interior of $\{\lambda: |\lambda - \lambda_0| < \varrho\}$ which is holomorphic in K'; that is, the relation*

$$\lambda(z) = \lambda_0 + \sum_{|\alpha|=1}^{\infty} \lambda_\alpha z^\alpha, \qquad z \in K',$$

holds. In this case, the projection $P(z)$, which is holomorphic in K, is the eigenprojection of $\lambda(z)$. Moreover, there exists a sufficiently small polydisc $K'' = \{z: |z_i| < r'' \leqq r',$ $i = 1, \ldots, n, r'' > 0\}$, such that there exists a holomorphic nonvanishing eigenvector $x(z)$ defined on K'', for example $x(z) = P(z)\, x_0$, where $A_0 x_0 = \lambda_0 x_0$, $x_0 \neq 0$.

Remark. According to the transformation (17), the investigation of the spectrum $\operatorname{spec}\big(A(z) \restriction P(z)\, X\big)$ and of the spectral objects of the operator $A(z)$ restricted to $P(z)\, X$ can be reduced to the corresponding investigation of the spectrum $\operatorname{spec}\{C(z) \restriction P_0 X\}$ and of the spectral objects of the operator $C(z)$ restricted to $P_0 X$. This means that all properties of spectral objects of $C(z) \restriction P_0 X$ can be reformulated as properties of corresponding spectral objects of $A(z) \restriction P(z) X$. Consequently, in the case $m(\lambda_0) = N < \infty$, the investigation of the spectral objects of $A(z) \restriction P(z)\, X$ can be reduced to the investigation of the spectral objects of an operator-valued function which is holomorphic in some polydisc and whose values are operators acting on a fixed complex linear space of finite dimension N.

In the following treatment it will therefore be assumed that X is finite-dimensional.

S 3. Holomorphic operator-valued functions of several complex variables whose values are operators acting on a finite-dimensional complex linear space

Let $\dim X = N < \infty$. Let G denote a region from $\mathbb{C}^n$. The points from G are denoted by $z = (z_1, z_2, \ldots, z_n)$. Furthermore, let $A(z) \in \mathscr{L}(X)$ be an operator-valued function which is holomorphic in G. For each $\zeta \in G$, the operator $A(\zeta)$ has a unique *spectral representation* and a *Jordan structure* (for these concepts, see Ch. 2, see also T. Kato

[1, p. 40]). Recall that the determination of the spectral representation appears as the first step towards the determination of the Jordan structure. In order to study the analytic properties of the Jordan structure of the operator-valued function $A(z)$, one first has to deal with the analytic properties of the spectral representation.

S 3.1. Spectral representation of $A(z)$

One has

$$\chi(\lambda; A(z)) = \det(\lambda - A(z)) = \lambda^N + a_1(z)\,\lambda^{N-1} + \cdots + a_N(z), \tag{1}$$

where $a_1(z), \ldots, a_N(z)$ denote certain holomorphic scalar functions defined on G. The set of all holomorphic functions on G is denoted by $\mathscr{H}_G$.

Lemma 1. *The set $\mathscr{H}_G$ is an integral domain (with identity) with respect to the operations $(f + g)(z) = f(z) + g(z)$ and $(f \cdot g)(z) = f(z) \cdot g(z)$, $f, g \in \mathscr{H}_G$.*

The proof of this lemma can be found, for example, in R. Gunning and H. Rossi [1, pp. 15—17]. To the integral domain $\mathscr{H}_G$ corresponds a quotient field. This field is denoted by $\mathscr{M}_G$ and is called the field of meromorphic functions on G. The elements of $\mathscr{M}_G$ are given by all quotients $p(z)/q(z)$, where p and q are holomorphic functions on G and where $q(z) \not\equiv 0$.

To the field $\mathscr{M}_G$ corresponds the ring of polynomials whose coefficients are from $\mathscr{M}_G$. This ring is denoted, as usual, by $\mathscr{M}_G[\lambda]$. It is a *Euclidean ring*, i.e., a ring where each element can be decomposed uniquely (up to units and ordering of the factors) into a product of prime factors (see, for example, B. L. van der Waerden [2, p. 72]).

A polynomial $p \in \mathscr{M}_G[\lambda]$ is called *normalized* if p has the form

$$p = \lambda^M + \alpha_1(z)\,\lambda^{M-1} + \cdots + \alpha_M(z), \qquad \alpha_\varrho(z) \in \mathscr{M}_G,$$

i.e., if the highest coefficient of the polynomial p is equal to 1. Each polynomial $p \neq 0$ can be normalized by multiplication by a suitable unit of the ring $\mathscr{M}_G[\lambda]$ (i.e., a non-vanishing meromorphic function).

The characteristic polynomial $\chi(\lambda; A(z))$ is a normalized polynomial of degree N with holomorphic coefficients. As usual, the ring of polynomials over $\mathscr{H}_G$ is denoted by $\mathscr{H}_G[\lambda]$. Therefore, χ is normalized, and $\chi \in \mathscr{H}_G[\lambda]$.

Lemma 2. *If $p, q, r \in \mathscr{M}_G[\lambda]$ are normalized polynomials, $r \in \mathscr{H}_G[\lambda]$ and $r = p \cdot q$, then $p \in \mathscr{H}_G[\lambda]$ and $q \in \mathscr{H}_G[\lambda]$.*

Proof. Since the coefficients $m(z)$ of the polynomials p, q are meromorphic, they have the form $m(z) = \alpha(z)/\beta(z)$, where $\alpha(z), \beta(z)$ are holomorphic in G. Thus one obtains a finite number of functions $\beta(z)$. The equation $\prod \beta(z) = 0$ defines an analytic manifold A in G (for analytic manifolds, see R. Gunning and H. Rossi [1, p. 112]). The manifold A is nowhere dense in G and $G \setminus A$ is connected (see R. Gunning and H. Rossi [1, p. 112]). Let $\zeta^0 \in A$. ζ^0 is a holomorphic point for all coefficients of the polynomial r. One has to show that ζ^0 is also a holomorphic point for all coefficients of the polynomials p and q. This will be done by induction. For $n = 1$, Lemma 2 is true (see Theorem 2.1 of the Appendix). Assume that Lemma 2 is true for $n - 1$.

Let

$$K = \{z : |z_i - \zeta_i^0| < r, i = 1, \ldots, n, r > 0\}$$

be a polydisc centered at ζ^0, where $K \subset G$. Fix $z_1 = \zeta_1$, $|\zeta_1 - \zeta_i^0| < r$, and consider the polydisc

$$K_{n-1} = \{(z_2, z_3, \ldots, z_n) : |z_i - \zeta_i^0| < r, i = 2, \ldots, n\}$$

of dimension $n - 1$. To this situation (polydisc K_{n-1} and the decomposition $r = pq$), one applies Lemma 2 (for $n - 1$ variables). Then one obtains: The coefficients $m(\zeta_1, z_2, z_3, \ldots, z_n)$ are holomorphic with respect to $(z_2, z_3, \ldots, z_n)$ on the polydisc K_{n-1}, i.e., $m(\zeta_1, z_2, \ldots, z_n) \in \mathscr{H}_{K_{n-1}}$ for all ζ_1 where $|\zeta_1 - \zeta_1^0| < r$. Fix an arbitrary point $(z_2, z_3, \ldots, z_n) = (\zeta_2, \zeta_3, \ldots, \zeta_n) \in K_{n-1}$ and consider the disc $K_1 = \{z_1 : |z_1 - \zeta_1^0| < r\}$ of dimension 1. Once more, one applies Lemma 2 (for $n = 1$) and obtains: the coefficients $m(z_1, \zeta_2, \ldots, \zeta_n)$ are holomorphic with respect to z_1 for all $(\zeta_2, \ldots, \zeta_n) \in K_{n-1}$. Therefore, the coefficients $m(z_1, z_2, \ldots, z_n)$ are holomorphic on the polydisc K; i.e., ζ^0 is a holomorphic point for all coefficients $m(z)$. ■

Since the ring $\mathscr{M}_G[\lambda]$ is Euclidean and Lemma 2 holds, one obtains

Corollary 1. *The characteristic polynomial $\chi(\lambda, A(z))$ can be decomposed uniquely (up to the ordering of the factors) into a product of prime polynomials of the form*

$$\chi = \prod_{\varrho=1}^{r} \pi_\varrho^{m_\varrho}, \tag{2}$$

where the polynomials π_ϱ are normalized irreducible and where $\pi_\varrho \in \mathscr{H}_G[\lambda]$, $\varrho = 1, \ldots, r$.

Let $p \in \mathscr{H}_G[\lambda]$ and assume that p is normalized. By $d_p(z)$ one denotes the ***discriminant*** of the polynomial p. One has

$$d_p(z) = \pm R(p, p'), \tag{3}$$

where R denotes the ***resultant*** of the polynomials p and $p' = \partial p/\partial\lambda$. (For the concepts of discriminant and resultant, see B. L. van der Waerden [2, p. 93].) Note that $d_p(z) \in \mathscr{H}_G$ and $d_p(z) \equiv 0$ if and only if p and p' have a nontrivial common divisor in $\mathscr{M}_G[\lambda]$. Furthermore, if $\zeta \in G$, then $d_p(\zeta) = 0$ if and only if the polynomial $p = p(\lambda; z)$ has at least one multiple root with respect to λ at $z = \zeta$.

If p is irreducible in $\mathscr{M}_G[\lambda]$, then $d_p(z) \not\equiv 0$, because p and p' do not have nontrivial common divisors. Thus, for an irreducible polynomial p the equation $d_p(z) = 0$ defines an analytic manifold A_p in G which is different from G. Moreover, if $z \in G \setminus A_p$ is fixed, then the polynomial $p(\lambda, z)$ has only simple roots with respect to λ.

Lemma 3. *Let χ denote the characteristic polynomial. In accordance with* (2), *one denotes the irreducible factors of the polynomial χ by $\pi_1, \pi_2, \ldots, \pi_r$. Then the discriminant $d_\pi(z) \in \mathscr{H}_G$ of the polynomial $\pi = \pi_1\pi_2 \cdots \pi_r$ satisfies the relation $d_\pi(z) \not\equiv 0$.*

Proof. One has

$$\pi' = \sum_{\varrho} \pi_1\pi_2 \cdots \pi_{\varrho-1}\pi_\varrho'\pi_{\varrho+1} \cdots \pi_r. \tag{4}$$

π and π' do not have any nontrivial common divisors in $\mathcal{M}_G[\lambda]$. In fact, assume the existence of such a divisor. Without loss of generality, one may assume that π_1 is such a divisor. Then it would follow from (4) that π_1 is a divisor of π_1', which cannot be true. ■

Consequently, an analytic manifold $A \neq G$ is defined in G by the equation $d_\pi(z) = 0$. A is closed with respect to G, nowhere dense in G, and $G \setminus A$ is a connected set; i.e., $G \setminus A$ is a region. If $\zeta \in G \setminus A$, then $\pi(\lambda; \zeta)$ has only simple roots with respect to λ. Let

$$\pi_\varrho(\lambda; \zeta) = \prod_{\sigma=1}^{g_\varrho} \big(\lambda - \lambda_{\varrho\sigma}(\zeta)\big), \qquad g_\varrho = \deg \pi_\varrho.$$

This implies that

$$\pi(\lambda; \zeta) = \prod_{\varrho=1}^{r} \prod_{\sigma=1}^{g_\varrho} \big(\lambda - \lambda_{\varrho\sigma}(\zeta)\big).$$

According to the implicit function theorem (see R. Gunning and H. Rossi [1, p. 26]), there exist a polydisc $K = \{z : |z_i - \zeta_i| < r, i = 1, \dots, n\}$ and holomorphic functions $\lambda_{\varrho\sigma}(z)$ on K such that

$$\pi(\lambda, z) = \prod_{\varrho=1}^{r} \prod_{\sigma=1}^{g_\varrho} \big(\lambda - \lambda_{\varrho\sigma}(z)\big), \qquad z \in K,$$

where $\lambda_{\varrho\sigma}(z) \neq \lambda_{\varrho'\sigma'}(z)$ for all $z \in K$ if $(\varrho, \sigma) \neq (\varrho', \sigma')$. In other words: In the neighborhood of $\big(\zeta, \lambda_{\varrho\sigma}(\zeta)\big)$, where $\lambda_{\varrho\sigma}(\zeta)$ is a simple root, the equation $\pi(\lambda, z) = 0$ is solved with respect to λ by a holomorphic function element $\big(z, \lambda_{\varrho\sigma}(z)\big)$. Consequently, over each point $\zeta \in G \setminus A$, one obtains exactly $\sum_{\varrho=1}^{r} g_\varrho$ different holomorphic function elements $\lambda_{\varrho\sigma}(z)$.

The points $z \in A$ are called *multiple points*, the points $z \in G \setminus A$ are called *simple points*.

Let $\mu = \mu(\lambda, z)$ be one of the irreducible polynomials $\pi_1, \pi_2, \dots, \pi_r$ and let $\mu(\lambda, z) = \prod_{\sigma=1}^{g} \big(\lambda - \lambda_\sigma(z)\big)$, where $z \in G \setminus A_\mu$. Then the holomorphic function elements $\lambda_1(z), \dots, \lambda_g(z)$ represent exactly the g branches of that analytic function $\mathfrak{P}$ over $G \setminus A_\mu$ which is uniquely defined by the polynomial μ. One has

Lemma 4. *To every irreducible polynomial $\mu \in \mathcal{H}_G[\lambda]$ corresponds a unique g-valued analytic function $\mathfrak{P}$ over $G \setminus A_\mu$, where $g = \deg \mu$. The g branches $\lambda_1(z), \dots, \lambda_g(z)$ of $\mathfrak{P}$ over $z \in G \setminus A_\mu$ satisfy the relation*

$$\mu(\lambda, z) = \prod_{\sigma=1}^{g} \big(\lambda - \lambda_\sigma(z)\big), \qquad z \in G \setminus A_\mu;$$

i.e.,

$$\mu\big(\lambda_\sigma(z), z\big) \equiv 0, \qquad \sigma = 1, \dots, g, \qquad z \in G \setminus A_\mu.$$

Proof. For this proof see, for example, B. A. Fuks [1, p. 119]. It is sufficient to show that the function elements $\lambda_1(z), \dots, \lambda_g(z)$, which, according to the implicit function theorem are defined locally over $G \setminus A_\mu$, belong to one and the same analytic function. This means that one has to show that all function elements can be obtained from an arbitrary element by analytic continuation in $G \setminus A_\mu$. For example, the element $\lambda_1(z)$ over $\zeta \in G \setminus A_\mu$ defines an analytic function, by analytic continuation, with the branches $\lambda_1(z), \lambda_2(z), \ldots, \lambda_l(z)$, $l \leqq g$ (without loss of generality). Consider the polynomial

$$\tilde{\mu}(\lambda, z) = \prod_{\varrho=1}^{l} \left(\lambda - \lambda_\varrho(z)\right) = \lambda^l + \tilde{a}_1(z)\, \lambda^{l-1} + \cdots + \tilde{a}_c(z).$$

The coefficients $\tilde{a}_1(z), \ldots, \tilde{a}_l(z)$ are *single-valued* and holomorphic in $G \setminus A_\mu$. Let $\zeta \in A_\mu$ and let $U \subset G$ be a neighborhood of ζ. Then the coefficients $\tilde{a}_1(z), \ldots, \tilde{a}_l(z)$ are bounded in $U \cap (G \setminus A_\mu)$. In fact, the coefficients $\tilde{a}_1(z), \ldots, \tilde{a}_l(z)$ are the elementary symmetric polynomials of the roots $\lambda_1(z), \ldots, \lambda_l(z)$. These roots satisfy the equation

$$\mu(\lambda, z) = \lambda^g + a_1(z)\, \lambda^{g-1} + \cdots + a_g(z) = 0.$$

The inverse roots $\tilde{\lambda} = 1/\lambda$ satisfy the equation

$$1 + a_1\tilde{\lambda} + \cdots + a_g\tilde{\lambda}^g = 0.$$

Consequently, these inverse roots cannot be arbitrarily small. Applying Riemann's theorem on the analytic continuation of holomorphic functions (see, for example, B. A. Fuks [1, p. 119]), one finds that the coefficients $\tilde{a}_1(z), \ldots, \tilde{a}_l(z)$ are holomorphic everywhere in G. Hence, $\tilde{\mu}$ is a divisor of μ in $\mathscr{M}_G[\lambda]$. Since $1 \leqq l \leqq g$, μ is irreducible and μ and $\tilde{\mu}$ are normalized; $\mu = \tilde{\mu}$ follows. One obtains $l = g$; i.e., each function element can be obtained by arbitrary analytic continuation of $\lambda_1(z)$. ■

Remark. According to Lemma 4, to every irreducible factor π_ϱ of the characteristic polynomial χ corresponds a unique analytic function $\mathfrak{P}_\varrho$ over $G \setminus A$. If

$$\pi_\varrho(\lambda, z) = \lambda^{g_\varrho} + a_{\varrho 1}(z)\, \lambda^{g_\varrho - 1} + \cdots + a_{\varrho g_\varrho}(z),$$

then a holomorphic function element $\lambda(z)$ of $\mathfrak{P}_\varrho$ satisfies the equation

$$\lambda(z)^{g_\varrho} + a_{\varrho 1}(z)\, \lambda(z)^{g_\varrho - 1} + \cdots + a_{\varrho g_\varrho}(z) \equiv 0$$

identically with respect to z.

Let ζ be a simple point and let $\lambda_{\varrho\sigma}(\zeta)$ denote one of the eigenvalues of $A(\zeta)$. Let Γ be a positively oriented sphere centered at $\lambda_{\varrho\sigma}(\zeta)$ and with a sufficiently small radius, such that Γ separates $\lambda_{\varrho\sigma}(\zeta)$ from the remaining part spec $A(\zeta) \setminus \{\lambda_{\varrho\sigma}(\zeta)\}$ of the spectrum spec $A(\zeta)$. Then there exists a polydisc

$$K = \{z \colon |z_i - \zeta_i| < r,\ i = 1, \ldots, n,\ r > 0\},$$

such that $\Gamma \subset \operatorname{res} A(z)$ for all $z \in K$. Moreover, the interior of Γ contains exactly one eigenvalue $\lambda_{\varrho\sigma}(z)$ of the operator $A(z)$ for all $z \in K$ and the projection

$$P_{\varrho\sigma}(z) = (2\pi i)^{-1} \int_\Gamma \left(\lambda - A(z)\right)^{-1} d\lambda, \qquad z \in K, \tag{5}$$

which is holomorphic in K, is the eigenprojection of $\lambda_{\varrho\sigma}(z)$ with respect to $A(z)$. One obtains exactly $\sum_{\varrho=1}^{r} g_\varrho$ different function elements for eigenprojections over $G \setminus A$. If two eigenvalues belong to different analytic functions, then the corresponding eigenprojections also belong to different analytic functions. One has

Lemma 5. *The eigenprojections $P_{\varrho 1}(z), P_{\varrho 2}(z), \ldots, P_{\varrho g_\varrho}(z)$ represent the g_ϱ branches of a unique analytic operator-valued function over $G \setminus A$ whose values are eigenprojections. Moreover, the equation*

$$\dim P_{\varrho\sigma}(z) = m_\varrho, \qquad \sigma = 1, 2, \ldots, g_\varrho, \tag{6}$$

holds.

Proof. Obviously, equation (6) is valid because, according to (2), for $z \in G \setminus A$ the algebraic multiplicity of the eigenvalue $\lambda_{\varrho\sigma}(z)$ equals m_ϱ.

Let $P_{\varrho\sigma}(z)$ and $P_{\varrho\sigma'}(z')$ be two eigenprojection function elements (5), where $\lambda_{\varrho\sigma}(z)$ and $\lambda_{\varrho\sigma'}(z')$ denote the corresponding function elements of the eigenvalues. Then the elements $\lambda_{\varrho\sigma}(z)$ and $\lambda_{\varrho\sigma'}(z')$ can be analytically continued along a path belonging to $G \setminus A$ which joins the points z and z' (see, for example, B. A. Fuks [1, p. 108]). Let K_1 and K_2 denote polydiscs which appear in this continuation, where $K_1 \cap K_2 \neq \emptyset$; let $\lambda_1(z)$ and $\lambda_2(z)$ denote the corresponding function elements of the eigenvalues, $P_1(z)$ and $P_2(z)$ the corresponding function elements for the eigenprojections. Then it is sufficient to show that $P_1(z) = P_2(z)$ if $z \in K_1 \cap K_2$. This identity is obvious, because the eigenprojection of an eigenvalue is uniquely determined. ∎

Remark. The following section contains the algebraic connection between $\lambda_{\varrho\sigma}(z)$ and $P_{\varrho\sigma}(z)$. This connection is completely analogous to the case $n = 1$ (see Sect. 5.2 and Subsect. 3.4.9). However, for $n > 1$ it seems to be difficult to deduce the algebraic connection from local properties of eigenvalues $\lambda_{\varrho\sigma}(z)$ and eigenprojections $P_{\varrho\sigma}(z)$ at the points of the exceptional manifold A. For this reason, another approach is preferred here. As usual, the eigennilpotent $D_{\varrho\sigma}(z)$ corresponding to $\lambda_{\varrho\sigma}(z)$ is given by

$$D_{\varrho\sigma}(z) = \big(A(z) - \lambda_{\varrho\sigma}(z)\big)\, P_{\varrho\sigma}(z), \qquad z \in G \setminus A. \tag{7}$$

Hence, for *simple points* one obtains the *spectral representation* of $A(z)$ in the form

$$A(z) = \sum_{\varrho=1}^{r} \sum_{\sigma=1}^{g_\varrho} \big(\lambda_{\varrho\sigma}(z)\, P_{\varrho\sigma}(z) + D_{\varrho\sigma}(z)\big), \qquad z \in G \setminus A, \tag{8}$$

which expresses some of its analytic properties.

S 3.2. Algebraic determination of the eigenprojections

Let μ be one of the irreducible factors of the characteristic polynomial χ and let $\mathfrak{P}$ be the corresponding analytic function over $G \setminus A$. For short, the pairs $\big(z^0, \lambda(z)\big)$, where $\lambda(z)$ is a function element of $\mathfrak{P}$ defined on some polydisc K centered at $z^0 \in G \setminus A$, are denoted by ζ. The point z^0 is called the *trace point* of ζ. Then the projection-valued function $P(\cdot)$ that corresponds to the analytic function $\mathfrak{P}$ can be considered as a function of ζ, $P = P(\zeta)$. Correspondingly, for $z \in G \setminus A$, the eigenvalue belonging to $\mathfrak{P}$

can be considered as a function $\lambda(\zeta)$ of ζ: if $\zeta = (z^0, \lambda(z))$, then $\lambda(\zeta) = \lambda(z^0)$. In this subsection, a global link between $\lambda(\zeta)$ and $P(\zeta)$ will be derived. For this purpose, one starts with some remarks:

I. By $\mathcal{M}_G(\mathfrak{P})$ one denotes the extension field of degree g over $\mathcal{M}_G$ obtained by adjunction of $\lambda(\zeta)$ to the field $\mathcal{M}_G$. The elements of $\mathcal{M}_G(\mathfrak{P})$ can be represented by expressions of the form

$$a_0(z) + a_1(z)\,\lambda(\zeta) + \cdots + a_{g-1}(z)\,\lambda(\zeta)^{g-1}, \tag{9}$$

where the coefficients $a_0(z), a_1(z), \ldots, a_{g-1}(z)$ are single-valued meromorphic functions and z denotes the trace point of ζ.

II. By $X(\mathcal{M}_G): \mathbb{C}$ one denotes the linear space (over $\mathbb{C}$) consisting of all meromorphic vector-valued functions from G into X.

Let $x_0(z), x_1(z), \ldots, x_{g-1}(z) \in X(\mathcal{M}_G): \mathbb{C}$. Consider the linear space over $\mathbb{C}$ consisting of all elements of the form

$$f(\zeta) = x_0(z) + x_1(z)\,\lambda(\zeta) + \cdots + x_{g-1}(z)\,\lambda(\zeta)^{g-1}. \tag{10}$$

Note that the meromorphic vector-valued functions $x_\varrho(z)$ are uniquely determined by the element $f(\zeta)$. This linear space is denoted by $X(\mathfrak{P})$ or $X(\mathfrak{P}): \mathbb{C}$. With respect to the linear spaces X, $X(\mathcal{M}_G)$ and $X(\mathfrak{P})$, one obviously has the inclusions

$$X \subset X(\mathcal{M}_G) \subset X(\mathfrak{P}). \tag{11}$$

III. The linear space $X(\mathfrak{P}): \mathbb{C}$ can also be considered as a linear space over the field $\mathcal{M}_G(\mathfrak{P})$. In this case, one uses the notation $X(\mathfrak{P}): \mathcal{M}_G(\mathfrak{P})$. A corresponding remark applies to $X(\mathcal{M}_G)$ and $\mathcal{M}_G$. Note that

$$\dim\big(X(\mathfrak{P}): \mathcal{M}_G(\mathfrak{P})\big) = \dim X : \mathbb{C} = N. \tag{12}$$

In fact, if $e_1, e_2, \ldots, e_N$ denotes a basis of X and $f_1, \ldots, f_N$ the corresponding adjoint basis in X^*, then

$$x_\varrho(z) = \sum_{\alpha=1}^{N} \big(x_\varrho(z), f_\alpha\big)\, e_\alpha, \qquad \varrho = 0, 1, \ldots, g-1, \qquad \big(x_\varrho(z), f_\alpha\big) \in \mathcal{M}_G.$$

Substituting this expression into (10), one obtains

$$f(\zeta) = \sum_{\alpha=1}^{N} e_\alpha \left(\sum_{\varrho=0}^{g-1} \big(x_\varrho(z), f_\alpha\big)\, \lambda(\zeta)^\varrho\right).$$

On the other hand, if

$$\sum_{\beta=1}^{N} \alpha_\beta(\zeta)\, e_\beta = 0, \qquad \alpha_\beta(\zeta) = \sum_{\varrho=0}^{g-1} \alpha_{\beta\varrho}(z)\, \lambda(\zeta)^\varrho,$$

then

$$\sum_{\varrho=1}^{g-1} \lambda(\zeta)^\varrho \left(\sum_{\beta=1}^{N} \alpha_{\beta\varrho}(z)\, e_\beta\right) = 0;$$

i.e.,

$$\sum_{\beta=1}^{N} \alpha_{\beta\varrho}(z)\, e_\beta = 0, \qquad \alpha_{\beta\varrho}(z) \in \mathscr{M}_G.$$

If z^0 denotes a holomorphic point with respect to all coefficients, then $\alpha_{\beta\varrho}(z^0) = 0$. This implies $\alpha_{\beta\varrho}(z) \equiv 0$; i.e., $\alpha_\beta(\zeta) = 0$.

IV. The operators from $\mathscr{L}\big(X(\mathfrak{P}) : \mathscr{M}_G(\mathfrak{P})\big)$ can be identified with operator-valued functions of the type

$$B(\zeta) = B_0(z) + B_1(z)\,\lambda(\zeta) + \cdots + B_{\varrho-1}(z)\,\lambda(\zeta)^{\varrho-1}, \tag{13}$$

where the coefficients $B_0(z), B_1(z), \ldots, B_{\varrho-1}(z)$ are meromorphic operator-valued functions from G into $\mathscr{L}(X)$, which are uniquely determined by the operator $B(\zeta)$.

The relation between $\lambda(\zeta)$ and $P(\zeta)$ can be obtained if the holomorphic operator-valued function $A(z)$ on G is considered as an operator from the space $\mathscr{L}\big(X(\mathfrak{P}) : \mathscr{M}_G(\mathfrak{P})\big)$. Over $\mathscr{M}_G$, in the decomposition of the characteristic polynomial χ of $A(z)$, the irreducible factor μ appears with multiplicity m. Over $\mathscr{M}_G(\mathfrak{P})$, in the decomposition of the polynomial χ the irreducible factor $\lambda - \lambda(\zeta)$ appears with the same multiplicity m; i.e., $\big(\lambda - \lambda(\zeta)\big)^m$ is the exact power of the irreducible factor $\lambda - \lambda(\zeta)$. Recall that to every exact power of an irreducible factor of the characteristic polynomial corresponds a unique reducing projection of $A(z)$ (see Sect. 4.2). This projection is denoted by $Q(\zeta)$ and its properties are collected in the following lemma.

Lemma 6. *The following relations are valid:*

I. $Q(\zeta) \in \mathscr{L}\big(X(\mathfrak{P}) : \mathscr{M}_G(\mathfrak{P})\big)$; *i.e., $Q(\zeta)$ can be reprented uniquely in the form*

$$Q(\zeta) = C_0(z) + C_1(z)\,\lambda(\zeta) + \cdots + C_{\varrho-1}(z)\,\lambda(\zeta)^{\varrho-1}, \tag{14}$$

where $C_0(z), C_1(z), \ldots, C_{\varrho-1}(z)$ are meromorphic operator-valued functions from G into $\mathscr{L}(X)$ and where z denotes the trace point of ζ.

II. *The function $Q(\zeta)$ is projection-valued, i.e., its values are projections.*

III. *$Q(\zeta)$ reduces $A(z)$; i.e.,*

$$Q(\zeta)\,A(z) = A(z)\,Q(\zeta). \tag{15}$$

IV. *The characteristic polynomial of $A(z) \restriction Q(\zeta)\,X$ is given by*

$$\chi\big(\lambda; A(z) \restriction Q(\zeta)\,X\big) = \big(\lambda - \lambda(\zeta)\big)^m. \tag{16}$$

V. *$Q(\zeta)$ is an operator polynomial in $A(z)$ with coefficients from $\mathscr{M}_G(\mathfrak{P})$; i.e., there exists a polynomial $q(\lambda) \in \mathscr{M}_G(\mathfrak{P})\,[\lambda]$, such that $Q(\zeta) = q\big(A(z)\big)$.*

Let z^0 be a holomorphic point with respect to all operator-valued functions $C_0(z)$, $\ldots, C_{\varrho-1}(z)$, $z^0 \in G \setminus A$ and let $U \subset G \setminus A$ be a neighborhood of z^0 consisting of holomorphic points only. Then, for each $z \in U$, the projection $Q(\zeta)$ is well-defined as a projection from $\mathscr{L}(X)$. It reduces $A(z)$, and (16) holds. The eigenprojection $P(\zeta)$ belonging to $\lambda(\zeta)$ also satisfies the equation

$$\chi\big(\lambda; A(z) \restriction P(\zeta)\,X\big) = \big(\lambda - \lambda(\zeta)\big)^m \tag{17}$$

for all $z \in U$. Consider the polynomial $\chi(\lambda; A(z))$ as a polynomial from $\mathbb{C}[\lambda]$, where $z \in U$ is arbitrary but fixed and where the eigenvalue $\lambda(\zeta)$ is fixed. Then the exact power of the irreducible factor $\lambda - \lambda(\zeta)$ of this polynomial is given by the number m. Hence, $Q(\zeta)$ and $P(\zeta)$ are reducing projections for $A(z)$ such that (16) and (17) hold, where the coinciding characteristic polynomials (16) and (17) are given by the exact power of the irreducible factor $\lambda - \lambda(\zeta)$ of the polynomial χ decomposed with respect to $\mathbb{C}[\lambda]$. Therefore, the equation $Q(\zeta) = P(\zeta)$ holds everywhere in U (for the proof, apply Lemma 4.2.2). Therefore, $Q(\zeta) \equiv P(\zeta)$ follows and one can state the following

Theorem 1. *Let $P(\zeta)$ be the (reducing) eigenprojection-valued function belonging to the eigenvalue function $\lambda(\zeta)$ with respect to the operator-valued function $A(z)$, which is holomorphic in G. Then $P(\zeta)$ can be uniquely represented in the form*

$$P(\zeta) = C_0(z) + C_1(z)\,\lambda(\zeta) + \cdots + C_{g-1}(z)\,\lambda(\zeta)^{g-1}; \tag{18}$$

where the meromorphic operator-valued functions $C_0(z), \ldots, C_{g-1}(z)$ are holomorphic in $G \setminus A$.

Proof. Only the last assertion that the coefficients are holomorphic in $G \setminus A$ remains to be proved. Let $z^0 \in G \setminus A$. By $\lambda_1(z), \lambda_2(z), \ldots, \lambda_g(z)$ one denotes the g branches of the function $\mathfrak{P}$ over z^0 and by $P_1(z), \ldots, P_g(z)$ the corresponding eigenprojections. Then one obtains

$$P_\alpha(z) = C_0(z) + C_1(z)\,\lambda_\alpha(z) + \cdots + C_{g-1}(z)\,\lambda_\alpha(z)^{g-1}, \qquad \alpha = 1, \ldots, g.$$

These g equations can be solved with respect to the g unknown functions $C_0(z), \ldots, C_{g-1}(z)$. According to Vandermonde's determinant theorem, one has

$$\det\{\lambda_\alpha(z)^\sigma\} = \prod_{\mu > \nu} \left(\lambda_\mu(z) - \lambda_\nu(z)\right), \qquad \alpha = 1, \ldots, g; \qquad \sigma = 0, \ldots, g-1.$$

This function is holomorphic at $z = z^0$ and, moreover, nonvanishing. Furthermore, the operators $P_1(z), \ldots, P_g(z)$ are holomorphic at $z = z^0$. Applying Cramer's rule, one obtains the coefficients $C_0(z), \ldots, C_{g-1}(z)$, which are holomorphic at z^0. ■

Remark. The sum

$$P_\varrho(z) = \sum_{\sigma=1}^{g_\varrho} P_{\varrho\sigma}(z), \qquad \varrho = 1, 2, \ldots, r, \tag{19}$$

is obviously single-valued in $G \setminus A$ and projection-valued. As in the proof of Theorem 1, one obtains that the projections $P_\varrho(z)$ are meromorphic projection-valued functions in G. Note that in this case one has to replace the field $\mathcal{M}_G(\mathfrak{P})$ by the field $\mathcal{M}_G$. Then one can first identify $P_\varrho(z)$ as a reducing projection that corresponds to the irreducible factor π_ϱ of the polynomial χ (see Sect. 4.2). This projection is uniquely determined by the equation

$$\chi\left(\lambda; A(z) \upharpoonright P_\varrho(z)\, X\right) = \pi_\varrho(\lambda; z)^{m_\varrho};$$

which implies (19).

Together with the eigenprojection $P(\zeta)$, the corresponding eigennilpotent $D(\zeta) = A(z) - \lambda(\zeta)\, P(\zeta)$ also belongs to $\mathscr{L}(X(\mathfrak{P}) : \mathscr{M}_G(\mathfrak{P}))$; i.e., like $P(\zeta)$, $D(\zeta)$ can be represented by (18). This representation of $D(\zeta)$ is the basis for the study of the refined structure of the eigennilpotent $D(\zeta)$.

S 3.3. Properties of the eigennilpotent

First of all, one has

$$D(\zeta)^m \equiv 0. \tag{20}$$

This identity implies

$$\chi(\lambda; D(\zeta)) = \lambda^N. \tag{21}$$

That is, the decomposition of $\chi(\lambda; D(\zeta))$ into a product of irreducible factors yields only one (linear) irreducible factor, namely λ. Hence (see Subsect. 4.4.3 and 4.4.4): all irreducible parts of $D(\zeta)$ are Jordan operators and $D(\zeta)$ can be decomposed (together with $P(\zeta)$) into a direct sum of irreducible (Jordan) operators (see Theorem 4.4.1). More precisely, one has

Lemma 7. *There exist uniquely determined numbers $m_1 \geqq m_2 \geqq \cdots \geqq m_l \geqq 1$, where $\sum_{\varrho=1}^{l} m_\varrho = m$, and Jordan projections $P_1(\zeta), P_2(\zeta), \ldots, P_l(\zeta)$, which, in general, are not uniquely determined, where $P_\varrho(\zeta) \in \mathscr{L}(X(\mathfrak{P}) : \mathscr{M}_G(\mathfrak{P}))$ and $\dim P_\varrho(\zeta) = m_\varrho$, such that the relations*

$$P(\zeta) = P_1(\zeta) + P_2(\zeta) + \cdots + P_l(\zeta), \tag{22}$$

$$P_\varrho(\zeta)\, P_\sigma(\zeta) = \delta_{\varrho\sigma} P_\varrho(\zeta), \tag{23}$$

$$P_\varrho(\zeta)\, D(\zeta) = D(\zeta)\, P_\varrho(\zeta), \qquad \varrho = 1, 2, \ldots, l, \tag{24}$$

hold. The matrix of the operator $D(\zeta) \restriction P_\varrho(\zeta)\, X(\mathfrak{P})$ is a Jordan block $\Gamma_{m_\varrho}(0)$ if one chooses a suitable (Jordan) basis $e_\varrho^\sigma(\zeta)$, $\sigma = 1, \ldots, m_\varrho$, in the subspace $P_\varrho(\zeta)\, X(\mathfrak{P})$. Then the relations

$$D(\zeta)\, e_\varrho^1(\zeta) = 0, \qquad D(\zeta)\, e_\varrho^2(\zeta) = e_\varrho^1(\zeta), \ldots, D(\zeta)\, e_\varrho^{m_\varrho}(\zeta) = e_\varrho^{m_\varrho - 1}(\zeta) \tag{25}$$

hold.

The following question arises: For which points ζ does the operator $D(\zeta)$, considered as an operator from $\mathscr{L}(X : \mathbb{C})$, possess the same Jordan structure as the operator-valued function $D(\zeta)$, considered as an element from $\mathscr{L}(X(\mathfrak{P}) : \mathscr{M}_G(\mathfrak{P}))$. To answer this question, the following lemma is proved.

Lemma 8. *There exists a point ζ^0, such that ζ^0 is holomorphic with respect to all projections $P_1(\zeta), P_2(\zeta), \ldots, P_l(\zeta), 1 - P(\zeta)$, and with respect to all vectors of the Jordan bases $e_\varrho^\sigma(\zeta)$. Moreover, the vectors $e_\varrho^\sigma(\zeta^0)$ form a Jordan basis in $P_\varrho(\zeta^0)\, X$ with respect to $D(\zeta^0)$.*

Proof. Obviously, there exists a point ζ^0 such that all objects $P_\varrho(\zeta)$, $1 - P(\zeta)$, $e_\varrho^\sigma(\zeta)$ are holomorphic at ζ^0. Let U denote a corresponding neighborhood of ζ^0 consisting of holomorphic points only. Let $P_K(\zeta)$ be one of the projections $P_\varrho(\zeta)$, where $e_K^1(\zeta), \ldots, e_K^{m_K}(\zeta)$ denotes the corresponding Jordan basis. Then ran $(e_K^1, e_K^2, \ldots, e_K^{m_K}) = m_K$ with respect to $\mathscr{M}_G(\mathfrak{P})$; hence, there exists an $m_K \times m_K$-matrix of coefficients of the vectors $e_K^1, \ldots, e_K^{m_K}$, such that the corresponding determinant $d_K(\zeta)$ does not vanish identically; i.e., $d_K(\zeta) \not\equiv 0$ in U. Therefore, except for the points ζ from a subset $A_K \subset U$, one obtains $d_K(\zeta) \neq 0$ for all points $\zeta \in U \setminus A_K$; i.e., the vectors $e_K^1, \ldots, e_K^{m_K}$ form a Jordan basis of $D(\zeta) \restriction P_K(\zeta) X$ for these values of ζ. Hence, the points ζ from $\bigcap_{K=1}^{l} (U \setminus A_K) \neq \emptyset$ have the following property: The vectors $e_\varrho^\sigma(\zeta)$, $\varrho = 1, \ldots, l$, form a Jordan basis of $P_\varrho(\zeta) X$, $\varrho = 1, \ldots, l$. But the existence of such points is asserted; hence, Lemma 8 is proved. ∎

Augment the basis $e_\varrho^\sigma(\zeta)$, which is adapted to $P_1(\zeta), \ldots, P_l(\zeta)$, by a basis of the subspace $\big(1 - P(\zeta)\big) X(\mathfrak{P})$, where one requires that the elements of this basis are holomorphic at ζ^0, and represent a basis of the subspace $\big(1 - P(\zeta^0)\big) X$ at ζ^0.

Then the vectors $e_\varrho^\sigma(\zeta)$, $\varrho = 0, 1, \ldots, l$, form a basis of $X(\mathfrak{P})$ which is adapted to the system $1 - P(\zeta), P_1(\zeta), \ldots, P_l(\zeta)$. Moreover, the vectors $e_\varrho^\sigma(\zeta^0)$ form a basis of X which is adapted to the system $1 - P(\zeta^0), P_1(\zeta^0), \ldots, P_l(\zeta^0)$; i.e., the latter vectors form a "constant" basis (in $X(\mathfrak{P})$) adapted to the corresponding "constant" system (in $\mathscr{L}\big(X(\mathfrak{P}) : \mathscr{M}_G(\mathfrak{P})\big)$. One has

Lemma 9. *There exists an operator $T(\zeta) \in \mathscr{L}\big(X(\mathfrak{P}) : \mathscr{M}_G(\mathfrak{P})\big)$, where $T(\zeta)^{-1}$ also exists and belongs to $\mathscr{L}\big(X(\mathfrak{P}) : \mathscr{M}_G(\mathfrak{P})\big)$, such that*

$$T(\zeta^0) = 1, \tag{26}$$

$$P_\varrho(\zeta) = T(\zeta) P_\varrho(\zeta^0) T(\zeta)^{-1}, \qquad \varrho = 1, \ldots, l, \tag{27}$$

$$D(\zeta) = T(\zeta) D(\zeta^0) T(\zeta)^{-1}; \tag{28}$$

i.e., $D(\zeta)$, considered as an operator from $\mathscr{L}\big(X(\mathfrak{P}) : \mathscr{M}_G(\mathfrak{P})\big)$, is similar to the "constant" operator $D(\zeta^0)$.

Proof. One defines the operator $T(\zeta)$ by

$$T(\zeta) e_\varrho^\sigma(\zeta^0) = e_\varrho^\sigma(\zeta).$$

Then the properties asserted above follow by a direct calculation. ∎

The operators $T(\zeta)$ and $T(\zeta)^{-1}$ can be represented uniquely, as operators from $\mathscr{L}\big(X(\mathfrak{P}) : \mathscr{M}_G(\mathfrak{P})\big)$, in the form

$$T(\zeta) = S_0(z) + S_1(z) \lambda(\zeta) + \cdots + S_{g-1}(z) \lambda(\zeta)^{g-1},$$

$$T(\zeta)^{-1} = R_0(z) + R_1(z) \lambda(\zeta) + \cdots + R_{g-1}(z) \lambda(\zeta)^{g-1}.$$

From these representations one obtains the existence of an analytic manifold B in G, $A \subseteq B \subset G$, with the following properties: the operators $T(\zeta)$ and $T(\zeta)^{-1}$ are well-defined as operators from $\mathscr{L}(X)$ for all points ζ except for those points whose trace points belong to B. Then one obtains

Lemma 10. *The operators $D(\zeta)$ are similar to the operator $D(\zeta^0)$ for all points ζ except for a set of points whose trace points belong to an analytic manifold B, where $A \subseteqq B \subset G$ In other words, $D(\zeta)$ has a fixed Jordan normal form for $\zeta \in G \setminus B$ and this Jordan normal form is the Jordan normal form of the operator $D(\zeta)$, considered as an operator from $\mathscr{L}(X(\mathfrak{P}) : \mathscr{M}_G(\mathfrak{P}))$.*

S 3.4. Jordan structure of $A(z)$

In Subsect. S 3.2 and S 3.3, the behavior of eigenvalues $\lambda(\zeta)$, eigenprojections $P(\zeta)$ and eigennilpotents $D(\zeta)$ of $A(z)$ was investigated, where these objects are assigned to a fixed irreducible factor μ of the characteristic polynomial χ. Since there are only finitely many irreducible factors $\pi_1, \pi_2, \ldots, \pi_r$, one can formulate the following theorems, whose proofs follow from Subsect. S 3.2 and S 3.3. Theorem 2 was established by the author in [6]. The symbol $\sim$ means similarity.

Theorem 2. *Let $A(z)$ be an operator-valued function,[1] holomorphic on $G \subset \mathbb{C}^n$, with values from $\mathscr{L}(X)$. Then there exists an analytic manifold B, $A \subseteqq B \subset G$, (where the analytic manifold A is defined by $d_\pi(z) = 0$; π denotes the discriminant of the polynomial $\pi = \pi_1\pi_2 \cdots \pi_r$; see Subsect. S 3.2) with the following properties: If*

$$D(z) = A(z) - \sum_{\alpha,\sigma} \lambda_{\alpha\sigma}(z)\, P_{\alpha\sigma}(z),$$

then $D(z)$ is single-valued in $G \setminus A$ and the relation

$$D(z) \sim D(z^0)$$

holds for all $z, z^0 \in G \setminus B$. Moreover, if $D(z^0) = \sum_{\alpha=1}^{r} \oplus \sum_{\sigma=1}^{g_\alpha} \oplus N_\alpha^\sigma$, where $N_\alpha^\sigma = D_{\alpha\sigma}(z^0) \upharpoonright P_{\alpha\sigma}(z^0)\, X$, then the relations

$$N_\alpha^\sigma \sim N_\alpha^1, \qquad \sigma = 1, 2, \ldots, g_\alpha; \qquad \alpha = 1, \ldots, r,$$

and

$$D_{\alpha\sigma}(z) \upharpoonright P_{\alpha\sigma}(z)\, X \sim N_\alpha^\sigma, \qquad z \in G \setminus B,$$

are valid.

Theorem 3. *If $z^0 \in G \setminus B$, then there exist a polydisc $K = \{z : |z_i - z_i^0| < \delta, i = 1, \ldots, n, \delta > 0\}$ and an operator-valued function $T(z)$, holomorphic in K, such that $T(z)^{-1}$ exists and is also holomorphic in K. Moreover, $T(z^0) = 1$ and*

$$D(z) = T(z)\, D(z^0)\, T(z)^{-1}, \qquad z \in K.$$

Therefore, the operator-valued function $A(z)$ has an "invariant Jordan structure" in $G \setminus B$. The nilpotent parts $D(z)$ of $A(z)$ are locally holomorphically similar for all $z \in G \setminus B$.

Appendix. Fundamentals of complex function theory and algebra

This Appendix covers most of the auxiliary material used in the main text, primarily from complex function theory and algebra. The theorems quoted in the text (especially in Ch. 3) will be proved in this Appendix.

Sect. A 1 summarizes the material necessary to define the concept of an n-valued algebroid analytic function over G (without proofs). Relevant references are the textbooks of S. Saks and A. Zygmund [1], of A. Hurwitz and R. Courant [1] and of H. Behnke and F. Sommer [1]). The other sections contain an overview of the theory of polynomials with meromorphic coefficients, their prime factor decompositions and their local factor decompositions (Weierstrass's preparation theorem for polynomials). The relation between irreducible polynomials and n-valued algebroid functions will be presented. Algebroid function fields will be mentioned. Finally, the Puiseux (Newton) diagram will be discussed.

Proofs are given from Sect. A 2 on.

A 1. n-Valued algebroid functions over a region G

A 1.1. Meromorphic elements

Let a be a point of the complex plane and let $f(z)$ be a single-valued function which is meromorphic at a. Then it has a Laurent expansion in the neighborhood of a,

$$f(z) = \sum_{n=-N}^{\infty} a_n (z-a)^n, \qquad a_{-N} \neq 0, \qquad 0 < |z-a| < \varrho, \qquad \varrho > 0,$$

which is convergent in some dotted disc $0 < |z-a| < \varrho$. $f(z)$ is meromorphic in the disc $|z-a| < \varrho$.

Let $|z-a| < r(a, f)$ denote the largest disc centered at a, such that $f(z)$ is meromorphic in this disc.

The ordered pair (a, f) consisting of the point a and the function $f(z)$ which is meromorphic in $|z-a| < r(a, f)$, is called a *meromorphic (function) element over the point* a with the corresponding disc $|z-a| < r(a, f)$. a is called the *trace point* and $r(a, f)$ the *radius* of the meromorphic element (a, f).

If $f(z)$ is meromorphic on a region G, then, at each point $a \in G$, the function $f(z)$ generates a meromorphic element (a, f). In order to exclude ambiguities, the function which is generated by $f(z)$ on the disc of the meromorphic element (a, f) will sometimes be denoted by $f_a(z)$.

Two meromorphic elements (a, f) and (b, g) are said to be equal, $(a, f) = (b, g)$, if $a = b$ and if the functions $f(z)$ and $g(z)$ coincide in some neighborhood of $a = b$. Then the two meromorphic elements have one and the same disc, and one has $f_a(z) \equiv g_a(z)$.

A 1.2. Analytic continuation

Let (a, f) be a meromorphic element with the corresponding disc $K = \{z: |z - a| < r\}$. Let $b \in K$. The meromorphic element (b, f_a) is said to be a *direct continuation* of (a, f). The element (a, f) can be directly continued to all points of K, but not to other points. For the radius $r(b, f_a)$ of the element (b, f_a) one obtains the estimate $r(b, f_a) \geqq r(a, f_a) - |b - a|$.

Let $\mathscr{C}$ be a closed piecewise smooth path in the complex plane, given by $z = z(t)$, $\alpha \leqq t \leqq \beta$, where the function $z(t)$ is piecewise smooth on $[\alpha, \beta]$. $\mathscr{C}$ is considered oriented. The same path with opposite orientation is denoted by $-\mathscr{C}$.

In the following discussion, meromorphic elements are also denoted by $P = P(a) = (a, f)$; i.e., $P(a)$ denotes a meromorphic element with trace point a.

Assume that to every point $z \in \mathscr{C}$ corresponds a unique meromorphic element $P(z)$.

Definition 1. The assignment $P(z)$, $z \in \mathscr{C}$, is said to be a *chain* (of meromorphic elements) *along* $\mathscr{C}$ if, to every $z \in \mathscr{C}$, corresponds an $\varepsilon = \varepsilon(z) > 0$, such that the meromorphic element $P(z + h)$ is always a direct continuation of $P(z)$, provided that $z + h \in \mathscr{C}$ and $|h| < \varepsilon$.

That is, if $P(z) = (z, f)$, then, for the points $z + h$ characterized in the definition, one always has $P(z + h) = (z + h, f)$.

Let $\mathscr{C}$ be a finite closed piecewise smooth path with initial point a and final point b. Let $P(a)$ be a fixed meromorphic element.

Theorem 1. *There exists at most one chain along $\mathscr{C}$ with a given initial element $P(a)$.*

This theorem provides the motivation for the following definition.

Definition 2. A meromorphic element $Q = Q(b)$ is said to be a *continuation* of the element $P = P(a)$ *along* $\mathscr{C}$ if there is a chain $P(z)$ along $\mathscr{C}$ such that $P = P(a)$ and $Q = P(b)$.

For short, one says that the chain $P(z)$ *joins* the elements P and Q. To every path $\mathscr{C}$ with initial point a and final point b and to every initial element $P = P(a)$ corresponds at most one element $Q = Q(b)$ that can be joined with P along $\mathscr{C}$.

Let G be a region and let $a, b \in G$.

Definition 3. The element $Q = Q(b)$ is said to be a *continuation* of $P = P(a)$ *inside* G if Q is a continuation of P along a path $\mathscr{C}$ belonging to G.

A 1.3. Analytic functions over G

Definition 4. A non-empty set $\mathfrak{P}$ of meromorphic elements whose trace points belong to G is said to be an *analytic function over* G if the following two conditions are satisfied:

I. To every two elements of $\mathfrak{P}$ corresponds a continuation inside G joining them.

II. If $P \in \mathfrak{P}$ and if Q is a continuation of G inside G, then $Q \in \mathfrak{P}$.

If P is any element with a trace point from G, then the set of all elements obtained by continuation of P inside G is an analytic function over G containing P. It is uniquely determined by P. Together with P, an analytic function $\mathfrak{P}$ also contains all direct continuations of P; their trace points belong to a suitable neighborhood of a.

Definition 5. Let $\mathfrak{P}$ be an analytic function over G and let H be a subregion of G, $H \subset G$. If P denotes an element of $\mathfrak{P}$ with a trace point from H, then P determines a unique analytic function $\mathfrak{H}$ over H. This function is called a *branch of* $\mathfrak{P}$ over H.

All elements of a branch $\mathfrak{H}$ of $\mathfrak{P}$ obviously belong to $\mathfrak{P}$. It is possible that there exist several different branches of $\mathfrak{P}$ over H.

Let $\mathfrak{P}$ be an analytic function over G. Then the number of different meromorphic elements of $\mathfrak{P}$ with fixed trace point $a \in G$ is at most denumerable (theorem of Poincaré-Volterra). We select especially those analytic functions over G whose number of different meromorphic elements over an arbitrary trace point does not exceed a fixed finite bound.

Definition 6. Let p be a natural number, $1 \leqq p < \infty$. Let $\mathfrak{P}$ be an analytic function over G. $\mathfrak{P}$ is said to be *p-valued* (over G) if every point $z \in G$ is a trace point of at most p different elements of $\mathfrak{P}$ and if there exists at least one point that is the trace point of exactly p different elements of $\mathfrak{P}$.

If $\mathfrak{P}$ is analytic over G and if $P = (a, f)$ denotes an element of $\mathfrak{P}$, then $f(a)$ (including ∞) is called a *value* of $\mathfrak{P}$ at the trace point a. To every trace point $a \in G$ corresponds a *value set* $\mathfrak{P}(a) = \{f(a): (a, f) \in \mathfrak{P}\}$. If $\mathfrak{P}$ is p-valued then at most p values are assigned to every trace point.

Let $\mathfrak{P}$ be p-valued and let D be the (open) subset of the trace set S (set of all trace points) consisting of all those trace points over which lie exactly p different elements of $\mathfrak{P}$. Denote the number of different values of $\mathfrak{P}$ over z by $n(z)$. The subset of S consisting of all those points z where $n(z) < p$ is isolated and closed in S.

If $f(z)$ is meromorphic in G, then the analytic function generated by $f(z)$ over G is 1-valued (single-valued). If $\mathfrak{P}$ is single-valued over G and if S denotes the trace set of $\mathfrak{P}$, then $\mathfrak{P}$ is a meromorphic function in S.

Definition 7. Let $\mathfrak{P}$ be analytic over G. If every $P(a) \in \mathfrak{P}$ can be continued along every path belonging to G and starting from a, then one says that $\mathfrak{P}$ can be *continued arbitrarily* in G.

Assume that $\mathfrak{P}$ can be continued arbitrarily in G and let $\mathfrak{P}$ be p-valued. Then to every point of G are assigned exactly p different elements of $\mathfrak{P}$ (i.e., $D = G$).

If G is simply connected, if $\mathfrak{P}$ is analytic over G and, furthermore, if $\mathfrak{P}$ can be continued arbitrarily in G, then $\mathfrak{P}$ is single-valued, i.e., meromorphic in G (monodromy theorem).

A 1.4. Singular points

Let S denote the trace set of an analytic function $\mathfrak{P}$ over G. The set $(S \cup \partial S) \cap G = M$ is divided into two classes: the class of regular points and the class of singular points.

Definition 7. A point $z \in M$ is called *regular* if there is a neighborhood U of z such that every branch of $\mathfrak{P}$ over U is meromorphic; i.e., every branch of $\mathfrak{P}$ over U represents a meromorphic function defined in U.

This means, that $z \in M$ is regular if there is a simply connected neighborhood U of z such that every branch of $\mathfrak{P}$ over U can be continued arbitrarily in G.

Definition 8. A point $z \in M$ is called *singular* if it is not regular.

The set R of all regular points is open. The set A of all singular points is closed in G. For p-valued functions $\mathfrak{P}$ one has $D \subsetneqq R$.

Let $\mathfrak{P}$ be analytic over G. Furthermore, let H be a subregion of G such that H contains a regular point, but no singular points. Then each point of H is regular. If $\mathfrak{H}$ is a branch of $\mathfrak{P}$ over H, then $\mathfrak{H}$ can be continued arbitrarily in H.

Definition 9. A singular point a is said to be *isolated* if there is a neighborhood U of a such that $U \setminus \{a\}$ contains no singular points.

If a is an isolated singular point of $\mathfrak{P}$ over G, then there is a neighborhood U of a such that $U \setminus \{a\}$ consists of regular points only. Then every branch of $\mathfrak{P}$ over $U \setminus \{a\}$ can be continued arbitrarily in $U \setminus \{a\}$.

Let $\mathfrak{P}$ be p-valued over G and let a be an isolated singularity of $\mathfrak{P}$. Furthermore, let U be a neighborhood of a such that $U \setminus \{a\}$ consists of regular points only. Then $\mathfrak{P}$ can be decomposed over $U \setminus \{a\}$ into finitely many different finite-valued branches. Let $\mathfrak{H}$ be one of these branches and let $\mathfrak{H}$ be n-valued. Then to every point of $U \setminus \{a\}$ are assigned exactly n different (meromorphic) elements of $\mathfrak{H}$.

If one denotes the elements of $\mathfrak{H}$ lying over $z \in U \setminus \{a\}$ by $f_1(z), f_2(z), \ldots, f_n(z)$, then the indices can be chosen such that the elements f_j undergo a cyclic permutation, $f_1 \to f_2 \to f_3 \to \cdots \to f_n \to f_1$ by a single positive winding of z around a.

The simplest illustration of the situation just described is given by the function

$$\zeta = (z - a)^{1/n}.$$

The point $z = a$ is singular; all other points are regular. This function is n-valued. To the point $z_0 \neq a$ are assigned the n different elements

$$f_i(z) = (z_0 - a)^{1/n} \sum_{\mu=0}^{\infty} \binom{1/n}{\mu} (z_0 - a)^{-\mu} (z - z_0)^{\mu}, \qquad |z - z_0| < |a - z_0|,$$

where for $(z_0 - a)^{1/n}$ one has to substitute the n different values of the root for $i = 1, 2, \ldots, n$, respectively. If $z - a = r\, e^{i\varphi}$ and $(z - a)^{1/n} = r'\, e^{i\varphi'}$, one obtains

$$r' = r^{1/n}, \qquad \varphi' = \frac{\varphi}{n} + \frac{2\pi}{n} k, \qquad k = 0, 1, 2, \ldots, n - 1,$$

and thereby the n different values of the root at the point z.

Let $\mathfrak{P}$ be p-valued over G. Assume that each singular point of $\mathfrak{P}$ is isolated; i.e., the set A of all isolated points is isolated in G. Since A is closed in G, the singular points can accumulate at most on ∂G. $G \setminus A$ is a subregion of G consisting of regular points only. $\mathfrak{P}$ can be continued arbitrarily in $G \setminus A$. To every point from $G \setminus A$ are assigned exactly p meromorphic elements of $\mathfrak{P}$.

A 1.5. Branching and algebroid branching

Let $\mathfrak{P}$ be p-valued over G and let A be isolated (and closed) in G. Furthermore, let a be singular and let U denote a neighborhood of a such that $U \setminus \{a\}$ consists of regular points only.

Definition 10. Let $\mathfrak{H}$ be a branch of $\mathfrak{P}$ over U of valuedness q. Then a is said to be a *branching point* of order q if $q \geqq 2$.

For convencience, also in the case $q = 1$, a is sometimes said to be a branching point of order 1.

Note that the concept of a branching point a always refers to a definite branch $\mathfrak{H}$ of $\mathfrak{P}$ over a sufficiently small neighborhood of a.

Let $\mathfrak{H}$ be n-valued over the neighborhood $|z - a| < r$ of a and let a be an (isolated) branching point of $\mathfrak{H}$ of order $n \geqq 2$. Assume that the points from $0 < |z - a| < r$ are regular. To every point z from $0 < |z - a| < r$ are assigned n different meromorphic elements. These n meromorphic elements are denoted as follows:

Let any branch of $\mathfrak{H}$ over the cut disc $0 < |z - a| < r$, $0 < \arg (z - a) < 2\pi$, be given. This branch is meromorphic and is denoted by $f_1(z)$. Let z encircle a once in the positive direction. Then, by analytic continuation, one obtains a meromorphic branch $f_2(z)$ over $0 < |z - a| < r$, $2\pi < \arg (z - a) < 4\pi$. Correspondingly, let $f_\varrho(z)$ denote the meromorphic branch, which is generated in the same way over

$$0 < |z - a| < r, \qquad (\varrho - 1)\, 2\pi < \arg (z - a) < \varrho\, 2\pi, \qquad \varrho = 1, 2, \ldots, n.$$

This notation is in accordance with the cyclic permutation property by analytic continuation by a single positive winding of z around a.

Next, consider the function $\zeta = (z - a)^{1/n}$. Since $|\zeta| = |z - a|^{1/n}$ and $\arg \zeta = n^{-1} \times \arg (z - a)$ the "ϱ-th sheet"

$$0 < |z - a| < r, \qquad (\varrho - 1)\, 2\pi \leqq \arg (z - a) \leqq \varrho\, 2\pi$$

is mapped by this function one-to-one onto the sector

$$S_\varrho\colon 0 < |\zeta| < r^{1/n}, \qquad (\varrho - 1)\, \frac{2\pi}{n} \leqq \arg \zeta \leqq \varrho\, \frac{2\pi}{n}.$$

The sectors $S_1, S_2, \ldots, S_n$ cover the disc $0 < |\zeta| < r^{1/n}$. Neighboring sectors have a (open) straight line in common, namely $\alpha\, e^{i\varrho \frac{2\pi}{n}}$, $0 < \alpha < r^{1/n}$, $\varrho = 0, 1, 2, \ldots, n - 1$.

On the dotted disc $0 < |\zeta| < r^{1/n}$ one defines the following single-valued function:

$$h(\zeta) = f_\varrho(a + \zeta^n), \qquad \zeta \in S_\varrho, \qquad \varrho = 1, 2, \ldots, n.$$

This definition is unambiguous, because at points ζ belonging to the sector boundaries just mentioned (for example, to the common boundary of S_ϱ and $S_{\varrho+1}$), the functions

$f_\varrho(z)$ and $f_{\varrho+1}(z)$ have one and the same value. The function $h(\zeta)$ is meromorphic at every point of each sector S_ϱ, because $f_\varrho(z)$ is meromorphic at those points z where $0 < |z - a| < r$. That is, $h(\zeta)$ is single-valued and meromorphic on $0 < |\zeta| < r^{1/n}$. Hence:

$\mathfrak{H}$ can be represented over $0 < |z - a| < r$ in the form $h\big((z - a)^{1/n}\big)$, where $h(\zeta)$ is single-valued and meromorphic on $0 < |\zeta| < r^{1/n}$.

The n values of $\mathfrak{H}$ over z are obtained by substitution of the n different values of $(z - a)^{1/n}$ into $h(\zeta)$. One obtains the n different elements of $\mathfrak{H}$ over a neighborhood of z in the same way.

Definition 11. The branching point a of $\mathfrak{P}$ with respect to $\mathfrak{H}$ is said to be *algebroid* if the function $h(\zeta)$ is meromorphic in $|\zeta| < r^{1/n}$.

This means that a is an algebroid branching point if $h(\zeta)$ is meromorphic at $\zeta = 0$.

Definition 12. Let a be a branching point of $\mathfrak{P}$ with respect to $\mathfrak{H}$ over $|z - a| < r$. The number α is called a *limit* of $\mathfrak{H}$ at a if, for any sequences $z_\varrho \to a$, $w(z_\varrho) \to \alpha$ for any value $w(z_\varrho)$ of $\mathfrak{H}$ over z_ϱ.

Let a be a branching point of $\mathfrak{P}$ with respect to $\mathfrak{H}$ over $|z - a| < r$. *Then a is an algebroid branching point if and only if $\mathfrak{H}$ has a limit at the point a.*

If a is an algebroid branching point for the n-valued branch $\mathfrak{H}$ over $0 < |z - a| < r$ then the corresponding function $h(\zeta)$ can be expanded into a Laurent series in a neighborhood of $\zeta = 0$,

$$h(\zeta) = \sum_{n=-N}^{\infty} a_n \zeta^n, \qquad a_{-N} \neq 0, \qquad 0 < |\zeta| < r_1^{1/n}, \qquad 0 < r_1 \leqq r.$$

Hence, the n values (branches) of $\mathfrak{H}$ over $0 < |z - a| < r$ are represented by the series expansions

$$\sum_{\varrho=-N}^{\infty} a_\varrho [(z - a)^{1/n}]^\varrho, \qquad 0 < |z - a| < r_1,$$

where each of the n different root values $(z - a)^{1/n}$ yields exactly one of the n values of $\mathfrak{H}$ over z. Let $\varkappa$ denote one of these root values and ξ a fixed primitive n-th root of unity. The n different root values are given by $\varkappa, \xi\varkappa, \xi^2\varkappa, \ldots, \xi^{n-1}\varkappa$ and, consequently one obtains the n series expansions

$$\sum_{\varrho=-N}^{\infty} a_\varrho (\xi^\mu \varkappa)^\varrho, \qquad \mu = 0, 1, 2, \ldots, n - 1.$$

The trace point a of an algebroid branching point for $\mathfrak{H}$ cannot be an accumulation point of trace points of α-points, i.e., of points z such that α is a value of $\mathfrak{H}$ over z.

A 1.6. p-Valued algebroid functions

Definition 13. A singular point a of $\mathfrak{P}$ over G is called *algebroid* if, for every branch $\mathfrak{H}$ of $\mathfrak{P}$ over a neighborhood of a, the point a is at most an algebroid branching point.

Note that an algebroid singularity is necessarily isolated.

Definition 14. A p-valued analytic function $\mathfrak{P}$ over G is called a *(p-valued) algebroid* function if its singular points are at most algebroid.

These algebroid singular points are isolated in G and they can accumulate at most on ∂G. A p-valued algebroid function has at most algebroid branching points.

If a is an algebroid singularity for $\mathfrak{P}$ and, moreover, a branching point of order n with respect to the branch $\mathfrak{H}$ of $\mathfrak{P}$, then $\mathfrak{H}$ can be represented over a sufficiently small neighborhood $|z - a| < r$ by a series expansion

$$f(z) = \sum_{\varrho=-N}^{\infty} a_\varrho (z - a)^{\varrho/n}, \qquad a_{-N} \neq 0, \qquad 0 < |z - a| < r. \tag{1.1}$$

Such a series expansion is called an *n-fold algebroid function element* at the point a and the corresponding ordered pair (a, f) an *n-fold algebroid element* of $\mathfrak{P}$. In particular, if $n = 1$, then one obtains a meromorphic element. These algebroid elements can be added to the analytic function $\mathfrak{P}$. Then the limit of the corresponding branching point is a value of $\mathfrak{P}$ at the point a.

A series expansion of the form (1.1) is called a *Laurent-Puiseux series*. In fact, it is a Laurent series formed by fractional powers of $z - a$. A power series formed by fractional powers of $z - a$ is called a *Puiseux series*.

An algebroid element $\sum\limits_{\varrho=-N}^{\infty} a_\varrho (z - a)^{\varrho/n}$ is called *formally n-fold* if the element is already m-fold, where m is a proper divisor of n; i.e., if $n = mp$ and if $a_\varrho \neq 0$ implies $\varrho = pk$, $k = 0, \pm 1, \pm 2, \ldots$ Thus,

$$\sum_\varrho a_\varrho (\xi^\mu \varkappa)^\varrho = \sum_k a_{pk} (\xi^\mu \varkappa)^{pk} = \sum_k a_{pk} (\xi^{\mu p} \varkappa^p)^k,$$

where ξ^p is a primitive m-th root of unity and $\varkappa^p$ is a fixed root $(z - a)^{1/m}$. Hence one obtains an m-fold element of the form $\sum \alpha_\varrho (z - a)^{\varrho/m}$. Each m-fold element is formally (mp)-fold for $p = 1, 2, 3, \ldots$

A 2. Polynomials with meromorphic coefficients

A 2.1. Basic concepts

Let G be a fixed region of the complex plane and let $f(z)$ and $g(z)$ denote meromorphic functions in G. Then $f(z) \pm g(z)$, $f(z) \cdot g(z)$ and also, provided that $g(z) \not\equiv 0$, $f(z)/g(z)$ are meromorphic functions in G. These operations satisfy the field axioms, and one has $f(z) \cdot g(z) = g(z) \cdot f(z)$. This field is called the (Abelian) *field* $\mathcal{M} \equiv \mathcal{M}_G$ *of meromorphic functions in* G.

Consider polynomials of the form

$$p(\lambda, z) = a_0(z)\, \lambda^m + a_1(z)\, \lambda^{m-1} + \cdots + a_{m-1}(z)\, \lambda + a_m(z), \tag{2.1}$$

whose coefficients $a_0(z), a_1(z), \ldots, a_m(z)$ are functions from $\mathcal{M}$. If $a_0(z) \not\equiv 0$, then $m = \deg p$ is called the *degree* of the polynomial p. Functions from $\mathcal{M}$ which do not vanish identically have the degree 0. The zero polynomial $p \equiv 0$ has no degree. Polynomials of the form (2.1) can be added and multiplied. The ring axioms are satisfied. This ring is called the *ring* $\mathscr{M}[\lambda] \equiv \mathscr{M}_G[\lambda]$ *of all polynomials over* $\mathscr{M}$ or, simply, the *polynomial ring* over $\mathscr{M}$. The meromorphic functions which do not vanish identically are the *units* of this ring. They have an inverse element which also belongs to the ring.

If a polynomial from $\mathscr{M}[\lambda]$ has the form

$$p(\lambda, z) = \lambda^n + a_1(z)\,\lambda^{n-1} + \cdots + a_n(z), \tag{2.2}$$

where the coefficients $a_1(z), a_2(z), \ldots, a_n(z)$ are holomorphic in G, then p is called an *entire polynomial.* The product of two entire polynomials is again entire. The set of all entire polynomials is denoted by $\mathscr{H}[\lambda] \equiv \mathscr{H}_G[\lambda]$.

Let $z_0 \in G$ and $p \in \mathscr{H}[\lambda]$; i.e., p has a representation of the form (2.2) with holomorphic coefficients. The set of all polynomials $p \in \mathscr{H}[\lambda]$ such that $a_i(z_0) = 0$ for all $i = 1, 2, \ldots, n$, is denoted by $\mathscr{H}^{z_0}_G[\lambda] \equiv \mathscr{H}^{z_0}[\lambda]$. The product of two polynomials from $\mathscr{H}^{z_0}[\lambda]$ again belongs to $\mathscr{H}^{z_0}[\lambda]$.

A 2.2. Divisors

If $p, q, r \in \mathscr{M}[\lambda]$ and if $r = pq$, then p and q are called *divisors* of r.

Every unit is a divisor of each polynomial, for, if p is a polynomial and α a unit then $p = \alpha(\alpha^{-1}p)$. Hence, units are always trivial divisors. The interesting questions with respect to divisibility are related to divisors q with $\deg q > 0$, i.e., to *proper* or *nontrivial divisors.* If one investigates a polynomial p with respect to divisors it is convenient to consider this polynomial in the *normalized* form

$$p(\lambda, z) = \lambda^n + a_1(z)\,\lambda^{n-1} + \cdots + a_n(z), \qquad a_i(z) \in \mathscr{M}. \tag{2.3}$$

Every polynomial can be normalized to the form (2.3) by suitable multiplication by a unit.

If p is a polynomial with $\deg p = n > 0$ and if p has a divisor q with $0 < \deg q < n$, then p is called *reducible*. Otherwise, p is called *irreducible* or a *prime polynomial.*

A 2.3. Division algorithm

Let p and q be normalized polynomials from $\mathscr{M}[\lambda]$,

$$p = \lambda^n + a_1\lambda^{n-1} + \cdots + a_n, \qquad q = \lambda^m + b_1\lambda^{m-1} + \cdots + b_m;$$

let $n \geqq m$. Then one can subtract a multiple of q from p such that the coefficient of the highest term vanishes. One forms

$$p - \lambda^{n-m}p = (a_1 - b_1)\,\lambda^{n-1} + \cdots.$$

If the degree of this polynomial is still not less than m, then, by subtracting a suitable multiple of q, one can make the coefficient of the highest term equal to zero. Continuation of this procedure finally leads to a remainder r the degree of which is less than m or which vanishes:

$$p - aq = r, \qquad \deg r < m \quad \text{or} \quad r \equiv 0.$$

This procedure is called the *division algorithm.*

A 2.4. Euclidean algorithm

By successive application of the division algorithm one obtains the ***Euclidean algorithm.*** One starts again with two normalized polynomials p and q with $\deg p \geqq \deg q$. For formal reasons, one writes

$$p = p_1, \qquad q = p_2, \qquad \deg p_1 \geqq \deg p_2.$$

Then there exist finitely many polynomials $p_3, p_4, \ldots, p_r$ with the following properties:

$$p_1 = q_1 p_2 + p_3, \qquad 0 < \deg p_3 < \deg p_2,$$

$$p_2 = q_2 p_3 + p_4, \qquad 0 < \deg p_4 < \deg p_3,$$

....

$$p_{r-2} = q_{r-2} p_{r-1} + p_r, \quad 0 \leqq \deg p_r < \deg p_{r-1},$$

$$p_{r-1} = q_{r-1} p_r.$$

The last equation indicates that the operation p_{r-1} divided by p_r leaves zero remainder. Of course, if $\deg p_r = 0$, i.e., if p_r is a meromorphic function (a unit), then the last equation is trivial. From the equations of the Euclidean algorithm the equation

$$p_r = \alpha p_1 + \beta p_2$$

follows, where α and β denote suitable polynomials.

A 2.5. Greatest common divisor

Let p and q be normalized polynomials with $\deg p \geqq \deg q$ and let g be a common divisor of p and q. Then g is also a divisor of p_r. Conversely, p_r is a common divisor of p and q. Hence, p_r is a common divisor of maximal degree. Normalizing p_r, one obtains a normalized common divisor of p and q of maximal degree. It is uniquely determined and called the ***greatest common divisor*** g of p and q. There are polynomials α and β such that

$$g = \alpha p + \beta q. \tag{2.4}$$

On writes $g = (p, q)$.

If $g = 1$, then p and q have no proper common divisor and they are said to be ***relatively prime.*** In this case one has

$$1 = \alpha p + \beta q$$

with suitable polynomials α and β.

If p, q, r denote three normalized polynomials, then $\big(p, (q, r)\big) = \big((p, q), r\big)$. This polynomial is called the ***greatest common divisor*** of p, q, r. One writes $g = (p, q, r)$. In the same way, one obtains the greatest common divisor $(p_1, p_2, \ldots, p_n)$ of finitely many normalized polynomials $p_1, p_2, \ldots, p_n$. If $(p_1, p_2, \ldots, p_n) = g$, then, according to the inductive definition of g, $g = \big(p_1, (p_2, p_3, \ldots, p_n)\big)$, equation (2.4) can be generalized and one obtains

$$g = \alpha_1 p_1 + \alpha_2 p_2 + \cdots + \alpha_n p_n$$

with suitable polynomials $\alpha_1, \ldots, \alpha_n$. In particular, if $(p_1, p_2, \ldots, p_n) = 1$, i.e., if the polynomials $p_1, \ldots, p_n$ are relatively prime, then one obtains $1 = \alpha_1 p_1 + \alpha_2 p_2 + \cdots + \alpha_n p_n$, which is a representation of unity.

A 2.6. Decomposition into prime factors

If p is a prime polynomial and if p is a divisor of ab, then p is a divisor of one of the factors a, b. To prove this, assume that p is not a divisor of a. Then p and a are relatively prime, hence $1 = \alpha a + \beta p$ with suitable polynomials α and β. Then $b = \alpha ab + \beta pb = (\alpha\lambda + \beta b)\, p$, where $ab = \lambda p$; hence, p is a divisor of b. One obtains

Theorem 1. *Let q be a normalized polynomial from $\mathscr{M}_G[\lambda]$. Then there exists a uniquely determined decomposition into prime factors*

$$q = \prod_{\varrho=1}^{r} p_\varrho^{m_\varrho} \qquad p_\varrho \in \mathscr{M}_G[\lambda], \tag{2.5}$$

where the polynomials $p_1, p_2, \ldots, p_r$ are normalized, irreducible and mutually different. Poles of the coefficients of $p_1, \ldots, p_r$ are necessarily poles of the coefficients of q.

If $q \in \mathscr{H}_G[\lambda]$, then all prime factors belong to $\mathscr{H}_G[\lambda]$; if $q \in \mathscr{H}_G^{z_0}[\lambda]$, then all prime factors belong to $\mathscr{H}_G^{z_0}[\lambda]$.

Proof. Obviously, a decomposition $q = q_1 q_2 \cdots q_m$ with irreducible polynomials $q_1, \ldots, q_m$ exists. Such a decomposition is unique. If there were two decompositions $q_1 q_2 \cdots q_m = q_1' q_2' \cdots q_n'$, then one would proceed as follows. First, q_1 would be a divisor of $q_1' \cdot (q_2' \cdots q_n')$, which implies $q_1 = q_1'$, or q_1 would be a divisor of $q_2' \cdots q_n'$. Then, after finitely many steps, $q_1 = q_{i_1}'$. Take the next polynomial q_2. In the same way one obtains $q_2 = q_{i_2}$. One would reach the same conclusion for any of the other factors. Hence, each normalized polynomial can be written uniquely (up to the order of the factors) as a product of powers (2.5) of normalized prime polynomials.

Next, one proves the assertion that poles of the coefficients of $p_1, \ldots, p_r$ are necessarily poles of the coefficients of q. For this purpose, one takes three normalized polynomials q, p, r,

$$p = \lambda^m + a_1(z)\,\lambda^{m-1} + \cdots + a_m(z),$$

$$q = \lambda^n + b_1(z)\,\lambda^{n-1} + \cdots + b_n(z),$$

$$r = \lambda^l + c_1(z)\,\lambda^{l-1} + \cdots + c^l(z),$$

such that $r = pq$. Then it is sufficient to prove the following statement: If z_0 is a pole of one of the coefficients $a_i(z)$, $b_j(z)$, then z_0 is also a pole of one of the coefficients $c_k(z)$. One has

$$\lambda^l + c_1(z)\,\lambda^{l-1} + \cdots = \left(\lambda^m + a_1(z)\,\lambda^{m-1} + \cdots\right)\left(\lambda^n + b_1(z)\,\lambda^{n-1} + \cdots\right). \tag{2.6}$$

Let z_0 be a pole of one of the coefficients $a_i(z)$, $b_j(z)$, but not a pole of the coefficients $c_1(z), \ldots, c_l(z)$; i.e., the functions $c_1(z), \ldots, c_l(z)$ are assumed to be holomorphic at the point $z = z_0$. Let h be the maximal order of z_0 as a pole of $a_1, a_2, \ldots, a_m$; let k be the maximal order of z_0 as a pole of $b_1, b_2, \ldots, b_n$. Then $\max(h, k) > 0$. Let

$$A_\varrho = \lim_{z \to z_0} (z - z_0)^h a_\varrho(z), \qquad B_\varrho = \lim_{z \to z_0} (z - z_0)^k b_\varrho(z).$$

Multiplying the equation (2.6) by $(z - z_0)^{h+k}$ and taking the limit $z \to z_0$, one obtains the equation

$$0 = (A_1 \lambda^{m-1} + \cdots)(B_1 \lambda^{n-1} + \cdots)$$

as an identity in λ. Since at least one of the numbers A_i and at least one of the numbers B_j are different from zero, one obtains a contradiction.

This implies: If $r \in \mathscr{H}_G[\lambda]$ and $r = pq$, where r, p, q are normalized, then $p, q \in \mathscr{H}_G[\lambda]$; i.e., the divisors of entire polynomials are again entire. Therefore, the prime factors of a polynomial $q \in \mathscr{H}_G[\lambda]$ also belong to $\mathscr{H}_G[\lambda]$.

Finally, let $r \in \mathscr{H}_G^{z_0}[\lambda]$ and $r = p \cdot q$ with $p, q \in \mathscr{H}_G[\lambda]$. Then (2.6) is valid: in particular, one obtains the equation

$$\lambda^l = \big(\lambda^m + a_1(z_0)\, \lambda^{m-1} + \cdots + a_m(z_0)\big)\big(\lambda^n + b_1(z_0)\, \lambda^{n-1} + \cdots + b_n(z_0)\big)$$

at the point $z = z_0$. The decomposition of λ^l into prime factors with respect to $\mathbb{C}[\lambda]$ is uniquely determined; hence, $a_i(z_0) = b_j(z_0) = 0$. Thus it is shown that the prime factors of polynomials from $\mathscr{H}_G^{z_0}[\lambda]$ again belong to $\mathscr{H}_G^{z_0}[\lambda]$. ■

A 3. Polynomials from $\mathscr{M}[\lambda]$ considered as polynomials from $\mathbb{C}[\lambda]$

A 3.1. Relatively prime polynomials

Let $p \in \mathscr{M}[\lambda]$ be a normalized polynomial. Let S_p denote the set of all points z from G such that at least one coefficient of p has a pole at z. S_p is isolated and closed in G. For each point $z \in G \setminus S_p$, the polynomial p is well-defined as a polynomial from $\mathbb{C}[\lambda]$.

Theorem 1. *Let p and q be normalized polynomials from $\mathscr{M}[\lambda]$ that are relatively prime in $\mathscr{M}[\lambda]$. Then, except for a set $H \subset G$, isolated and closed in G, the polynomials p and q are relatively prime in $\mathbb{C}[\lambda]$ for all $z \in G \setminus (S_p \cup S_q)$; i.e., for all $z \in G \setminus (S_p \cup S_q \cup H)$ the polynomials have no common root.*

Proof. Let $\deg p \geqq \deg q > 0$. Applying the Euclidean algorithm to p and q, where the last remainder is p_r, one obtains the equation $\deg p_r = 0$, because the polynomials p and q are relatively prime. That is, one has $p_r = f(z)$, where $f(z)$ is a not identically vanishing meromorphic function in G. Let E denote the union of all poles of all coefficients of the polynomials p_ϱ, q_ϱ generated by the Euclidean algorithm. E is isolated and closed in G (note that $S_p \cup S_q \subseteqq E$). Furthermore, the set N of all roots of $f(z)$ is isolated and closed in G.

Let $z_0 \in G \setminus E$ and let λ_0 denote a common root of $p(\lambda, z_0)$ and $q(\lambda, z_0)$; i.e., one has $p(\lambda_0, z_0) = q(\lambda_0, z_0) = 0$. Then, from the equations of the Euclidean algorithm it follows that

$$p_1(\lambda_0, z_0) = p_2(\lambda_0, z_0) = p_3(\lambda_0, z_0) = \cdots = p_{r-1}(\lambda_0, z_0) = f(z_0) = 0;$$

i.e., z_0 is necessarily a root of $f(z)$. Hence, for all $z \in G \setminus (E \cup N)$, the polynomials p and q have no common root. ■

A 3.2. Behavior of irreducible polynomials

Theorem 2. *Let p be a normalized and irreducible polynomial from $\mathscr{M}[\lambda]$. The set of all poles of all coefficients of p is denoted by S_p. Then, for all $z_0 \in G \setminus S_p$, the polynomial $p = p(\lambda, z_0)$ is well-defined as a polynomial of $\mathbb{C}[\lambda]$. Moreover, except for a subset of G, isolated and closed in G, $p(\lambda, z_0)$ has only simple roots for all $z_0 \in G \setminus S_p$.*

Proof. The assertion is obvious for $\deg p = 1$. Let $\deg p > 1$ and $q(\lambda, z) = \frac{\partial p}{\partial \lambda}(\lambda, z)$. Since p is irreducible, the identity $(p, q) = 1$ follows; i.e., q and p are relatively prime. According to Theorem 1, p and q have no common root, except for a subset that is isolated and closed in G. In other words, except for the mentioned subset, the roots of $p(\lambda, z)$ are simple for all $z \in G \setminus S_p$. ■

A 4. Local factor decompositions of polynomials from $\mathscr{M}[\lambda]$

This section deals with a special case of the Weierstrass preparation theorem.

A 4.1. Factor decompositions of normalized polynomials

Let $p \in \mathscr{M}_G[\lambda]$. Then p can be decomposed into a product of prime factors with respect to $\mathscr{M}_G[\lambda]$. Transition to a subregion $H \subset G$ could change the prime factor decomposition of p in $\mathscr{M}_H[\lambda]$; that is, it is possible that irreducible polynomials in $\mathscr{M}_G[\lambda]$ become reducible with respect to $\mathscr{M}_H[\lambda]$. If $z_0 \in G \setminus S_p$, then, in particular, one can ask for the prime factor decomposition of p with respect to $\mathscr{M}_K[\lambda]$, where K is a suitable sufficiently small neighborhood of z_0. To answer this question, one uses the Weierstrass preparation theorem (for polynomials).

Theorem 1. *Let $p \in \mathscr{M}_G[\lambda]$ and let p be normalized. Furthermore, let $z_0 \in G \setminus S_p$ and $p(\lambda, z_0) = \prod_{\varrho=1}^{r} (\lambda - \lambda_\varrho)^{k_\varrho}$. Then there exists a neighborhood of z_0, say a disc $K = \{z: |z - z_0| < \varepsilon\}$, $\varepsilon > 0$, such that the following proposition holds; There exist polynomials $q_1(\mu, z)$, $q_2(\mu, z), \dots, q_r(\mu, z)$ belonging to $\mathscr{H}^{z_0}_K[\mu]$ and with $\deg q_\varrho = k_\varrho$, $\varrho = 1, 2, \dots, r$, such that the relation*

$$p(\lambda, z) = \prod_{\varrho=1}^{r} q_\varrho(\lambda - \lambda_\varrho, z), \qquad z \in K, \tag{4.1}$$

is valid. (4.1) *is a decomposition of p with respect to $\mathscr{H}_K[\lambda]$.*

Proof. Let $q(\lambda) = \prod_{\varrho=2}^{r} (\lambda - \lambda_\varrho)^{k_\varrho}$. Then, $p(\lambda, z_0) = (\lambda - \lambda_1)^{k_1} q(\lambda)$ and $q(\lambda_1) \neq 0$. Let L be a suitable disc centered at λ_1 such that $p(\lambda, z_0)$ has no root in $(L \cup \partial L) \setminus \{\lambda_1\}$. Let $\beta = \min_{\lambda \in \partial L} |p(\lambda, z_0)|$. The polynomial $p(\lambda, z)$ converges to $p(\lambda, z_0)$ as $z \to z_0$ uniformly with respect to $\lambda \in \partial L$. Hence, there exists a disc $K = \{z: |z - z_0| < \delta\}$, such that the coefficients of $p(\lambda, z)$ are holomorphic in K and

$$\sup_{\lambda \in \partial L} |p(\lambda, z) - p(\lambda, z_0)| < \beta, \qquad |z - z_0| < \delta.$$

Next, one can apply Rouche's theorem to $p(\lambda, z)$, where $z \in K$. According to this theorem, $p(\lambda, z)$ and $p(\lambda, z_0)$ have exactly the same number of roots (their multiplicities taken into account) in the interior of L. That is, $p(\lambda, z)$ has exactly k_1 roots in the interior of L. They are denoted by $\lambda_1 + \mu_1(z), \lambda_1 + \mu_2(z), \ldots, \lambda_1 + \mu_{k_1}(z)$ (their multiplicities taken into account). Then $\mu_1(z_0) = \mu_2(z_0) = \cdots = \mu_{k_1}(z_0) = 0$. Consider the polynomial

$$q(\lambda, z) = \prod_{\varrho=1}^{k_1} [(\lambda - \lambda_1) - \mu_\varrho(z)] = (\lambda - \lambda_1)^{k_1} + A_1(z)\,(\lambda - \lambda_1)^{k_1-1} + \cdots + A_{k_1}(z).$$

One has $A_1(z_0) = A_2(z_0) = \cdots = A_{k_1}(z_0) = 0$. The assertion is that these coefficients $A_\varrho(z)$ are holomorphic in K.

From the calculus of residues one obtains

$$S_j(z) = \sum_{\varrho=1}^{k_1} (\mu_\varrho(z))^j = \frac{1}{2\pi i} \int_{\partial L} (\lambda_1 - \lambda)^j \frac{\frac{\partial p}{\partial \lambda}(\lambda, z)}{p(\lambda, z)}\, d\lambda, \qquad j = 1, 2, \ldots$$

For short,

$$g_j(\lambda, z) = (\lambda - \lambda_1)^j \frac{\frac{\partial p}{\partial \lambda}(\lambda, z)}{p(\lambda, z)}, \qquad \lambda \in \partial L, \qquad z \in K.$$

The function $g_j(\lambda, z)$ is continuous in $\partial L \times K$ and so is $\partial g_j/\partial z$. Hence, S_j' exists and

$$S_j'(z) = \frac{1}{2\pi i} \int_{\partial L} \frac{\partial g_j}{\partial z}(\lambda, z)\, d\lambda,$$

which implies that $S_j(z)$ is holomorphic in K.

As is well-known, the coefficients $A_\varrho(z)$, $\varrho = 1, \ldots, k_1$, are the elementary symmetric polynomials of the roots $\mu_1(z), \ldots, \mu_{k_1}(z)$. They can also be written as polynomials with respect to the functions $S_j(z)$. This can be seen as follows. For short, let $\lambda - \lambda_1 = \mu$. Then,

$$q(\lambda, z) = \mu^{k_1} + A_1\mu^{k_1-1} + \cdots + A_{k_1}.$$

Let

$$\frac{q(\lambda, z)}{\mu - \mu_j} = \mu^{k_1-1} + B_1^{(j)}\mu^{k_1-2} + \cdots + B_{k_1-1}^{(j)}.$$

Comparing the coefficients, one obtains the questions

$$A_i = B_i^{(j)} - \mu_j B_{i-1}^{(j)}, \qquad i = 1, 2, 3, \ldots, k_1 - 1,$$

where $B_0^{(j)} = 1$. If one puts $A_0 = 1$, then these equations can be uniquely solved,

$$B_1^{(j)} = A_1 + A_0\mu_j,$$
$$B_2^{(j)} = A_2 + A_1\mu_j + A_0\mu_j^2,$$
$$B_3^{(j)} = A_3 + A_2\mu_j + A_1\mu_j^2 + A_0\mu_j^3,$$
$$\cdots\cdots\cdots\cdots\cdots\cdots\cdots\cdots\cdots\cdots$$
$$B_{k_1-1}^{(j)} = A_{k_1-1} + A_{k_1-2}\mu_j + \cdots + A_1\mu_j^{k_1-2} + A_0\mu_j^{k_1-1}.$$

Hence,

$$\frac{\partial}{\partial\mu} q(\lambda, z) = \sum_{j=1}^{k_1} \frac{q(\lambda, z)}{\mu - \mu_j} = A_0 S_0 \mu^{k_1-1} + (S_0 A_1 + S_1 A_0)\mu^{k_1-2}$$
$$+ (S_0 A_2 + S_1 A_1 + S_2 A_0)\, \mu^{k_1-3} + \cdots$$
$$+ (S_0 A_{k_1-1} + S_1 A_{k_1-2} + \cdots + S_{k_1-1} A_0),$$

where $S_j = \sum_\varrho \mu_\varrho^j$. Comparing the coefficients, one obtains the equations

$$(k_1 - 1)\, A_1 = S_0 A_1 + S_1 A_0,$$
$$(k_1 - 2)\, A_2 = S_0 A_2 + S_1 A_1 + S_2 A_0,$$
$$\cdots\cdots\cdots\cdots\cdots\cdots$$
$$A_{k_1-1} = S_0 A_{k_1-1} + S_1 A_{k_1-2} + \cdots + S_{k_1-1} A_0.$$

Because $A_0 = 1$ and $S_0 = k_1$, it follows from these equations that $A_1, A_2, \ldots, A_{k_1-1}$ can be represented as polynomials in $S_1, S_2, \ldots, S_{k_1-1}$. The last coefficient A_{k_1} can be obtained from the equation

$$\sum_{j=1}^{k_1} q(\lambda_1 + \mu_j, z) = S_{k_1} + S_{k_1-1} A_1 + S_{k_1-2} A_2 + \cdots + k_1 A_{k_1} = 0.$$

Therefore, the coefficients $A_j(z)$ are holomorphic in the disc K.

Hence, the polynomial $q(\lambda, z)$ belongs to $\mathscr{H}_K^{z_0}[\mu]$; in particular, it belongs to $\mathscr{H}_K[\lambda]$. The original polynomial $p(\lambda, z)$ from $\mathscr{M}_G[\lambda]$ belongs to $\mathscr{H}_K[\lambda]$, too.

Obviously, $q(\lambda, z)$ is a divisor of $p(\lambda, z)$ in $\mathscr{H}_K[\lambda]$. To prove this assertion, it is sufficient to show that for every $z \in K$ each root of $q(\lambda, z)$ is also a root of $p(\lambda, z)$, which is obvious. The polynomial $r = p/q$ belongs to $\mathscr{H}_K[\lambda]$ and is different from zero for each $(\lambda, z) \in L \times K$. For, if there were a pair $(\lambda, z) \in L \times K$ with $r(\lambda, z) = 0$, then the sum of all multiplicities of roots of $p(\lambda, z)$ at the point z would be at least equal to $k_1 + 1$, which is a contradiction. Hence,

$$p(\lambda, z) = [(\lambda - \lambda_1)^{k_1} + A_1(z)\, (\lambda - \lambda_1)^{k_1-1} + \cdots + A_{k_1}(z)]\, r(\lambda, z), \tag{4.2}$$

where all coefficients $A_i(z)$ are holomorphic in K, $A_i(z_0) = 0$, $r \in \mathscr{H}_K[\lambda]$ and $r(\lambda, z) \neq 0$ in $L \times K$.

A root of $p(\lambda, z_0)$, which is different from λ_1, for example λ_2, is necessarily a root of $r(\lambda, z_0)$. Hence, the considerations just described can again be applied to λ_2 and $r(\lambda, z)$. Then, one obtains a disc K' centered at z_0 (possibly one smaller than K) and a disc L' centered at λ_2, such that

$$r(\lambda, z) = s(\lambda, z)\, t(\lambda, z) = [(\lambda - \lambda_2)^{k_2} + B_1(z)\, (\lambda - \lambda_2)^{k_2-1} + \cdots + B_{k_2}(z)]\, t(\lambda, z),$$

where $s, t \in \mathscr{H}_{K'}[\lambda]$ and $t(\lambda, z) \neq 0$ for $(\lambda, z) \in L' \times K'$. This procedure terminates after finitely many steps and thus Theorem 1 has been proved. ■

A 4.2. Simple roots

If, for example, λ_1 is a simple root of $p(\lambda, z)$ at the point z_0, i.e., if $p(\lambda_1, z_0) = 0$ and $(\partial p/\partial\lambda)(\lambda_1, z_0) \neq 0$, then the exponent k_1 in the decomposition (4.2) is equal to 1 and one obtains

$$p(\lambda, z) = \big(\lambda - \lambda_1 + A_1(z)\big)\, r(\lambda, z).$$

Hence, the equations $p(\lambda, z) = 0$ and $\lambda - \lambda_1 + A_1(z) = 0$ are equivalent in $L \times K$; i.e., the set of all pairs (λ, z) from $L \times K$ with $p(\lambda, z) = 0$, which is the zero manifold of p in $L \times K$, is given by the function $\lambda(z) = \lambda_1 - A_1(z)$, which is holomorphic in K. That is, in the neighborhood of a simple root the equation $p(\lambda, z) = 0$ is solved with respect to λ by a holomorphic function element $\lambda(z)$; i.e. the equation $p\big(\lambda(z), z\big) \equiv 0$ holds identically in K and $\lambda(z)$ is the only function element in $L \times K$ which solves this equation.

A 4.3. Total decomposition of a normalized polynomial

Let $p \in \mathcal{M}_G[\lambda]$ be normalized. Then p has a prime factor decomposition with respect to $\mathcal{M}_G[\lambda]$, $p = \prod\limits_{\alpha=1}^{n} p_\alpha^{m_\alpha}$, $\deg p_\alpha = g_\alpha$, p_α irreducible in $\mathcal{M}_G[\lambda]$. The zeros of $p_\alpha(\lambda, z)$ at $z = z_0$ are denoted by $\lambda_{\alpha\tau}$, $\tau = 1, 2, \ldots, r_\alpha$, and the corresponding multiplicities by $\nu_{\alpha\tau}$. One has $r_\alpha \leqq g_\alpha$ (note that multiplicities $\nu_{\alpha\tau} > 1$ can occur at most for points z_0 from an isolated and closed set in G). Then, in a suitable neighborhood $K = K(z_0)$, there exist decompositions

$$p_\alpha(\lambda, z) = \prod_{\tau=1}^{r_\alpha} q_{\alpha\tau}(\lambda - \lambda_{\alpha\tau}, z), \qquad \deg q_{\alpha\tau} = \nu_{\alpha\tau}, \qquad q_{\alpha\tau}(\cdot, z) \in \mathcal{H}_K^{z_0}[\cdot].$$

The polynomials $q_{\alpha\tau}$ are not necessarily irreducible with respect to $\mathcal{H}_K^{z_0}[\cdot]$. However, in the corresponding prime factor decomposition there are no multiple factors, because otherwise the polynomial $q_{\alpha\tau}(\cdot, z)$ would have multiple roots for every $z \in K$, which is not true. Therefore, one has

$$q_{\alpha\tau}(\mu, z) = \prod_{\varkappa=1}^{a_{\alpha\tau}} q_{\alpha\tau\varkappa}(\mu, z), \qquad q_{\alpha\tau\varkappa} \in \mathcal{H}_K^{z_0}[\mu],$$

where the polynomials $q_{\alpha\tau\varkappa}$ are irreducible with respect to $\mathcal{H}_K^{z_0}[\mu]$.

A 5. Irreducible polynomials and finite-valued algebroid functions

A 5.1. Preliminaries

Let $p \in \mathcal{M}_G[\lambda]$ be a normalized polynomial and let (z_0, f) be a meromorphic element over $z_0 \in G$. *The element* (z_0, f) *satisfies the equation* $p(\lambda, z) = 0$ *if the equation* $p\big(f(z), z\big) = 0$ *holds identically with respect to* z*, where* z *belongs to a neighborhood of* z_0.

Let $\mathfrak{P}$ denote an analytic function over G. $\mathfrak{P}$ *satisfies the equation* $p(\lambda, z) = 0$ *if every meromorphic element of* $\mathfrak{P}$ *satisfies the equation* $p(\lambda, z) = 0$.

An analytic function $\mathfrak{P}$ over G satisfies the equation $p(\lambda, z) = 0$ if and only if a single meromorphic element of $\mathfrak{P}$ satisfies the equation $p(\lambda, z) = 0$.

The *zero manifold* of p over G is defined by the set of all ordered pairs (λ, z) with $z \in G$, such that $p(\lambda, z) = 0$. The zero manifold can be described by analytic functions satisfying the equation $p = 0$. The investigation of the zero manifold of a given p can be reduced to the corresponding investigation for the prime factors of p with respect to $\mathcal{M}_G[\lambda]$. Hence, without loss of generality, p is assumed to be irreducible in the following discussion.

A 5.2. n-Valued algebroid functions over G and their irreducible equations

Theorem 1. *Let* $\mathfrak{P}$ *be an n-valued algebroid function over G. Then there exists exactly one normalized polynomial $p(\lambda, z)$ of degree n from $\mathcal{M}_G[\lambda]$ such that* $\mathfrak{P}$ *satisfies the equation $p(\lambda, z) = 0$. This polynomial is irreducible and at most the poles and singularities of* $\mathfrak{P}$ *are poles of its coefficients.*

Proof. Let A be the set of all singular points of $\mathfrak{P}$ and let $a \in G \setminus A$. Then a is the trace point of exactly n meromorphic elements of $\mathfrak{P}$. These meromorphic elements are denoted by $(a, f_a^{(1)}), (a, f_a^{(2)}), \ldots, (a, f_a^{(n)})$. Let K_a be a disc contained in $G \setminus A$ centered at a, such that all discs belonging to the n meromorphic elements over a contain K_a. Furthermore, let $S(x_1, x_2, \ldots, x_n)$ denote any symmetric polynomial. Consider in K_a the meromorphic function $S_a(z) = S\big(f_a^{(1)}(z), f_a^{(2)}(z), \ldots, f_a^{(n)}(z)\big)$. The assertion is that these meromorphic functions $S_a(z)$ define a single meromorphic function $S(z)$ in the whole region $G \setminus A$. To prove this assertion, it is sufficient to show that the two functions $S_a(z)$ and $S_b(z)$ coincide for all points $z \in K_a \cap K_b$, where $a, b \in G \setminus A$ are any two points; i.e., one has to show $S_a(z) = S_b(z)$ for $z \in K_a \cap K_b$. Let $z \in K_a \cap K_b$. Then the elements $(z, f_a^{(1)}), \ldots, (z, f_a^{(n)})$ form a set of n different meromorphic elements from $\mathfrak{P}$ over the trace point z; correspondingly, the elements $(z, f_b^{(1)}), \ldots, (z, f_b^{(n)})$ form a set of n different meromorphic elements from $\mathfrak{P}$ over the trace point z. That is, these two sets coincide (up to the order of the elements). Hence, the asserted equation

$$S_a(z) = S\big(f_a^{(1)}(z), \ldots, f_a^{(n)}(z)\big) = S\big(f_b^{(1)}(z), \ldots, f_b^{(n)}(z)\big) = S_b(z)$$

follows, because $S(x_1, \ldots, x_n)$ is symmetric. If one puts $S(z) = S_z(z)$, one obtains a single-valued meromorphic function in $G \setminus A$. Poles of $S(z)$ are necessarily poles of branches of $\mathfrak{P}$.

If $a \in A$, i.e., if a is a singular point of $\mathfrak{P}$, then a is necessarily algebroid singular. Hence, if $\mathfrak{H}$ denotes a branch of $\mathfrak{P}$ in the neighborhood of a and $f(z)$ a meromorphic element of $\mathfrak{H}$, then $\lim\limits_{z \to a} f(z)$ exists and is independent of the choice of the element $f(z)$ from $\mathfrak{H}$ with the trace point z (this limit can be different for different branches $\mathfrak{H}$). Then, $\lim\limits_{z \to a} S(z)$ also exists, because the limits exist for the occurring arguments. Hence, a cannot be an essential singularity for $S(z)$; i.e., $S(z)$ is meromorphic in the whole region G.

Let $S_1, S_2, \ldots, S_n$ denote the elementary symmetric polynomials of $x_1, x_2, \ldots, x_n$.

Then,

$$\prod_{i=1}^{n} (\lambda - x_i) = \lambda^n + S_1\lambda^{n-1} + S_2\lambda^{n-2} + \cdots + S_n.$$

According to the foregoing consideration, the functions $S_1, \ldots, S_n$ yield n meromorphic functions $\tilde{S}_1(z), \ldots, \tilde{S}_n(z)$ in G. One considers the polynomial

$$p(\lambda, z) = \lambda^n + \tilde{S}_1(z)\, \lambda^{n-1} + \cdots + \tilde{S}_n(z)$$

from $\mathcal{M}_G[\lambda]$ and asserts: The analytic function $\mathfrak{P}$ satisfies the equation $p(\lambda, z) = 0$. Let a be regular and let $f_1(z), \ldots, f_n(z)$ denote the meromorphic elements of $\mathfrak{P}$ over a. Then one obtains in a neighborhood U of a the identity

$$\prod_{i=1}^{n} \big(\lambda - f_i(z)\big) = p(\lambda, z), \qquad z \in U.$$

Hence, $p\big(f_1(z), z\big) \equiv 0$ for $z \in U$. That is, the element (a, f_1) satisfies the equation $p = 0$. Therefore, $\mathfrak{P}$ satisfies the equation $p = 0$.

If $q(\lambda, z)$ is a normalized polynomial such that $\mathfrak{P}$ satisfies the equation $q = 0$, then the degree of q is at least n, because there are trace points z_0 such that n different values $\lambda_1, \lambda_2, \ldots, \lambda_n$ of $\mathfrak{P}$ lie over z_0, but these numbers are necessarily zeros of $q(\lambda, z_0)$.

From this remark it follows that $p(\lambda, z)$ is irreducible with respect to $\mathcal{M}_G[\lambda]$. Let $p = q_1 \cdot q_2$, where q_1, q_2 are normalized polynomials from $\mathcal{M}_G[\lambda]$ such that $0 < \deg q_1 < n$, $0 < \deg q_2 < n$. Furthermore, let (a, f) be a holomorphic element of $\mathfrak{P}$. Then, $p\big(f(z), z\big) \equiv 0$ for $z \in U$, where U is a suitable neighborhood of a. Consequently, for every z, one obtains the equation $q_1\big(f(z), z\big) = 0$ or the equation $q_2\big(f(z), z\big) = 0$. Hence, there is a sequence $z_\varrho \to a$ such that $q_1\big(f(z_\varrho), z_\varrho\big) = 0$ (or correspondingly for q_2). Therefore, one necessarily obtains $q_1\big(f(z), z\big) \equiv 0$ for $z \in U$. Since $\deg q_1 < n$, this conclusion is a contradiction. Hence, p is irreducible.

Let $q(\lambda, z)$ be a second normalized polynomial of degree n such that $\mathfrak{P}$ satisfies the equation $q = 0$. Furthermore, let (a, f) be a holomorphic element of $\mathfrak{P}$. Then the polynomials $q(\lambda, z)$ and $p(\lambda, z)$ have a common zero for every z from a suitable neighborhood of a, namely $f(z)$. According to Theorem 3.1, this result implies the existence of a nontrivial common divisor of p and q with respect to $\mathcal{M}_G[\lambda]$. Since p is irreducible, one obtains: p is a divisor of q. Now both polynomials are normalized and of degree n. Therefore, $p = q$. ■

A 5.3. Definition of n-valued algebroid functions by irreducible polynomials

According to Subsect. *A* 5.2, to every n-valued algebroid function $\mathfrak{P}$ over G corresponds a unique normalized irreducible polynomial p of degree n such that $\mathfrak{P}$ satisfies the equation $p = 0$.

The inverse relation is also true: To every irreducible polynomial $p \in \mathcal{M}_G[\lambda]$ of degree n corresponds uniquely an n-valued algebroid function $\mathfrak{P}$ ever G such that $\mathfrak{P}$ satisfies the equation $p = 0$.

Theorem 2. *Let p be a normalized irreducible polynomial of degree n from $\mathcal{M}_G[\lambda]$. Then there exists exactly one n-valued algebroid function $\mathfrak{P}$ over G such that $\mathfrak{P}$ satisfies the* *equation $p = 0$. Poles or singularities of $\mathfrak{P}$ are necessarily poles of coefficients of p or the points z_0 where $p(\lambda, z_0)$ has at least one multiple root.*

Proof. Let A denote the set consisting of all poles of coefficients of $p(\lambda, z)$ and of all points z where $p(\lambda, z)$ has a multiple root. Since p is irreducible, A is isolated and closed in G.

Let $z_0 \in G \setminus A$. Then the zeros $\lambda_1, \lambda_2, \ldots, \lambda_n$ of $p(\lambda, z_0)$ are simple and, according to Subsect. A 4.2, there is a neighborhood U of z_0 such that $U \subset G \setminus A$ and

$$p(\lambda, z) = \big(\lambda - \lambda_1(z)\big)\big(\lambda - \lambda_2(z)\big) \cdots \big(\lambda - \lambda_n(z)\big), \qquad z \in U,$$

where the functions $\lambda_1(z), \ldots, \lambda_n(z)$ are holomorphic in U and $\lambda_\varrho = \lambda_\varrho(0)$, $\varrho = 1, \ldots, n$. That is, the holomorphic elements $(z_0, \lambda_1), \ldots, (z_0, \lambda_n)$ satisfy the equation $p(\lambda, z) = 0$ in U.

Hence, the element (z_0, λ_1) defines an analytic function $\mathfrak{P}$ over G which satisfies the equation $p = 0$. If (z_1, g) denotes a meromorphic element of $\mathfrak{P}$ with a trace point $z_1 \in G \setminus A$, then (z_i, g) necessarily coincides with one of the elements (z_1, λ_i), $i = 1, \ldots, n$. Hence, poles of $\mathfrak{P}$ necessarily belong to A.

Let $a \in S \cup \partial S$ with respect to $\mathfrak{P}$ and, furthermore, let $a \in G \setminus A$. If U denotes a sufficiently small neighborhood of a, then $U \setminus \{a\}$ contains at least trace points $z_1 \in G \setminus A$ from $\mathfrak{P}$. A corresponding element (z_1, g) of $\mathfrak{P}$ is holomorphic and it also has to coincide with one of the elements (a, λ_i) in a neighborhood of z_1, hence in U. These properties imply that a is a trace point, $a \in S$, and, moreover, that only holomorphic elements of $\mathfrak{P}$ lie over U; i.e., a is regular with respect to $\mathfrak{P}$. Therefore, singularities of $\mathfrak{P}$ necessarily belong to A. Moreover, it follows that all points of $G \setminus A$ are regular and $\mathfrak{P}$ can be continued arbitrarily in $G \setminus A$. Moreover, $\mathfrak{P}$ is at most n-valued, because an element (z_1, g) of $\mathfrak{P}$ with $z_1 \in G \setminus A$ necessarily coincides with one of the n elements (z_i, λ_i), $i = 1, \ldots, n$.

Furthermore, let $a \in A$. Let U denote a neighborhood of a with $U \setminus \{a\} \subset G \setminus A$. Every branch $\mathfrak{H}$ of $\mathfrak{P}$ over $U \setminus \{a\}$ can be continued arbitrarily and its valuedness h is finite. Moreover, in a sufficiently small disc $0 < |z - a| < r$, the branch $\mathfrak{H}$ can be represented in the form $g\big((z - a)^{1/h}\big)$, where $g(\zeta)$ is single-valued and holomorphic in $0 < |\zeta| < r^{1/h}$. Note that $z = a + \zeta^h$. The fact that the elements of $\mathfrak{H}$ over $0 < |z - a| < r$ satisfy the equation $p(\lambda, z) = 0$ can also be written in the form

$$p\big(g(\zeta), a + \zeta^h\big) \equiv 0, \qquad 0 < |\zeta| < r^{1/h}. \tag{5.1}$$

Consider sequences $\zeta_\varrho \to 0$ such that the corresponding sequence $g(\zeta_\varrho)$ has a finite limit $\lim\limits_{\varrho \to \infty} g(\zeta_\varrho) = b$. Recall that the point a could be a pole of coefficients of $p(\lambda, z)$. The maximal order of these poles is denoted by k. Then the coefficients of the polynomial $(z - a)^k\, p(\lambda, z) = q(\lambda, z)$ are holomorphic at $z = a$ and one has $q(\lambda, a) \not\equiv 0$; i.e., at least one coefficient is different from zero. From (5.1), the relation

$$q\big(g(\zeta), a + \zeta^h\big) \equiv 0, \qquad 0 < |\zeta| < r^{1/h},$$

follows and, taking the limit $\zeta_\varrho \to 0$, one obtains $q(b, a) = 0$. Since this equation has at most n different solutions, there are at most n different finite limits for sequences $g(\zeta_\varrho)$ with $\zeta_\varrho \to 0$. Hence, according to a well-known theorem of Casorati-Weierstrass, the point $\zeta = 0$ cannot be an essential singularity for $g(\zeta)$. Therefore, $\zeta = 0$ is at most a pole of $g(\zeta)$, which implies that a is at most an algebroid singularity for $\mathfrak{P}$.

These results imply the existence of an analytic function over G which is at most n-valued, algebroid, whose poles and (algebroid) singularities belong necessarily to A and which satisfies the equation $p = 0$.

The valuedness of $\mathfrak{P}$ is denoted by m, where $m \leqq n$. One applies Theorem 1. According to this theorem, to the function $\mathfrak{P}$ corresponds a unique normalized irreducible polynomial q of degree m such that $\mathfrak{P}$ satisfies the equation $q = 0$. That is, $\mathfrak{P}$ satisfies the equations $p = 0$ and $q = 0$. Hence, these two polynomials necessarily have common roots for sequences z_ϱ which accumulate in the interior of G. Therefore, according to Theorem 3.1, they have nontrivial common divisors in $\mathcal{M}_G[\lambda]$. Since both polynomials are normalized and irreducible, $p = q$. Therefore, $\mathfrak{P}$ is n-valued and satisfies the equation $p = 0$. $\mathfrak{P}$ is uniquely determined by p, because $\mathfrak{P}$ is n-valued. This implies that every meromorphic element (z, λ) satisfying the equation $p = 0$ necessarily belongs to $\mathfrak{P}$. ■

A 5.4. Consequences

Theorems 1 and 2 establish a one-to-one correspondence between the normalized irreducible polynomials p of $\mathcal{M}_G[\lambda]$ of degree n and the n-valued algebroid functions $\mathfrak{P}$ over G such that $\mathfrak{P}$ satisfies the equation $p = 0$. A special case of this correspondence should be emphasized.

Theorem 3. *Let $p(\mu, z) \in \mathcal{H}_K^{z_0}[\mu]$, where K denotes a disc $|z - z_0| < \varepsilon$, $\deg p = g$ and p is irreducible with respect to $\mathcal{H}_K^{z_0}[\mu]$ (hence, also irreducible with respect to $\mathcal{H}_K[\lambda]$). Furthermore, it is assumed that $p(\mu, z)$ has only simple roots with respect to μ for every $z \in K$ with $z \neq z_0$.*

Then p uniquely defines a g-valued algebroid function $\mathfrak{P}$ over K such that z_0 is the only singularity of $\mathfrak{P}$ over K, namely a branching point of order g. The limit of $\mathfrak{P}$ as $z \to z_0$ is unique and equal to zero.

$\mathfrak{P}$ is represented over K by a convergent Puiseux series of the form

$$\sum_{\varrho=0}^{\infty} a_\varrho[(z - z_0)^{1/g}]^\varrho, \qquad a_0 = 0; \qquad |z - z| < \varepsilon; \tag{5.2}$$

i.e., if for $z \neq z_0$, one denotes the values of $\mathfrak{P}$ over z by $\lambda_1(z), \lambda_2(z), \ldots, \lambda_g(z)$, then (after a suitable ordering)

$$\lambda_\sigma(z) = \sum_{\varrho=0}^{\infty} a_\varrho[\zeta^\sigma(z - z_0)^{1/g}]^\varrho, \qquad \sigma = 1, 2, \ldots, g,$$

where the root $(z - z_0)^{1/g}$ is fixed and where ζ denotes a primitive g-th root of unity.

Proof. First one applies Theorem 2. The assigned analytic function $\mathfrak{P}$ over K is g-valued, where z_0 is the only singularity which is necessarily a branching point of $\mathfrak{P}$ of order g. According to Subsect. A 1.5 one obtains the series expansion (5.2). ■

Let p be normalized and irreducible from $\mathcal{M}_G[\lambda]$. Let $\mathfrak{P}$ denote the assigned function and let $z_0 \in G \setminus S_p$. The behavior of $\mathfrak{P}$ over a neighborhood U of z_0 corresponds to the decomposition behavior of p with respect to $\mathcal{M}_U[\lambda]$. Let U be chosen such that the polynomial $p(\lambda, z)$ has only simple roots with respect to λ for $z \in U \setminus \{z_0\}$ and, moreover, the coefficients of p are holomorphic in U. Let $\lambda_1, \ldots, \lambda_r$ and $\nu_1, \ldots, \nu_r$ denote the zeros of $p(\lambda, z_0)$ and their multiplicities, respectively. Then there exist a decomposition

$$p(\lambda, z) = \prod_{\tau=1}^{r} q_\tau(\lambda - \lambda_\tau, z), \qquad \deg q_\tau = \nu_\tau, \qquad q_\tau(\cdot, z) \in \mathcal{H}_U^{z_0}[\cdot],$$

and a decomposition $q_\tau(\mu, z) = \prod_{\varkappa=1}^{a_\tau} q_{\tau\varkappa}(\mu, z)$, $q_{\tau\varkappa}(\mu, z) \in \mathcal{H}_U^{z_0}[\mu]$. Note that the polynomials $q_{\tau\varkappa}(\mu, z)$ are irreducible in $\mathcal{H}_U[\mu]$ and necessarily mutually different. Let $\deg q_{\tau\varkappa} = g_{\tau\varkappa}$.

According to Theorem 3, a unique $g_{\tau\varkappa}$-valued function $\mathfrak{P}_{\tau\varkappa}$ is defined by $q_{\tau\varkappa}$ over U, with only one singularity z_0, which is a branching point of order $g_{\tau\varkappa}$. The function $\mathfrak{P}_{\tau\varkappa}$ has a limit λ_τ as $z \to z_0$.

It is impossible that the polynomials $q_{\tau\varkappa}(\mu, z)$ become reducible if one passes to a smaller neighborhood U, for, if so, such a polynomial $q_{\tau\varkappa}$ would generate two distinct analytic functions over a neighborhood of z_0.

Hence, $\mathfrak{P}$ can be decomposed over U in groups of branches $\mathfrak{P}_{\tau\varkappa}$, $\tau = 1, \ldots, r$; $\varkappa = 1, 2, \ldots, a_\tau$, where the functions $\mathfrak{P}_{\tau\varkappa}$ with fixed τ have one and the same limit λ_τ as $z \to z_0$. z_0 is a branching point of order $g_{\tau\varkappa}$ for $\mathfrak{P}_{\tau\varkappa}$.

A corresponding decomposition can be obtained for the points $z_0 \in S_p$.

If $\mathfrak{P}$ is n-valued and algebroid over G and if a is singular with respect to $\mathfrak{P}$, then there exists a neighborhood $0 < |z - a| < \varepsilon$ of a, such that there are exactly n different values of $\mathfrak{P}$ over z, where z runs through all points of this neighborhood.

Otherwise, there would be a sequence $z_\varrho \to a$ such that $p(\lambda, z_\varrho)$ had multiple roots with respect to λ for every z_ϱ, which is not true.

The zero manifold of a normalized irreducible polynomial p is completely described over $G \setminus S_p$ by $\mathfrak{P}$. If $z_0 \in G \setminus S_p$ is regular, then the *values* of $\mathfrak{P}$ over z_0 yield all pairs (z_0, λ) of the zero manifold; if $z_0 \in G \setminus S_p$ is singular, then the *limits* of all branches of $\mathfrak{P}$, as $z \to z_0$, yield all pairs (z_0, λ) of the zero manifold. In this way, all pairs (z_0, λ) over $G \setminus S_p$ are taken into account, where $p(\lambda, z_0) = 0$.

A 6. Algebroid function fields

A 6.1. Definition of analytic functions on $\mathfrak{P}$

Let $\mathfrak{P}$ be an n-valued algebroid function over G and let p denote the corresponding normalized irreducible polynomial. Let $\tilde{\mathfrak{P}}$ denote a subset of $\mathfrak{P}$ such that the meromorphic elements of $\tilde{\mathfrak{P}}$ can be mutually continued within $\tilde{\mathfrak{P}}$.

Theorem 1. *Assume that to every meromorphic element $P(z)$ of $\tilde{\mathfrak{P}}$ with the trace point z corresponds a meromorphic element $Q(z)$ with the same trace point z. Furthermore, assume that if $P'(z')$ is a direct continuation of $P(z)$ and if $Q'(z')$ denotes the element assigned to*

$P'(z')$, then $Q'(z')$ is a direct continuation of $Q(z)$. Then the set $\tilde{\mathfrak{Q}}$ of all elements $\{Q(z)\}$ defines a single analytic function over G.

Proof. One merely has to show that the assigned elements $Q(z)$ can be mutually continued. Let $Q_1(z_1)$, $Q_2(z_2)$ denote two such elements, where $P_1(z_1)$, $P_2(z_2)$ are corresponding inverse images from $\tilde{\mathfrak{P}}$. These inverse images can be mutually analytically continued within $\tilde{\mathfrak{P}}$. Let $\mathscr{C}$ be a corresponding path from z_1 to z_2 and let $P(z)$ denote a chain defined on $\mathscr{C}$, belonging to $\tilde{\mathfrak{P}}$, where $P(z_1) = P_1(z_1)$, $P(z_2) = P_2(z_2)$. To every $P(z)$ of the chain corresponds an element $Q(z)$. By assumption, these elements $Q(z)$ also form a chain. ■

If one identifies, as is often done, $\mathfrak{P}$ with its Riemann surface, then, according to Theorem 1, one can interpret *the analytic function $\mathfrak{Q}$, defined uniquely by $\tilde{\mathfrak{Q}}$, as an analytic function defined on the Riemann surface* $\mathfrak{P}$. More precisely, the "part" $\tilde{\mathfrak{Q}}$ of $\mathfrak{Q}$ is defined on the "part" $\tilde{\mathfrak{P}}$ of $\mathfrak{P}$. Note that the characteristic property of the assignment $P(z) \to Q(z)$ for meromorphic elements of $\tilde{\mathfrak{P}}$ and $\tilde{\mathfrak{Q}}$ is the *compatibility with the property "direct continuation"* of the meromorphic elements; i.e., a direct continuation of $P(z)$ always has to correspond with the corresponding direct continuation of $Q(z)$.

A 6.2. Field of algebroid functions on $\mathfrak{P}$

If $\mathfrak{G}$ is a finite-valued algebroid function over G, and if to every meromorphic element (z, f) of $\mathfrak{P}$ corresponds a unique meromorphic element (z, g) of $\mathfrak{G}$ such that to a direct continuation of (z, f) corresponds the corresponding direct continuation of (z, g), then one calls $\mathfrak{G}$ an *algebroid function* on $\mathfrak{P}$.

Each meromorphic element $(z, g) \in \mathfrak{G}$ appears as an image, because by analytic continuation of images one again obtains images.

If (a, f) is an n-fold element of $\mathfrak{P}$, then the meromorphic elements assigned to the meromorphic elements of this n-fold element over $U(a) \setminus \{a\}$ form a branch of $\mathfrak{Q}$ over $U(a) \setminus \{a\}$. Its valuedness m is a divisor of n and corresponds to an m-fold element (a, g) over $\mathfrak{G}$. This m-fold element (a, g) is assigned to the n-fold element (a, f). Thus, each m-fold element (a, g) of $\mathfrak{G}$ appears as an image.

For algebroid functions $\mathfrak{G}$ on $\mathfrak{P}$, the elementary arithmetical operations are defined in a natural way. Let $\mathfrak{G}_1$ and $\mathfrak{G}_2$ denote two algebroid functions on $\mathfrak{P}$. Let (z, g_1) be a meromorphic element of $\mathfrak{G}_1$, (z, g_2) a corresponding one of $\mathfrak{G}_2$, both assigned to the element (z, f) of $\mathfrak{P}$. Then one considers the meromorphic elements

$$(z, g_1 \pm g_2);\ (z, g_1 \cdot g_2);\ (z, g_1/g_2), \qquad g_2 \not\equiv 0.$$

They define algebroid functions on $\mathfrak{P}$, denoted by

$$\mathfrak{G}_1 \pm \mathfrak{G}_2, \qquad \mathfrak{G}_1 \cdot \mathfrak{G}_2, \qquad \mathfrak{G}_1/\mathfrak{G}_2, \tag{6.1}$$

respectively, because they are assigned uniquely to the *functions* $\mathfrak{G}_1$ and $\mathfrak{G}_2$.

For example, let (z', g_1') and (z', g_2') denote other meromorphic elements of $\mathfrak{G}_1$ and $\mathfrak{G}_2$ assigned to the element (z', f'). Then $(z', g_1' \cdot g_2')$ is a continuation of $(z, g_1 g_2)$, namely along a path which is suitable for the continuation of (z, f) into (z', f').

The operations (6.1) satisfy the field axioms. Hence, the set of all algebroid functions on $\mathfrak{P}$ forms an (Abelian) field. It is denoted by $A(\mathfrak{P})$.

Since arithmetical operations have been introduced for algebroid functions on $\mathfrak{P}$, the fact that $\mathfrak{P}$ satisfies the equation $p(\lambda, z) = 0$ can be expressed by the equation $p(\mathfrak{P}, z) = 0$; i.e., if $p(\lambda, z) = \lambda^n + a_1(z)\,\lambda^{n-1} + \cdots + a_n(z)$, then $\mathfrak{P}^n + a_1(z)\,\mathfrak{P}^{n-1} + \cdots + a_n(z) \equiv 0$.

A 6.3. Characterization of the function field $A(\mathfrak{P})$

Let p be normalized and irreducible in $\mathcal{M}_G[\lambda]$. Let $\deg p = g$. Then one can form, in the algebraic sense, the *extension field* of $\mathcal{M}_G$ by *adjunction of* $\mathfrak{P}$ to $\mathcal{M}_G$. As usual, this field is denoted by $\mathcal{M}_G(\mathfrak{P})$. It has the degree g over $\mathcal{M}_G$ and is formed by the set of all elements

$$a_0(z) + a_1(z)\,\mathfrak{P} + a_2(z)\,\mathfrak{P}^2 + \cdots + a_{g-1}(z)\,\mathfrak{P}^{m-1}, \tag{6.2}$$

where the coefficients $a_0(z), \ldots, a_{g-1}(z)$ belong to $\mathcal{M}_G$. That is, the functions $1, \mathfrak{P}, \mathfrak{P}^2, \ldots, \mathfrak{P}^{g-1}$ form a field basis of $\mathcal{M}_G(\mathfrak{P})$ over $\mathcal{M}_G$.

To prove this property one first has to show that the product of two functions (6.2) again has the form (6.2). It is sufficient to show that any polynomial

$$q(\mathfrak{P}, z) = b_0(z) + b_1(z)\,\mathfrak{P} + \cdots + b_m(z)\,\mathfrak{P}^m, \qquad m \geqq g,$$

always has the form (6.2). Applying the division algorithm to q and p, one obtains

$$q = sp + r, \qquad \deg r < g,$$

and by substitution of $\mathfrak{P}$, one obtains $q(\mathfrak{P}, z) = r(\mathfrak{P}, z)$.

Moreover, the quotient $1/r(\mathfrak{P}, z)$ has the form (6.2), where $r(\lambda, z)$ denotes a non-vanishing polynomial from $\mathcal{M}_G[\lambda]$ of degree $m < g$. The polynomials $r(\lambda, z)$ and $p(\lambda, z)$ are relatively prime, because p is irreducible. Hence, 1 can be represented by $1 = ar + bp$, where $a, b \in \mathcal{M}_G[\lambda]$. Substituting $\mathfrak{P}$ in this equation, one obtains $1 = a(\mathfrak{P}, z) \times r(\mathfrak{P}, z)$ or $1/r(\mathfrak{P}, z) = a(\mathfrak{P}, z)$. As a polynomial in $\mathfrak{P}$, $a(\mathfrak{P}, z)$ can be transformed again into the form (6.2).

The fields $\mathcal{M}_G(\mathfrak{P})$ *and* $A(\mathfrak{P})$ *coincide.* One has

Theorem 2. *Let* $\mathfrak{P}$ *be algebroid and n-valued over* G. *Then the function* $\mathfrak{F}$ *is algebroid on* $\mathfrak{P}$ *if and only if it can be represented in the form*

$$\mathfrak{F} = a_0(z) + a_1(z)\,\mathfrak{P} + \cdots + a_{g-1}(z)\,\mathfrak{P}^{g-1}, \tag{6.3}$$

where the coefficients $a_0(z), \ldots, a_{g-1}(z)$ *are meromorphic in* G.

Proof. Obviously, $\mathcal{M}_G(\mathfrak{P}) \subseteqq A(\mathfrak{P})$, because each polynomial (6.3) represents an algebroid function on $\mathfrak{P}$. One must prove $A(\mathfrak{P}) \subseteqq \mathcal{M}_G(\mathfrak{P})$. Let $\mathfrak{H} \in A(\mathfrak{P})$ and let z be a regular point of $\mathfrak{P}$. The corresponding meromorphic elements of $\mathfrak{P}$ are denoted by $(z, f_1), (z, f_2), \ldots, (z, f_g)$. The assigned meromorphic elements of $\mathfrak{H}$ are denoted by (z, h_1),

$(z, h_2), \ldots, (z, h_g)$ (they need not be mutually different). One forms the functions

$$\left.\begin{aligned} r_0(z) &= h_1(z) + h_2(z) + \cdots + h_g(z), \\ r_1(z) &= h_1 f_1 + h_2 f_2 + \cdots + h_g f_g, \\ r_2(z) &= h_1 f_1^2 + h_2 f_2^2 + \cdots + h_g f_g^2, \\ &\cdots\cdots\cdots\cdots\cdots\cdots\cdots\cdots\cdots \\ r_{g-1}(z) &= h_1 f_1^{g-1} + h_2 f_2^{g-1} + \cdots + h_g f_g^{g-1}. \end{aligned}\right\} \tag{6.4}$$

By analytic continuation of $r_0, r_1, \ldots, r_{g-1}$ in the whole region G, one first shows that they are single-valued over G and then that they are meromorphic in G. That is, $r_0, r_1, \ldots, r_{g-1} \in \mathcal{M}_G$. The determinant of the linear system of equations (6.4) is given by Vandermonde's determinant

$$D(z) = \begin{vmatrix} 1 & 1 & 1 & \cdots 1 \\ f_1 & f_2 & f_3 & \cdots f_g \\ f_1^2 & f_2^2 & f_3^2 & \cdots f_g^2 \\ \cdots & \cdots & \cdots & \cdots \\ f_1^{g-1} & f_2^{g-1} & f_3^{g-1} & \cdots f_g^{g-1} \end{vmatrix} = \prod_{i>k} (f_i - f_k).$$

Therefore, $D(z)^2$ is single-valued, hence meromorphic on G. By solving the system, for example, with respect to $h_1(z)$, one obtains

$$h_1(z) = \frac{1}{D(z)^2} \begin{vmatrix} r_0 & 1 & \cdots 1 \\ r_1 & f_2 & \cdots f_g \\ \cdots & \cdots & \cdots \\ r_{g-1} & f_2^{g-1} & \cdots f_g^{g-1} \end{vmatrix} \cdot \begin{vmatrix} 1 & 1 & \cdots 1 \\ f_1 & f_2 & \cdots f_g \\ \cdots & \cdots & \cdots \\ f_1^{g-1} & f_2^{g-1} & \cdots f_g^{g-1} \end{vmatrix} = \frac{1}{D(z)^2} \det\big(a_{ij}(z)\big),$$

and, furthermore,

$$a_{ij}(z) = r_{i-1} f_1^{j-1} + f_2^{i+j-2} + \cdots + f_g^{i+j-2} = r_{i-1} f_1^{j-1} - f_1^{i+j-2} + \sum_{\varrho=1}^{g} f_\varrho^{i+j-2}.$$

The function $\sum_{\varrho=1}^{g} f_\varrho^{i+j-2}$ is meromorphic on G. Therefore,

$$h_1(z) = q\big(f_1(z), z\big),$$

where $q(\lambda, z)$ is a polynomial from $\mathcal{M}_G[\lambda]$. Hence, $\mathfrak{H} = q(\mathfrak{P}, z)$. ■

A 6.4. Algebroid vector- and operator-valued functions

Let $x(z)$ be a holomorphic vector-valued function defined in a neighborhood of z_0. Then $x(z)$ defines an algebroid vector-valued function over G if the scalar holomorphic function $\big(x(z), f\big)$ defines an algebroid function over G for each $f \in X^*$.

Let $e_1, \ldots, e_N$ be a basis in X; let $f_1, \ldots, f_N$ denote the corresponding adjoint basis. Assume that $x(z)$ defines an algebroid vector-valued function over G. One has

$$x(z) = \sum_{\varrho=1}^{N} \big(x(z), f_\varrho\big)\, e_\varrho.$$

$(x(z), f_\varrho)$, $\varrho = 1, \ldots, N$, defines an algebroid function over G. Assume that this algebroid function belongs to $\mathcal{M}(\mathfrak{P})$ for each ϱ, where $\mathfrak{P}$ is a fixed algebroid function over G. Then,

$$(x(z), f_\varrho) = \sum_{\sigma=0}^{g-1} d_{\varrho\sigma}(z)\, \mathfrak{P}^\sigma,$$

and one obtains

$$x(z) = \sum_{\varrho=1}^{N} \sum_{\sigma=0}^{g-1} \alpha_{\varrho\sigma} \mathfrak{P}^\sigma e_\varrho = \sum_{\sigma=0}^{g-1} \left(\sum_{\varrho=1}^{N} \alpha_{\varrho\sigma}(z)\, e_\varrho \right) \mathfrak{P}^\sigma.$$

The functions $\sum_{\varrho=1}^{N} \alpha_{\varrho\sigma}(z)\, e_\varrho = a_\sigma(z)$ are meromorphic vector-valued functions in G. For $x(z)$, one obtains the representation

$$\sum_{\sigma=0}^{g-1} a_\sigma(z)\, \mathfrak{P}^\sigma.$$

These special algebroid vector-valued functions over G are called *algebroid vector-valued functions on* $\mathfrak{P}$. They form a linear space over $\mathbb{C}$.

In the same way, one defines algebroid operator-valued functions over G and *algebroid operator-valued functions on* $\mathfrak{P}$.

A 7. Puiseux-Newton diagram

A 7.1. Problem

If p is a normalized polynomial from $\mathcal{M}_G[\lambda]$ and $z_0 \in G$, then the corresponding zero manifold can be represented, over a suitable neighborhood U of z_0, by finitely many analytic functions $\mathfrak{H}_1, \ldots, \mathfrak{H}_r$ over U, where at most z_0 can appear as an algebroid singularity. The function $\mathfrak{H}_i$ can be represented over U by a Puiseux series or by a Laurent-Puiseux series. The next problem is actually calculating the possible Laurent-Puiseux series if p and z_0 are given.

One can solve this problem by formal "Ansätze", by using power series with respect to fractional powers of $(z - z_0)$ and comparing the corresponding coefficients. The necessary calculations and considerations are rather complicated but they can be somewhat simplified with the help of the Puiseux-Newton diagram.

A 7.2. Diagram

For the Puiseux-Newton diagram, for example, see M. M. Vainberg and V. A. Trenogin [1, p. 34].

Let $p(\lambda, z) = \lambda^n + a_1(z)\, \lambda^{n-1} + \cdots + a_n(z)$. Let k denote the maximum order of poles of its coefficients at z_0. One considers $(z - z_0)^k\, p(\lambda, z)$. The coefficients of this polynomial are holomorphic at $z = z_0$; at least one of them does not vanish at $z = z_0$.

Without loss of generality, one takes $z_0 = 0$. Then the polynomial

$$q(\lambda, z) = z^k\lambda^n + b_1(z)\,\lambda^{n-1} + \cdots + b_n(z)$$

must be considered in the neighborhood of $z = 0$, where the functions $b_i(z)$ are holomorphic at $z = 0$. Consider the equation

$$q(\lambda, z) = (b_0 z^{a_0} + \cdots) + (b_1 z^{a_1} + \cdots)\,\lambda + \cdots$$
$$+ (b_{n-1} z^{a_{n-1}} + \cdots)\,\lambda^{n-1} + z^{a_n}\lambda^n = 0,$$

where $\cdots$ indicates higher powers of z. The Ansatz for a solution is

$$\lambda = e_1 z^{\varepsilon_1} + e_2 z^{\varepsilon_2} + \cdots,$$

where $e_1, e_2, \ldots$ are nonvanishing coefficients and $\varepsilon_1 < \varepsilon_2 < \cdots$. One obtains

$$0 = (b_0 z^{a_0} + \cdots) + (b_1 z^{a_1} + \cdots)(e_1 z^{\varepsilon_1} + \cdots) + \cdots$$
$$+ (b_{n-1} z^{a_{n-1}} + \cdots)(e_1^{n-1} z^{(n-1)\varepsilon_1} + \cdots) + z^{a_n}(e_1^n z^{n\varepsilon_1} + \cdots).$$

Reordering the right-hand side according to powers of z, the coefficients of all powers of z must vanish, in particular, the coefficient of the smallest power of z. The terms of the smallest power of z are to be found among the terms

$$b_0 z^{a_0},\ b_1 e_1 z^{a_1+\varepsilon_1},\ b_2 e_1^2 z^{a_2+2\varepsilon_1},\ \ldots,\ b_{n-1} e_1^{n-1} z^{a_{n-1}+(n-1)\varepsilon_1},\ e_1^n z^{a_n+n\varepsilon_1}.$$

For the terms of the smallest power of z to cancel mutually, it is necessary that at least two of them are present. Hence one has to determine ε_1 such that, among the numbers

$$a_0,\ a_1 + \varepsilon_1,\ a_2 + 2\varepsilon_1,\ \ldots,\ a_{n-1} + (n-1)\,\varepsilon_1,\ a_n + n\varepsilon_1,$$

the smallest one occurs at least twice. This condition yields finitely many possibilities for ε_1. The equation expressing the vanishing of the coefficient of the smallest power of z then yields a condition for the corresponding coefficients e_1. Thus, one obtains finitely many possibilities for $e_1 z^{\varepsilon_1}$. Starting from each of them, one has to repeat the procedure for the next coefficient $e_2 z^{\varepsilon_2}$.

To determine the smallest exponent ε_1, one considers the points

$$A_0 = (0, a_0),\ A_1 = (1, a_1),\ A_2 = (2, a_2),\ \ldots,\ A_{n-1} = (n-1, a_{n-1}),\ A_n = (n, a_n)$$

with respect to a Cartesian (x, y)-plane. If the coefficient of λ^k, $k = 0, 1, \ldots, n$, vanishes identically, i.e., if the term $b_k(z)\,\lambda^k$ is missing, then the point A_k will be disregarded. Fig. 6 shows an example of such a diagram. If $\tan\tau = \varepsilon_1$, where τ is measured with respect to the *negative* horizontal axis, then

$$a_k + k\varepsilon_1 = a_k + k\tan\tau \tag{7.1}$$

is the segment of the vertical axis cut off by the straight line through A_k defined by the angle τ.

In order to obtain a coincidence of two of the numbers (7.1), one has to choose τ such that two of the points A_k lie on the same straight line. Moreover, in order to ensure that none of the numbers (7.1) is smaller than the two coinciding ones, this straight line is restricted by the condition that none of the points A_k lies "below" it.

Therefore, the straight lines in question form the "lower" boundary of the convex hull of the points $A_1, \ldots, A_k$.

The slopes of the straight lines are denoted by $\tau_1, \ldots, \tau_p$. Then the possible values of ε_1 are given by $\varepsilon_1^{(\varrho)} = \tan \tau_\varrho$. If an "admissible" straight line joins the points $A_\nu = (\nu, a_\nu)$ and $A_\mu = (\mu, a_\mu)$, then the corresponding ε_1 is given by (see Fig. 7)

$$\varepsilon_1 = \frac{a_\nu - a_\mu}{\mu - \nu}.$$

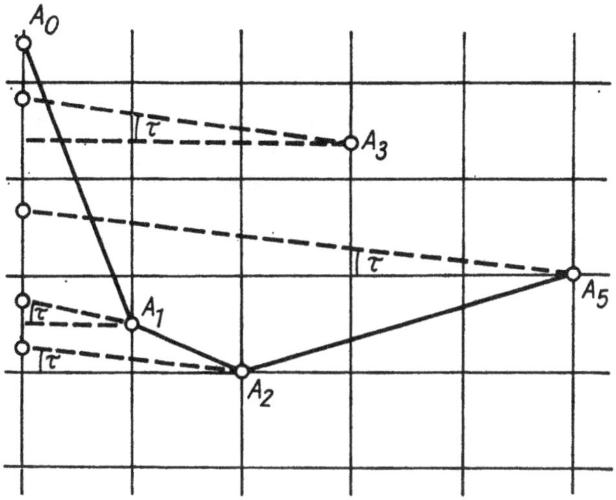

Fig. 6

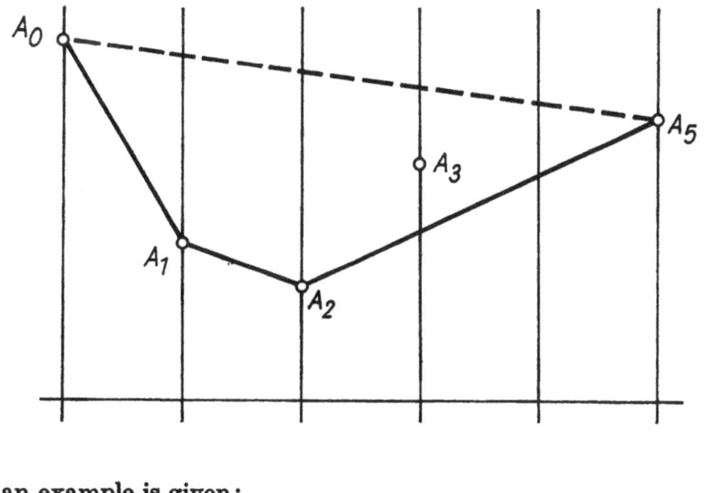

Fig. 7

Finally, an example is given:

$$q(\lambda, z) = z^2 + 2z^3 - (z + z^2)\,\lambda + z\lambda^2 + \lambda^3 = 0.$$

The diagram takes the form of Fig. 8. One obtains

$$\varepsilon_1 = 1; \qquad \varepsilon_1 = 1/2.$$

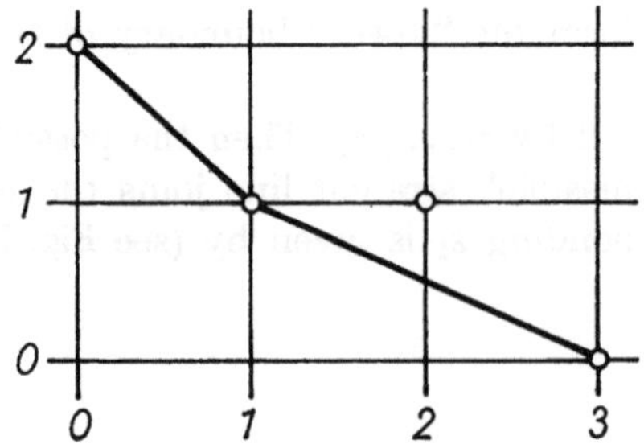

Fig. 8

If $\varepsilon_1 = 1$, then the corresponding straight line contains two points. Hence, two terms coincide: $1 - e_1 = 0$; therefore, $e_1 = 1$. If $\varepsilon_1 = 1/2$, then again two terms coincide: $-e_1 + e_1^3 = 0$. Since $e_1 \neq 0$, $e_1 = \pm 1$. Therefore, one obtains three different Puiseux series starting with the terms $\lambda_1 = z + \cdots$, $\lambda_2 = z^{1/2} + \cdots$, $\lambda_3 = -z^{1/2} + \cdots$. The procedure for the calculation of the higher terms is similar. Note that, necessarily, $\varepsilon_2 > \varepsilon_1$.

Bibliography

Articles

BART, H., M. A. KAASHOEK, and D. C. LAY

[1] Stability properties of finite meromorphic operator functions I. Koninkl. Nederl. Akademie van Wetenschappen — Amsterdam Proceedings, Series A, 77, No. 3.

[2] Relative Inverses of Meromorphic Operator Functions and Associated Holomorphic Projection Functions. Math. Ann. **218** (1975), 199—210.

BAUMGÄRTEL, H.

[1] Zur Störungstheorie beschränkter linearer Operatoren eines Banachschen Raumes. Math. Nachr. **26** (1964), 361—179.

[2] Analytische Störung isolierter Eigenwerte endlicher algebraischer Vielfachheit von nicht-selbstadjungierten Operatoren. Mber. Dt. Akad. Wiss. Berlin **10** (1968), 250—258.

[3] Ein Reduktionsprozeß für analytische Störungen nichthalbeinfacher Eigenwerte. Mber. Dt. Akad. Wiss. Berlin **11** (1969), 81—89.

[4] Jordansche Normalform holomorpher Matrizen. Mber. Dt. Akad. Wiss. Berlin **11** (1969), 23—24.

[5] Zur Abschätzung der Konvergenzradien von Störungsreihen. Mber. Dt. Akad. Wiss. Berlin **11** (1969), 556—572.

[6] Einige Probleme und Ergebnisse der Störungstheorie linearer Operatoren. Mitt. Math. Ges. DDR (1970), 70—83.

[7] Analytic perturbation theory for linear operators depending on several complex variables. Mat. Issl. **9** (1974) vyp. 1 (31), 17—39 (*Russian*).

BIRKHOFF, G.

[1] Three observations on linear algebra. Univ. Nac. Tucuman Rev. Ser. A **5** (1946), 147—151.

BIRNBAUM, S.

[1] Perturbation of eigenvalues — with an engineering application. SIAM Rev. **7** (1965), 13—30.

BROWNELL, F. H.

[1] Explicit Perturbation Formulas and Convergence Theorems. J. Math. Anal. and Appl. **6** (1963), 190—206.

BUTLER, J.

[1] Perturbation Series for Eigenvalues of Analytic Non-Symmetric Operators. Arch. Math. **10** (1959), 21—27.

DAVIS, C.

[1] The Rotation of Eigenvectors by a Perturbation I. J. Math. Anal. and Appl. 6 (1963), 159 to 173.
[2] The Rotation of Eigenvectors by a Perturbation II. J. Math. Anal. and Appl. 11 (1965), 20–27.

DUNFORD, N.

[1] Spectral Theory. Bull. Amer. Math. Soc. 49 (1943), 637–651.
[2] Spectral Theory I. Convergence to Projections. Trans. Amer. Math. Soc. 54 (1943), 185–217.

ENI, V. M.

[1] Multiplicity of eigenvectors and of characteristic numbers for polynomial operators. Izv. Akad. Nauk Mold. SSR 7 (1965), 94–97 (*Russian*).
[2] Analytic perturbations of characteristic numbers and of eigenvectors for polynomial operators. Mat. Issl. 1 (1966), 189–195 (*Russian*).
[3] Stability of the root number of polynomial operators. Izv. Akad. Nauk Mold. SSR 4 (1966), 78–81 (*Russian*).
[4] Stability of the root number of an analytic operator-valued function and perturbation of its characteristic numbers and eigenvectors. Dokl. Akad. Nauk SSSR 173 (1967), 1251–1254 (*Russian*).
[5] Semistability of the root number of a linear operator. Mat. Issl. 2 (1967), 73–84 (*Russian*).

ERDMANN, A.

[1] Über eine Normalform für polynomiale Operatorfunktionen auf endlichdimensionalen Räumen und deren Abhängigkeit von holomorphen Störungen. Preprint AdW der DDR, Okt. 1977.

FÖRSTER, K.-H.

[1] Über lineare, abgeschlossene Operatoren, die analytisch von einem Parameter abhängen. Math. Z. 95 (1967), 251–258.

FRIEDRICHS, K. O.

[1] On the perturbation of continuous spectra. Comm. Pure Appl. Math. 1 (1948), 361–406.

GAVURIN, M. K.

[1] Eigenvalues of operators depending on a parameter. Vestnik Leningr. Univ. 9 (1952), 77–95 (*Russian*).
[2] Estimates for eigenvalues and eigenvectors of perturbed operators. Dokl. Akad. Nauk SSSR 96 (1954), 1093–1095 (*Russian*).

GREENLEE, W. M.

[1] Perturbation of Eigenvalues with an Engineering Application. J. Math. Anal. and Appl. 27 (1969), 1–20.
[2] Perturbation of Eigenvalues with an Engineering Application — A Supplement. J. Math. Anal. and Appl. 28 (1969), 674–680.

GOCHBERG, I. C.

[1] Linear operators depending analytically on a parameter. Dokl. Akad. Nauk SSSR 78 (1951), 629–632 (*Russian*).

GOCHBERG, I. C., and M. A. KAASHOEK

[1] Unsolved Problems in Matrix and Operator Theory I. Partial Multiplicities and Additive Perturbations. Integral Equations and Operator Theory 1/2 (1978), 278–283.
[2] Unsolved Problems in Matrix and Operator Theory II. Partial Multiplicities for Products. Integral Equations and Operator Theory 2/1 (1979), 116–120.

GOCHBERG, I. C., and M. G. KREIN

[1] The basic propositions on defect numbers, root numbers and indices of linear operators Uspechi Mat. Nauk 12 (1957), 43–118 (*Russian*).

GOCHBERG, I. C., P. LANCASTER and L. RODMAN

[1] Representations and Divisibility of Operator Polynomials. Canad. J. Math. **30** (1978) No. 5, 1045–1069.
[2] Perturbation Theory for Divisors of Operator Polynomials. SIAM J. Math. Anal. **10** (1979) No. 6, 1161–1183.
[3] Spectral analysis of selfadjoint matrix polynomials. Ann. of Math. **112** (1980), 33–71.

KATO, T.

[1] On the convergence of the perturbation method I, II. Progr. Theoret. Phys. 4 (1949), 514 to 523; 5 (1950), 95–101; 207–212.
[2] On the adiabatic theorem of quantum mechanics. J. Phys. Soc. Japan 5 (1950), 435–439.
[3] On the convergence of the perturbation method. J. Fac. Sci. Univ. Tokyo Sect. I, 6 (1951), 145–226.
[4] On the perturbation theory of closed linear operators. J. Math. Soc. Japan 4 (1952), 323–337

KELDYŠ, M. V.

[1] On eigenvalues and eigenfunctions for some classes of nonselfadjoint equations. Dokl. Akad. Nauk SSSR 77 (1951), 11–14 (*Russian*).

KREIN, S. G., and V. P. TROFIMOV

[1] On holomorphic operator-valued functions depending on several complex variables. Func. Anal. i Prilož. 2 (1969), 85–86 (*Russian*).

LANCASTER, P.

[1] On eigenvalues of matrices dependent on a parameter. Num. Math. 6 (1964), 377–387.
[2] On Analytic Perturbation Theory. Research Paper No. 19, May 1967. Dep. of Math. Univ. of Calgary, Alberta, Canada.
[3] A Fundamental Theorem on Lambda-Matrices with Applications — I. Ordinary Differential Equations with Constant Coefficients. Linear Algebra and its Appl. 18 (1977), 189–211.
[4] A Fundamental Theorem on Lambda-Matrices with Applications — II. Difference Equations with Constant Coefficients. Linear Algebra and its Appl. 18 (1977), 213–222.

LANGER, H.

[1] Über Lancaster's Zerlegung von Matrizen-Scharen. Arch. Rat. Mech. and Anal. **29** (1968). 75–80.

LICHTENSTEIN, L.

[1] Untersuchungen über die Gleichgewichtsfiguren rotierender Flüssigkeiten, deren Teilchen einander nach dem Newtonschen Gesetze anziehen. 2. Abh. Stabilitätsbetrachtungen. Math. Z. 7 (1920), 126–231.

LIDSKII, V. B.

[1] On the perturbation theory for nonselfadjoint operators. Ž vycisl. Mat. i Mat. Fiz. **6** (1966)' 52—60 (*Russian*).
[2] Eigenvalues for sums and products of symmetric matrices. Dokl. Akad. Nauk SSSR **75** (1950), 769—772 (*Russian*).

LIVŠIC, B. L.

[1] Perturbation theory for an operator of simple structure. Dokl. Akad. Nauk SSSR **133** (1960), 800—803 (*Russian*).

LOGINOV, B. V.

[1] Error estimates for the perturbation method. Taškent. Gos. Univ. Naučn. Trudy Vyp. **245** (1964), 58—65 (*Russian*).

LOGINOV, B. V., and V. E. POSPEEV

[1] Eigenvalues and eigenvectors of a perturbed operator. Izv. Akad. Nauk Uz. SSR, Ser. Fiz.-Mat. Nauk 11 (1967), 29—35 (*Russian*).

LOUD, W. S.

[1] An eigenvalue problem related to stability questions in perturbed autonomous systems. J. Math. Mech. **11** (1962), 139—151.

MIRSKY, L.

[1] Proofs of two Theorems on Doubly Stochastic Matrices. Proc. Amer. Math. Soc. **9** (1958), 371—374.

MOTZKIN, T. S., and O. TAUSSKY

[1] Pairs of matrices with property L. Trans. Amer. Math. Soc. **73** (1952), 108—114.
[2] Pairs of matrices with property L, II. Trans. Amer. Math. Soc. **80** (1955), 387—401.

PALLASKE, U.

[1] Iterative Störungsrechnung bei mehrfachen und benachbarten Eigenwerten. Dissertation 1969, Universität Köln.

PATTABHIRAMAN, M. V., and P. LANCASTER

[1] Spectral Properties of a Polynomial Operator. Numer. Math. **13** (1969), 247—259.

POINCARÉ, H.

[1] Sur les groupes continus. Oeuvres, vol. 3, 173—212.

PORATH, G.

[1] Störungstheorie der isolierten Eigenwerte für abgeschlossene lineare Transformationen im Banachschen Raum. Math. Nachr. **20** (1959), 175—230.
[2] Störungstheorie für lineare Transformationen im Banachschen Raum. Wiss. Z. Techn. Hochsch. Dresden **9** (1959/60), 1121—1125.

RELLICH, F.

[1] Störungstheorie der Spektralzerlegung, I. Mitteilung. Analytische Störung der isolierten Punkteigenwerte eines beschränkten Operators. Math. Ann. **113** (1937), 600—619.
[2] Störungstheorie der Spektralzerlegung, II. Math. Ann. **113** (1937), 677—685.
[3] Störungstheorie der Spektralzerlegung, III. Math. Ann. **116** (1939), 555—570.
[4] Störungstheorie der Spektralzerlegung, IV. Math. Ann. **117** (1940), 356—382.
[5] Störungstheorie der Spektralzerlegung. V. Math. Ann. **118** (1942), 462—484.
[6] Störungstheorie der Spektralzerlegung. Proceedings of the International Congress of Mathematicians 1950, vol. I, 606—613.

RIBARIČ, N., and I. VIDAV

[1] Analytic Properties of the Inverse $A(z)^{-1}$ of an Analytic Linear Operator Valued Function $A(z)$. Arch. rat. Mech. Anal. **32** (1969), 298—310.

ROSENBLOOM, P.

[1] Perturbation of linear Operators in Banach Spaces. Arch. Math. **6** (1955), 89—101.

SABIROV, T.

[1] On perturbations of eigenvalues and eigenvectors for linear operators. Dokl. Akad. Nauk Tadžik. SSR **9** (1966), 3—6 (*Russian*).

SCHÄFKE, F. W.

[1] Über Eigenwertprobleme mit zwei Parametern. Math. Nachr. **6** (1951), 109—124.
[2] Verbesserte Konvergenz- und Fehlerabschätzungen für die Störungsrechnung. Z. angew. Math. Mech. **33** (1953), 255—259.
[3] Zur Störungstheorie der Spektralzerlegung. Math. Ann. **133** (1957), 219—234.

SCHMIDT, E.

[1] Zur Theorie der linearen und nichtlinearen Integralgleichungen, Teil 3. Über die Auflösungen der nichtlinearen Integralgleichungen und die Verzweigung ihrer Lösungen. Math. Ann. **65** (1908), 370—399.

SCHRÖDER, J.

[1] Fehlerabschätzungen zur Störungsrechnung bei linearen Eigenwertproblemen mit Operatoren eines Hilbertschen Raumes. Math. Nachr. **10** (1953), 113—128.
[2] Fehlerabschätzungen zur Störungsrechnung für lineare Eigenwertprobleme bei gewöhnlichen Differentialgleichungen. Z. angew. Math. Mech. **34** (1954), 140—149.
[3] Störungsrechnung bei Eigenwert- und Verzweigungsaufgaben. Arch. rat. Mech. Anal. **1** (1957/58), 436—468.

SCHRÖDINGER, E.

[1] Quantisierung als Eigenwertproblem, III. Störungstheorie, mit Anwendung auf den Starkeffekt der Balmer-Linien. Ann. Phys. **80** (1926), 437—490.

STEINBERG, S.

[1] Meromorphic Families of Compact Operators. Arch. rat. Mech. Anal. **31** (1968), 372—379.

v. Sz.-Nagy, B.

[1] Perturbations des transformations autoadjointes dans l'espace de Hilbert. Comment. Math. Helv. 19 (1946/47), 347–366.
[2] Perturbations des transformations linéaires fermées. Acta Sci. Math. Szeged 14 (1951), 125 to 137.

Taussky, O.

[1] A weak property L for pairs of matrices. Math. Z. 71 (1959), 463–465.

Terray, J., and P. Lancaster

[1] On the Numerical Calculation of Eigenvalues and Eigenvectors of Operator Polynomials. J. Math. Anal. and Appl. 60 (1977), 370–378.

Trenogin, V. A.

[1] Perturbation of the eigenvalues and eigenelements of linear operators. Dokl. Akad. Nauk SSSR 167 (1966), 519–522 (*Russian*).

Turner, R. E. L.

[1] Perturbation of Compact Spectral Operators. Comm. Pure Appl. Math. 18 (1965), 519–541.

Višik, M. I., and L. A. Ljusternik

[1] Perturbation of the eigenvalues and eigenvectors for some nonselfadjoint operators. Dokl. Akad. Nauk SSSR 130 (1960), 251–253 (*Russian*).
[2] Solution of perturbation problems for matrices and selfadjoint and nonselfadjoint differential equations. Uspechi Mat. Nauk 15 (93) (1960), 3–80 (*Russian*).

Volterra, V.

[1] Sui fondamenti della teoria delle equazioni differenziali lineari. Mem. Soc. Ital. Sci. (3) 6 (1887), 1–104; (3) 12 (1902), 3–68.

Wolf, F.

[1] Analytic Perturbation of Operators in Banach Spaces. Math. Ann. 124 (1952), 317–333.

Books and monographs

Behnke, H., and F. Sommer

[1] Theorie der analytischen Funktionen einer komplexen Veränderlichen. Springer-Verlag, Berlin–Göttingen–Heidelberg 1955.

Bellman, R.

[1] Introduction to Matrix Analysis. McGraw Hill, New York–Toronto–London 1960. Nauka, Moscow 1969 (*Russian transl.*).

Bieberbach, L.

[1] Lehrbuch der Funktionentheorie I. Leipzig und Berlin 1930.

Boseck, H.

[1] Einführung in die Theorie der linearen Vektorräume. 2nd ed. VEB Deutscher Verlag der Wissenschaften, Berlin 1967.

Coddington, E. A., and N. Levinson

[1] Theory of Ordinary Differential Equations. McGraw Hill, New York—Toronto—London 1955.

Courant, R., and D. Hilbert

[1] Methoden der Mathematischen Physik I. 3rd ed. Springer-Verlag, Berlin—Heidelberg—New York 1968.

Dunford, N., and J. T. Schwartz

[1] Linear Operators. Part I. General Theory. John Wiley, New York—London 1958.

Fick, E.

[1] Einführung in die Grundlagen der Quantentheorie. Geest & Portig, Leipzig 1968.

Friedrichs, K. O.

[2] Perturbation of Spectra in Hilbert Space. Amer. Math. Soc., Providence (R. I.) 1965. (Moscow 1969. *Russian transl.*).

Fuks, B. A.

[1] Introduction to the theory of functions of several complex variables. Gos. Izd. Fiz.-Mat. Lit. Moscow 1962 (*Russian*).

Gantmacher, F. R.

[1] Matrizenrechnung I. VEB Deutscher Verlag der Wissenschaften, Berlin 1958 (*Transl. from Russian*).

[2] Matrizenrechnung II. VEB Deutscher Verlag der Wissenschaften, Berlin 1966 (*Transl. from Russian*).

Gelfand, I. M.

[1] Lectures on Linear Algebra. Nauka, Moscow 1966 (*Russian*).

Glazman, I. M., and Ju. I. Ljubič

[1] Finite-dimensional linear analysis. Nauka, Moscow 1969 (*Russian*).

Gochberg, I. C., and M. G. Krein

[2] Introduction to the theory of linear nonselfadjoint operators. Nauka, Moscow (*Russian*).

Graeub, W.

[1] Linear Algebra. Springer-Verlag, Berlin—Heidelberg—New York 1958.

Gunning, R. C., and H. Rossi

[1] Analytic functions of several complex variables. Prentice-Hall, Englewood Cliffs (NJ) 1965.

HOUSEHOLDER, A. S.

[1] The Theory of Matrices in Numerical Analysis. New York—Toronto—London 1964.

HURWITZ, A., and R. COURANT

[1] Funktionentheorie (mit einem Anhang von H. RÖHRL). 4th ed. Springer-Verlag, Berlin—Heidelberg—New York 1964.

KATO, T.

[1] Perturbation theory for linear operators. Springer-Verlag, Berlin—Heidelberg—New York 1966.

KOCHENDÖRFFER, R.

[1] Determinanten und Matrizen. BSB B. G. Teubner-Verlagsgesellschaft, Leipzig 1961.

LANCASTER, P.

[5] Theory of Matrices. New York—London 1969.

LJUBARSKI, G. I.

[1] Anwendungen der Gruppentheorie in der Physik. VEB Deutscher Verlag der Wissenschaften, Berlin 1962 (*Transl. from Russian*).

LUDWIG, G.

[1] Die Grundlagen der Quantenmechanik. Springer-Verlag, Berlin—Göttingen—Heidelberg 1954.

RAYLEIGH, L.

[1] The Theory of Sound, I. London 1894 (2nd ed. London 1927).

RIESZ, F.

[1] Les systèmes d'équations linéaires à une infinité d'inconnues. Paris 1913.

RIESZ, F., and B. v. SZ.-NAGY

[1] Vorlesungen über Funktionalanalysis. VEB Deutscher Verlag der Wissenschaften, Berlin 1956 (*Transl. from French according to the 2nd and 3rd ed.*).

RELLICH, F.

[7] Perturbation theory of eigenvalue problems. Gordon and Breach, New York—London—Paris 1969.

SAKS, S., and A. ZYGMUND

[1] Analytic Functions. Warsaw—Wroclaw 1952.

SCHRÖDINGER, E.
[2] Collected papers on wave mechanics. London and Glasgow 1928.

VAINBERG, M. M., and V. A. TRENOGIN

[1] Bifurcation theory for solutions of nonlinear equations. Nauka, Moscow 1969 (*Russian*).

VOLTERRA, V., and B. HOSTINSKY

[1] Opérations infinitesimales linéaires. Paris 1938.

van der Waerden, B. L.

[1] Die gruppentheoretische Methode in der Quantenmechanik. Springer-Verlag, Berlin 1932.
[2] Algebra II. 5th ed. Springer-Verlag, Berlin—Göttingen—Heidelberg 1967.
[3] Moderne Algebra I. 3rd ed. Springer-Verlag, Berlin—Göttingen—Heidelberg 1950.

Weyl, H.

[1] Gruppentheorie und Quantenmechanik. 2nd ed. S. Hirzel, Leipzig 1931.

Wigner, E.

[1] Gruppentheorie und ihre Anwendungen auf die Quantenmechanik der Atome. Berlin 1931.

Wilkinson, J. H.

[1] The Algebraic Eigenvalue Problem. Oxford 1965.

Subject index

Bei Fragen zur Produktsicherheit wenden Sie sich bitte an:
If you have any questions regarding product safety,
please contact:

Walter de Gruyter GmbH
Genthiner Straße 13
10785 Berlin
productsafety@degruyterbrill.com